TREATISE ON ANALYTICAL CHEMISTRY

PART I

THEORY AND PRACTICE

SECOND EDITION

TREATISE ON
ANALYTICAL CHEMISTRY

PART I

THEORY AND PRACTICE

SECOND EDITION

VOLUME 12

Edited by PHILIP J. ELVING

Department of Chemistry, University of Michigan

Associate Editor: C. B. MURPHY

Xerox Corporation, Rochester, New York

Editor Emeritus: I. M. KOLTHOFF

School of Chemistry, University of Minnesota

AN INTERSCIENCE® PUBLICATION

JOHN WILEY & SONS New York—Chichester—Brisbane—Toronto—Singapore

An Interscience® Publication
Copyright © 1983 by John Wiley & Sons, Inc.

Library of Congress Catalog Number: 78–1707

ISBN 0-471-89653-5

Printed in the United States of America

10 9 8 7 6 5 4 3 2 1

jwl   9-1-83

TREATISE ON ANALYTICAL CHEMISTRY

PART I

THEORY AND PRACTICE

VOLUME 12: *SECTION J*
Thermal Methods   *Chapters 1–9*

AUTHORS OF VOLUME 12

PRONOY K. CHATTERJEE

ROGER M. HART

LEE D. HANSEN

HORST G. LANGER

R. C. MACKENZIE

J. L. MARGRAVE

JOHN C. MELCHER

ROBERT L. MONTGOMERY

M. MURAT

DONALD ROBERTSON

DONALD A. SEANOR

DONALD D. WAGMAN

Authors of Volume 12

Dr. Pronoy K. Chatterjee

Personal Products, Milltown, New Jersey, Chapter 9

Dr. Lee D. Hansen

Brigham Young University, Thermochemical Institute, 267 FB, Provo, Utah, Chapter 4

Mr. Roger M. Hart

Hart Scientific, Provo, Utah, Chapter 4

Dr. Horst G. Langer

*Dow Chemical U. S. A., Central Re-
search—New England Laboratory, Post
Office Box 400, Wayland, Massachusetts, Chapter 6*

Dr. R. C. Mackenzie

*The Macaulay Institute for Soil Re-
search, Craigiebuckler, Aberdeen AB9
2QJ, SCOTLAND, Chapter 1*

Dr. J. L. Margrave

*Department of Chemistry, Rice Univer-
sity, Houston, Texas, Chapter 5*

Mr. John C. Melcher

*871 Lesley Road, Villanova, Pennsylva-
nia, Chapter 3*

Dr. Robert L. Montgomery

Chemistry Department, Rice University, Houston, Texas, Chapter 5

Dr. M. Murat

Laboratoire de Chimie Applique, INSA, 20 Avenue Albert Einstein, 69621 Villeurbanne, FRANCE, Chapter 7

C. B. Murphy

42 Clarke's Crossing, P.O. Box 631, Fairport, New York 14450

Dr. Donald Robertson

619 McKean Road, RD-1, Ambler, Pennsylvania, Chapter 3

Dr. Donald A. Seanor

Xerox Corporation, Joseph C. Wilson Center for Technology, Rochester, New York, Chapter 8

Dr. Donald D. Wagman

Blaasstrasze 25/9, A-1190 Vienna, AUSTRIA, Chapter 2

Foreword

The division of chapters between Volumes 12 and 13, which will contain the section on "Thermal Methods of Analysis," was dictated in part by unforeseen delays in the preparation of some of the individual chapters, e.g., authors who originally undertook the preparation of chapters were not able to fulfill their commitments and other experts had to be enlisted to prepare the chapters in question. In all fairness to the authors whose manuscripts were ready for publication and to the users of the *Treatise*, it was decided to proceed with the present volume. Volume 13 will contain, *inter alia*, chapters on differential thermal analysis, differential scanning calorimetery, and thermogravimetry.

The editors are pleased to acknowledge the cooperation and assistance given in preparation of the section on "Thermal Methods of Analysis" by the Nomenclature Committee of International Confederation for Thermal Analysis, in particular by the Committee Chairman, Dr. R. C. Mackenzie of the Macaulay Institute for Soil Research, Aberdeen, Scotland. The members of the Committee concerned with the chapters in this volume were T. Daniels, C. J. Keattch, D. Dollimore, and F. W. Wilburn. The Committee helped to review manuscripts submitted for the section on "Thermal Methods" primarily to ensure that proper nomenclature was consistently maintained, as this would be to the advantage of the reader and would, additionally, encourage the standardization of thermal analysis nomenclature. Any deviations from the use of optimum nomenclature are the responsibility of the authors and editors of the *Treatise on Analytical Chemistry* and not of the ICTA Committee.

Preface to the Second Edition of the Treatise

In the mid 1950s, the plan ripened to edit a "Treatise on Analytical Chemistry" with the objective of presenting a comprehensive treatment of the theoretical fundamentals of analytical chemistry and their implementation (Part I) as well as of the practice of inorganic and organic analysis (Part II); an introduction to the utilization of analytical chemistry in industry (Part III) was also considered. Before starting this ambitious undertaking, the editors discussed it with many colleagues who were experts in the theory and/or practice of analytical chemistry. The uniform reaction was most skeptical; it was not thought possible to do justice to the many facets of analytical chemistry. Over several years, the editors spent days and weeks in discussion in order to define not only the aims and objectives of the Treatise but, more specifically, the order of presentation of the many topics in the form of a table of contents and the tentative scope of each chapter. In 1959, Volume 1 of Part I was published. The reviews of this volume and of the many other volumes of Part I as well as of those of Parts II and III have been uniformly favorable, and the first edition has become recognized as a contribution of classical value.

Even though analytical chemistry still has the same objectives as in the 1950s or even a century ago, the practice of analytical chemistry has been greatly expanded. Classically, qualitative and quantitative analysis have been practiced mainly as "solution chemistry." Since the 1950s, "solution analysis" has involved to an ever increasing extent physicochemical and physical methods of analysis, and automated analysis is finding more and more application, for example, its extensive utilization in clinical analysis and production control. The accomplishments resulting from automation are recognized even by laymen, who marvel at the knowledge gained by automated instruments in the analysis of the surfaces of the moon and of Mars. The computer is playing an ever increasing role in analysis and particularly in analytical research. This revolutionary development of analytical methodology is catalyzed by the demands made on analytical chemists, not only industrially and academically but also by society. Analytical chemistry has always played an important role in the development of inorganic, organic, and physical chemistry and biochemistry, as well as in that of other areas of the natural sciences such as mineralogy and geochemistry. In recent years, analytical chemistry—often of a rather sophisticated nature—has become increasingly important in the medical and biological sciences, as well as in the solving of such social problems as environmental pollution, the tracing of toxins, and the dating of art and archaeological objects, to mention only a few. In the area of atmospheric science, ozone reactivity and persistence in the stratosphere is presently a topic of great priority; extensive analysis is required both for monitoring atmospheric constituents and for investigating model systems.

One example of the increasing demands being made on analytical chemists is the growing need for speciation in characterizing chemical species. For example, in reporting that lake water contains dissolved mercury, it is necessary to report in which oxidation state it is present, whether as an inorganic salt or complex, or in an organic form and in which form.

As a result of the more or less revolutionary developments in analytical chemistry, portions of the first edition of the Treatise are becoming—and, to some extent, have become—out-of-date, and a revised, more up-to-date edition must take its place. In recognition of the extensive development and because of the increased specialization of analytical chemists, the editors have fortunately secured for the new edition the cooperation of experts as coeditors for various specific fields.

In essence, it is the objective of the second edition of the Treatise, as it was of the first edition (whose preface follows this one), to do justice to the theory and practice of contemporary analytical chemistry. It is a revision of Part I, which mirrors the development of analytical chemistry. Like the first edition, the second edition is not an extensive textbook; it attempts to present a thorough introduction to the methods of analytical chemistry and to provide the background for detailed evaluation of each topic.

Minneapolis, Minnesota   I. M. KOLTHOFF
Ann Arbor, Michigan   P. J. ELVING

Preface to the First Edition of the Treatise

The aims and objectives of this Treatise are to present a concise, critical, comprehensive, and systematic, but not exhaustive, treatment of all aspects of classical and modern analytical chemistry. The Treatise is designed to be a valuable source of information to all analytical chemists, to stimulate fundamental research in pure and applied analytical chemistry, and to illustrate the close relationship between academic and industrial analytical chemistry.

The general level sought in the Treatise is such that, while it may be profitably read by the chemist with the background equivalent to a bachelor's degree, it will at the same time be a guide to the advanced and experienced chemist—be he in industry or university—in the solution of his problems in analytical chemistry, whether of a routine or of a research character.

The progress and development of analytical chemistry during most of the first half of this century has generally been satisfactorily covered in modern textbooks and monographs. However, during the last fifteen or twenty years, there has been a tremendous expansion of analytical chemistry. Many new nuclear, subatomic, atomic, and molecular properties have been discovered, several of which have already found analytical application. In the development of techniques for measuring these and also the more classical properties, the revolutionary progress in the field of instrumentation has played a tremendous role.

It has been difficult, if not impossible, for anyone to digest this expansion of analytical chemistry. One of the objectives of the present Treatise is not only to describe these new properties, their measurement, and their analytical applicability, but also to classify them within the framework of the older classifications of analytical chemistry.

Theory and practice of analytical chemistry are closely interwoven. In solving an analytical chemical problem, a thorough understanding of the theory of analytical chemistry and of the fundamentals of its techniques, combined with a knowledge of and practical experience with chemical and physical methods, is essential. The Treatise as a whole is intended to be a unified, critical, and stimulating treatment of the theory of analytical chemistry, of our knowledge of analytically useful properties, of the theoretical and practical fundamentals of the techniques for their measurement, and of the ways in which they are applied to solving specific analytical problems. To achieve this purpose, the Treatise is divided into three parts: I, analytical chemistry and its methods; II, analytical chemistry of the elements; and III, the analytical chemistry of industrial materials.

Each chapter in Part I of the Treatise illustrates how analytical chemistry draws on the fundamentals of chemistry as well as on those of other sciences; it stresses for its particular topic the fundamental theoretical basis insofar as it affects the analytical approach, the methodology and practical fundamentals used both for the development of analytical methods and for their implementation for analytical service, and the critical factors in their application to both organic and inorganic materials. In general, the practical discussion is confined to fundamentals and to the analytical interpretation of the results obtained. Obviously then, the Treatise does not intend to take the place of the great number of existing and exhaustive monographs on specific subjects, but its intent is to serve as an introduction and guide to the efficient utilization of these specialized monographs. The emphasis is on the analytical significance of properties and of their measurement. In order to accomplish the above aims, the editors have invited authors who are not only recognized experts for the particular topics, but who are also personally acquainted with and vitally interested in the analytical applications. Only in this way can the Treatise attain the analytical flavor which is one of its principal objectives.

Part II is intended to be very specific and to review critically the analytical chemistry of the elements. Each chapter, written by experts in the field, contains in addition to a critical and concise treatment of its subject, critically selected procedures for the determination of the element in its various forms. The same critical treatment is contemplated for Part III. Enough information is presented to enable the analyst both to analyze and to evaluate a product.

The response in connection with the preparation of the Treatise from all colleagues has been most enthusiastic and gratifying to the editors. It is obvious that it would have been impossible to accomplish the aims and objectives cited in the preface without the wholehearted cooperation of the large number of distinguished authors whose work appears in this and future volumes of the Treatise. To them and to our many friends who have encouraged us we express our sincere appreciation and gratitude. In particular, considering that the Treatise aims to cover all of the aspects of analytical chemistry, the editors have found it desirable to solicit the advice of some colleagues in the preparation of certain sections of the various parts of the Treatise. They would like at this time to acknowledge their indebtedness to Professor Ernest B. Sandell of the University of Minnesota for his interest and active cooperation in the organizing and detailed planning of the Treatise.

Minneapolis, Minnesota　　　　　　　　　　　　　　　　　　　I. M. KOLTHOFF
Ann Arbor, Michigan　　　　　　　　　　　　　　　　　　　　P. J. ELVING

Acknowledgment

In view of the wide scope of the Treatise, it has been considered essential to have the advice and aid of experts in various areas of analytical chemistry. For the section on "Thermal Methods of Analysis," the editor has been fortunate to have the cooperation of Dr. C. B. Murphy of the Xerox Corporation as Associate Editor; his collaboration is acknowledged with gratitude.

P.J.E.

PART I. THEORY AND PRACTICE

CONTENTS—VOLUME 12

SECTION J.   Thermal Methods

3. Principles of Thermometry. By *John C. Melcher and Donald*

TREATISE ON ANALYTICAL CHEMISTRY

PART I

THEORY AND PRACTICE

SECOND EDITION

SECTION J: Thermal Methods

Part I
Section J

Chapter 1

NOMENCLATURE IN THERMAL ANALYSIS

By R. C. MACKENZIE, *The Macaulay Institute for Soil Research, Craigiebuckler, Aberdeen, Scotland*

Contents

I. INTRODUCTION

Discipline in nomenclature is essential in all branches of science as, in its absence, communication would be greatly restricted. In some subjects, a slow development has enabled nomenclature to develop naturally over a period of time, but in others a period of great activity has caused a sudden inrush of new and often confusing names. In these latter circumstances, standardization of nomenclature eventually becomes essential.

Before nomenclature can be standardized, all terms that have been used in the designated field must be surveyed and then correlated and classified to

1

bring out similarities and differences: although somewhat tedious, this procedure is essential for the development of a credible system. Subsequent work involves decision on the validity of synonyms and selection of names to be recommended for adoption. The criteria used in the last stage may vary in each instance. For example, in some circumstances choice can be based on chronology—i.e., selection of the earliest recorded name—whereas in others a fairly recent term is preferable systematically. Occasionally, no available term is suitable and a new name has to be coined. This procedure must not be adopted lightly, however, as people are essentially conservative, and powerful reasons have to be adduced for rejecting any name in current use even if it is not highly entrenched. For names that are well entrenched, a new term is most unlikely to be acceptable no matter how logical it may be; thus, it is unlikely that *differential thermometry* (43), despite its inherent accuracy, would ever replace *differential thermal analysis*. In this last stage, therefore, considerable judgement is required and a psychological approach is essential.

The somewhat rapid growth of thermal analysis in the two decades preceding the mid-1960s led to a confusing miscellany of terms, some of which were in everyday use and some of which were esoteric. Moreover, the same technique could be known by several names or, conversely, one name or abbreviation could be used for two quite different techniques. The time was therefore ripe for assessment of the position and the First International Conference on Thermal Analysis in 1965 set up a Nomenclature Committee that has continued to function under the aegis of the International Confederation for Thermal Analysis (ICTA) and in close collaboration with the Thermal Methods Group of the Analytical Division of the Royal Society of Chemistry. The reader is referred to an earlier publication (35) for detail on the *modus operandi* of this Committee.

So far, five reports on nomenclature in thermal analysis have been prepared and published by this Committee. The first (34), which deals with names and definitions of major techniques as well as general aspects, has been adopted by the International Union of Pure and Applied Chemistry (IUPAC) (17), the International Standards Organization (ISO) (14), and the American Society for Testing and Materials (ASTM) (1), and it and the second (36), which considers apparatus, technique, and curve nomenclature for differential thermal analysis (DTA) and thermogravimetry (TG), by Association Française de Normalisation (AFNOR) (2) and the Soil Science Society of America (40). An integrated combination of the second with the third (37) report has now been adopted by IUPAC (18). The fourth report (38), which sets out a new and logical system of classification and definition and extends the first report to techniques that have come into prominence since 1969, is being considered by IUPAC (19). The fifth report (39) deals solely with symbols. The material in all five reports is presented as a unified system in the following account, which also incorporates a few so far unpublished minor modifications.

All these reports refer to, and are definitive for, the English language only; however, various subcommittees are currently considering nomenclature in other languages. The Chinese-, French-, Italian- and Japanese-language subcommittees have already published definitive documents dealing with the matter in the various reports (5–7, 9–13, 16, 20–23, 25) and translations* of some have appeared in Chinese (30), Czech (3), Polish (27–29, 41), Portuguese (15), Rumanian (42), Russian (26), Slovenian (24), and Slovak (8).

II. GENERAL RULES

1. *Thermal analysis* and not "thermography" is accepted as the name in English, since the latter has at least two other meanings in this language, the major one being medical (4). The adjective is then *thermoanalytical* (cf. physical chemistry and physicochemical); the term "thermoanalysis" should not be used.

2. *Differential* is the adjectival form of difference; *derivative* is used for the first derivative (mathematical) of any curve.

3. The term "analysis" is to be avoided as far as possible since the methods considered do not comprise analysis as generally understood chemically: some terms that include the word *analysis* are too widely accepted, however, to be changed.

4. The term *curve* is preferred to "thermogram" for the following reasons. (a) "Thermogram" is used for the results obtained by the medical technique of thermography [see 1]. (b) If applied to certain curves (e.g., thermogravimetric curves), "thermogram" would not be consistent with the dictionary definition. (c) For clarity there would have to be frequent use of terms such as differential thermal thermogram, thermogravimetric thermogram, etc., which are not only cumbersome but also confusing.

5. In multiple techniques, *simultaneous* is used for the application of two or more techniques to the same sample at the same time (for formal definitions see Section IV.C); *combined* then indicates the use of separate samples for each technique.

6. Technique names based on proprietary names of commercial instruments are, as far as possible, avoided, since such names (a) give undue prominence to individual manufacturers and (b) become obsolete should the instruments involved be superseded.

III. TERMINOLOGY AND ABBREVIATIONS

The names and abbreviations adopted for some techniques considered at an early stage in the program, together with terms that were for various

---

* The gross errors that can arise through translation via one or two foreign languages back to the original have been outlined in a note in *J. Therm. Anal.*, **13**, 163 (1978).

TABLE 1.I
Recommended and Rejected Terminology for Some Thermoanalytical Techniques
(34, with modifications)

Acceptable name	Acceptable abbreviation	Rejected name(s)
A. General		
Thermal analysis		Thermography
		Thermoanalysis
B. Individual techniques		
Thermogravimetry	TG	Thermogravimetric analysis
		Dynamic thermogravimetric analysis
Derivative thermogravimetry	DTG	Differential thermogravimetry
		Differential thermogravimetric analysis
		Derivative thermogravimetric analysis
Isobaric mass-change determination		Dehydration curves
Evolved gas detection	EGD	Effluent gas detection
Evolved gas analysis[a]	EGA	Effluent gas analysis
		Gas effluent analysis
		Thermovaporimetric analysis
Heating curve determination[b]		Thermal analysis
Heating-rate curve determination[b]		Derivative thermal analysis
Inverse heating-rate curve determination[b]		
Differential thermal analysis	DTA	Dynamic differential calorimetry
Differential scanning calorimetry	DSC	Differential enthalpic analysis
		Enthalpography
C. Multiple Techniques		
Simultaneous TG-DTA, etc.		DATA (Differential and thermogravimetric analysis)
		Derivatography
		Derivatographic analysis

[a] Abbreviations such as MTA (mass-spectrometric thermal analysis) and MDTA (mass spectrometry and differential thermal analysis) should be avoided.

[b] When determinations are performed during the cooling cycle, these become *cooling curve determination, cooling-rate curve determination,* and *inverse cooling-rate curve determination,* respectively.

reasons rejected (either at the time or later), are listed in Table 1.I. **It is important to note** that approved abbreviations have been kept to the minimum and should always be in capital letters without periods. A suggestion that the limited number of abbreviations approved should be in the form given, irrespective of language, while it has much to recommend it, has not as yet been universally adopted in practice. Although only a limited number of abbreviations are officially approved, it may be convenient for brevity to use others in an individual paper **provided the technique is first named in full with the abbreviation adopted in brackets.**

No pronouncements are made on techniques that are considered border-line—e.g., calorimetry and thermometric titrimetry—and that may well be the subject of study by other bodies.

IV. DEFINITIONS AND CONVENTIONS

A. GENERAL

The only general definition is that of thermal analysis itself, and this poses considerable problems as it seems impossible to find a form of words that covers all generally recognized thermoanalytical techniques while at the same time excluding those that are nonthermoanalytical. The nearest seems to be the following:

Thermal analysis. A group of techniques in which a physical property of a substance* is measured as a function of temperature while the substance is subjected to a controlled temperature program.

B. INDIVIDUAL TECHNIQUES

1. Classification

A recent survey of about 100 proposed thermoanalytical techniques has led to selection of a certain number that have been, are, or may soon be of considerable general value. These have been classified in such a manner as to bring out interrelationships, and the resulting arrangement (Table 1.II) can readily be modified to accommodate additional physical properties and/or techniques as required. In a general classification such as this, one can only attain a certain level of distinction, beyond which the specialist must take over. Thus, for example, derivative techniques are not listed in Table 1.II, since derivative curves can be calculated or recorded for most measurements. Although most derivative techniques are distinguished by the adjective *derivative,* some, like *heating-rate curve determination,* have individual names and some, like *derivative thermogravimetry,* have an importance virtually equal to that of the parent technique. For these reasons, a few derivative techniques are separately defined below. In some definitions, conventions for standard plotting of curves are indicated.

2. Definitions

Thermogravimetry (TG). A technique in which the mass of a substance is measured as a function of temperature while the substance is subjected to a controlled temperature program.

---

* Throughout this text, *substance* is to be understood in the sense of *substance and/or its reaction product(s).*

TABLE 1.II

A Flexible and Logical Classification System for Thermoanalytical Techniques (38), with Examples of Techniques in Current Use

Physical property	Derived technique(s)	Acceptable abbreviation
Mass	Thermogravimetry	TG
	Isobaric mass-change determination	
	Evolved gas detection	EGD
	Evolved gas analysis	EGA
	Emanation thermal analysis	
	Thermoparticulate analysis	
Temperature	Heating curve determination	
	Differential thermal analysis	DTA
Enthalpy	Differential scanning calorimetry[a]	DSC
Dimensions	Thermodilatometry	
Mechanical characteristics	Thermomechanical measurement	
	Dynamic thermomechanical measurement	
Acoustic characteristics	Thermosonimetry	
	Thermoacoustimetry	
Optical characteristics	Thermoptometry	
Electrical characteristics	Thermoelectrometry	
Magnetic characteristics	Thermomagnetometry	

[a] The confusion that has arisen about this term seems best resolved by separating two modes *(power-compensation DSC* and *heat-flux DSC),* as described in the definition given in the text.

The record is the thermogravimetric or TG curve; the mass should be plotted on the ordinate decreasing downwards and temperature (*T*) or time (*t*) on the abscissa increasing from left to right.

Derivative thermogravimetry (DTG). A technique yielding the first derivative of the thermogravimetric curve with respect to either temperature or time.

The record is the derivative thermogravimetric or DTG curve; the derivative should be plotted on the ordinate with mass losses downwards and temperature or time on the abscissa increasing from left to right.

Isobaric mass-change determination. A technique in which the equilibrium mass of a substance at constant partial pressure of the volatile product(s) is measured as a function of temperature while the substance is subjected to a controlled temperature program.

The record is the isobaric mass-change curve; the mass should be plotted on the ordinate decreasing downwards and temperature on the abscissa increasing from left to right.

Evolved gas detection (EGD). A technique in which the evolution of gas from a substance is detected as a function of temperature while the substance is subjected to a controlled temperature program.

Evolved gas analysis (EGA). A technique in which the nature and/or amount of volatile product(s) released by a substance are/is measured as a function of temperature while the substance is subjected to a controlled temperature program.

The method of analysis should always be clearly stated.

Emanation thermal analysis. A technique in which the release of radioactive emanation from a substance is measured as a function of temperature while the substance is subjected to a controlled temperature program.

Thermoparticulate analysis. A technique in which the release of particulate matter from a substance is measured as a function of temperature while the substance is subjected to a controlled temperature program.

Heating curve determination. A technique in which the temperature of a substance is measured as a function of the programmed temperature while the substance is subjected to a controlled temperature program in the heating mode.

Sample temperature should be plotted on the ordinate increasing upwards and programmed temperature or time on the abscissa increasing from left to right.

Heating-rate curve determination. A technique yielding the first derivative of the heating curve with respect to time (i.e., dT/dt) plotted against temperature or time.

The function dT/dt should be plotted on the ordinate and temperature or time on the abscissa increasing from left to right.

Inverse heating-rate curve determination. A technique yielding the first derivative of the heating curve with respect to temperature (i.e. dt/dT) plotted against either temperature or time.

The function dt/dT should be plotted on the ordinate and temperature or time on the abscissa increasing from left to right.

Differential thermal analysis (DTA). A technique in which the temperature difference between a substance and a reference material is measured as a function of temperature while the substance and reference material are subjected to a controlled temperature program.

The record is the differential thermal or DTA curve; the temperature difference (ΔT) should be plotted on the ordinate with endothermic reactions downwards and temperature or time on the abscissa increasing from left to right.

Quantitative differential thermal analysis (quantitative DTA*). This term covers those uses of DTA where the equipment is designed to produce quantitative results in terms of energy and/or any other physical parameter.

---

* The abbreviation QDTA is considered not to be warranted.

*Differential thermal analysis (DTA) in an isothermal environment.** A variant of DTA in which the temperature difference between a substance and a reference material is continuously recorded against time as the two specimens are maintained in a nominally isothermal environment.

Differential scanning calorimetry (DSC). A technique in which the difference in energy inputs into a substance and a reference material is measured as a function of temperature while the substance and reference material are subjected to a controlled temperature program.

Two modes, *power-compensation differential scanning calorimetry (power-compensation DSC)* and *heat-flux differential scanning calorimetry (heat-flux DSC),†* can be distinguished depending on the method of measurement used.

Thermodilatometry. A technique in which a dimension of a substance under negligible load is measured as a function of temperature while the substance is subjected to a controlled temperature program.

The record is the thermodilatometric curve; the dimension should be plotted on the ordinate increasing upwards and temperature or time on the abscissa increasing from left to right.

Linear thermodilatometry and *volume thermodilatometry* are distinguished on the basis of the dimension measured.

Thermomechanical measurement. A technique in which the deformation of a substance under nonoscillatory load is measured as a function of temperature while the substance is subjected to a controlled temperature program.

The mode, as determined by the type of stress applied (compression, tension, flexure, or torsion), should always be stated.

Dynamic thermomechanical measurement. A technique in which the dynamic modulus and/or damping of a substance under oscillatory load is measured as a function of temperature while the substance is subjected to a controlled temperature program.

Torsional braid measurement is a particular case of dynamic thermomechanical measurement in which the material is supported on a braid.

Thermosonimetry. A technique in which the sound emitted by a substance is measured as a function of temperature while the substance is subjected to a controlled temperature program.

Thermoacoustimetry. A technique in which the characteristics of imposed acoustic waves are measured as a function of temperature after passing

---

* The term "isothermal DTA" is, from the definition of differential thermal analysis, incorrect.

† The ICTA Nomenclature Committee considers a system with multiple sensors (e.g., a Calvet-type arrangement) or with a controlled heat leak (Boersma-type arrangement) would be heat-flux DSC, whereas systems without these or equivalent arrangements would be quantitative DTA.

through a substance while the substance is subjected to a controlled temperature program.

Thermoptometry. A technique in which an optical characteristic of a substance is measured as a function of temperature while the substance is subjected to a controlled temperature program.

Measurement of total light, light of specific wavelength(s), refractive index, and luminescence lead to *thermophotometry, thermospectrometry, thermorefractometry,* and *thermoluminescence,* respectively; observation under the microscope leads to *thermomicroscopy.* Other terms may have to be added.

Thermoelectrometry. A technique in which an electrical characteristic of a substance is measured as a function of temperature while the substance is subjected to a controlled temperature program.

The most common measurements are of resistance, conductance, or capacitance.

Thermomagnetometry. A technique in which a magnetic characteristic of a substance is measured as a function of temperature while the substance is subjected to a controlled temperature program.

C. MULTIPLE TECHNIQUES

Simultaneous techniques are being used increasingly, particularly in industrial applications where time and cost have to be kept to the minimum. The following definitions apply:

Simultaneous techniques. The application of two or more techniques to the same sample at the same time—e.g., simultaneous thermogravimetry and differential thermal analysis.

In writing, the names of simultaneous techniques, when used in full, should be separated by *and* (see above) and, when abbreviated acceptably, by a hyphen—e.g., simultaneous TG-DTA. All abbreviations should be in capital letters without periods, unless this is contrary to established practice for any technique outside the field of thermal analysis.

Coupled simultaneous techniques. The application of two or more techniques to the same sample when the two instruments involved are connected through an interface*—e.g., simultaneous differential thermal analysis and mass spectrometry.

In coupled simultaneous techniques, as in discontinuous simultaneous techniques (below), the first technique to be mentioned is that by which the first, in time, measurement is made—e.g., when a DTA instrument and a mass spectrometer are connected through an interface, DTA-MS is the correct form and **not** MS-DTA.

---

* A specific piece of equipment that enables two instruments to be joined together.

Discontinuous simultaneous techniques. The application of coupled techniques to the same sample when sampling for the second technique is discontinuous—e.g., discontinuous simultaneous differential thermal analysis and gas chromatography (discontinuous DTA-GC) when discrete portions of evolved volatile(s) are collected from the sample situated in the instrument used for the first technique for examination by the second.

D. DTA AND TG APPARATUS AND TECHNIQUE

In selecting appropriate terms for apparatus and technique, certain arbitrary choices had to be made as there is sometimes little to choose between alternative terms. In the definitions below, it is assumed that the thermo-sensing device is a thermocouple; should another sensor be used its name should replace *thermocouple* throughout.

1. DTA

The *sample* is the actual material investigated, whether diluted or undiluted.

The *reference material* is a known substance, usually inactive thermally over the temperature range of interest.

The *specimens* are the sample and reference material.

The *sample holder* is the container or support for the sample.

The *reference holder* is the container or support for the reference material.

The *specimen-holder assembly* is the complete assembly in which the specimens are housed. When the heating or cooling source is incorporated in one unit with the containers or supports for the sample and reference material, this would be regarded as part of the specimen-holder assembly.

A *block* is a type of specimen-holder assembly in which a relatively large mass of material is in intimate contact with the specimens or specimen holders.

The *differential* (or ΔT) *thermocouple* is the thermocouple system used to measure temperature difference.

2. TG

A *thermobalance* is an apparatus for weighing a sample continuously while it is being heated or cooled.

The *sample* is the actual material investigated, whether diluted or undiluted.

   In TG, samples are normally not diluted, but in simultaneous TG-DTA diluted samples are sometimes used.

The *sample holder* is the container or support for the sample.

3. DTA and TG

The *temperature* (or *T*) *thermocouple* is the thermocouple system used to measure temperature; its position with respect to the sample should always be stated.

The *heating rate* is the rate of temperature increase, which is customarily quoted in degrees per minute (on the Celsius or kelvin scales). Correspondingly, the *cooling rate* is the rate of temperature decrease. The heating or cooling rate is said to be *constant* when the temperature/time curve is linear.

In simultaneous TG-DTA, definitions follow from those given for TG and DTA separately.

E. DTA AND TG CURVES

The reasons for rejection of the imprecise term "thermogram," which regrettably still appears in the literature, have already been given (Section II). The conventions to be followed in reporting DTA and TG results, as specified by the Standardization Committee of ICTA (31), and the conventions for curves given above (Section IV.B) must be borne in mind in reading the definitions below.

1. DTA

In DTA **it must be remembered** that, although the ordinate is conventionally labeled ΔT, the measurement recorded is usually the output from the sensor used. For thermocouples, and indeed some other sensors, output can vary markedly with temperature so that any scale on the ΔT axis will be valid for one temperature only, and this temperature should be clearly stated.

All definitions refer to a single peak such as that shown in Fig. 1.1: multiple peak systems showing shoulders or more than one maximum or minimum can be considered to result from the superposition of single peaks. Several ways of interpolating the base line have been proposed and that used in Fig. 1.1 is purely exemplary. It must be noted, however, that the location of points B and D depends on the method of interpolation.

The *base line* (AB and DE, Fig. 1.1) corresponds to the portion or portions of the DTA curve for which ΔT is approximately zero.

A *peak* (BCD, Fig. 1.1) is that portion of the DTA curve that departs from and subsequently returns to the base line.

An *endothermic peak* or *endotherm* is a peak where the temperature of the sample falls below that of the reference material; that is, ΔT is negative.

An *exothermic peak* or *exotherm* is a peak where the temperature of the sample rises above that of the reference material; that is, ΔT is positive.

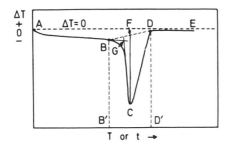

Fig. 1.1. Formalized DTA curve.

Peak width (B'D', Fig. 1.1) is the temperature or time interval between the points of departure from and return to the base line.

Peak height (CF, Fig. 1.1) is the distance, vertical to the temperature or time axis, between the interpolated base line and the peak tip (C, Fig. 1.1).

Peak area (BCDB, Fig. 1.1) is the area enclosed between the peak and the interpolated base line.

The *extrapolated onset* (G, Fig. 1.1) is the point of intersection of the tangent (GC, Fig. 1.1) drawn at the point of greatest slope on the leading edge of the peak with the extrapolated base line (BG, Fig. 1.1).

2. TG

All definitions refer to a single-stage process such as that shown in Fig. 1.2: multistage processes can be considered as resulting from a series of single-stage processes.

A *plateau* (AB, Fig. 1.2) is that part of the TG curve where the mass is essentially constant.

The *initial temperature, T_i,* (B, Fig. 1.2) is that temperature (on the Celsius or kelvin scales) at which the cumulative mass change reaches a magnitude that the thermobalance can detect.

The *final temperature, T_f,* (C, Fig. 1.2) is that temperature (on the Celsius or kelvin scales) at which the cumulative mass change reaches a maximum.

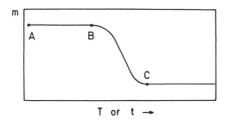

Fig. 1.2. Formalized TG curve.

The *reaction interval* is the temperature difference between T_f and T_i, as defined above.

F. SYMBOLS

The following recommendations relate to symbols employed in connection with TG, DTG, DTA, and DSC—currently the most widely used techniques:

1. The international system of units (SI units) should be adhered to, except in rare instances where recommended symbols conflict with long-established practice.

2. The use of symbols with superscripts, such as $\dot{T}$, should be avoided.

3. The use of double subscripts, such as T_{sp} or T_{pd}, should be avoided. If such symbols are deemed necessary, they must be clearly defined on first introduction in the publication.

4. Notwithstanding recommendation 1 above, the symbol T should be used for temperature whether expressed in degrees Celsius (°C) or in kelvin (K). For temperature interval the symbol K or °C can be used in accordance with Resolution 3 of the 13th General Conference of Weights and Measures (CGPM) (see *Compt. Rend. 13th CGPM,* 1967–68, p. 104).

5. The symbol t should be used for time, whether expressed in seconds (s), minutes (min) or hours (h).

6. The heating rate can be expressed either as dT/dt, when a true derivative is intended, or as β in K min^{-1} (see recommendation 4 above). The heating rate so expressed need not be constant and can be positive or negative.

7. The symbols m for mass and W for weight* are recommended.

8. The symbol α is recommended for the fraction reacted.

9. The ordinate in DTA should be expressed in terms of ΔT, the difference in temperature between the sample and the reference material.

10. The ordinate in DSC should be expressed in terms of dQ/dT or dQ/dt rather than dH/dT or dH/dt, since Q represents quantity of heat or electricity whereas H represents enthalpy.

11. The following rules should be observed for subscripts: (a) Where the subscript relates to an object, it should be a capital letter, e.g., m_S represents the mass of the sample; T_R represents the temperature of the reference material. (b) Where the subscript relates to a phenomenon occurring, it should be in lower case, e.g., T_g represents the glass transition temperature; T_c represents the temperature of crystallization; T_m represents the tempera-

---

* A quantity of the same nature as a force—i.e., the product of mass and the acceleration due to gravity (see *Compt. Rend. 3rd CGPM,* 1901, p. 70).

ture of melting; T_σ represents the temperature of a solid-state transition.* (c) Where the subscript relates to a specific point in time or to a point on the curve, it should be in lower case or in figures, e.g., T_i represents the initial temperature; m_f represents the final mass; $t_{0.5}$ represents the time at which the fraction reacted is 0.5; $T_{0.3}$ represents the temperature at which the fraction reacted is 0.3; T_p represents the temperature of the peak; T_e represents the temperature of the extrapolated onset (as defined in IV.E.1 above).

V. CONCLUSIONS

The definitions and guidelines listed above form a fairly comprehensive and logical system of nomenclature for the main thermoanalytical techniques. They should always be read, however, in conjunction with the recommendations of the Standardization Committee of ICTA on DTA and TG (31), EGA (32), and thermomechanical measurements (33). New techniques invariably emerge from time to time and the Nomenclature Committee of ICTA is always willing to consider and recommend appropriate terminology for any such brought to its notice. It also keeps a watching brief on thermal analysis nomenclature in scientific journals and publications and draws the attention of authors and editors, where necessary, to international standards. This service should eventually lead to an acceptable code of practice being universally adopted.

ACKNOWLEDGMENTS

The author wishes to express his thanks to the Nomenclature Committee of ICTA for permission to use material as yet unpublished. As Chairman of this Committee, he also acknowledges his indebtedness to all who have, from the inception of this program in 1965 to its completion in 1980, served in any capacity—namely, (the late) Prof. L. G. Berg (USSR), Dr. D. T. Y. Chen‡ (Hong Kong), Dr. T. Daniels† (UK, Vice-Chairman), Dr. D. Dollimore† (UK), Mr. J. A. Forrester (UK), Dr. J. H. Flynn† (USA), Prof. P. D. Garn† (USA), Dr. B. O. Haglund† (Scandinavia), Dr. M. Harmelin (France), Dr. J. O. Hill† (Australia), Dr. A. A. Hodgson (UK), Prof. H. Kambe†‡ (Japan), (the late) Dr. M. D. Karkhanavala (India), Dr. C. J. Keattch† (UK, Secretary since 1965), Dr. G. M. Kline† (USA), Prof. D. Krug‡ (German Federal Republic), Prof. G. Lombardi†‡ (Italy), Dr. H. G. McAdie† (Canada), Dr. C. B. Murphy† (USA), Prof. H. R. Oswald (Switzerland), Dr. J. P. Redfern† (UK), Dr. J. Rouquerol†‡ (France), Dr. J. Šesták† (Czechoslovakia), Dr. R. Setton (France), Dr. J. H. Sharp† (UK), Dr. O. T. Sørensen† (Denmark), Dr. H. G. Wiedemann (Switzerland), and Dr. F. W. Wilburn† (UK). All of these have freely given their time and expertise to advance the aims of the Committee and to ensure the successful outcome of its work.

* The greek σ is recommended as subscript here since T_s could possibly be confused with T_S and double subscripts are to be avoided (see recommendation 3).

† Members of Committee in 1980.
‡ Organizers of language subcommittees in 1980.

REFERENCES

1. American Society for Testing and Materials, *Annual Book of ASTM Standards*, ASTM, Philadelphia, Pt. 41, E473-73, 1973 (see also E473-82, 1982).
2. Association Française de Normalisation, *Fascicule de Documentation*, NF T01-021, 1974.
3. Bárta, R., *Silikáty*, **12**, 377 (1968).
4. Cade, C. M., and B. V. Barlow, *Sci. Progr. (London)*, **55**, 167 (1967).
5. Chen, D. T. Y., *Chemistry, Taiwan*, No. 1, 26 (1977).
6. Chen, D. T. Y., *Chemistry, Taiwan*, No. 2, B1 (1979).
7. Chen, D. T. Y., *Chemistry, Taiwan*, No. 2, A18 (1980).
8. Fajnor, V. S., "Návrh Slovenského Názvoslovia Termickej Analýzy" in M. Vaniš, *Zborn. VIII Celoštát, Konf. Term. Analýze, Vysoké Tatry, 1979*, SVST, Bratislava, 1979, p. 49.
9. French Sub-Committee on Nomenclature in Thermal Analysis, *Analusis*, **2**, 459 (1973).
10. French Sub-Committee on Nomenclature in Thermal Analysis, *Actualité Chim. (Paris)*, No. 4, 35 (1973).
11. French Sub-Committee on Nomenclature in Thermal Analysis, *J. Therm. Anal.*, **6**, 241 (1974).
12. French Sub-Committee on Nomenclature in Thermal Analysis, *Thermochim. Acta*, **8**, 325 (1974).
13. French Sub-Committee on Nomenclature in Thermal Analysis, *Analusis*, **3**, 236 (1975).
14. International Standards Organisation, "Plastics—Definition of Terms," *ISO Recommendations R472-1969: Addendum 4*, August 1975.
15. Ionashiro, M. and I. Giolito, *Cerâmica, Brasil*, **26**, 17 (1980).
16. Italian Sub-Committee on Nomenclature in Thermal Analysis, *Notiz. Ital. ICTA Gruppo Italiano*, No. 9, 3 (1972); No. 10, 2 (1972).
17. IUPAC, Analytical Chemistry Division, Commission on Analytical Nomenclature, *Pure Appl. Chem.*, **37**, 439 (1974).
18. IUPAC, Analytical Chemistry Division, Commission on Analytical Nomenclature, and ICTA Nomenclature Committee *Pure Appl. Chem.*, **52**, 2385 (1980).
19. IUPAC, Analytical Chemistry Division, Commission on Analytical Nomenclature, and ICTA Nomenclature Committee, *Pure Appl. Chem.*, **53**, 1597 (1981).
20. Japanese Sub-Committee on Nomenclature in Thermal Analysis, *Calorim. Therm. Anal. Newsl. (Tokyo)*, **1**, 22 (1970).
21. Japanese Sub-Committee on Nomenclature in Thermal Analysis, *Calorim. Therm. Anal. Newsl. (Tokyo)*, **2**, 35 (1971).
22. Japanese Sub-Committee on Nomenclature in Thermal Analysis, *Calorim. Therm. Anal. Newsl. (Tokyo)*, **2**, 62 (1971).
23. Japanese Sub-Committee on Nomenclature in Thermal Analysis, *Calorim. Therm. Anal. (Tokyo)*, **2**, 53 (1975).
24. Jernejčič, J., *Vest. Sloven. Kem. Drustva*, **20**, 51 (1973).
25. Kambe, H., *Calorim. Therm. Anal. (Tokyo)*, **5**, 167 (1978).
26. Kashik, D. S., "Terminologiya v Termicheskom Analize," in G. E. Domburg *et al.*, *Termicheskii Analiz. Tezisy Dokladov VII Vsesoyuznogo Soveshchaniya* [Thermal Analysis. Summaries of Papers Presented at the VII All-Union Conference.], Zinatne, Riga, 1979, Vol. 2, p. 139.
27. Langier-Kuzniarowa, A., *Przeglad Geol.*, **21**, 42 (1973).
28. Langier-Kuzniarowa, A., and L. Stoch, *Mineral. Polon.*, **4**, 97 (1973).

29. Langier-Kuzniarowa, A., and L. Stoch, *Chem. Anal. (Warsaw)*, **20,** 669 (1975).

30. Liu, Zhen-Hai, *Hua Hsueh Tung Pao*, No. 4, 235 (1981).

31. McAdie, H. G., *Anal. Chem.*, **39,** 543 (1967).

32. McAdie, H. G., *Anal. Chem.*, **44,** 640 (1972).

33. McAdie, H. G., *Anal. Chem.*, **46,** 1146 (1974).

34. Mackenzie, R. C., *Talanta,* **16,** 1227 (1969).

35. Mackenzie, R. C., *J. Therm. Anal.,* **4,** 215 (1972).

36. Mackenzie, R. C., C. J. Keattch, D. Dollimore, J. A. Forrester, A. A. Hodgson, and J. P. Redfern, *Talanta,* **19,** 1079 (1972).

37. Mackenzie, R. C., C. J. Keattch, T. Daniels, D. Dollimore, J. A. Forrester, J. P. Redfern, and J. H. Sharp, *J. Therm. Anal.,* **8,** 197 (1975).

38. Mackenzie, R. C., *Thermochim. Acta,* **28,** 1 (1979).

39. Mackenzie, R. C., *Thermochim. Acta,* **46,** 333 (1981).

40. Soil Science Society of America, *Glossary of Soil Science Terms,* Soil Science Society of America, Madison, Wisconsin, 1978, p. 22.

41. Stoch, L., *Mineral. Polon.* **6,** 101 (1976).

42. Todor, D. N., *Rev. Chem. (Bucharest),* **24,** 822 (1973).

43. Wendlandt, W. W., *Thermal Methods of Analysis,* 2nd ed., Wiley, New York, 1974, p. 134.

ELEMENTS OF CHEMICAL THERMODYNAMICS: INTRODUCTION TO THERMAL METHODS

By Donald D. Wagman, *Chemical Thermodynamics Division, National Bureau of Standards, Washington, D.C.*

Contents

I. INTRODUCTION

As one studies the fundamental concepts and principles of the various branches of chemistry, certain basic ideas and rules are found to be common to all of them. Some of these relate to the structural concepts of nature as exemplified by the developments of atomic theory, quantum mechanics, and statistical mechanics; others involve the concepts of the transformations between heat and work and the relations between these transformations and chemical or physical changes in systems. The general study of these transformations is the province of chemical thermodynamics.

The origins of thermodynamics are phenomenological, and its history is interwoven with the developments of engineering and the study of heat engines. This early evolution led to the postulation of two simple laws of very general applicability. The further logical development of the ideas based on these laws provided the unifying principles for the rational understanding of observations in many branches of chemistry, including problems in chemical affinity, chemical metallurgy, electrochemistry, and colloidal and surface chemistry.

It is not our purpose here to give a detailed and complete treatment of the concepts of chemical thermodynamics. There are currently available a number of textbooks of differing degrees of complexity and rigor devoted to that aim (9,12,15). Nor shall we attempt to describe and discuss the many ways in which chemical thermodynamics relates to all areas of analytical chemistry. An analysis of the application of thermodynamics to general systems in chemical equilibrium has been presented in an earlier chapter (8); the reader will find there a presentation of the fundamental laws of thermodynamics. In the following chapters, we will be concerned with the application of techniques of analysis that involve the measurement of temperature changes as an indicator of the amount of chemical reaction or as an indication of the chemical composition of a system existing in multiphase equilibrium. In some applications, such as enthalpimetric titrations, the evolution of heat— or more strictly, the variation in heat evolution—may be used to measure quantitatively the extent of a reaction which does not lend itself to measurement by more traditional techniques. In other techniques, such as cryoscopy or ebulliometry, the temperature at which two phases, usually a solid and liquid or a liquid and vapor, exist at equilibrium is related to the (thermodynámic) composition of the phases. We may also be concerned with the manner in which the temperature of the equilibrium varies with the composition of one or both phases.

The first class of methods may be considered first law methods. They involve thermal measurements, in which the quantity of heat is directly proportional to the amount of reaction. The second class of techniques may be considered as second law methods, in which an experimentally determined change, as in temperature, is related essentially to a logarithmic variation in a concentration variable. In this chapter we shall endeavor only to show how these methods derive from the fundamental laws of thermodynamics. Detailed discussions of the various techniques will be given in the following chapters.

II. THE FIRST LAW OF THERMODYNAMICS

The phenomenological foundation of the first law is the principle of conservation of energy. If we consider a system of constant mass, in which q equals the heat absorbed, w is the mechanical (PV) energy absorbed, and u equals all other forms of energy absorbed, then we may write for any process going from state A to state B

$$\Delta E = E_B - E_A = \Sigma q + \Sigma w + \Sigma u \qquad (1)$$

For a process with only PV work of expansion in which $\Sigma w = - \int p \, dV$, equation 1 becomes

$$\Delta E = q - \int p \, dV \tag{2}$$

For a constant pressure process, $\Delta E = q - p\Delta V$ or

$$(\Delta E + p\Delta V) = q; \quad dP = 0 \tag{3}$$

If we define the enthalpy $H = E + PV$, equation 3 becomes

$$H_B - H_A = \Delta H = q; \quad (dP = 0)$$

The quantity $(dH/dT)_P = C_p$ defines the quantity called the heat capacity at constant pressure; similarly the quantity $(dE/dT)_V = C_v$ defines the heat capacity at constant volume.

III. THE SECOND LAW OF THERMODYNAMICS

Every system, if undisturbed by external forces, will change spontaneously until it approaches a final state of rest or equilibrium. Two examples of natural processes are the free expansion of a gas from a high-pressure region to one of low pressure and the flow of heat from a region of high temperature to one of lower temperature. It is also a fact of experience that no system will spontaneously change in the opposite direction, that is, away from equilibrium. For such an unnatural process to occur, some external force is required. This force may be said to increase the capacity for spontaneous change of the system. We can associate with this capacity for spontaneous change a thermodynamic property called the entropy S. This quantity is defined for a *reversible* process as the quantity dq/T, in which the quantity dq is the measure of the amount of heat entering the system from the surroundings. At the same time, the entropy of the surroundings decreases by the same amount.

The definition of entropy change given here is the classical one based on the existence of reversible processes. To determine the entropy change for an irreversible process, it is convenient to select a series of reversible processes that will take the system of interest from the initial state to the desired final state and compute the entropy change as the sum of the changes for the individual steps, each step being computed from the relation

$$\Delta S = \int (dq/T)_{\text{rev}} \tag{4}$$

While the irreversible process and the reversible steps connect the same initial and final states of the system, this is not true for the surroundings. The second law of thermodynamics tells us that if we include all entropy changes, for the reversible processes,

$$\Delta S_{\text{surroundings}} + \Delta S_{\text{system}} = 0 \tag{5}$$

and for the irreversible processes,

$$\Delta S_{\text{surroundings}} + \Delta S_{\text{system}} > 0 \tag{6}$$

If we confine our attention to reversible processes in which only mechanical work is done, we can write the first law of thermodynamics in differential form as

$$dE = dq + dw$$
$$= T dS - p dV \tag{7}$$

Hence at constant volume we can write

$$(dE/dT)_V = T(dS/dT)_V \tag{8}$$
$$(dS/dT)_V = C_V/T \tag{9}$$

Also from our starting equation (7), by adding the equality $d(PV) = P dV + V dP$ we obtain

$$dE + d(PV) = T dS + V dP$$

or

$$d(E + PV) = dH = T dS + V dP \tag{10}$$

For a constant pressure process

$$(dH)_P = T(dS)_P$$
$$(dH/dT)_P = C_P = T(dS/dT)_P$$
$$(dS/dT)_P = C_P/T \tag{11}$$

Certain other combinations of the fundamental thermodynamic variables are also of importance. Thus, if we subtract the identity $d(TS) = T dS + S dT$ from both sides of equation (7) we obtain

$$dE - d(TS) = -p dV - S dT$$
$$d(E - TS) = dA = -p dV - S dT$$

The quantity A is frequently referred to as the Helmholtz energy.

By subtracting the same identity $d(TS) = T dS + S dT$ from equation (10) we obtain

$$dH - d(TS) = dG = V dP - S dT \tag{12}$$

The quantity $G \equiv H - TS$, now called the Gibbs energy or free enthalpy, is the quantity formerly called the free energy and designated by the symbol F in the earlier traditional American notation. Present-day international convention has now assigned this quantity the symbol and name used here (10).

IV. THERMODYNAMIC CRITERIA FOR EQUILIBRIUM

The first two laws of thermodynamics have served to define two fundamental quantities E and S; combined with our experimental variables P, V,

T, we have five basic properties. We have also seen that for some purposes certain special combinations of these quantities are sufficiently important to merit being given special designations, such as H and G. We shall examine briefly the reasons for some of these choices.

If we write equation 1 in a differential notation we have

$$dE = dq + dw' - P\,dV$$

in which dw' represents all energy absorbed by the system other than heat or PV work. If now we let $-dw'$ represent work done by the system, we can write (replacing dq by $T\,dS$)

$$-dw' = T\,dS - P\,dV - dE \qquad (13)$$

$$= -(dE + P\,dV - T\,dS)$$

Any system undergoing a spontaneous change can be made to do useful work, at least in principle. Hence, for such a system the capacity for spontaneous change as measured by the amount of useful work that it can perform is expressible in terms of the fundamental properties E, S, P, V, and T by the relation

$$-dw' = -(dE + P\,dV - T\,dS) > 0 \qquad (14)$$

For a system at equilibrium, which is in a state of rest and hence cannot perform work, we have the criterion

$$-dw' = -(dE + P\,dV - T\,dS) = 0 \qquad (15)$$

The usefulness of this criterion can be illustrated by applying it to certain special conditions. For instance, we may write equation 13 as

$$-dw' = -[dE + d(PV) - V\,dP - d(TS) + S\,dT]$$

and if we set the experimental variables P and T as constant, i.e., $dT = dP = 0$, our equation becomes

$$-dw' = -d(E + PV - TS) = -dG$$

and the criterion for a spontaneous reaction under the conditions of constant T and P becomes

$$dG < 0 \qquad (16)$$

and for equilibrium

$$dG = 0 \qquad (17)$$

Similarly for conditions of constant T and V, the criteria become

$$d(E - TS) = dA < 0 \quad \text{for a spontaneous reaction} \qquad (18)$$

$$= dA = 0 \quad \text{for equilibrium}$$

The general expression 13 will take various other forms depending on what conditions we may apply, but the ones given above are of most applicability to the usual problems of chemical thermodynamics.

Before going ahead with the application of thermodynamics to systems of varying chemical composition, let us examine some of the relationships that exist among the thermodynamic quantities. If we start with the general equation applicable to a reversible process with only work of expansion (PdV) being done

$$dE = TdS - PdV \qquad (19)$$

and make use of the definitions for $H(= E + PV)$, $A(= E - TS)$, and $G(= H - TS)$ we obtain

$$dH = TdS + VdP \qquad (20)$$

$$dA = -SdT - PdV \qquad (21)$$

$$dG = VdP - SdT \qquad (22)$$

All of these fundamental relations are of the general form

$$dX = xdu + y\,dv$$

and result in partial differential relationships of the form

$$(dX/du)_v = x$$

$$(dX/dv)_u = y$$

Thus,

$$(dE/dS)_V = T$$

$$(dG/dP)_T = V$$

$$(dG/dT)_P = -S$$

Furthermore, for equations of this form we have the general criterion

$$(dx/dv)_u = (dy/du)_v$$

Each fundamental equation yields a relation of the above form, thus from equation 22

$$-(dS/dP)_T = (dV/dT)_P \qquad (23)$$

and from equation 21

$$(dS/dV)_T = (dP/dT)_V \qquad (24)$$

These relations among the various differential coefficients of the thermodynamic functions represent one of the most important tools in the successful application of thermodynamic principles to actual chemical and physical systems.

V. THE THIRD LAW OF THERMODYNAMICS

The first and second laws of thermodynamics have defined the quantities E, H, and S in terms of experimental measurements of the changes in these quantities. However, they do not provide any basis for a determination of the absolute values of the quantities. The heat content of a substance is measured relative to the value in an arbitrary reference state; heats of reaction represent the difference in heat content between the final product states and the initial reactant states. It would be possible to define the state consisting of isolated electrons and nuclei all at 0 K as an absolute zero, or even the isolated gaseous atoms at 0 K (although the gaseous ionic species would then represent negative energy values), but this would introduce the inconvenience of large numerical quantities into thermochemical and thermodynamic data with no corresponding gain in rigor. So for purposes of tabulation of thermochemical (heat of reaction) data it is customary to select as zero the values for the elements in their normal states (gaseous elements as ideal gases) at the appropriate temperature. The heat content or energy content of any substance, when determined as a function of temperature, is usually measured relative to the value for that substance at 0 K. The reference state for each substance must thus be reached by some suitable extrapolation from the lowest temperature of measurement, but the value at that temperature is represented by the symbol H_0 to which no absolute numerical value may be assigned. Thus we may write

$$H_T - H_0 = \int_{T=0}^{T=T_i} C_{\text{theor}} \, dT + \int_{T_i}^{T} C_{\text{exp}} \, dT \tag{25}$$

The first term on the right-hand side of the equation represents the necessary theoretical contribution from 0 K to the lowest temperature of experiment, and the second term represents the experimentally measured contribution.

With respect to the calculations of entropy for individual substances, the second law provides a similar relation, i.e.,

$$S_T - S_0 = \int_0^{T_i} (C_p/T)_{\text{theor}} \, dT + \int_{T_i}^{T} (C_p/T)_{\text{exp}} \, dT \tag{26}$$

The second law says nothing about the evaluation of S_0. However, the third law of thermodynamics, which may be expressed in many different ways, states that for any perfect crystalline substance, element, or compound, the value of the entropy at 0 K may be set equal to zero. For nonperfect crystalline materials the entropy will be finite and positive. This enables us to assign values to the constant S_0 and thus to obtain an absolute scale of values for entropy.

The developments in the field of quantum statistical mechanics have now expanded and generalized our understanding of the significance of entropy and the third law. In these terms, the thermodynamic state of zero entropy is defined as that in which the system is in the lowest quantum states compatible with the distribution law appropriate for the system. This more general description enables us to assign absolute entropy values to hypothetical ideal gases as well as to condensed phases. It also introduces entropy contributions from such things as nuclear spins. As these contributions ordinarily remain unchanged in the course of chemical reactions, it is customary in thermodynamics to list values of "virtual" entropies, i.e., absolute entropies minus the contributions due to nuclear spin or isotopic mixing.

VI. EQUILIBRIUM BETWEEN DIFFERENT PHASES OF A PURE COMPONENT

For any pure substance existing in a single phase, only two variables such as T and P need be specified to determine the values of all the other thermodynamic properties of the system. Such a system may be said to have two degrees of freedom.

Consider the case of two phases of a single substance coexisting at equilibrium at a given T and P. If we designate the phases as 1 and 2, respectively, the criterion for equilibrium is (equation 17)

$$G_1(T, P) = G_2(T, P)$$

If we shift the T and P to a slightly different set of values so as to maintain equilibrium

$$G_1 + dG_1 = G_2 + dG_2$$

or

$$dG_1 = dG_2$$

But by equation 22

$$dG = (dG/dP)_T\, dP + (dG/dT)_P\, dT = V\, dP - S\, dT$$

Hence we may write

$$V_1\, dP - S_1\, dT = V_2\, dP - S_2\, dT$$

or

$$dP/dT = (S_2 - S_1)/(V_2 - V_1) = \Delta S/\Delta V \qquad (27)$$

For this system the changes in the variables P and T are related by equation 27 so that only one variable may be independently assigned. Such a system has one degree of freedom. A solid and liquid or a liquid and its saturated vapor represent examples of a univariant system.

If a third phase were present, another relation similar to equation (27) would be established. Under these circumstances, the only solution would be that dP and dT are identically zero, i.e., that no variations of T and P are allowed. Such a system is described as nonvariant or invariant.

VII. FUGACITY, STANDARD STATES, AND ACTIVITY

The Gibbs energy change for any substance as a function of pressure is given by the relation

$$(dG/dP)_T = V \tag{28}$$

For an ideal gas $V_i = RT/P$ and, at constant T,

$$G_i(P_2) - G_i(P_1) = RT \ln(P_2/P_1) \tag{29}$$

For any real gas one can write

$$V_r = RT/P - \beta = V_i - \beta \tag{30}$$

Experimentally, a real gas approaches the properties and behavior of an ideal gas as the pressure of the gas approaches zero. Therefore, as $P \to 0$, $V_r \to V_i$ and $G_r \to G_i$. Integrating equation 28 for the real gas, using equation 30 we obtain

$$G_r(P_2) - G_r(P_1) = RT \ln(P_2/P_1) - \int_{P_1}^{P_2} \beta \, dP \tag{31}$$

Combining equations 29 and 31

$$(G_r - G_i)_{P_2} - (G_r - G_i)_{P_1} = - \int_{P_1}^{P_2} \beta \, dP \tag{32}$$

If now $P_1 \to 0$, and $P_2 = P$,

$$(G_r - G_i)_P = - \int_0^P \beta \, dP$$

But $G_i(P) = G_i(P = 1) + RT \ln P$, hence

$$G_r(P) = G_i(P = 1) + RT \ln P - \int_0^P \beta \, dP \tag{33}$$

The fugacity f is defined by the relation

$$RT \ln f = RT \ln P - \int_0^P \beta \, dP \tag{34}$$

and consequently

$$G_r(P) = G_i(P = 1) + RT \ln f \tag{35}$$

From the derivation of equation 35, the state corresponding to $G_i(P = 1)$ is reached by expansion of the real gas to a very low pressure such that the real gas behaves as an ideal gas, followed by compression, as an ideal gas, to 1 atmosphere. This state is called the "standard state" and is designated by the superscript ° applied to the appropriate thermodynamic symbol. Equation 35 becomes at a given temperature and pressure

$$G(P,T) = G°(T) + RT \ln f \tag{36}$$

The fugacity of any substance in a condensed phase is equal to the fugacity of that substance in the gas phase in equilibrium with the condensed phase. Under ordinary conditions, the vapor pressure of the condensed phase (the pressure of the gas phase in equilibrium with the condensed phase) is low enough that it may be assumed that the fugacity is equal to the pressure. In such cases, the fugacity of the condensed phase is equal to the vapor pressure.

For convenience, the standard state for condensed phases is chosen as the substance, liquid or solid, at 1 atmosphere external pressure. The state of unit fugacity used for gases is not suitable for condensed phases because the fugacity of condensed phases is usually low and the state of unit fugacity is not easily attained.

The ratio of the fugacity in any given state to its value in the standard state is defined as the activity in the given state, and is designated by the symbol a. Thus,

$$G - G^\circ = RT \ln (f/f^\circ) = RT \ln a \tag{37}$$

For gases, since $f^\circ = 1$ atm, $f = a$, and $a^\circ = 1$. For condensed phases, although $f^\circ \neq 1$, $a^\circ = 1$. Furthermore, slight changes in external pressure will produce an insignificant change in the activity of condensed phases.

We have been considering phase changes in a system in which the variables change in a manner such that equilibrium is maintained. We shall now examine, for a one-component system in two-phase equilibrium, the effect of a change in only one of the variables, maintaining the other experimental variable constant. For the equilibrium conditions, described by the subscripted variables T_e, P_e, we can write, following our definition in equation 12:

$$(G_2 - G_1)_e = (H_2 - H_1)_e - T_e(S_2 - S_1)_e$$
$$\Delta G_e = \Delta H_e - T_e \Delta S_e \tag{38}$$

If we change to a new temperature T not far removed from T_e, $T = T_e + \delta$, the system will no longer be at equilibrium. However, we may still write for the isothermal phase change at this new temperature

$$\Delta G_T = \Delta H_T - T \Delta S_T$$

To a first approximation, we may assume that $\Delta H_T = \Delta H_e$ and $\Delta S_T = \Delta S_e = \Delta H_e/T_e$. Hence, we may write

$$\Delta G_T = \Delta H_e - T(\Delta H_e/T_e) = \Delta H_e(1 - T/T_e) \tag{39}$$

Thus, for a process in which at equilibrium the value of ΔH_e is positive (heat is absorbed), an increase in temperature will result in a negative value for ΔG, that is, the process will tend to go spontaneously. Similarly, if the temperature is decreased the reverse process will be thermodynamically favored.

If we do not assume that ΔH_e and ΔS_e are constant but are some simple function of T, we can use our equations 11 to write:

$$\Delta H_T = \Delta H_e + \Delta C_P(T - T_e) \qquad (40)$$

$$\Delta S_T = \Delta S_e + \Delta C_P \ln(T/T_e)$$

in which we have assumed that the heat capacity difference between the two phases, ΔC_P, is constant over the temperature range between T and T_e. Then

$$\Delta G_T = \Delta H_T - T \Delta S_T$$

$$= \Delta H_e + \delta \Delta C_P - (T_e + \delta)\Delta C_P\ln(T/T_e) - (T_e + \delta)\Delta S_e$$

If we approximate $\ln(T/T_e) = \ln(1 + \delta/T_e) \approx \delta/T_e$ this reduces to

$$\Delta G_T = \Delta H_e - T_e\Delta S_e - \delta\Delta S_e - \delta^2 \Delta C_P/T_e$$

But for the equilibrium conditions,

$$\Delta G_e = \Delta H_e - T_e\Delta S_e = 0$$

Also,

$$\Delta S_e = \Delta H_e/T_e$$

Hence,

$$\Delta G_T = -(\delta/T_e)(\Delta H_e + \delta \Delta C_P) \qquad (41)$$

The relative behavior of the Gibbs energies for the various phases of a pure substance in the regions of equilibrium phase change (fusion, vaporization) is shown in Fig. 2.1 and Table 2.I for the element Br_2. The values for

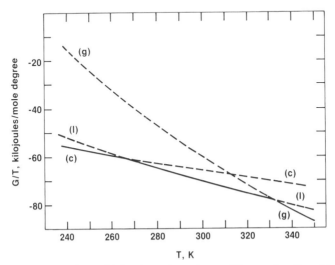

Fig. 2.1. Plot of the function (G/T) for the various phases of Br_2 as a function of temperature. The intersections at 265.9 K and 332.7 K are the normal melting point and boiling point respectively. The intersection of the (c) and (g) curves at 311.9 K represents the temperature at which the vapor pressure of the solid phase equals 1 atm.

TABLE 2.I

Gibbs Energy Changes for the Phase Transitions for Br_2

T, K	$-G°$(J/mol)			$\Delta G°$(J/mol)			
				$l - c$		$g - l$	
	c	l	g	obs	calc	obs	calc
240	13,405	12,481	3,678	924	1020		
250	14,485	13,853	6,063	632	629		
260	15,493	15,255	8,460	238	234		
265.9	16,104	16,103	9,883	1	0		
270	16,529	16,689	10,870	− 160	− 163		
280	17,585	18,146	13,297	− 561	− 563		
290	18,669	19,638	15,732	− 969	− 967		
300	19,769	21,152	18,179	− 1383	− 1372		
310	20,899	22,694	20,644	− 1795	− 1781		
320	22,045	24,255	23,117			1138	1147
330	23,217	25,846	25,598			248	240
332.7	23,573	26,272	26,271			1	0
340	24,414	27,456	28,096			− 640	− 642
350	25,627	29,089	30,602			− 1513	− 1501

crystalline and liquid Br_2 are calculated from the heat capacity and enthalpy measurements in reference 5; extrapolation into the metastable regions is shown by dotted lines in the Fig. 2.1. The values for the gas at 1 atmosphere have been calculated by standard statistical-mechanical relations (9). The values of G for each phase at various temperatures are given in Table 2.I, each value being calculated relative to $G = 0$ for Br_2(c) at 0 K. The values for ΔG_T for the crystal-to-liquid and liquid-to-gas transitions calculated from these values are listed under "obs;" the corresponding values calculated from equation 41 are listed under "calc." The temperatures 265.9 K and 332.7 K correspond to the normal fusion and boiling temperatures (at 1 atm).

VIII. THERMODYNAMIC PROPERTIES OF SOLUTIONS: THE PHASE RULE

Up to now we have been considering systems of one component in which only two variables need be considered, usually temperature and pressure. With a phase consisting of more than one substance, the system may be thermodynamically defined by specifying the composition as well as the temperature and pressure of each phase present.

Consider any thermodynamic property Z of a homogeneous phase (solution) containing n_1 moles of component 1, n_2 moles of component 2, etc. The value of Z is a function of P, T, n_1, n_2, . . .

$$Z = Z(P,T,n_i) \tag{42}$$

Differentiating under conditions of constant T and P, we obtain

$$dZ = (dZ/dn_1)_{P,T,n_2...}dn_1 + (dZ/dn_2)_{P,T,n_1,n_3...}dn_2 + \ldots$$

or

$$dZ = \bar{Z}_1 dn_1 + \bar{Z}_2 dn_2 + \ldots$$

$$= \sum_1^n \bar{Z}_i dn_i \tag{43}$$

The quantity

$$\bar{Z}_i \equiv (dZ/dn_i)_{P,T,n_j}$$

is called the partial molal property. Since the value of $\bar{Z}_i$ is itself a function of P, T, and the composition of the phase, equation 43 may be integrated at constant P, T and composition to yield the significant relation

$$Z = \sum_1^n \bar{Z}_i n_i \tag{44}$$

Thus, for solutions the partial molal property $\bar{Z}_i$ plays the same role as the molal value Z for a pure substance.

Equation 43 is of particular importance when applied to the Gibbs energy G. The general form of this equation becomes

$$dG = (dG/dP)_{T,n_i} dP + (dG/dT)_{P,n_i} dT + \Sigma(dG/dn_i)_{T,P} dn_i \tag{45}$$

$$= V dP - S dT + \Sigma \bar{G}_i dn_i$$

Integrating at constant P, T and composition

$$G = \Sigma \bar{G}_i n_i \tag{46}$$

Differentiation of 42 and combination with 45 yields the relation

$$0 = S dT - V dP + \Sigma n_i d\bar{G}_i \tag{47}$$

From the parallel roles of the quantity $\bar{G}_i$ to the mechanical potential P and the thermal potential T, the partial molal Gibbs energy is frequently called the chemical potential.

Consider a system of c components existing in p phases at equilibrium at a given T and P. If we change the conditions of temperature and pressure slightly, we can apply equation 47 to each phase. The conditions for thermal, mechanical, and chemical equilibrium in the system require that dP, dT and each $d\bar{G}_i$ be the same for all phases. Thus we obtain a set of p equations containing the $c + 2$ unknowns dP, dT, $d\bar{G}_i$, $d\bar{G}_2$, ... $d\bar{G}_c$.

If the number of phases is equal to the number of unknowns, i.e., if $p = c + 2$, each of the variables must be zero, and the system is invariant. If $p < c + 2$, some of the variables may be arbitrarily selected. The number of nonzero variables which may be arbitrarily chosen is called the degree of freedom of the system f and is given by the relation

$$f = c + 2 - p \qquad (48)$$

The number 2 occurs in equation 48 because of our limitation to T and P as external variables. Had magnetic, electrical, or gravitational energies been considered, a different number would have occurred in equation 48. In general, had we considered c components, p phases, and r physical variables, the expression of the phase rule would be

$$f = c - p + r \qquad (49)$$

If we consider a solution of a salt in water, we have a two-component system. If the solution is unsaturated with respect to the salt, we have two phases present—the liquid solution and the vapor phase of pure $H_2O(g)$ at pressure P. Applying the phase rule, we find $c + 2 - p = 2$. Any two of the three variables T, P, and composition of the solution may be varied over a range; the remaining variable will be fixed. If we increase the concentration of the solution to the saturation limit (at a given temperature) another phase, the solid salt, is present, and the phase rule predicts only one degree of freedom. At the given (fixed) temperature the concentration of the solution is fixed and so is the vapor pressure of $H_2O(g)$ over the solution. The particular value of the pressure depends on the particular salt used and the temperature of the system. However, for a given set of conditions, such a system can be useful in maintaining constant humidity even where water vapor is being lost from the system, since the pressure remains fixed as long as the three phases (salt-solution-vapor) are present. A similar situation exists with a salt and its hydrate, in which case the phases would be the hydrated salt, the anhydrous salt or some lower-hydrated form, and $H_2O(g)$. Again the phase rule indicates that the pressure of $H_2O(g)$ over the two solid phases (the equilibrium decomposition pressure of the hydrate) will be determined only by the temperature.

IX. THE IDEAL SOLUTION

Just as it is convenient in considering the properties of gases to define a hypothetical substance called the ideal gas, so in dealing with solutions it is useful to define a system called an ideal solution. Many equivalent definitions may be given; we shall define it as a solution in which the fugacity of each component is proportional to its mole fraction over the entire range of composition.

$$f_i = k N_i \qquad (50)$$

Since it holds at all concentrations it holds at $N_i = 1$ and hence $k = f_i^*$, where f_i^* is the fugacity of the pure substance in the state of the solution (gas, liquid, or solid) at the given temperature and pressure.

Let us consider some of the properties of the ideal solution. The partial molal Gibbs energy of component i in solution may be equated to its value in the gas phase.

$$\tilde{G}_i = G_i^\circ(g) + RT \ln f_i$$

For the pure component

$$\tilde{G}_i^* = G_i^\circ(g) + RT \ln f_i^*$$

Hence,

$$\tilde{G}_i - \tilde{G}_i^* = RT \ln (f/f^*) = RT \ln N_i \tag{51}$$

If we differentiate equation 51 with respect to P at constant T, we obtain

$$(d\tilde{G}_i/dP)_T - (d\tilde{G}_i^*/dP)_T = 0 \tag{52}$$

$$\bar{V}_i = \bar{V}_i^*$$

Since the partial molal volume of each component is equal to the molal volume of the pure component, there is no volume change on forming an ideal solution.

If we divide equation 51 by T and then differentiate with respect to T at constant P, we obtain

$$(d(\tilde{G}_i/T)/dT)_P - (d(\tilde{G}_i^*/T)/dT)_P = 0$$

$$\bar{H}_i/T^2 = \bar{H}_i^*/T^2 \tag{53}$$

Because the partial molal heat content for each component in an ideal solution is equal to that for the pure component, there is no heat of mixing for an ideal solution.

If we differentiate equation 51 with respect to N_i, bearing in mind that G_i^* refers to the pure component and $(dG_i^*/dN_i)_{T,P} = 0$, then

$$d\tilde{G}_i/dN_i = RT/N_i \tag{54}$$

Let us now consider the equilibrium between a pure solid and an ideal liquid solution. This is the situation that exists for a saturated solution in equilibrium with excess solid or for the freezing curve for a component separating out from an ideal solution. Let the solution consist of two components, 1 and 2, and the solid phase consist of pure 1. Then at equilibrium

$$G_1(c) = \tilde{G}_1(\text{soln})$$

For equilibrium to be maintained under a slight change

$$dG_1(c) = d\tilde{G}_1(\text{soln})$$

But

$$dG_1(c) = V_1(c)dP - S_1(c)dT$$

$$d\tilde{G}_1(\text{soln}) = \bar{V}_1 dP - \bar{S}_1 dT + (RT/N_1)dN_1 \tag{55}$$

If the change is made at constant T

$$V_1(c)\,dP = \bar{V}_1\,dP + RT\,d\ln N_1$$

$$(d\ln N_1/dP)_T = (V_1(c) - \bar{V}_1)/RT = (V_1(c) - \bar{V}_1^*)/RT$$

Since the term $\bar{V}_1^* - V_1(c) = \Delta Vm$ represents the volume change on melting for pure component 1, we may express the change in the equilibrium composition of the liquid solution with change in pressure as

$$[(d\ln N_1)/dP]_T = -\Delta Vm/RT \tag{56}$$

(The quantities P, ΔVm, and R must all be in consistent units.)

If in the first approximation we assume that ΔVm is a constant independent of pressure, we may integrate equation 56 between limits and obtain

$$\Delta\ln N_1 = \ln(N_1/N_1') = -(\Delta Vm/RT)(P - P') \tag{57}$$

If the equilibrium shift occurs at constant pressure, equation 55 becomes

$$-S_1(c)\,dT = -\bar{S}_1\,dT + RT\,d\ln N_1$$

$$[(d\ln N_1)/dT]_P = [\bar{S}_1 - S_1(c)]/RT \tag{58}$$

Since component 1 is in equilibrium, we may replace $\bar{S}_1 - S_1(c)$ by $(\bar{H}_1 - H_1(c))/T$ and obtain

$$[(d\ln N_1)/dT]_P = [\bar{H}_1 - H_1(c)]/RT^2 \tag{59}$$

But for the ideal solution, $\bar{H}_1 = \bar{H}_1^*$ and $\bar{H}_1^* - H_1(c)$ is the heat of fusion of pure component 1, which we may write as ΔHm. Thus,

$$(d\ln N_1/dT)_P = \Delta Hm/RT^2 \tag{60}$$

Let us consider this equation in more detail. It may be integrated between limits, choosing as the lower limit of integration the composition $N_1 = 1$, with the corresponding temperature $T = T^*$, the freezing temperature of pure component 1. Then

$$\int_{N_1=1}^{N_1} d\ln N_1 = \int_{T^*}^{T} (\Delta Hm/RT^2)\,dT \tag{61}$$

If we assume that ΔHm is independent of T over the range T^* to T, equation 61 may be integrated to yield

$$\ln N_1 = (\Delta Hm/R)(1/T^* - 1/T) = (\Delta Hm/R)((T - T^*)/TT^*) \tag{62}$$

If the temperature range is small and the upper limit N_1 is close to unity,

$$TT^* \approx (T^*)^2, \quad \ln N_1 = \ln(1 - N_2) \approx -N_2$$

and equation 62 becomes

$$N_2 = -(\Delta Hm/RT^{*2})(T - T^*) \tag{63}$$

The quantity $\Delta Hm/RT^{*2}$ is called the cryoscopic constant. Its value is a function only of the properties of the solvent (component 1) and is independent of the other substances in the ideal solution. Since ΔHm is always positive, the cryoscopic constant is positive and consequently $T < T^*$, that is, the equilibrium temperature between a pure solid and a liquid solution is lower than that between pure solid and pure liquid. Furthermore, when the solution approaches pure component $1(N_1 \rightarrow 1)$, the lowering of the freezing point is directly proportional to the mole fraction of other components (N_2) in the solution.

Equation 61 may also be integrated under not quite so severe restrictions. Let us assume first that ΔHm may be expressed as a linear function of the temperature in the form

$$\Delta Hm = \Delta Hm^* + \Delta C_P (T - T^*)$$

If we now integrate as before between the limits T and T^* (N_1 and $N_1 = 1$) we obtain

$$\ln N_1 = -(\Delta H^*/R)(1/T - 1/T^*) + (\Delta C_P/R)[\ln(T/T^*) + T^*(1/T - 1/T^*)] \quad (64)$$

Let

$$T = T^* - \Delta T = T^*(1 - \Delta T/T^*)$$

$$1/T - 1/T^* = (T^* - T)/TT^* = (\Delta T/T^{*2})[1 - (\Delta T/T^*)]^{-1}$$

$$= (\Delta T/T^{*2})(1 + (\Delta T/T^*) + (\Delta T/T^*)^2 + \ldots)$$

$$\ln (T/T^*) = \ln (1 - \Delta T/T^*) = -\Delta T/T^* - \tfrac{1}{2}(\Delta T/T^*)^2 - \ldots$$

$$\ln N_1 = \ln (1 - N_2) = -N_2 - \tfrac{1}{2}N_2^2 - \ldots$$

Making the appropriate substitutions and neglecting terms in powers of $(\Delta T)^3$ and higher, the integrated equation becomes

$$-N_2 - \tfrac{1}{2}N_2^2 = -(\Delta Hm^*/RT^{*2})\Delta T$$
$$[1 + (\Delta T/T^*) - \tfrac{1}{2}(\Delta C_P/\Delta Hm^*)\Delta T + \ldots] \quad (65)$$

If we define the quantities

$$A = \Delta Hm^*/RT^{*2}$$

$$B = 1/T^* - \tfrac{1}{2}(\Delta C_P/\Delta Hm^*)$$

equation 65 may be written

$$N_2 + \tfrac{1}{2}N_2^2 = A\Delta T(1 + B\Delta T + \ldots) \quad (66)$$

As before, the mole fraction of the impurity component or solute is expressed in terms of the properties of the pure solvent and the change in the equilibrium temperature between the solid and liquid phases of the solvent component. There is no requirement that there be only one "impurity" component present. The quantity N_2 represents the total mole fraction of solute components present. In the simple approximation represented by equation 63, one can assume that each component contributes a lowering of the solvent freezing temperature independently of the other impurities present, but the simple additivity will not hold for the more extended region covered by equation 66.

The preceding relations lead to another method for the calculation of the amount of impurity in a substance based on measurements on the system in the region just below the melting point. As the substance cools through the melting range, the concentration of impurity in the liquid fraction increases, the relation between temperature and concentration being given by equation 63. At any given point on the freezing curve, we represent the mole fraction of impurity in the liquid phase $N_2 = n_2/(n_1 + n_2) \approx n_2/n_1$, and the mole fraction of impurity in the initial system as $N_2^* = n_2/(n_1^* + n_2) \approx n_2/n_1^*$. If the fraction of sample melted $F = n_1/n_1^* = N_2^*/N_2$, we may rewrite equation 63 as

$$N_2^*/F = A(T^* - T)$$

or

$$T = T^* - (N_2^*/A)(1/F) \tag{67}$$

Hence, a plot of the equilibrium temperature T versus the reciprocal of the fraction of the sample in the liquid phase at T yields a straight line with a slope determined by the ratio of mole fraction of impurity to the cryoscopic constant. This relation is the basis of one of the most precise thermodynamic methods for the determination of purity, as described in a later chapter.

This relation may also be extended to apply to heat capacity measurements in this range. In these measurements, heat energy is supplied to the system and the temperature increase is measured. The energy increment produces a rise in temperature and also, as shown by equation 67, a change in the fraction of solvent melted. We may express the total increase in enthalpy as

$$dH = (dH/dT)_F \, dT + (dH/dF)_T \, dF$$

If we differentiate equation 67, we obtain

$$dF = (N_2^*/A)(T^* - T)^{-2} dT$$

Now $(dH/dF)_T = \Delta Hm$, the heat of fusion of the solvent in an ideal solution; hence,

$$dH = C_P \, dT + N_2^* R T^{*2} (T^* - T)^{-2} dT$$

The total derivative (dH/dT) represents the observed heat capacity; hence, the anomalous increase in heat capacity at any temperature, called the "premelting heat capacity," is given by

$$\Delta C_P' = N_2^* R T^{*2} (T^* - T)^{-2} \tag{68}$$

As the thermodynamic melting temperature T^* is approached, $\Delta C'_P$ increases very rapidly, producing a significant deviation from what might be considered the normal heat capacity curve. However because of the difficulty in establishing the "true" curve precisely, this method is seldom applicable for precise measurements of purity.

If we consider the equilibrium that exists between an ideal liquid solution and a vapor phase consisting of only one component, we can proceed as before to write

$$dG_1(g) = d\bar{G}_1(soln, N_1)$$

and obtain as the final differential relations

$$(d \ln N_1/dP)_T = [V_1(g) - V_1^*(liq)]/RT \tag{69}$$

$$(d \ln N_1/dT)_P = [H_1^*(g) - H_1^* (liq)]/RT^2 = \Delta Hv/RT^2 \tag{70}$$

where ΔHv is the enthalpy of vaporization of pure component 1.
On the other hand, if both components of the liquid solution are volatile and form an ideal gaseous solution, the fundamental equation becomes, using the prime to designate the vapor phase,

$$d\bar{G}_1' (g, N_1') = d\bar{G}_1(liq, N_1)$$

$$\bar{V}_1' dP - \bar{S}_1'dT + (RT/N_1') dN_1' = \bar{V}_1 dP - \bar{S}_1 dT + (RT/N_1) dN_1$$

and at constant pressure

$$RT \, d \ln(N_1'/N) = (H_1^{*'} - H_1^*)/T) \, dT \tag{71}$$

X. THE EQUILIBRIUM CONSTANT

Consider a chemical system that has reacted according to the equation

$$c \, C + d \, D + \ldots = l \, L + m \, M + \ldots \tag{72}$$

If the system has come to equilibrium, we may apply equations 17 and 43 to obtain

$$(dG)_{T,P} = \sum_i \bar{G}_i \, dn_i = 0 \tag{73}$$

For a system of constant total mass, the stoichiometric coefficients of the reaction relate the various dn_i such that equation 73 becomes, when applied to reaction 72

$$(l \, \bar{G}_L + m \, \bar{G}_M + \ldots.) - (c \, \bar{G}_C + d \, \bar{G}_D + \ldots) = 0 \tag{74}$$

The general expression for the molal Gibbs energy for any component of the solution is given by equation 37

$$\bar{G}_i - \bar{G}_i^\circ = RT \ln (f_i/f_i^\circ) = RT \ln a_i$$

Applying this relation to equation 74 yields

$$(l\bar{G}_L^\circ + mG_M^\circ + \ldots) - (c\bar{G}_C^\circ + \bar{d}G_D^\circ + \ldots) + RT \ln (a_L^l a_M^m)/(a_C^c a_D^d) = 0$$

or

$$\Delta G^\circ = - RT \ln (a_L^l a_M^m)/(a_C^c a_D^d) \qquad (75)$$

The ratio of activities on the right-hand side of equation 75 defines the quantity called the thermodynamic equilibrium constant K for the reaction 72. For gases, we may replace the activity by the fugacity (as $f^\circ = 1$), and if the pressure is low, $f = P$. Similarly, for pure solid phases at ordinary pressures $f = f^\circ$, and the activity may be set equal to unity.

Equation 75 relates the equilibrium constant K, a function of the activities of the various species in the equilibrium system, to the total change in the Gibbs energy when all reactants in their standard states are converted completely into the products in their respective standard states. Since the standard states at any temperature are defined at a given pressure, ΔG° and K are not functions of the pressure, although the individual values of the activities at equilibrium may be functions of the pressure. The variation of K with temperature is obtained from equation 75:

$$[d(\Delta G^\circ/T)/dT]_P = -R(d \ln K)/dT = - \Delta H^\circ/T^2$$

$$(d \ln K)/dT = \Delta H^\circ/RT^2 \qquad (76)$$

If we integrate equation 76, assuming ΔH° is constant, we obtain the relation

$$\ln K = (-\Delta H^\circ/RT) + A = -\Delta G^\circ/RT$$

Let us apply this to the decomposition of a salt hydrate such as was mentioned in the discussion of the phase rule. For the decomposition of a dihydrate, such as $NaI \cdot 2H_2O$, the equation is

$$NaI \cdot 2H_2O(c) = NaI(c) + 2H_2O(g) \qquad (77)$$

At equilibrium, the phases present are the dihydrate, the anhydrous salt, and water vapor at pressure P. The activities of the pure solid crystals may be set equal to unity, for $H_2O(g)$ the activity may be set equal to the water vapor pressure P. Then the equilibrium constant

$$K = \frac{a(NaI) \times a^2(H_2O)}{a(NaI \cdot 2H_2O)} = P^2$$

The variation of P with temperature will be given by the relation

$$d \ln K/dT = \Delta H^\circ/RT^2 = 2(d \ln P/dT)$$

If we integrate with respect to T, again assuming ΔH° is constant, we obtain the relation

$$\ln P = (-\Delta H^\circ/2RT) + \text{constant}$$

Note that this equation has exactly the same form as the usual equation for the vapor pressure of a pure substance as a function of temperature, in which the coefficient of 1/T is related to the enthalpy of vaporization of 1 mole of vapor and the constant is related to the entropy difference between the gaseous and condensed phases, each in its appropriate standard state. Thus, from a knowledge of the enthalpy change of the decomposition equation 77 and the entropies of each of the substances involved, we can calculate $\Delta G°$ for the reaction and from this calculate K and P, the equilibrium pressure of $H_2O(g)$ over the hydrate. Calculations of this sort can be used to estimate and compare relative efficiencies of various substances as drying agents.

XI. APPLICATIONS TO ANALYSIS

As was mentioned earlier, the methods of analysis to be treated in this volume may be divided into first law and second law methods, depending on the nature of the measurement relation involved. In a first law method, use is made of the fact that a chemical reaction is characterized by the liberation or absorption of a quantity of heat that is directly proportional to the amount of reaction taking place. Strictly speaking, exact proportionality will hold only for a very limited class of reactions. If the reaction under consideration consists of pure individual phases reacting to form new pure phases, or if the solution phases are ideal solutions in which the heat of mixing is zero (see equation 53), the heat of reaction is strictly proportional to the amount of reaction. In most practical systems, heats of dilution and mixing, stirring, and other factors produce problems that can only be solved by careful design of the calorimetric experiments. The use of concentrated reagents, sensitive thermistors, and essentially adiabatic conditions makes it possible to isolate these deviations so that the pure heat of reaction effects may be readily obtained from the measurements.

Consider the titration of an acid with a base. If both acid and base are strong, the heat of neutralization of very dilute solutions represents the heat of combination of aqueous H^+ and OH^- ions to form H_2O:

$$H^+ + OH^- \rightarrow H_2O(liq); \qquad \Delta H = -13.35 \text{ kcal/mole at } 298.15°K \quad (16)$$

It is generally necessary to use concentrated solutions as titrant in order to reduce the volume changes and energy equivalent changes in the system during the titration, hence the actual heat effect may differ by as much as a kilocalorie. For instance, in the heat of neutralization of 1 molal KOH with 1 molal HCl, $\Delta H = -13.90$ kcal/mol (16). In addition to this heat of neutralization, there is a relatively small heat of dilution effect produced by the added volume of titrating solution. When the neutralization reaction is completed, this small effect becomes the sole heat of reaction. Hence it is possible to locate the equivalence point as that point at which the rate of heat production in the system falls off sharply (it might even change sign). Since both rates will be linear with respect to added titrant within a few per cent, the location of the equivalence point as the intersection of the lines can be made with good precision.

In the case of titration of a weak acid (or base), the situation is somewhat more complicated theoretically but in practice one may obtain results that are comparable in precision to those for a strong acid. The reaction of neutralization may be considered as the sum of two reactions: (1) the ionization of the acid: $HA \rightarrow H^+ + A^-$ and (2) the neutralization of the H^+ ion: $H^+ + OH^- \rightarrow H_2O$. The heat of ionization of most weak acids (step 1) varies from about -1 to $+3$ kcal/mol in aqueous solutions at 25°C (the temperature coefficient ranges from -0.035 to -0.060 kcal/mol deg). Formic acid, with an ionization constant $= 1.77 \times 10^{-4}$, has a value of ΔH of ionization $= -0.10$ kcal/mol; the corresponding values for HOCl(aq) are 3×10^{-8} and $+3.3$ kcal/mol. For these acids, the net heat of neutralization will be greater than 10 kcal/mol, a value large enough to be suitable for calorimetry.

The existence of a significant hydrolysis reaction is a complicating factor. This represents a reversal of the neutralization reaction and thus tends to decrease the measured heat of reaction. However, calculations show that hydrolysis does not become significant until the equivalence point is approached. For example, in the neutralization by a strong base of 0.01 molal acid having an ionization constant $K_a = 1 \times 10^{-10}$, the degree of hydrolysis present when 50% of stoichiometric equivalence is reached is less than 2%, and at 80% of equivalence the hydrolysis amounts to 4%. Thus, 80% of the neutralization reaction of this weak acid yields a straight line (within 4%) for the rate of heat evolution as a function of amount of base added. During the remainder of the neutralization, the rate of heat evolution will change rapidly, going over to what might be considered the linear "after-rate." Although the curvature may be pronounced at the stoichiometric equivalence point, linear extrapolations from the major portion of the neutralization and from the after-rate enable one to fix the end point to within a few per cent.

These same acid–base systems are also susceptible to second law analysis methods. These methods involve relationships deriving ultimately from the relation $\Delta G° = -RT \ln K$ and hence relate a logarithmic concentration change to some measure of the change in Gibbs free energy of the system. In the usual potentiometric titration methods, the concentration (activity) of the hydrogen ion is measured by the electromotive force generated by the system when combined with a reference electrode. When a strong acid is titrated with a strong base, the equivalence point is indicated by a sharp change in emf resulting from a very rapid change in concentration of H^+(aq). The break in emf occurs over a very narrow increment in titrant added. Because of the logarithmic relation between H^+ concentration and emf, the first 90% of the usual neutralization reaction produces little change in pH and emf and contributes little information toward the location of the equivalence point.

When a weak acid is titrated potentiometrically, the pH and the corresponding emf change in a gradual fashion and it becomes difficult to locate the point of inflection corresponding to equivalence. For this system, the enthalpimetric titration method proves most valuable. Here the titration curve for the first portions of the reaction can be used to provide the necessary information to establish equivalence. Thus, for the weak acid described above, the first 80% of the measured reaction heat can be used to extrapolate for the remaining 20%, whereas in potentiometric titrations the first portions of the neutralization contribute no useful information. Detailed applications of enthalpy titration procedures to various specialized analysis applications are given in a later chapter.

Other second law methods of analysis to be treated in this volume derive from the relations based on equations 66, 67, and 71. Subject to the limitations indicated in the derivations, these methods can be used to determine small amounts of impurity with precision. In general they give no information about the chemical identity of the impurity, but on the other hand little specific knowledge of the species is required for the method to be applicable. Hence these methods, especially freezing point methods based on equation 66 or calorimetric methods based on equation 67, have proven to be extremely useful as methods to establish purity for substances that melt in the operating range of precision calorimeters.

Although the above equations do not discriminate among the substances composing the impurity, cryoscopic methods based on equation 66 have been used on occasion as a tool for the analysis of mixtures. Consider a mixture of two isomers A and B, the relative amounts of A and B being unknown. If a known amount of the mixture is added to a known larger amount of solvent C of known cryoscopic constant and the freezing point depression of C determined, the molecular weight of the isomers may be determined. If, however, the solvent or major component is chosen to be the pure substance A (of known cryogenic properties), then only the component B of the mixture acts as the impurity, the added substance A no longer being an impurity. Similarly, the pure substance B may be used as the solvent, in which case only the component A will act as impurity to depress the equilibrium freezing temperature.

One of the earliest calorimetric methods for the quantitative measurement of impurity in a substance that could be conveniently melted involved the use of an adiabatic calorimeter to determine the fractional melting of a sample (from measurement of the heat input to the sample as a fraction of the heat of melting) and the equilibrium temperature (see equation 67). Measurements of this type required large sample sizes and long measurement times. Recently, differential scanning calorimetry (DSC) has been used for this purpose, taking advantage of the rapid measurement capability and small sample size required. In a typical DSC measurement, the melting range of the sample appears as a peak on the temperature plot, with the shape of the peak (height and breadth) being a function of the sample purity.

A review of the application of DSC to purity determinations is given in reference 11.

An interesting application of quantitative thermochemistry to analysis of the amount of unsaturation in organic compounds makes use of the fact that the addition of 1 mol of hydrogen to a carbon–carbon double bond liberates almost 120 kJ per mol of bonds saturated or per mol of H_2 absorbed. Measurements of this sort can frequently be used to provide evidence as to possible molecular structures. This type of measurement has been called "enthalpimetry" (14).

A recent development in the application of thermochemistry to biological and biochemical analysis has come with the development of the heat-conduction microcalorimeters. These calorimeters are small, having volumes less than 1 ml, and able to detect thermal power down to 1 microwatt (1 μW) and total energies of less than 1 mJ. In the conduction calorimeter, thermal contact between the reaction vessel and a heat "sink" is established by thermocouples; the voltage generated by the thermopile is proportional to the rate of heat flow from the reaction cell. Biological process are not particularly low-energy processes, on a molar basis, but they usually occur in aqueous systems at very low concentrations, and clinical samples may contain less than one μmol of material in very dilute solution. However microcalorimetry has been used successfully in the assay of glucose in blood serum, using the enzyme catalyzed phosphorylation of glucose reaction: glucose (in solution) + adenosine triphosphate (in solution)

$\xrightarrow[\text{enzyme}]{\text{hexokinase}}$ (adenosine diphosphate + glucose-6-phosphate) (in solution). The enthalpy change for this reaction is -61.4 kJ/mol of glucose. Measurements for glucose content made on samples of blood serum yielded results that compared very satisfactorily with analyses made by more traditional chemical methods. The calorimetric reaction is complete within 15 min, so that the entire analysis can be accomplished within a few hours. However the method must assume that one and only one reaction occurs in the calorimeter cell. In the case of biological systems of the type considered here, this usually requires the use of very specific (i.e., pure) enzymes. Thus, one limitation on this method of analysis relates to the availability of appropriate (enzyme) catalysts of high purity. More details can be found in reference 13.

In using the relations developed above, one must bear in mind the conditions that have been assumed. The liquid phase has been assumed to obey the ideal solution law. This condition will usually be satisfied over some small range of impurity concentrations, although the exact upper limit may not be predicted. It is also assumed that the solid phase consists only of the major component. If the impurity proves to be soluble in the solid phase of the solvent, the problem becomes much more complex. Note that this situation is not the same as that existing at the simple eutectic point at which both the solute and solvent separate as pure solid phases.

XII. SOURCES OF CHEMICAL THERMODYNAMIC DATA

The numerical data needed for various applications of thermal methods of analysis include such quantities as heats of reaction, melting temperatures, heats of fusion, vapor pressure–temperature data. The basic equation for enthalpimetric titrations may be written as $\Delta T = (N \cdot \Delta H)/(EE)$ in which the measured temperature rise is proportional to N, the number of moles of reaction taking place and the molal heat of reaction ΔH. The quantity EE represents the energy equivalent or "effective heat capacity" of the calorimetric system. In general, the most effective method of determining the energy equivalent of the system is by an enthalpimetric titration of a system for which ΔH and N are accurately known. It is true that the energy values for chemical reactions ultimately must be referred to electrical energy measurements as the fundamental working standards for energy, but when used for analysis purposes, a standardizing reaction similar in nature to the one being studied will tend to reduce some of the systematic bias characteristic of the particular calorimeter system.

Similarly the cryoscopic constants A and B defined in equation 66 require a knowledge of melting points, heats of fusion, and ΔCp of fusion, and ebulliometric methods require knowledge of the corresponding vapor pressure–temperature equilibria. The use of equation 67 requires the use of an adiabatic system in which the fraction melted at the temperature T is calculated from a knowledge of the increments of heat energy added and the total heat of fusion of the sample.

Frequently, the numerical thermodynamic data needed for the applications of thermal analyses may be obtained directly from the analytical procedures, as described in the following chapters. Values so obtained may be the best to use in these circumstances although they may not agree with the most precise data available. As mentioned before, these factors tend to compensate for systematic deviations of the measurement system.

Numerical tabulations of data on the chemical thermodynamic properties of systems are usually prepared to serve various purposes. They may simply tabulate values of certain properties under specific conditions. Tables of solubilities, heat capacities, boiling and freezing points, and vapor pressure tables are instances of such compilations. They may be designed to make possible the rapid accurate calculation of the energies of chemical reactions, or the calculation of reaction equilibria under varying experimental conditions. In these instances they are usually expressed as values of the heats and Gibbs energies of formation for the chemical compounds and mixtures as a function of temperature.

The formation reaction is defined to mean the reaction in which the given substance in a designated standard state (solid, liquid, etc.) is formed isothermally from the constituent elements, each in a specified standard state called the standard reference state. The reference state for the elements is taken to be that state which is stable at the reference temperature and at 1 atmosphere pressure. The symbol f is usually added to the thermodynamic property to indicate the formation reaction. By definition then, the values of $\Delta Hf°$ and of $\Delta Gf°$ for the elements in their appropriate reference states are zero. It follows from the additivity of thermodynamic equations that for any given reaction the change in any thermodynamic property will be the algebraic sum of the values for the formation reaction of all participating substances:

$$\Delta H° = \sum_{\text{products}} \Delta Hf° - \sum_{\text{reactants}} \Delta Hf°$$

Knowing the values of the thermodynamic properties for the formation reaction of all substances thus enables us to calculate the values for any other reaction involving these substances.

To simplify the presentation of data for aqueous electrolyte solutions, the values for the heat and Gibbs energies ($\Delta H°$ and $\Delta G°$) of the ionization process for strong electrolytes are taken to be zero. In addition the convention is adopted that $\Delta Hf°$ and $\Delta Gf°$ for $H^+(aq) = 0$, that is,

$$\tfrac{1}{2}H_2 \,(g, f = 1 \text{ atm}) = H^+(aq, a = 1) + e^- \qquad \Delta Hf° = \Delta Gf° = 0$$

The separate convention is also adopted that the standard entropy of $H^+(aq)$ shall be zero. (Note that this implies a value of $\Delta Sf°$ for e^-. However, no problems arise when the requirement of electrical neutrality for a total reaction is maintained.) The properties of a neutral strong electrolyte in aqueous solution in the standard state are thus equal to the algebraic sum of these values for the appropriate kinds and number of ions assumed to constitute the molecule of the given electrolyte.

Subject to these conventions, it is possible to summarize the available experimental data on heats of reaction and chemical equilibria in terms of tables of values of $\Delta Hf°$, $\Delta Gf°$, and $S°$ for chemical substances, solutions, and other systems. The proper preparation of such tables requires that careful analysis and evaluation be made of many types of experimental measurement. Some of the various types of data that must be considered in a complete evaluation of thermochemical data and the properties derived therefrom are given in Table 2.II.

The final selected values must represent the best fit to the experimental data and their interrelations as expressed by the equations:

$$\Delta H° = \Delta G° + T \Delta S°$$

$$\Delta G°/T = -R \ln K$$

$$d(\Delta G°/T)/dT = -\Delta H°/T^2$$

To attain this and to maintain the internal self-consistency of the tables, great care must be taken in the application of corrections for energy units, temperature scale, side reactions, etc. Also all auxiliary data used in the

TABLE 2.II
Thermodynamic Measurements and Derived Properties

Measurements	Used for
Heats of reaction Heats of solution and dilution Heats of transition, fusion, vaporization	ΔH
Heat capacity, 0–298°K Heat capacity, $T > 298°$K	$S°$ and correction of ΔH and ΔG for temperature
Chemical equilibria and variation with T Vapor pressures and variation with T Electromotive force and variation with T Solubility and variation with T	ΔG and ΔH
Activity coefficients in solution PVT measurements	ΔG
Ionization potentials Appearance potentials Molecular dissociation energies	ΔH
Molecular structures, bond distances, and angles Vibrational and rotational spectra Electronic spectra	$S°$

calculations must be constant throughout the tables. Among such auxiliary data, of course, are values of the fundamental physical constants which have recently been recommended by CODATA, the Committee on Data for Science and Technology (2). Some of the values from that set, useful in chemical thermodynamics, are given in Table 2.III. Conversion factors for various units of molecular energy have been calculated and are presented in Table 2.IV. These conversion factors are based on the defining relations indicated by the italicized values in the table.

Establishing a complete table of thermochemical values is at present a sequential process in which values of substances are calculated using only reactions involving elements (having defined values of $\Delta Hf°$ and $\Delta Gf°$) or substances whose thermodynamic constants have been previously established in the evaluation process. It is also necessary to select an initial set of values for certain substances that serve as fundamental building blocks for all the other substances in the tables. These values then become part of the basic structure of the final tables and must be considered in determining questions of interconsistency with other tables.

Although a given compilation of numerical data may be completely self-consistent, it may not be consistent with numerical data in other compilations. This may present serious (and often unrecognized) problems to a user of such tabular data. One of the aims of the recently created (1963) National Standard Reference Data System under the U. S. National Bureau of Standards is to promote the compilation of critically evaluated and mutually interconsistent tables of numerical data on physical and chemical properties

TABLE 2.III

Fundamental Constants of Thermodynamics

Absolute temperature of the triple point of water	273.16 K
Standard atmosphere	101325 N/m^2
Standard acceleration of gravity, g	9.80665 m/sec^2
Thermochemical calorie	4.1840 J
Velocity of light in vacuum, c	2.9979246 $\times$ 10^8 m/sec
Avogadro number, N	6.022045 $\times$ 10^{23} mole^{-1}
Faraday constant, F	96484.56 C/equiv.
Gas constant, R	8.31441 J/deg · mole
Boltzmann constant, k	1.380662 $\times$ 10^{-23} J/deg
Second radiation constant, hc/k	1.43879 deg/cm^{-1}

of materials. This includes agreement on problems in nomenclature, symbols, basic numerical data, etc., as well as methods of analysis and evaluation of data.

Recent advances in large-scale computer processing and handling of chemical information have had a tremendous impact on the ability of scientists to evaluate, store, and retrieve thermodynamic information rapidly. The developments have proceeded along two principal lines. In one the availability of very large, rapid-access on-line memories has led to the growth of a number of computer-accessible data banks that may be used for identification of substances. Data banks related to spectral properties such as infrared and mass spectra are a principal example of these. In addition there are several banks, such as EROICA at the University of Tokyo, Japan, AVESTA in Kiev, U.S.S.R., TRL at Washington University, St. Louis, Missouri, KEYDATA located at Cambridge, England, which store numerical values of various physical and thermodynamic properties for organic and some inorganic substances. In addition, there are several data systems that provide correlation equations to calculate the desired properties based on various evaluated theoretical and semi-empirical relationships. Many of these systems operate as part of a larger system designed to carry out elaborate chemical manufacturing process design simulations. A compilation of these various data systems has been prepared recently by Hilsenrath (6).

The other significant application of large-scale computers has been in the development of methods for the processing and evaluation of masses of experimental data to obtain reliable numerical values. The area of metallurgical thermodynamics has proven to be specially fertile for the growth of automated data storage and processing. In one application, evaluated phase diagram data are stored in a large data base from which the detailed diagrams with associated thermodynamic data, such as transition, eutectic, and other temperatures can be reproduced using appropriate computer graphics. Such a data base is being prepared under a program sponsored jointly by the National Bureau of Standards and the American Society for Metals. The results of the evaluations are published in the Bulletin of Alloy Phase Diagrams (1).

TABLE 2.IV

Conversion Factors for Units of Molecular Energy[a]

	J/mole	cal/mole	cm³ atm/mole	kW · hr/mole	Btu/lb mole	cm⁻¹/molecule	e.v./molecule
1 J/mole =	1	2.390057×10^{-1}	9.869223	2.77778×10^{-7}	0.429923	8.359345×10^{-2}	1.036435×10^{-5}
1 cal/mole =	*4.18400*	1	41.2929	1.162222×10^{-6}	1.798796	3.49755×10^{-1}	4.33644×10^{-5}
1 cm³ atm/mole =	*0.1013250*	2.42173×10^{-2}	1	2.81458×10^{-8}	4.35619×10^{-2}	8.47011×10^{-3}	1.050168×10^{-6}
1 kW · hr/mole =	*3,600,000*	860,421	3.55292×10^{7}	1	1,547,721	300,936	37.3117
1 Btu/lb mole =	*2.32600*	5.55927×10^{-1}	22.9558	6.46111×10^{-7}	1	1.944384×10^{-1}	2.41075×10^{-5}
1 cm⁻¹/molecule =	11.96266	2.85914	118.0623	3.32296×10^{-6}	5.14302	1	1.239852×10^{-4}
1 e.v./molecule =	*96484.56*	23060.4	952.229	2.68013×10^{-2}	41480.9	8065.479	1

[a] The italic numbers represent the fundamental values used in deriving this table. The remaining factors were obtained by applying the relationships: n_{ij} = $n_{ik}\, n_{kj}$ and $n_{ii} = n_{ik}\, n_{ki} = 1$.

An alternate approach has been to store the thermodynamic data for the Gibbs energies for the various metallic elements as functions of temperature and pressure (including the various crystalline modifications). These data are then combined with appropriate solution theory relations to locate the equilibrium phases at each composition and temperature. Detailed discussion and description of this method of calculation is given in (7).

A different use of large computers has been their application in solving very large networks of thermodynamic equations. As has already been described, a measurement of an enthalpy or Gibbs energy change in a reaction can be expressed as a linear combination of the enthalpies or Gibbs energies of formation of each substance participating in the reaction. Similarly the entropy change for a reaction is the algebraic sum of the entropies of the products less that of the reactants. Hence, the total set of all such experimental measurements made over the years constitutes in fact a very large set of approximately 40,000 linear equations with about 10,000 unknowns. Not all of the unknowns are independent because of the thermodynamic constraint equation $G = H - TS$ which serves to connect the values of the enthalpy, Gibbs energy, and entropy of formation of the individual compounds. Recent techniques in evaluating these large networks of data have been developed in a joint program at the U. S. National Bureau of Standards and the University of Sussex, England. In this approach, which deals only with data corrected to 298 K and 1 atm pressure, a set of equations, involving compounds containing a particular element, is solved by a procedure involving both least sums and least squares criteria to obtain a set of "best" values of $\Delta Hf°$, $\Delta Gf°$, and $S°$ for each substance in the network. Only a portion of the total set of equations is treated at any time because of computer limitations. As each portion of the network is solved, the results become a part of the constant auxiliary data used in the evaluation of subsequent portions of the network. A detailed description of the method has been presented in (3).

A more elaborate procedure, designed to treat data for a smaller number of compounds, but including data over a large range of temperatures, has been developed by the U. S. Geological Survey. In this procedure, the heat capacity for each unknown substance is expressed as an arbitrary polynomial in T and from this relation empirical expressions are derived for $\Delta Hf°$, $\Delta Gf°$, and $S°$ for each substance as a function of temperature. All temperature-dependent measurements such as heat capacities, vapor pressures, etc., are then included in the reaction set to be solved. A description of this procedure is given in (4).

A number of compilations of thermodynamic data present useful summaries of needed numerical values. Although new experimental data make frequent revisions necessary, full-scale complete revisions are at present very time-consuming. Partial summaries and revisions represent useful attempts to keep current with new results, but the problems of mutual consistency require great caution in the use of such material. A partial list of sources of comprehensive tables of chemical thermodynamic data and the properties they include is given below:

1. "Selected Values of Chemical Thermodynamic Properties," *Natl. Bur. Std. Circ.* 500, Government Printing Office, Washington, D. C. (1952). Contains values of $\Delta Hf°$, $\Delta Gf°$ at 25°C, heats and temperatures of transition where known for all inorganic substances. This circular has now been revised, the revisions being published as a series of Notes in the *NBS Technical Notes* Series, NBS-TN 270–3 to 270–8.

2. *Thermochemical Constants of Substances*, V. P. Glushko, Ed., Academy of Sciences, U.S.S.R., Moscow (1965). This is being published in a set of 10 volumes and is similar in content to *NBS Circular* 500.

3. "Selected Values of Physical and Thermodynamic Properties of Hydrocarbons and Related Compounds," American Petroleum Institute Research Project 44, Carnegie Institute of Technology, Pittsburgh, Pa., 1953. Contains values for heats of formation, vaporization, fusion, vapor pressures, density, index of refraction, viscosity, etc., for various classes of hydrocarbon compounds, including experimental and correlated values.

4. *Thermochemistry of Aqueous Uniunivalent Electrolyte Solutions*, by V. B. Parker, NSRDS-NBS 2, Government Printing Office, Washington, D. C., 1964. Contains heat of solution, dilution, and heat capacity tables for 1-1 electrolytes in water.

5. *Contributions to the Data on Theoretical Metallurgy*, Vols. II, XIII, XIV, XV, by K. K. Kelley, U. S. Bureau of Mines Bulletins 371, 584, 592, 601, Government Printing Office, Washington, D. C. Heats of fusion, vapor pressure data, heat capacity, and entropy data on inorganic compounds.

6. *Selected Values of the Thermodynamic Properties of the Elements*, by R. Hultgren, P. D. Desai, D. T. Hawkins, M. Gleiser, K. K. Kelley, and D. D. Wagman, American Society for Metals, Metals Park, Ohio, 1973. Heat capacities, entropy, heats of sublimation and fusion, and thermal functions from 0 K to high temperatures.

7. *Physico-Chemical Constants of Pure Organic Compounds*, by J. Timmermans, Elsevier, New York. Vol I, 1950; Vol. II, 1965. Contains vapor and liquid densities, vapor pressures, freezing points, heat capacities, etc., of pure organic compounds. Volume II is a supplement to Volume I.

8. *Oxidation States of the Elements and their Potentials in Aqueous Solutions*, 2nd ed., by W. M. Latimer, Prentice-Hall, New York, N. Y., 1952. Tables of Gibbs free energies, ionization, and oxidation–reduction equilibria for inorganic compounds.

9. *Atlas of Electrochemical Equilibria at 25°C*, by M. Pourbaix, Gauthier-Villars, Paris, 1963. Tables and diagrams representing equilibria involving the elements, their oxides and hydrides in aqueous solution. An English language edition has recently (1966) been published by Pergamon Press Ltd., London.

10. *Stability Constants of Metal-Ion Complexes*, by L. G. Sillén and A. E. Martell, *Spec. Publ.* No. 17, The Chemical Society, London, England, 1964. A compilation of reported values for the equilibria involving the formation of complex ions of all elements, and derived thermodynamic constants (ΔS, ΔH). Additional supplements have been published by R. M. Smith and A. E. Martell as *Critical Stability Constants*. These have been published by Plenum Press, N. Y.

11. *Solubilities of Inorganic and Metal-Organic Compounds*, 4th ed., by A. Seidell and W. F. Linke. Volume I (1958) was published by Van Nostrand, Princeton, N. J.; Volume II (1966) was published by the American Chemical Society, Washington, D. C. A compilation of published measurements of the solubilities of all inorganic and metal-organic compounds in various solvents. The 3rd Edition volume covering the data on organic molecules will presumably be revised separately.

12. *A Reference Book on Solubilities*, V. V. Kafarov, Ed., Academy of Sciences, U.S.S.R., Moscow, 1961. Contains tables of selected values of solubilities and solution densities as a function of temperature for salts in water and other solvents. An English edition, edited by H. Stephen and T. Stephen, has been published (1963) by Pergamon Press Ltd., London.

13. *Physical Chemistry of Electrolyte Solutions*, 3rd ed., by H. S. Harned and B. B. Owen. *A.C.S. Monograph Series* No. 137, published by Reinhold, New York, 1958. Contains tables of activity coefficients, ionization equilibria, dielectric constants, etc.

14. *Landolt-Börnstein Tables of Numerical Data and Functions of Physics, Chemistry, Astronomy, Geophysics, and Technology*, 6th ed., Springer, Berlin, 1961. Volume II, Part 4, by K. Schäfer and E. Lax contains tables of calorimetric data, including heats of formation, fusion, vaporization, heat capacities.

REFERENCES

1. *Bulletin of Alloy Phase Diagrams*, published quarterly by the American Society of Metals, Metals Park, Ohio.

2. CODATA Task Group on Fundamental Constants, *Recommended Consistent Values of the Fundamental Physical Constants, 1973*, CODATA Bulletin 11, Paris, 1973.

3. Garvin, D., V. B. Parker, D. D. Wagman, and W. H. Evans, "A Combined Least Sums and Least Squares Approach to the Evaluation of Thermodynamic Data Networks", NBSIR 76–1147, U. S. National Bureau of Standards, Washington, D. C., 1976.

4. Haas, J. L., Jr. and J. R. Fisher, *Am. J. Sci.* **276,** 525–545 (1976).

5. Hildenbrand, D. L., W. R. Kramer, R. A. McDonald, and D. R. Stull, *J. Am. Chem. Soc.* **80,** 4129–4132 (1958).

6. Hilsenrath, J., Natl. Bur. Stand. Tech. Note 1122, "Summary of On-Line or Interactive Physico-Chemical Numerical Data Systems", U. S. Government Printing Office, Washington, D. C. (1980).

7. Kaufman, L., and H. Bernstein, "Computer Calculation of Phase Diagrams," in *Refractory Materials*, A. M. Alper, Ed., Vol. 4, Academic Press, N. Y., 1976.

8. Lee, T. S., and O. Popovych, "Chemical Equilibrium and the Thermodynamics of Reactions," this Treatise, Part I, 2nd ed., Vol. 1, Chap. 9.

9. Lewis, G. N., M. Randall, K. S. Pitzer, and L. Brewer, *Thermodynamics*, McGraw-Hill, New York, 1961.

10. McGlashan, M. L., and M. A. Paul, *Manual of Symbols and Terminology for Physicochemical Quantities and Unitis*, IUPAC Additional Publication, Butterworths, London, 1975.

11. Palermo, E. F., and J. Chiu, *Thermochim. Acta* **14**, 1–12 (1976).

12. Paul, M. A., *Principles of Chemical Thermodynamics*, McGraw-Hill, New York, 1951.

13. Prosen, E. J., and R. N. Goldberg, NBSIR 73–180, U. S. National Bureau of Standards, Washington, D. C. (1973).

14. Rogers, D. W., *Am. Lab.* **12**, 18–24 (1980).

15. Rossini, F. D., *Chemical Thermodynamics*, Wiley, New York, 1950.

16. Wagman, D. D., W. H. Evans, V. B. Parker, I. Halow, S. M. Bailey, and R. H. Schumm, *Selected Values of Chemical Thermodynamic Properties*, Natl. Bur. Stand. Tech. Note 270–3, U. S. Government Printing Office, Washington, D. C. 20402, 1968.

PRINCIPLES OF THERMOMETRY

By John C. Melcher and Donald Robertson

Contents

I. INTRODUCTION

A. DEFINITION OF TEMPERATURE

Temperature may be defined as the condition of a body that determines its ability to transfer heat to or receive heat from other bodies. It is a measure of the molecular kinetic energy in the mass of the body.

Thermometry is the science and art of measuring temperature as needed to serve the needs of scientific, technological, and social disciplines. Thermometers for measuring temperature use thermally induced changes of some other property of a material as a measure of temperature. Volume, length, pressure, electrical resistance, thermoelectric voltage, and other properties, when properly calibrated, will measure temperature changes. These can be read out as length of liquid in a capillary tube, rotation of a dial gage, an electric voltage, current, or frequency, etc. Sensors and readout display devices are commercially available to cover substantially all of the temperature scale from 0–4000 K, and to make reasonable estimates of the very high temperatures involved in nuclear and plasma reactions.

B. LIST OF SYMBOLS AND ABBREVIATIONS

A	Ampere(s)
a	Specific heat ratio
a	Absorption coefficient of a surface for Radiation
ACS	American Chemical Society
AIP	American Institute of Physics
ANSI	American National Standards Institute
API	American Petroleum Institute
ASM	American Society for Metals
ASTM	American Society for Testing and Materials
BIPM	Bureau Internationale des Poids et Mesures
Btu	British thermal unit
°C	Celsius temperature, Kelvin temperature $-$ 273.15
c_1	First radiation constant = $3.7405\ 10^{-16}\ \text{W/m}^2$
c_2	Second radiation constant = 0.014388 mK
e	Base of natural log = 2.718
eV	Electron Volt
E	Electromotive force
I	Electric current
IEEE	Institute of Electrical and Electronic Engineers.
IPTS	International Practical Temperature Scale—followed by year of adoption (19)48 or (19)68
J	Joules
Hz	Frequency of alternating current, cycles per sec
K	Absolute temperature, Kelvins
k	Boltzmann's constant = $1.30854\ 10^{-23}\ \text{J/K}$
L	Luminous intensity, luminosity
m	Meters of length
M	Molecular weight of a gas
MKS or MKSA	Meter, kilogram, second, ampere Metric System (Prior to 1962)
N	Newtons, force
n	Power of radiation thermometer curve
NBS	National Bureau of Standards (USA)
NPL	National Physical Laboratory (UK)
NRC	National Research Council (Canada)

p	Pressure
PTB	Physikalische Techniche Bundstanstalt (West Germany)
PMC	Process measurement and control section of SAMA
Q	Ratio of signals from spectral bands of a color ratio pyrometer
R	Electrical resistance, or gas constant
r	Reflection constant of a surface for Radiation
PRT	Platinum resistance thermometer, precision type
RTD	Resistance temperature detector
SAMA	Scientific Apparatus Makers Association
SI	Systeme Internationale, Metric System, adopted 1962
T	Kelvin or absolute thermodynamic temperature
t	Celsius temperature
T	Time
TC	Thermocouple
v	Velocity of sound
V	Volume
V	Volt
W	Watts
W	Work or energy
ϵ	Emissivity coefficient of a surface for radiation
λ	Wavelength of radiation in meters
Π	Peltier coefficient of thermoelectricity
Ω	Electrical resistance in ohms

II. EARLY DEVELOPMENT OF THERMOMETERS

Heat processing of food and ceramics and the comfort of heated shelter were known to primitive man. As civilization advanced, glass manufacture, smelting and refining of metals, textile processing, and fermentation were added to the human activities requiring attention to temperature as part of the heating process quality control. But there was no quantification of a temperature scale. The skill of the craftsman to infer the correct temperature state from responses of his body to radiant or ambient heat and to observe physical changes visible at elevated temperatures were the "measurements" that made heat processing successful.

Galileo, in 1594, invented the first known thermometer. His "thermoscope" was a glass bulb, "about the size of a hen's egg," blown on the end of a small glass tube. It was dipped, bulb up, into a cup of water or wine. The bulb was heated to expand the air. As the bulb cooled, water rose in the tube. As temperature changed, the liquid level would rise and fall with changes of temperature. There is no record of the basis for calibration. Since the cup was open to the atmosphere, it responded also to changes in atmospheric pressure. Following Torricelli's invention of the barometer in 1643, Ferdinand II, Grand Duke of Tuscany, developed the sealed bulb Florentine thermometer, using alcohol expanding from a bulb into a sealed capillary stem. Degrees, intended to represent thousandths of the volume of the bulb, were marked by glass beads fused onto the stem. These thermometers were accepted, despite lack of calibration standards, in meteorological, scientific, and body temperature measurements.

Fahrenheit, (31) in 1714, demonstrated the prototype of the modern liquid in glass thermometer. He selected mercury as the expanding fluid, because of its low freezing point and high boiling point. The thermometer was sealed, but how should it be scaled? Newton, about 1701, had suggested a scale of 0° at the temperature of melting ice, and 12° at the blood heat of the human body. Fahrenheit did not want winter air temperatures reading below zero, and he had no confidence in the accuracy of the ice point. Therefore, he selected the lowest temperature attainable in a mixture of ice and common salt as 0° on his temperature scale, and 12° as blood heat. To attain finer divisions, the scale was revised to use 96° as blood heat. By experiment, he and others determined that the ice point was, indeed, constant at 32°; previous uncertainties had been due to supercooling. Boiling point experiments established 212° as the boiling point of water under standard barometric pressure. Refinement in techniques for body temperature measurement corrected the blood heat temperature to about 98.6°. This revised Fahrenheit scale, refined as the fixed reference points of melting ice and boiling water have been more precisely measured, continues to be the customary scale in popular use in the United States. The Celsius scale will replace the Fahrenheit scale in the United States when convenience, education, and custom permit.

With the fundamental temperature interval between the ice point and the boiling point of water established as a natural constant, scales based upon it were proposed. deReaumer, about 1730, proposed an alcohol thermometer scale of 80° for the fundamental interval, making each degree equal to a thousandth part expansion of the alcohol in the bulb—at last a standardized scale for those Florentine thermometers! The Reaumer scale received some acceptance for many years, but it has no legal acceptance today. Some Reaumer thermometers still exist.

Celsius, (24) in 1742, proposed his Centigrade scale, with the fundamental interval divided into 100 degrees. This scale, supported at the time by many leading scientists, and now named the Celsius scale, is the legal basis of the International Practical Temperature Scale of 1968 used throughout the world. Bolton (16) describes these early thermometers in detail.

Mercury in glass thermometers, with accepted and reproducible temperature scales, became the standard means for measuring temperature over the range $-39°C$ to about 300°C during the 18th century. As the range was extended upward to about 500°C through use of better glass and introduction of gas under pressure in the capillary to prevent boiling of the mercury, applicability increased. The mercury-in-glass thermometer remains today the most common thermometer for scientific and industrial use. Extension of the liquid-in-glass range downward, using alcohol and other liquids with lower freezing points than mercury, adds to the usefulness of the liquid in glass family of thermometers, discussed more fully in section V.A.

Hero of Alexandria demonstrated before 200 A. D. that gases expand when heated, but there is no record that expanding gas was used for temperature measurement until Galileo devised his thermoscope. With liquid-in-glass thermometers available to quantify temperature measurements, 17th century physicists were able to conduct the experiments in pressure, volume, and temperature relationships which are basic to classical thermodynamics. Boyle (17), in 1660, demonstrated the basic law,

$$P/V = R \tag{1}$$

when temperature is constant. But what was the temperature? And what was the general case for the temperature, volume, pressure relationship? Boyle's law, the publications and encouragement of Newton, and the standard Centigrade and Fahrenheit temperature scales encouraged experiments with thermal expansion of gases as a basis for temperature measurement. Various arrangements of constant-volume gas thermometers using pressure change as the readout display and constant-pressure gas thermometers with ingenious means for determining volume change were devised during the 17th and 18th centuries.

Marriotte, in 1784, confirmed that Boyle's law was valid over a wide range of standard temperatures. Charles, in 1787, demonstrated a practical "thermometric hydrometer," with expanding air as the temperature sensor and a mercury column as the temperature display. Gay-Lussac (33) put it all together in 1802 with the combined gas law:

$$pV = RT \tag{2}$$

III. THERMODYNAMIC BASIS OF TEMPERATURE SCALES

A. GAS THERMOMETRY

Heat engines, using the expansion of steam released from a pressure vessel, were demonstrated by Hero of Alexandria about A.D. 100. These remained curious toys for 16 centuries. Newcomen, in 1705, commercialized the first piston steam engine to drive pumps lifting water from mines. Watt, in 1769, invented the vastly superior engine with boiler, piston, and cylinder array and condenser separated. Adding a crankshaft driven by the piston rod

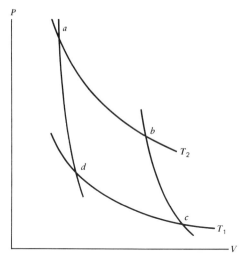

Fig. 3.1. Carnot thermodynamic cycle.

produced the rotary motion that could drive industrial machinery and steam-boats. Fuel efficiency was minimal, probably less than 5% under the most favorable operating conditions. Could the emerging sciences of calorimetry and thermometry provide the information needed by engineers to design more efficient heat engines?

LaPlace and Lavoisier, by 1780, were obtaining caloric data on heats of combusion, vaporization, and fusion. Watt, in 1782, patented the steam engine indicator, which allowed study of the pressure, volume, and stroke relationships in a piston engine, and some deductions about the work output of the steam supply.

Sadi Carnot, in 1824, wrote *Reflections on the Motive Power of Heat* (23), which was the first scientific response to the efficiency problem. Carnot's principles, now known as the second law of thermodynamics, are fully discussed in Chapter 1. The gas thermometry and temperature relationships described in this chapter develop from the specific pressure, volume, temperature, and work relationships in a reversible heat engine using heated gas or vapor as motive power.

To outline this development, we must consider the cyclic operation of a reversible heat engine, the variable parameters being the temperature (as yet lacking a quantitative definition) and two related variables, pressure and volume of a gas. The cycle is shown schematically as the loop *abcda* in which *ab* and *dc* are isothermals and *ad* and *bc* are adiabatics. In traversing the cycle, let Q_2 be the heat absorbed in going from *a* to *b* and Q_1 the heat given up in going from *c* to *d*. By definition, no heat exchanges occur from *b* to *c* and from *d* to *a*. Now it can be shown by qualitative arguments including the second law that Q_2/Q_1 is a function only of the temperatures, T_1 and T_2, and is independent of the nature of the engine, the properties of the working medium, the variables P and V, etc. Thus, we have a basis for the desired absolute or thermodynamic scale of temperature.

Joule, during the period 1840–1850, determined that the value of the mechanical equivalent of heat was 772 ft · lb/Btu, and that the value was substantially constant for all energy conversions, whether the measured value be kinetic, thermal, electrical, chemical, or some other form of energy. Conservation of energy was confirmed as the first law of thermodynamics. Joule's name is honored as the unit of work in the SI measuring system. His value has been revised upward by about a percent as accuracy of measurement has improved. Q in the Carnot cycle could be quantified. Thus, the basic laws of thermodynamics were complete.

Experiments by many scientists were producing data about thermodynamic energy conversions useful in heat engine design. But what numbers should be assigned to the temperatures in the Carnot equation?

William Thomson (not yet Lord Kelvin; he was only 24 years old) joined with Joule in a long series of experiments to establish a scale of absolute temperature. Together they reasoned that the efficiency of the Carnot cycle was directly measured by the temperature ratio if an absolute temperature zero, where a gas had zero thermal energy, could be established.

Experimental observations by Charles and Gay-Lussac had established that all gases expand at the rate of about 1/273 of their volume at 0°C for each degree of temperature rise at constant pressure, and contract at the same rate with falling temperature. This provided a provisional temperature scale in Celsius degrees for a gas thermometer. Joule and Thompson (39), working together for many years, sought a scale for their proposed Absolute temperature. Their experiments included the classic porous-plug adiabatic expansion of gas under pressure, released to atmospheric pressure, in which temperature change measured the work done by the change in pressure. This Joule-Thomson effect was key to the establishment of a gas thermometer with an absolute temperature scale. Experiments with many gases, over as wide temperature ranges as could be achieved at the time, demonstrated that atmospheric air and other commonly available gases followed the gas law very well. Hydrogen was found to be the perfect gas, within the limits of their experimental precision. In 1862, Thomson proposed the Absolute temperature scale, dimensioned in Celsius degrees, with the ice point at 273.1° Absolute, and the steam point at 373.1° Absolute (76). Thus, a scale, based upon the fact that energy of molecular motion ceases at absolute zero, was quantified. Later research has refined the ice point value to 273.15 K. The Absolute temperature scale is now known as the Kelvin scale, reflecting both Thomson's original contribution and Queen Victoria's honor bestowed in 1892. For the convenience of English and American engineers designing steam engines and related apparatus, Rankine used an absolute temperature scale dimensioned in Fahrenheit degrees from the same absolute zero.

Precise gas thermometers, capable of millidegree accuracy over the range from about 100 K to about 400 K, with accuracy decreasing to about 0.2 K at

1338 K are now used mainly by the National Bureau of Standards and other national and international metrology laboratories for occasional experiments calibrating other reference temperature-measuring standards. These gas thermometers usually have glass bulbs filled with helium for the lower temperatures. Platinum bulbs filled with nitrogen are used at the higher temperatures. They are usually operated at constant volume, and pressure is read out with precise mercury manometers.

Gas-filled thermometers with industrial accuracy are discussed in Section V.B.

B. RADIATION THERMOMETRY

Thermodynamic theory evolved from the special case of the Carnot cycle, the concept of absolute temperature, and the realization of a thermodynamic temperature scale for a gas thermometer. Helmoltz, Gibbs, Maxwell, and Stephen achieved major steps in the progress of thermodynamic theory from gas theory toward the general case relationships among temperature, energy transfer, energy conversion, etc. (see Chapter 1). These paragraphs will discuss the development of radiation thermometers and their scales based upon the fundamentals of thermodynamics.

Helmholtz, about 1854, while examining the sources and effects of solar energy, established the principle that radiant energy, received at the surface of an opaque body, is converted into heat that will warm the receiving body. The amount of heating depends upon the proportion of the radiation absorbed. The sum of the absorbed energy plus the reflected energy equals the total radiation in the beam. Furthermore, the heated body would radiate heat, at a rate that increased with temperature, and that was some function of the reflective and absorptive character of the surface of the body. Reflective bodies, bright metal for example, were poor radiators and poor receivers. Absorptive bodies, black-painted or oxidized-surface metals, for example, were good radiators, as well as good receivers.

In 1860, Kirchhoff conceived the idea of a complete radiator, called a "blackbody," that would emit the maximum radiatiation possible for a given temperature and would absorb completely any incident radiation. This is realized experimentally as an isothermal hollow enclosure that has a small opening and is insulated on its outside surfaces. Let another body, with any surface characteristics, be introduced into the cavity, but separated from the inner surface of the cavity to prevent conductive heat transfer from the cavity walls, and it will exchange radiation with the cavity walls and eventually come to the same temperature as the cavity walls. New vistas for quantifying the exchange of radiant energy! Heating of the inserted body would reduce the temperature of the black body in proportion to the respective masses and specific heats. Alternatively, heat could be added to the black body to maintain a constant temperature while the inserted body was heating up. The process was reversible—an inserted body at higher temperature

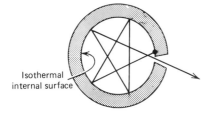

Isothermal
internal surface

Fig. 3.2. Experimental black body. Energy emitted from any point of the internal surface picks up additional energy by multiple reflection to satisfy Kirchhoff's Law, $\epsilon + r = 1$, and finally escapes through the small opening.

than the black body would increase the temperature of the black body. Incident radiation entering the black body through the hole would increase its temperature. Radiation emitted through the hole would be a function of the temperature. The hole could be closed with an imperfect absorber, which will absorb a fraction of the radiation within the cavity, and it will emit as much as it absorbs. Kirchoff's conclusion was that the radiation emission coefficient was equal to its absorption coefficient, and that the sum of absorbtion coefficient and reflection coefficient is equal to unity.

Maxwell published in 1873 the equations of electromagnetic theory, relating the density of radiant energy to the pressure of radiation and in 1879 Stephan established by experiment that the total radiation of a hot body was proportional to the fourth power of its absolute temperature. Boltzmann, in 1884, demonstrated that radiation was analogous to a fluid in thermodynamic transfer of heat. This is stated as the Stephan-Boltzmann law:

$$W_2/W_1 = (T_2/T_1)^4 \tag{3}$$

where T_1 is the higher temperature and T_2 is the lower temperature condition.

Commercial total radiation thermometers, which approach the requirements for a perfect thermodynamic radiation thermometer, are available and are discussed in Section VIII.A.

While theory and practice of the principles of electromagnetic radiation and thermodynamic applications were developing, great strides were made in the understanding of the infrared, visible, and ultraviolet spectra. Photometry was added to available methods of physical measurement. The change in the color distribution of radiant energy as temperature increases had been observed for centuries. At low temperatures, the energy from any radiator is all in the infrared range. At about 600°C, dull red radiant color is visible. As temperature increases, the dominant color changes through orange and straw colors to a dazzling white. Wein (77) established experimentally, with black-body radiators over the then available range of attainable temperatures, that the wavelengths of radiation emitted reached a peak value, thus producing the dominant color, at a wavelength intermediate between the lowest and highest values observed. The dominant color shifted inversely with wavelength toward the violet end as temperature is increased. This shift of dominant color with temperature is Wein's displacement law

$$\lambda_m = \frac{2897.9}{T} \tag{4}$$

This basic work on black-body radiation earned Wein the Nobel prize for physics in 1911.

The concept of using spectral radiance to measure temperature appeared in the 1860s. A photometric device, which used an oil lamp as a standard source and a red glass in the eyepiece to limit response to the red region, was used by Le Chatelier in 1892. But the calibration was empirical, the standard lamp was not reliable, and the photometric comparison was difficult. The instrument had limited use, except to stimulate improvements leading toward the development of successful optical pyrometers.

Planck (57) examined Wein's displacement law and his equations in the course of developing his *Treatise on the Theory of Radiation* in which he introduced the quantum theory. This established the Planck equation for radiation

$$L_\lambda = \frac{C_1}{\lambda^5 \pi \exp\left(\dfrac{C_2}{\lambda T} - 1\right)} \tag{5}$$

where L_λ is the luminous intensity of black-body radiation (spectral radiance) at wavelength λ in meters, T is the absolute temperature, C_1 is a function of Planck's constant, and C_2 is a function of the Stephan Boltzmann constant. Figure 3.3 illustrates graphically Wein's displacement law, with spectral distribution calculated according to Planck's equation. Planck's equation quantifies the temperature and energy relationships in total radiation from a black body more precisely than the Stephan-Boltzmann fourth-power relationship. But that did not solve the technical deficiencies in radiation sensors, which cannot respond as black-body receivers with adequate sensitivity. It does, through defining the energy density at a selected wavelength as a function of temperature, provide the basis for thermodynamic temperature measurements based upon spectral radiance.

Consider the *ratio* of luminosity L at a known temperature and wavelength to L at an unknown temperature at the same wavelength. The ratio, derived from Planck's equation is:

$$\frac{L_2}{L_1} = \frac{\exp\left(\dfrac{C_2}{\lambda T_1} - 1\right)}{\exp\left(\dfrac{C_2}{\lambda T_2} - 1\right)} \tag{6}$$

Let T_1 be the melting point of gold, which has been established by measurement with a gas thermometer, and we have a thermodynamic temperature scale that can extend upward as far as we like, to measure temperature of black bodies.

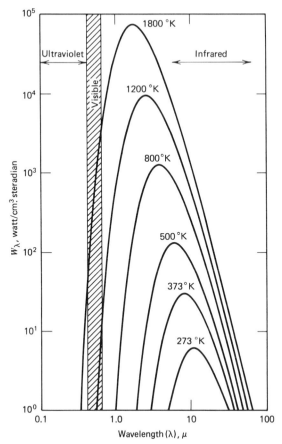

Fig. 3.3. Spectral blackbody radiation as a function of wavelength. (Courtesy NBS.)

The brightness ratio remained difficult to determine by photometric techniques based upon the standard oil or amyl acetate lamps and crude red filters available at the time. Edison's 1879 invention of the electric lamp led to the E. F. Morse invention, in 1902 (50) of the disappearing filament optical pyrometer. Improvements made possible by advances in technology make this the present standard for realizing the thermodynamic temperature scale over the range above the freezing point of gold.

The disappearing filament optical pyrometer is shown schematically in Fig. 3.4. The telescope is focused on the radiant source. The filament of a calibrated lamp is located in the focal plane of the objective lens. A red filter, in conjunction with the long wavelength cutoff of the human eye, provides nearly monochromatic brightness match. The current through the pyrometer lamp is adjusted until the filament disappears in the target image and is measured. The temperature of the lamp filament will increase approximately as the square of the current, it will glow visibly at about 700°C and reach a

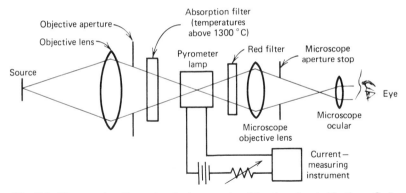

Fig. 3.4. Disappearing filament optical pyrometer. (Courtesy Leeds Northrup Co.)

maximum practicable temperature at about 1250°C. Measured lamp current is converted into apparent temperature from the known calibration characteristic of the lamp. Absorbing filters with calibrated transmission extend the range upward.

The details of optical pyrometer construction and use are discussed in Section VIII.B.

All of the sensors used in thermometry depend to some extent upon the laws of thermodynamics for the response they provide to temperature changes. All of the sensors depend upon the laws of thermodynamics to transfer heat energy from the object under test to the sensor. These thermodynamic aspects of sensor applications are discussed in the sections describing the available sensors.

IV. INTERNATIONAL PRACTICAL TEMPERATURE SCALE OF 1968

As commerce and trade developed in the civilized world, each country established legal standards for measuring mass, volume, length, time, etc., within its domain. Responding to the needs of increasing international commerce and navigation, a treaty was signed in 1875 establishing an International Bureau of Weights and Measures (BIPM) on international territory at Sèvres, near Paris. BIPM was charged with the unification and improvement of the MKS metric measuring system and with determining the conversions from metric units to customary nonmetric units used by individual nations. Temperature standards were added to BIPM responsibilities in 1879. Electrical, chemical, and other standards of physical measurement were added from time to time. In 1962, the present metric system, designated Systeme Internationale de Poids et Mesures (SI) was formally adopted by 40 signatory nations. The basis of the standards has gradually changed from prototype artifacts such as meter bars, kilogram weights, and pendulum clocks to the present basic standards defined as wavelengths of krypton light, cesium frequency, magnetic force of electric current, thermodynamic temperature on the Kelvin scale, mass of C_{12}, and luminous intensity of freezing platinum. The National Bureau of Standards (54) has published complete information on SI measurements, including the basic units, the derived units, the physical constants, nomenclature, and conversions from SI to U.S. customary units. Another NBS publication (64) covers the recommended procedures and practices for the changeover from customary to SI units that is now in progress.

BIPM, NBS, the National Physical Laboratory (NPL) in England, and the Reichsanstalt (now Bundsanstalt) (PTB) in Germany conducted long series of experiments, using gas thermometers, radiation thermometers, thermoelectric and electrical resistance thermometers, and mercury thermometers to determine the best methods to use for the practical realization of the thermodynamic temperature scale in science and industry. The objectives included the development of precise equations for the response of electrical, radiation, and gas thermometers. Equally important was precise measurement of freezing points, boiling points, and other thermodynamic transitions at fixed temperatures that could be useful in the calibration of temperature sensors and readout devices. The melting point of gold was recognized as the key temperature through which to couple the gas thermometer scale to the optical pyrometer and thermocouple scales. Day and Sosman (28) at Carnegie Institution of Washington concentrated on high-temperature gas thermometry, reaching the melting point of palladium, reported as 1549.2°C, and establishing values of melting points for zinc, antimony, silver, copper, gold, nickel, and cobalt. Measurement of these melting points with platinum–platinum rhodium thermocouples, and of the higher temperature points with an optical pyrometer, joined thermocouple sensors to the thermodynamic temperature scale. Callendar (20) established the calibration equation matching the platinum-resistance thermometer to the gas thermometer scale. Others in national laboratories, industry, and academia were actively inventing, developing, improving, and promoting the many forms of sensors and readout instruments described and discussed in the following sections of this chapter.

In 1927, following many national and international symposia on temperature measurement and negotiation by the technical and diplomatic representatives to BIPM, 31 nations at the general conference of weights and measures adopted an International Temperature Scale. Five fixed points were defined between boiling oxygen and the gold freezing point. Interpolating means were defined, and the scale was defined in Celsius degrees over the range from -200°C to 3000°C. In 1948, the scale was renamed as the International Practical Temperature Scale of 1948, some of the defining fixed points were adjusted in value, and some of the interpolating equations were modified. The advancing precision and accuracy of physical measurements, which led to the SI improvements over the MKSA metric system, encouraged extensive review and experimentation in thermometry. Better readout instruments for electrical measurements were available. Metals and alloys for electrical sensors could be refined to higher purity and more carefully annealed. Computers could establish higher-order polynomials to fit calibration curves of sensors to the thermodynamic scale of temperature. Provisional scales for the cryogenic region below 100 K had been tested over a period of years and found satisfactory. Optical pyrometry and high-temperature thermocouple thermometry had been examined in depth. These developments led to the new International Practical Temperature Scale (IPTS) of 1968, with technical and diplomatic acceptance by 40 nations, including the United States and other major developed nations of the free world, as well as Russia and the major developed Communist nations. A Consultative Committee on Thermometry (CCT) coordinates the work on temperature scales, thermometric fixed points, and related work in the field of precise thermometry. CCT reports and recommends improvements in the IPTS.

IPTS 68 (48) is defined over the range 12 K to 4000 K by 13 fixed points and the defining equations of the Carnot cycle,

$$\frac{T - (T_0 + .01)}{T} = \frac{W}{Q} \tag{7}$$

the gas law,

$$\frac{T}{T_0 + .01} = \frac{PV}{P_0 V_0} \tag{8}$$

and Planck's radiation law

$$J_\lambda = \frac{C_1}{\lambda^5 \exp\left(\frac{C_2}{\lambda T} - 1\right)} \tag{9}$$

which constitute the thermodynamic scale. The platinum resistance thermometer is of specified high quality, and readout with a Wheatstone bridge of high accuracy is the interpolating instrument for the range below 900 K. The interpolating equation is a polynomial

$$\frac{R_t}{R_0} = 1 + at + bt^2 + ct^3 \ldots \tag{10}$$

The platinum–platinum 10% rhodium (type S) thermocouple of specified high quality, with readout by a precision potentiometer, is the interpolating instrument for the range 900–1337 K. The interpolating equation is

$$E = a + bt + ct^2 \tag{11}$$

The optical pyrometer is the interpolating instrument above 1337 K. The interpolating equation is

$$\frac{L_\lambda(T_{68})}{L_\lambda(T_{68(A_4)})} = \frac{\exp\left[\dfrac{C_2}{\lambda T_{68(AV)}}\right]}{\exp\left[\dfrac{C_2}{\lambda T_{68}}\right]} \tag{12}$$

The thermocouple provides necessary overlaps below the high end of the resistance thermometer range and above the low end of the optical pyrometer range. Constants for interpolating equations for a particular sensor are determined by calibration at the defining fixed points.

Table 3.I displays IPTS 68 arranged from 0 K at the top. Vertical columns show assigned value of the significant fixed reference points in T K, $t°$C. Defining points and defined fixed points are so designated. Other points are mainly those for which NBS can furnish calibrated samples, useful in calibrating standard and working temperature sensors.

TABLE 3.I
Thermodynamic Temperature and IPTS 68

Significant point	K	°C	Defining IPTS 68	Interpolating Instrument
Absolute 0	0	− 273.15	Yes	—
B.P. ^{4}He	4.222	− 268.928	No	Not agreed upon
T.P. equil H_2	13.81	− 259.34	Yes	PRT
T.P. normal H_2	13.956	− 259.194	No	PRT
B.P. normal H_2 25/76 STP Atm	17.042	− 266.108	Yes	PRT
B.P. equil H_2	20.28	− 252.87	Yes	PRT
B.P. normal H_2	20.397	− 252.753	No	PRT
T.P. Ne	24.561	− 248.569	No	PRT
B.P. Ne	27.102	− 246.048	Yes	PRT
T.P. O_2	54.361	− 218.759	Yes	PRT
T.P. Ne	63.146	− 210.004	No	PRT
B.P. N_2	77.344	− 195.806	No	PRT
B.P. O_2	90.188	− 172.962	Yes	PRT
Subl. pt CO_2	194.674	− 78.476	No	PRT
F.P. Hg	234.314	− 38.836	No	PRT
F.P. H_2O	273.15	0	No	PRT
T.P. H_2O	273.16	0.01	Yes	PRT
T.P. phenoxy benzene	300.02	26.87	No	PRT
Blood heat	310	37	No	
B.P. H_2O	373.15	100	Yes	PRT
T.P. benzoic acid	395.52	122.37	No	PRT
F.P. In	429.784	156.634	No	PRT
F.P. Sn	505.118	231.968	Yes	PRT
F.P. Bi	544.592	271.442	No	PRT
F.P. Cd	594.258	321.108	No	PRT

TABLE 3.1 *Continued*

Significant point	K	°C	Defining IPTS 68	Interpolating Instrument
F.P. Pb	600.652	327.502	No	PRT
B.P. Hg	629.81	356.66	No	PRT
F.P. Zn	692.73	419.58	Yes	PRT
B.P. S	717.824	444.671	No	PRT$_1$
F.P. CuAl eutectic	821.38	548.23	No	PRT$_1$
F.P. Sb	903.905	630.755	No	Type S TC PRT
F.P. Al	933.61	660.46	No	Type S TC
F.P. Ag	1235.08	961.93	Yes	Type S TC
F.P. Au	1337.58	1064.43	Yes	Type S TC Optical Pyrometer
F.P. Cu	1358.03	1084.88	No	Type S TC
F.P. Ni	1728	1455	No	Optical Pyrometer
F.P. Co	1768	1495	No	Optical Pyrometer
F.P. Pd	1827	1554	No	Optical Pyrometer
F.P. Pt	2042	1769	No	Optical Pyrometer
F.P. Rh	2236	1963	No	Optical Pyrometer
M.F. H_2O3	2327	2054		Optical Pyrometer
F.P. Tr	2720	2447	No	Optical Pyrometer
M.P. Nb	2750	2477	No	Optical Pyrometer
M.P. Mo	2896	2623	No	Optical Pyrometer
M.P. W	2695	3422	No	Optical Pyrometer
+ crater carbonarc	3787	3514	No	Optical Pyrometer
	4000 K	Upper limit of	IPTS 68	Optical Pyrometer

Assignment of accuracy to these figures is not possible at present; errors may be as high as several units in the last digit shown.

B. CRYOGENIC TEMPERATURES

Cryogenics is roughly defined as the science and technology of reactions and phenomena occurring at temperatures below about 100 K. The region is scientifically and technologically important because of such broadly applicable technologies as liquefaction of gases and electrical superconductivities. It is also a frontier area for fundamental research in physical, chemical, and life sciences. Analytical techniques in this temperature region are discussed in Chapter 17. Cryogenic temperature-measuring techniques are discussed in Section VII.B.

Thermal energy increments for small temperature changes are very significant at these low temperatures, requiring precise calibration data for sensors. Heat leaks can result in substantial errors in temperature measurement. Linkage to IPTS 68 and to a true thermodynamic temperature scale is difficult, since the triple point of hydrogen, 13.81 K is the lowest assigned T_{68} temperature.

Several national laboratories and the CCT have cooperated in the comparison of their provisional low temperature scales, using T_{68} fixed points, gas and acoustic thermometers, and various interpolating sensors to develop an international provisional temperature scale, EPT-76, for the region 0.5 K to 30 K. This scale (55,36) is continuous with IPTS 68 at the Ne boiling point, 27.102 K. It is a few mK lower than T_{68} down to the H triple point. Below 13.8 K it ij defined by assigned T_{76} values at superconducting transition points and the boiling point of ^{4}He. Germanium, carbon, and NMR ther-

TABLE 3.1b
EPT-76 Provisional Temperature Scale for 0.5–30°K

Significant point	Assigned temperature T_{76}	Corresponding temperature T_{68}	Interpolating instrument[a]
Superconducting transition—Cd	0.519 K	0.517 K	GeRT or NMR
Superconducting transition—Zn	0.851	0.848	GeRT or NMR
Superconducting transition—Al	1.1796	1.1764	GeRT or NMR
Superconducting transition—In	3.4145	3.4077	GeRT or NMR
B.P. ^{4}He, p_0 = 1 std. atmosphere	4.2221	4.215	Acoustic or He thermometer
Superconducting transition—Pb	7.1999		Acoustic or He thermometer
T.P. Equil H$_2$	13.8044	13.810	PRT
B.P. Equil H$_2$ 25/76 std. atm.	17.0373	17.042	PRT
B.P. Equil H$_2$ 1 std. atm.	20.2734	20.280	PRT
T.P. Ne	24.5591	24.5612	PRT
B.P. Ne, 1 std. atm.	27.102	27.102	PRT

[a] Interpolating equations for GeRT and NMR thermometers are determined by calibration; they are not defined in EPT-76.

mometers are suitable interpolating sensors for work requiring precision of 5 mK. Thermocouples and PRTs are useful, but usually less sensitive. Table 3.Ib displays EPT-76 in the format of Table 3.I.

EPT-76 and the results of ongoing research to improve the fit of T_{68} to a true thermodynamic temperature will be published from time to time in *Metrologia* and as text revisions to IPTS 68. A new IPTS 8? will be established during the next few years when research results justify improvement (37).

C. TEMPERATURE-MEASURING SENSORS FOR GENERAL USE

Table 3.II displays the sensors available for temperature measurement, for guidance in selecting units useful in a particular experiment. The lowest temperature measured on the Kelvin scale determines the location in the table. Precision means the ability of the measuring system to detect small changes in temperature or the scatter in successive measurements of the same temperature. Accuracy means the limit of error in reproducing IPTS 68. Usual precision and accuracy is that obtained by using apparatus according to the manufacturers instructions, without correcting for systematic errors. Best accuracy and precision requires state-of-the-art quality in sensors and readout instruments. Instruments must be calibrated to achieve NBS or equivalent correction data. Corrections must be applied to data taken during experiments using careful measurement techniques. For contact sensors, precision and accuracy refer to the temperature state of the *sensor*. For radiation thermometers, precision and accuracy are based upon *black-body* sighting conditions. More information about the commonly used and readily available sensors will be found in the designated sections of this chapter. References are cited for the newer and less-established devices. References for the commonly used thermometers, supplementing the brief information included in a particular section, are cited in the text of the section.

Proceedings of the symposia, *Temperature, Its Measurement and Control in Science and Industry* (1–4) contain tutorial, review, and research papers covering the whole spectrum of temperature measurement technology and applications. These occasional meetings, sponsored by American Institute of Physics in cooperation with other technical societies and NBS, review state of the art in sensors, readout and auxiliary apparatus, and applications. They are excellent reference sources for detailed information on unusual problems in thermometry. The text books (7,11,26) contain detailed information on many of the devices and applications described briefly in this chapter.

Figure 3.5 displays the capability of NBS to reproduce IPTS 68, and the uncertainty which NBS assigns to their reports of calibration on sensors submitted to them. For compression, temperature is plotted on a logarithmic scale. Uncertainty is expressed as parts of the Kelvin temperature readings.

TABLE 3.II
Temperature Sensor Guide

Name	Physical quantity	Sensor form	Associated instruments and display	Temp. range K	Usual application	Usual precision	Best precision	Usual accuracy	Best accuracy	More information
Magnetic thermometer	Magnetic susceptance	Crystal of cerous magnesium nitrate	Mutual inductance bridge	0.01 to 15 K	Laboratory measurement	0.001°K	Not defined			(36)
Noise thermometer	Johnson noise	Superconducting quantum interference device—SQUID	Electronic measurement of magnetic flux	0.01–4 + K	Laboratory measurement		Not defined			(38)
Nuclear magnetic resonance thermometer	Free induction decay of Cu^{63} and Cu^{65}	Material placed within an inductive field	Signal generator, amplifier detector, lock in amplifier, RMS voltmeter	005–about 40	Laboratory measurement		Not defined			(38)
Acoustic thermometer	Sound velocity in cavity	Acoustical resonant cavity	Electro-acoustical transponder and length measuring instruments	0.1–50	Laboratory measurement		0.001 K			Section X
Vapor thermometer	Vapor pressure of fluid	Metal bulb filled with saturated vapor, capillary connection to readout	^{3}He and precision manometer	0.2–2	Laboratory measurement		0.001 K			(53), Part I p.37
			^{4}He and precision manometer	1–5.2	Laboratory measurement		0.001 k			
			Other vapors, with Bourdon gauge or other pressure readout device	4–400	Industrial or laboratory measurement and control	0.5% of temperature span	0.25% of span	1% of span	0.5% of span	Section X.B
Germanium thermometer	Electrical resistance	Chip of doped germanium, with four terminals	Potentiometer or Kelvin bridge	1.5–100	Measurement and control in laboratory	0.1 K	0.0001 K	1 K	0.001 K	Section VII.B
Carbon thermometer	Electrical resistance	1/8 or 1/4 W carbon resistor	Wheatstone bridge	1.5–100	Measurement and control in laboratory	0.1 K	0.001 K	1 K	0.01 K	Section VII.B
Gas thermometer	Gas pressure constant volume	Gas-filled bulb, capillary connection to readout manometer	Precision manometer for high precision—precise measurement of bulb volume	4–1338	Reproduce thermodynamic temperature scale in national metrology laboratories		0.001 K at 4 K, 0.2 K at 1338 K		same as precision	Section III.A
			Industrial form has Bourdon gauge attached to capillary		Measurement and control in laboratory and industry	0.5% of span	0.2% of span	1% of span	0.5% of span	Section V.B

Resistance thermometer	Electrical resistance	Strain-free platinum wire (PRT) with four leads	Four-terminal bridge or potentiometer and standard resistor	4–900	Temperature standard laboratory measurement	0.001 K	0.00001 K	0.01 K	0.0001 K	Section VII.A
		Platinum or base metal wire or film, encapsulated—usually three leads (RTD)	Wheatstone bridge, arranged for balanced or voltmeter readout—chart recorders—digital meters	10–1338	Measurement and control in laboratory and industry	0.1% of span	0.01% of span	0.5% of span	0.02% of span	Section VII.A
Thermocouple	Thermal EMF	Pt-Pt 10% Rh wires, with measuring and reference junctions	dc potentiometer	300–1338	Temperature standard, laboratory measurement	0.5 K	0.1 K	1 K	0.2 K	Section VI
		Other noble, base and exotic metal thermocouples	Potentiometer, digital meter, millivoltmeter, chart recorder	4–2400	Measurement and controls in laboratory and industry	ANSI standards	Varies among T.C. types	ANSI standards	Varies among T.C. types	Section VI
Quartz thermometer	Frequency of piezo electric oscillation	Y-cut quartz crystal	Oscillator, with counter and timer	30–500	Laboratory and industrial measurement	0.001 K	0.0001 K	0.02 K	0.0002 K	Section IX
Thermistor	Electrical resistance	Metal oxide chip or bead with 2 leads	Wheatstone bridge, digital indicator or chart recorder	30–600	Laboratory and industrial measurement and control, control of consumer appliances	0.5 K	0.001 K	1% of span	0.1 K	Section VII.B
Acoustic thermometer	Sound velocity	Metal rod	Pulse generator, magnetostrictive transducer; time interval apparatus	30–3000 Depending on metal	Laboratory measurement	1% of reading		2% of reading		
Nuclear quadrupole resonance thermometer	Frequency of magnetic field	Potassium chlorate specimen in tank coil of oscillator	Oscillator with feedback circuit to clamp at resonant frequency—display on recorder	30–300	Laboratory measurement	Not yet determined				(56)
Liquid in glass thermometer	Thermal expansion	Glass bulb filled with mercury or other liquid	Capillary graduated in ° of temperature is an integral part of the device	125–900	Universal in laboratory, industry, home	0.5% of span	0.001% of span	1% of span	0.005% of span	Section II.V.A
Liquid filled thermometer	Thermal expansion	Metal bulb filled with mercury or other liquid	Capillary connecting to Bourdon gage or pressure switch	150–900	Indicating, recording, and control in laboratory, industry, and home	0.5% of span	0.1% of span	1% of span	0.25% of span	Section V.B
Bimetallic thermometer	Differential thermal expansion	Adherent metal strips coiled as spiral or helix	Dil gauge with pointer, or an electric switch	150–650	Indicating, recording and control in laboratory, industry, and home	1% of span	0.5% of span	2% of span	1% of span	Section V.C

TABLE 3.II (*Continued*)

Name	Physical quantity	Sensor form	Associated instruments and display	Temp. range K	Usual application	Usual precision	Best precision	Usual accuracy	Best accuracy	More information
Noise thermometer	Johnson noise voltage	Platinum filament resistor	Low-noise amplifier and peak voltage comparator	273–1338	Laboratory measurement	Not yet determined				
Liquid crystal thermometer	Optical color and transmission changes	Various cholesterol and other organic compounds		Near room & body temps.						Section XI
Specific temperature indicators	Melting point or phase change	Pellets, cones, paints, crayons	Color change or mechanical deformation	293–2000	"Tattle tale" to show temperature exposure	1% of nominal value				Section XII
Total radiation thermometer	Total radiant energy	Thermopile, bolometer, or other photon detector	Optical system, potentiometer or other device for measuring detector output	273–5000 and up	Measurement and control in laboratory and industry	Varies with temperature and ambient conditions				Section VIII.A
Spectrally selective radiation thermometers	Radiant energy in selected wave band	Photon detector	Optical system, including filter for selected wave band—device for measuring electrical output of detector	278–5000 and up	Measurement and control in laboratory and industry	Varies with temperature and ambient conditions				Section VIII.B.4.
Monochromatic optical pyrometers	Spectral radiance in narrow band	Human eye in manual form	Calibrated filament lamp, telescope, filter, absorption screen, device for current measurement	1025–5000 and up	IPTS 68 interpolations laboratory and industrial measurement	2 K or 2% of reading	1 K or 1% of reading	Varies with temperature, emissivity		Section III.B.2.
		Photoelectric detector in automatic form	Same as manual form, with addition of electronic servo balancing system, can have recording display	Same as manual form	IPTS 68 interpolations laboratory and industrial measurement and control	0.5° or 0.51%	0.2° or 0.2%	Conditions		Section VIII.B.2.
Color radio pyrometers	Ratio of spectral radiance in two (or three) narrow bands	Photon detector	Telescope, interference filters, voltage ratio meter	875–3000	Laboratory and industrial measurement and control under gray body conditions	Varies with temperature and gray body conditions				Section XII.B.4

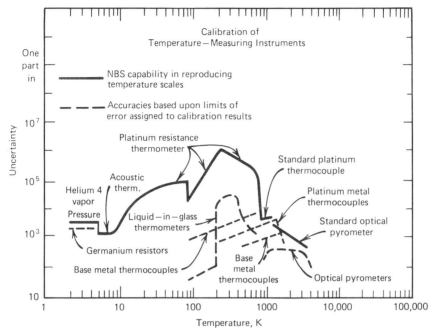

Fig. 3.5. Calibration of temperature-measuring instruments.

V. EXPANSION THERMOMETERS

A. LIQUID-IN-GLASS THERMOMETERS

1. Description of Commercially Available Thermometers

Liquid-in-glass thermometers are by far the most numerous sensors for measurement of temperature. Three centuries of manufacture and use have produced inexpensive, reliable devices. Even the mass-produced thermometers for household use are usually accurate within a percent of scale span over a long period of time. In the chemical laboratory, mercury thermometers, inert to most chemicals, are the workhorse instruments for the range −39 to +550°C, which covers most of the processes of interest to analytical chemists.

High-quality mercury thermometers are very carefully made. Accurately proportioned bulbs blown onto uniform bore capillary tubing are carefully annealed, filled with the purest mercury, and sealed off with suitable pressure of nitrogen in the capillary space. After aging the unit through temperature cycles, calibration proceeds by exposing it to successively higher temperatures and marking the stem at the mercury column position for each calibration point. Intermediate divisions are engine divided and etched or fused to the stem, along with the numbers. Accuracy is typically 0.5% of

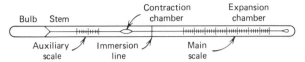

Fig. 3.6. Principle features of a solid-stem liquid-in-glass thermometer. (Courtesy NBS Monograph 150.)

scale span or one scale division when the unit is properly used. Scale corrections, determined by calibration against a more precise standard, are applied when more accurate readings are needed. Fig. 3.6 illustrates the features of mercury-in-glass thermometers. Not all of the features shown would necessarily be incorporated in any one thermometer.

Manufacturers and scientific-apparatus supply houses can furnish thermometers in lengths from about 10 cm to about 60 cm. Temperature spans can be as short as 5°C and as long as 500°C. When measurements below −39°C are required, the fluid can be mercury–thallium eutectic for minimum temperature about −56°C, or organic fluids for temperatures to about −200°C. Limited-purpose thermometers with scales and mounting arrangements suitably selected are available for most of the test procedures sponsored by the ACS, API, ASTM, and other sponsors of specific temperature tests and analyses. Thermometers constructed of very high-temperature glass or fused silica, with mercury or gallium as the working fluid, can be used at temperatures higher than 550°C, but thermocouples (Section VI) and other electrical (Section VII) or radiation thermometers (Section VIII) are usually chosen for these higher temperatures.

Maximum reading thermometers (the clinical thermometer is a common example) include a constriction of the capillary immediately above the bulb. This allows the mercury to rise when heated, but prevents the column from receding until it is shaken down by centrifugal force. Armor cases and pocket cases are available to protect the glass element. Industrial thermometers enclose the bulb in a metal well suitable for threading into a pipe tap on a pressure vessel, with the emergent stem enclosed and protected. Sling psychrometers mount a wetting device on the wet bulb: the wet and dry bulb thermometers are mounted on an arm that is spun rapidly to make a reading. Cases with cups enclosing the bulb trap a fluid under test, a convenience for gauging temperature in tanks. Ingenious design of capillary tubing assists vision in reading the mercury column. Lenticular shapes for magnification, red reflectors and white backgrounds for contrast, and specially ribbed tubes to minimize squinting are among the vision aids offered.

The Beckman differential thermometer is a special form of limited range, maximum reading thermometer that pushes the sensitivity of mercury thermometers to the practical limit of about 0.002°C. It is especially useful in the analytical laboratory to measure differential temperatures; i.e., rise or fall in calorimetry, small differences in freezing or boiling points, etc. Essential components are a large bulb of mercury, a large bore capillary with a range such as 0–150°C, a small bore differential capillary with a range such as 5°C, and a mercury reservoir which permits the starting point of the differential range to be adjusted for the experiment at hand. A particular thermometer will be constructed to read the maximum of a rising temperature or the minimum of a falling temperature in the differential range. Manipulating the amount of stored mercury to change the starting point of the differential range requires patience and skill.

2. Application Notes and Caveats

Liquid-in-glass thermometers approach the temperature of the test medium slowly. Allow at least 2 min in stirred liquid, and 5 min or more in low-velocity gases, for equilibrium to be established. Check readings, made several seconds apart, will establish stability. If process temperature changes faster than 0.5°/min., and you need better than 1° precision of measurement, then use a faster sensor, such as a fine gauge thermocouple. Lags of temperature readings in calorimetry are a special case, described by White (78).

Immerse the thermometer in the test material to the maximum reading that is expected, thus avoiding emergent stem errors due to the temperature of the stem and the mercury in the capillary. Stem correction for mercury in glass thermometers is approximately $0.00016n \, (t_1 - t)$ where n is the number of °C emergent, t_1 is bath temperature, and t is stem temperature. For organic liquids in glass, the correction constant is about 0.001, not 0.00016. This approaches 40°C error at 500°C when the bulb only is immersed. Tie a small thermometer to the stem if need be to measure the approximate stem temperature. *Exception*: When the thermometer is calibrated for a specific immersion, immerse to that depth, which should be marked on the stem.

When a thermometer is warmed, then cooled, the volume of the bulb does not immediately return to its original volume at the lower temperature. This hysteresis results in slightly lower readings. A check reading at the ice point (Section VI.B) will establish the amount of ice-point depression that needs to be added to the measurement. Thermometers exposed to temperatures up to about 100°C will seldom exhibit ice-point depression greater than 0.1°C. Thermometers exposed to temperatures above about 150°C need an ice-point check and corrections after every experimental run. Thermometers exposed continuously to temperatures above about 370°C, or intermittently to temperatures above about 405°C, may exhibit significant permanent changes in bulb volume. This results from strain deformation as the glass approaches the strain point. Erratic ice-point corrections that are large compared with required precision of reading signal the need for a replacement thermometer.

In calorimetry, when the range of temperature is small, and the interval between measurements is short, heat the thermometer to the highest expected temperature *before* starting a series of experiments. Ice-point depression will take place before the observations, and it will have no effect on temperature change readings that are the required data from the experiment.

Fill test wells in vessels, solids, or pipelines with mercury, oil, or other fluid suitable to the temperature range to reduce time lag and errors from heat conduction along the thermometer stem.

Support thermometers on stands, in stoppers, or other suitable ways to avoid strain on the bulb and errors due to local hot spots in a vessel. Try to keep the stem vertical and the bulb down. Stoppers used to mount thermometers are expendable. Cut the stopper so that the thermometer can be inserted and removed safely.

Measure freezing points as the liquid cools slowly from the liquid state. The liquid will supercool, and the release of the heat of fusion will quickly reheat to the true freezing point as crystals of solid form. Cooling rate should be slow enough to sustain the freeze at least 15 min in accurate work. Melting points are not reliable because the solid usually superheats before melting, and the equilibrium temperature of solid and liquid phases is difficult to establish. An exception to this rule is the use of shaved ice packed in a Dewar flask to establish 0°C. (See Section XIV.B.)

Shake down maximum reading thermometers with a rotary snap motion. Grasp the thermometer, bulb up, swing it through about 90° using the wrist as an axis, stopping abruptly. The action is repeated as necessary until the mercury column stands substantially lower than the expected reading. Join a separated mercury column bay grasping the thermometer firmly, bulb downward in the hand. Strike the hand sharply from below. Repeat as needed to join the column. For difficult cases of separated columns, see (79).

Beckman thermometers are set by transferring mercury in a series of shaking-down rotations and column-joining taps. Follow the manufacturer's directions precisely. The expansion bulb must be full. The mercury in the differential capillary must stand at the temperature expected at the start of the experiment. Surplus mercury must be in the middle of the reserve reservoir. Calibration is usually based on a setting of 20°C. Setting factors must be applied when the setting is changed.

Measure boiling points with the thermometer in the vapor over the boiling liquid. (Section XIV.B). The liquid superheats in the vessel, and equilibrium is established in the vapor at the boiling point at the ambient pressure. Allow plenty of time—30 min is not unreasonable—for purging air and establishing equilibrium in the vapor chamber. Correct the measurement to standard pressure.

In calorimetry, account for the heat capacity of the thermometer. If data are not available, measure the volume of the bulb, and the volume of the immersed stem, in a partially filled graduate. Heat capacity is about 2.1 J/ml for the glass, and about 1.9 J/ml for the mercury.

Store thermometers, bulbs down, in cushioned racks or containers to minimize breakage and separation of the liquid column.

B. FILLED-SYSTEM EXPANSION THERMOMETERS

Fluid expansion thermometers provide for remote reading of temperatures. A metal bulb, filled with mercury or other fluid, is connected by capillary tubing to a Bourdon tube pressure gauge; as the fluid expands the pressure will increase. The gauge dial is calibrated in temperature. These thermometers are widely used in industry, and to a lesser extent in the laboratory, when remote display of a measurement is required. A filled-system thermometer is shown in Fig. 3.7.

When the fluid is mercury, the scale is linear and ranges between −39°C and 550°C can be covered. The shortest practicable span is about 50°C. Transmission distances to about 70 m are practicable. The capillary tubing connecting the bulb to the remote gauge acts like an emergent stem on the mercury-in-glass thermometer, requiring correction or compensation for the ambient temperature effect.

When the fluid is nitrogen or other gas under pressure, the scale is linear, and ranges between −150°C and 500°C can be covered. The shortest practicable span is about 100°C. Since the ambient effect on the transmission tubing is small, compensation for this is usually not required.

When the fluid is alcohol, n-butane, propane, or other fluids having a low boiling point, the scale is nonlinear, according to the vapor pressure-temperature relationship of the fluid. Fluid is selected for the desired operating range. The pressure inside the thermometer is determined by the temperature at a free surface of the liquid. The lower limit of range may be about

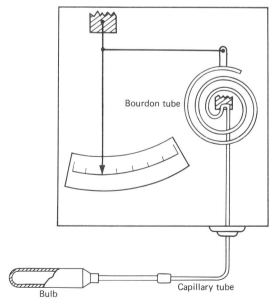

Fig. 3.7. Filled-system thermometer. (Courtesy SAMA.)

− 50°C, the upper limit of range may be about 400°C, and the useful span for a particular instrument will be greater than 40°C and less than 200°C. The ambient temperature of the capillary does not affect the calibration directly, but a cross-ambient effect associated with the transfer of fluid will delay operation of the remote indicator when the measured temperature at the bulb crosses the ambient temperature of the capillary. If the remote indicator is not at the same elevation as the thermometer bulb, a calibration for the difference in these elevations should be made. Otherwise, the head of liquid in the capillary will add to or subtract from the vapor pressure in the system to cause an error. The liquid-vapor interface must be inside the sensitive bulb.

C. BIMETALLIC EXPANSION THERMOMETERS

1. Description of Available Instruments

When two strips of metal, having different coefficients of thermal expansion, are joined together, fastened at one end, and heated, the bimetal strip will bend to form an arc of a circle. The highly expansive or active metal will be on the outside, and the low expansive or dead metal will be on the inside. Early applications of this principle depending upon rivets, solder, or cement to join the two materials into a bimetal strip provided rugged temperature sensors with unreliable calibration.

Direct high-temperature–high-pressure-welded lamination of metals in ingot form, a process that was developed originally to provide rolled plate for the jewelry industry, provides strong and mechanically inseparable bimetal laminations. These can be rolled and annealed to the desirable size of strip needed for reliable bimetal thermometry (45).

The thermal activity or flexivity of a bimetal strip can be measured as a function of radius change, temperature change, and thickness of strip. Choice of materials and strip thickness are the basic controls used to determine desired characteristics.

Simple bimetal thermometers, widely used as air temperature indicators, are flat hair springs. The outside of the spring is anchored to a post. The inner end of the spring drives a pointer along a temperature scale as it rotates with changes of temperature. These simple forms are not suitable for immersion in fluids, installation in pipelines, or many other laboratory purposes.

A multiple helix arrangement of the bimetal spring, developed by Weston Instruments, Inc., about 1940 (45), provided needed compactness, with a sensitive portion comparable in size with the bulb of a mercury-in-glass thermometer. Thus, the inherent ruggedness of the bimetallic elements was introduced in the form of a rotary pointer dial thermometer with an enclosed, immersible, small-diameter metal tube, as shown in Fig. 3.8.

These thermometers are available for temperatures in the region from − 80°C to 500°C, with temperature spans as short as 50°C. Dial diameters

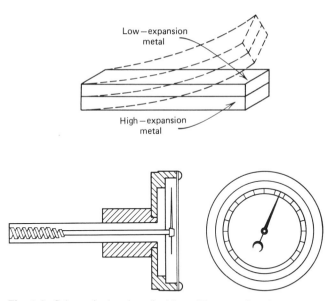

Fig. 3.8. Schematic drawing of a bimetallic expansion thermometer.

range from about 2 cm to about 10 cm. Mounting arrangements permit immersion in liquids or gases, insertion in soft solids, and insertion in wells for pipelines and pressure vessels.

Bimetal dial thermometers can be accurate to 0.5% of range. Safe operation at overswings of 10% or more above the top of the calibrated range and at underswings up to 50% below the bottom of the range is useful in many applications.

2. APPLICATION NOTES AND CAVEATS

Choose thermometer locations for good thermal contact with the hot body. In general, location requirements are similar to those for a liquid-in-glass thermometer.

The sensitive helix, which is 2–10 cm long, must be fully immersed.

Check and adjust thermometer zero at the ice point as required to correct for set in the helix.

Immerse the thermometer deep into the hot body to reduce error from conduction of heat along the emergent tube. There are no stem corrections, but conduction errors, which may be significant, cannot readily be estimated.

Allow sufficient immersion time to equalize temperature with the hot body; 1 min in stirred liquid is appropriate. Allow a longer time if the element is enclosed in a protection tube, or if temperatures of semisolids or gases are measured.

Tap the dial gently before reading to minimize effects from frictional drag on the readings.

VI. THERMOELECTRIC THERMOMETERS

A. THEORY AND PRINCIPLES OF THERMOELECTRICITY

In 1821, Seebeck (70) discovered that when the ends of two wires of different metals are joined to form a closed circuit, and the junctions are maintained at different temperatures, an electric current flows around the circuit. Peltier, in 1834, observed that when an electric current passes through the junction of two dissimilar metals, heat is released or absorbed at the junction depending upon the direction of current flow. Heat is proportional to the current, and to the Peltier coefficient

$$\Pi = t\,\frac{dV}{dt} \tag{13}$$

The coefficient, determined experimentally, is a function of the metals chosen and the temperature. Both the Seebeck and the Peltier effects are reversible. Thus, heat is absorbed at the hot junction and released at the cold junction to drive electric current through the circuit. Conversely, an electric current from an external source driven through the circuit will heat one junction and cool the other. Thus, we have the equivalent of a Carnot engine for conversion of heat to electricity.

William Thomson, in 1851, applied the laws of thermodynamics to the thermoelectric circuit. He found that an electric current produces different thermal effects as it passes from hot to cold, or cold to hot in the same metal. The Thomson effect has a coefficient that varies widely among metals, alloys, and semiconductors. Since thermoelectric thermometry uses very small currents, it is largely dependent upon the Peltier effect, and the Thomson effect can be neglected.

The source of the thermoelectric potential is the tendency for electrons or holes carrying an electric charge to diffuse away from the heated portion of an electrical conductor. If conductors of different materials are joined, and the junction is heated, electrons flow from the material with the least number of electrons to the material with the most. The electron flow will reverse if the junction is cooled. The heat generated or absorbed as current passes through a thermocouple junction equals the energy difference between the conducting electrons (or holes) across the junction. Metallic (electron) conductors are used in thermoelectric thermometry. Thermoelectric generators and refrigerators pair hole conductors with electron conductors to maximize the potential available and the heat released or absorbed at the junctions.

Bacquerel, in 1830, tried to use platinum–palladium, and other metal pairs, as temperature-measuring devices. Lack of sensitive voltage-measuring instruments prevented development by him of a practical thermoelectric thermometer. Le Chatelier (47), in 1885, needed a thermometer to measure temperatures in the calcining of hydraulic cement. He developed the platinum–platinum 10% rhodium thermocouple, and used a d'Arsonval galvanometer with a high series resistance as a millivoltmeter. The method was successful. Calibration was established by comparison with gas thermometers. The freezing points of sulfur and gold were established as reference check points. Le Chatelier's thermocouples and millivoltmeters were soon adopted for high-temperature research in metallurgical, ceramic, and glass industries, as well as the cement industry. Application to real-time process measurement in industry, based upon Le Chatelier's work, led to the development of the array of thermocouple materials, readout instruments, and accessory hardware available in today's market. The Le Chatelier thermocouple, Type S in American National Standards Institute MC 96.1 (6), is the interpolating device for IPTS 68 from 630°C to 1064°C.

The successful use of thermocouples in the measurement of temperature depends upon three experimentally established laws. The law of homegeneous circuit states that an electric current cannot be sustained in a circuit of a single homogeneous metal by applying heat along it. The law of intermediate metals, analogous to the second law of thermodynamics, states that the algebraic sum of the electromotive forces in a circuit composed of any number of dissimilar metals is zero if all of the circuit is at uniform temperature. The law of successive or intermediate temperature states that two dissimilar homogeneous metals producing thermal emf, E_1, when junctions are at temperatures t_1 and t_2, and E_2 when the junctions are at t_2 and t_3, will produce emf $E_1 + E_2$ when junctions are at t_1 and t_3. Combination of the three laws gives the following statement: The algebraic sum of the thermoelectric emf generated in any given circuit containing any number of dissimilar homogeneous metals is a function of the temperatures at the junctions only.

In the discussion of thermoelectricity, it was explained that the amount of thermoelectric effect in a circuit of two dissimilar materials depends upon the temperature difference between the two junctions. To measure temperature by exposing one of the junctions to an unknown temperature, it is necessary to know the temperature of the other junction.

It is usual to refer to the junction exposed to the temperature being measured as the "measuring junction" while the other junction is known as the reference junction.

The reference junction can either be held at a known constant temperature, and the readout calibrated accordingly, or may be allowed to float with ambient temperature.

The melting ice pack (Section XIV.B) is frequently used to provide a fixed reference-junction temperature. While it is simple, inexpensive, and accurate, it requires considerable maintenance. The ice supply must be replenished at frequent intervals to assure proper precision.

Equipment is available to provide an accurately controlled cavity for maintaining isothermal reference temperature. It is usually designed to control at the ice point, although some devices are available that maintain higher temperature. The calibration of the readout instrument must be based upon the reference temperature selected.

When the reference junction is allowed to float with ambient temperature, a properly controlled bias voltage must be provided in the measuring instrument. This is accomplished by incorporating a temperature-sensitive element in good thermal contact with the reference junction. This element appropriately adjusts the bias voltage to compensate for the temperature change of the reference junction.

Thermoelectric materials are available to cover temperatures ranging from about 2 K to above 3000 K. The voltage generated is low, ranging from less than 1 mV to about 7 mV for each 100°C of temperature span. The thermocouple, with an established or measured reference-junction temperature, when connected to an emf measuring device becomes a thermoelectric thermometer. Readout devices are discussed in Section VI.B, Thermoelectric materials, and applications are discussed in the Sections VI.D and VI.E. More detailed information will be found in ANSI, ASTM, and NBS publications (6,8,10,9,27,61,65,67).

B. THERMOCOUPLE READOUT INSTRUMENTS

The readout system of a thermoelectric thermometer may be a deflection-type, high-resistance millivoltmeter, a sensitive-digital voltmeter, or a potentiometer (Fig 3.9).

Deflection millivoltmeters (Fig. 3.9a) make measurements with an accuracy of about 1%. Thermoelectric current will range up to a few hundred

---

Fig. 3.9. Thermoelectric thermometer schematics. (a) Thermocouple, AB, and millivoltmeter, M, measure temperature t, with reference junction at t_r. $E_{AB} = f(t - t_r)$. Meter current, $I_M = E_{AB}/(R_M + R_{AB})$. For direct reading of t, offset meter zero $f(t_r)$. R_M = resistance of meter, in Ω, R_{AB} = resistance of thermocouple and leads, in Ω. (b) Thermocouple, AB, and potentiometer, P, measure temperature t with reference junction at t_r. $E_{AB} = f(t - t_r)$. At potentiometer balance, $I_a = 0$ and $I_p R_S = E_{AB}$, where I_p = potentiometer current, and R_S = portion of slide wire resistance to balance. For direct reading of t, offset potentiometer zero $f(t_r)$. (c) Differential thermocouple, BAB, and measuring instrument, M, measure the difference between t_1 and t_2. $E_{BB} = f(t_1 - t_2)$. M may be millivoltmeter or potentiometer; t_{r_1} and t_{r_2} must be equal. (d) N thermocouples with voltages adding in series are a thermopile. $\sum_1^N E_{AB_N} = Nf(t - t_r)$. M may be a millivoltmeter or a potentiometer. For direct reading of T_1, offset zero $f(Nt_r)$. (e) N thermocouples in parallel measure average temperature at n junctions. $E_{AB} = f\left(\sum_1^n \frac{(t_1 + t_2 + ..t_n) - t_r}{n}\right)$.

M may be a millivoltmeter or a potentiometer. Resistances of all thermocouples must be equal. For direct reading, offset zero of M = $f(t_r)$. (f) Block diagram of linearized Digital Voltmeter, alternate for millivoltmeter or potentiometer with any thermocouple arrangement. (Courtesy of Leeds & Northrup. Co.)

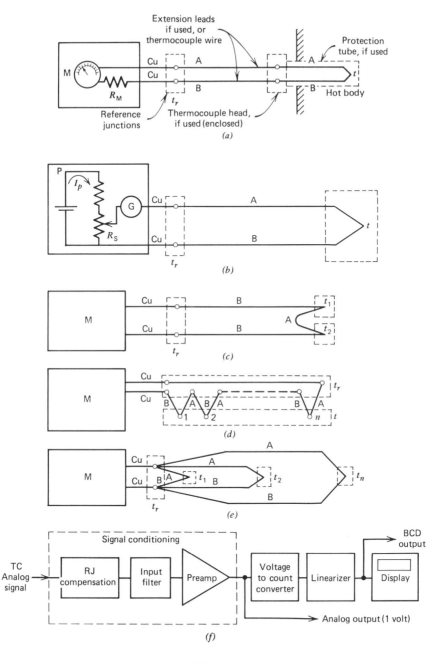

Fig. 3.9.

microamperes. Resistance of the meter is in proportion to the millivolt span to be covered. The scale may be calibrated in temperature degrees or in millivolts. The reference junction is at the instrument terminals. The reference-junction temperature is measured by a bimetal thermometer coupled to the zero adjustment of the meter to obtain compensation. Compensation for the resistance of the thermocouple is usually made by manual adjustment of a series resistance.

Potentiometer systems (Fig. 3.9*b*) are also used for measuring thermocouple signals and can provide the best accuracy and precision. Here a resistance network carrying a known current is tapped to provide a known voltage to balance against the voltage of the thermocouple. A galvanometer or electronic null detector identifies the balance point. Reference junctions may be immersed in a bath of melting ice for 0°C reference, or the reference-junction temperature may be measured and compensated. The potentiometer may be scaled in millivolts or may be direct reading in temperature. Portable indicators and automatically balanced indicators and recorders for thermocouples are available. They are accurate to about 0.25% of range span, direct reading in temperature, and automatically compensated for reference-junction temperatures. Laboratory potentiometers, such as the type-K potentiometer, and more precise instruments are used when accuracy of 0.01% or better is required.

Digital-voltmeter instruments (Fig. 3.9*f*) are well suited for measuring thermocouple signals. Meters with 10-μV measurement capability are adequate for industrial measurements using base metal thermocouples. For precise measurements, meters having a measurement sensitivity of 1 μV or less are recommended.

Readout may be displayed in voltage for conversion into temperature from tables, or directly in temperature. Direct reading meters have a built-in linearizing system, and automatic reference-junction compensation.

Voltage-indicating meters require the use of constant temperature reference-junction devices such as ice baths, etc. Direct-reading instruments provide internal bias voltage systems to compensate for reference-junction temperature variations.

C. AVAILABLE THERMOCOUPLES AND THERMOCOUPLE ASSEMBLIES

Table 3.III tabulates the thermocouples in common use that are readily available commercially. Those designated by ANSI type letters are required to meet the initial calibration tolerances stated in ANSI C 96.1, (6) and listed in the table. Elements meeting the standard tolerances are satisfactory for most laboratory and industrial applications. Selected tolerances, available at a small premium price, apply to the best wire available from a mill run. In addition to having better accuracy, the selected wire will be more homogeneous from end-to-end in a given spool. These thermocouple materials are available as bare wires in a wide range of sizes, as made-up thermocouples in

TABLE 3.III

Characteristics of Thermocouples

Material	Ni-10% Cr + Cu-Ni –	Cu – Cu-Ni –	Iron + Cu-Ni –	Ni-10%Cr + Ni-5%Al-Si –	Pt-10%Rh + Pt –	Pt-13%Rh + Pt –	70%Pt-30%Rh + 94%Pt-6%Rh –	W + W-26%Re –	Ni-Cr-Si + Ni-Si –
ANSI type	E	T	J	K	S	R	B		
Temp. range	0–900°C	0–350° C	0–750°C	0–1250°C	0–1450°C	0–1450°C	870–1700°C	No rating	0–1250°C
Tolerance, initial calibration ±	1.7°C or 0.5%	1°C or 0.75%	2.2°C or 0.75%	2.2°C or 0.75%	1.5°C or 0.25%	1.5°C or 0.25%	0.5%	No rating	2.2°C or 0.75%
Selected tolerance	1°C or 0.4%	0.5°C or 0.4%	1.1°C or 0.4%	1.1°C or 0.4%	0.6°C or 0.1%	0.6°C or 0.1%	not available	not available	not available
Cryogenic range	–200–0°C	–200–0°C	N.A.	–200–0°C	not available for cryogenic temperatures				
Tolerance, initial calibration ±	1.7°C or 1%	1°C or 1.5%	not specified	2.2°C or 2%					
Melting point	1280°C	1083°C	1280	1375°C	1760°C	1760°C	1810°C	3200°C	1400°C
Useful low temperature	–270°C	–270°C	–210°C	–270°C	0°C	0°C	870°C	700°C	0°C
Useful high temp – 8 AWG	870°C		760°C	1260°C					1300°C
Useful high temp – 14 AWG	650°C	370°C	590°C	1090°C				2300°C	1200°C
Useful high temp – 20 AWG	540°C	260°C	480°C	980°C	1480°C (#24 Ga)	1480°C (#24 Ga)	1700°C (#24 Ga)	2000°C	1000°C
Useful high temp 30 AWG	430°	200°C	370°C	870°C					1000°C
Average output	70 μV/°C	50 μV/°C	55 μV/°C	40 μV/°C	10 μV/°C	8 μV/°C	16 μV/°C	16 μV/°C	40 μV/°C
Oxidizing atmosphere	Yes	Protect	Protect	Yes	Yes	Yes	Yes	Protect	Yes
Reducing atmosphere	Yes	Yes	Yes	Protect	Protect	Protect	Protect	Yes	Yes
Fabrication method	Arc weld Neutral flame Gas weld Silver solder Crimp	Soft solder Silver solder crimp	Braze Arc weld Neutral flame Gas weld Crimp	Neutral flame Gas weld Crimp Arc weld	Arc weld	Arc weld	Arc weld	Arc weld	Arc weld Reducing flame Gas weld

85

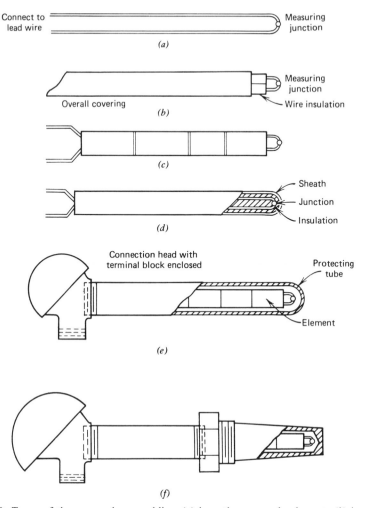

Fig. 3.10. Types of thermocouple assemblies: (a) bare thermocouple element, (b) insulated wire thermocouple element, (c) element with double bore insulators, (d) element with compacted ceramic insulation, (e) thermocouple in protection tube, (f) thermocouple in well for pressure vessel.

Fig. 3.11. (a) Thermal emf's of common thermocouple materials *vs.* platinum. ($T_0 = 0°C.$) (b) Thermoelectric power, dE/dT, of common thermocouples.

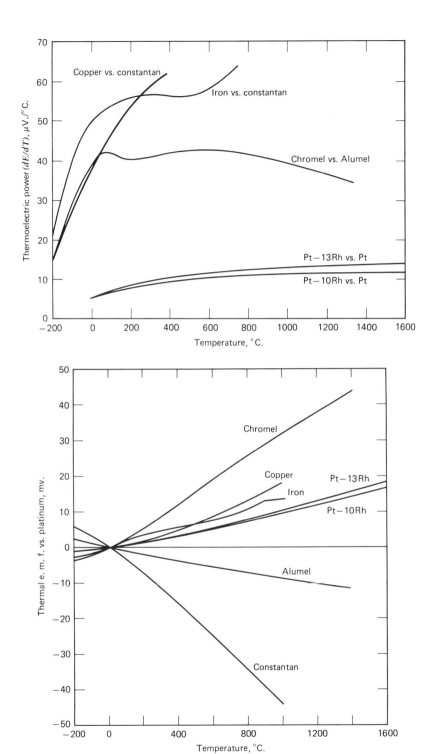

Fig. 3.11.

a wide range of lengths, as insulated wires and insulated thermocouple pairs, and as protected thermocouples mounted in many ways. These are illustrated in Fig. 3.10. Thermoelectric output is displayed in Fig. 3.11. Comment on the selection of thermocouples and protection tubes follows in section VI.D.

D. SELECTION GUIDES FOR THERMOCOUPLES AND PROTECTION TUBES

Type T, copper +, copper nickel −, and Type E, chromium nickel +, copper nickel −, are the thermocouples of choice for a wide range of general laboratory measurements where the scientist must make up his own thermocouples to fit a particular experimental need. Both types are available as individual wires, matched pairs, metal sheathed insulated pairs and assembled thermocouples, in wire sizes ranging from 14 to 50 AWG. Insulations available include enamel, enamel and wrap or braid of natural, synthetic, or mineral fiber, extruded vinyl, nylon, or Teflon plastic. Type-T junctions can be soft soldered; hence, they may be easier to fabricate. The positive element is pure copper, not likely to have localized nonhomogeneity. Accuracy at temperatures below 0°C is better. Upper limit is imposed by the rapid oxidation of the copper wire. Type E, chromium nickel +, copper nickel −, has about 75% greater emf/°C, an advantage for precision in measurement and control. Both elements oxidize slowly, and the oxide film adheres to afford some protection to the underlying metal; higher temperatures can be measured. Since both wires are alloys, local nonhomogeneity may introduce errors.

Type-J, iron +, copper nickel −, thermocouples also have a broad spectrum of application in laboratory and industrial use. The variety of wire sizes, insulation, sheathing, and protection tubes parallels those available for types E and T. emf output is intermediate between types E and T. Type-J thermocouples are uniquely resistant to deterioration from reducing atmospheres. Useful range extends upward to about 980°C in reducing atmospheres. Oxidation that causes failure will also shift the calibration.

Type-K, Chromium-nickel +, nickel, aluminum, silicon − thermocouples are the low-cost choice for use in oxidizing and neutral atmospheres over the range 400–1250°C. Hydrogen, carbon monoxide, and sulfur-bearing gases will destroy the calibration and cause early failure by embrittlement at temperatures above 500°C. They may be operated for short periods of time at temperatures up to 1300°C in safe atmospheres.

NiCrSi and NiSi are two thermoelectrically stable nickel alloys recently developed under the leadership of the Materials Research Laboratories of the Australian department of Defence and standardized by the National Bureau of Standards of the U. S. Dept of Commerce (19a). When paired as a thermocouple, these materials form a stable temperature sensor. In comparison with type-K thermocouples, stability in the area of 500°C and at the upper end of the useful temperature range is reported to be much better. Thermocouple materials are available from several alloy manufacturers, and thermocouple wire, completed couples, and instruments, both recording and controlling, are available from at least one instrument company.

Type S, Le Chatelier's platinum, platinum 10% rhodium thermocouple has an unique position as the original practical thermocouple and as the interpolating sensor for a portion of IPTS 68. Calibration reports, precise to 0.1 and accurate to 0.2°C, can be obtained from NBS at fixed points. Type R, platinum–platinum 13% rhodium, has about 20% higher output. Type B has about 20% lower output, but can measure higher temperatures. The platinum materials are stable in oxidizing atmospheres, but are subject to contamination and embrittlement in reducing atmospheres. Grain growth, alloy diffusion at the junction, and contamination by reducible oxides all cause drift of calibration and eventual failure at the junction. Reduced silica from the insulators can be a problem.

Tungsten–rhenium alloys are strictly for use at very high temperatures, in vacuum or reducing atmospheres. Pure tungsten is brittle; adding rhenium to both elements reduces brittleness, with minor change in thermal emf. Trace quantities of oxygen will quickly destroy the thermocouple.

Other thermocouple pairs, platinum 15% iridium–palladium, iridium–iridium 40% rhenium, and a number of refractory semiconductor pairs are being used experimentally in effort to extend the upper limit for thermocouple temperature measurement.

Gold–cobalt versus silver is being studied for cryogenic application. It has about three times the thermoelectric power of copper constantan at temperatures below 100 K, an attractive characteristic.

E. APPLICATION NOTES AND CAVEATS

Thermocouples, like all other contact temperature sensors, measure their own temperature. Hence, it is imperative to locate the measuring junction to attain, as nearly as possible, the temperature of the material under test. The output of the thermocouple depends upon the temperature at the point where the wires separate (51). Hence, a small junction, with good heat transfer from the test material, is essential. Heat will leak along the thermocouple wires. Hence, immersion of several centimeters into the test material is desirable. Any protection sheath or well surrounding the thermocouple will add to the heat leak along the wires. Heat transfer to a thermocouple is a "thin rod" type of problem.

Time lag in measuring the temperature of the test object depends upon the heat capacity of the thermocouple compared with the heat transfer capability of the test body. Time constants, the time to achieve 62% of temperature change, will range from a few milliseconds for a bare junction of fine gauge wire immersed in flowing liquid to several minutes for sheathed or protected junctions immersed in gas flowing at 20 m/sec. Heat transfer from gases to thermocouples may be poor. In general, choose the smallest thermocouple and the smallest sheath or well that can meet the experimental requirements for pressure, temperature, corrosion resistance, and operating time at temperature. Small thermocouples must be used to follow rapid fluctuations and to control processes subject to major upsets. Larger thermocouples may be used in large-scale apparatus where the system approaches a steady thermal state. In a system of low heat capacity, the temperature must be measured by the smallest thermocouple that can endure the experiment. Discarding short-lived thermocouples frequently is a small price to pay for better data or better temperature control.

The ideal thermocouple and its production tube or sheath will resist expected oxidation, reduction, and corrosion in the worst-case hostile temperature or other condition. When the thermocouple enters a hostile environment, the protection tube must be leak proof at operating pressure and temperature. Purging the tube with a flow of air or inert gas, not sufficient to affect the temperature reading, is often useful.

Thermocouples in hot gases, especially above 300°C, may receive radiant heat from flames or other heat sources, or lose heat by radiation to the cooler walls of confining ducts. These errors can be reduced by using deeply immersed small thermocouples. Open-ended tubes, with the junctions protruding a few centimeters, are suitable for support when needed. Radiation shields can be helpful. Aspiration to increase gas velocity and hence heat transfer at the junction may be required. Commercial forms of gas thermocouples mounted in stagnation-type probes, originally developed for testing jet engines, are available and useful when gas velocity exceeds 30 m/sec.

An important advantage of thermocouples, used with potentiometers as the measuring instruments, is that the emf is transmitted without error over substantial distances. Electrical interference can, unfortunately, introduce stray voltages and noise to degrade the measurement. Never locate thermocouple leads in conduits containing electric power circuits. Well-designed measuring instruments will reject both the common mode (ground loop) and magnetic (transverse mode) interference associated with heaters, motors, etc., in the thermal system. Electrical insulation degrades at high temperatures, sometimes permitting leakage current to enter thermocouple circuits. Thermionic current may be present at high temperatures. Electrical shielding and guarding may be needed to overcome these. Thermocouples may be grounded, if the readout system will permit it, but only at one point. A second ground, at a different temperature from the first, will cause circulating currents between the two grounds through the thermocouple leg. The IR drop from the current in the thermocouple leg will cause a voltage bias in the thermocouple to produce an error.

Surface temperatures of stationary objects can best be measured with the wires peened into shallow holes drilled close together at the test point. Welding or soldering wires to the surface, or locating the thermocouple on the surface under an insulating pad, introduce heat transfer errors. With any of these arrangements, the emergent leads must run along the surface of the object for several centimeters to minimize the heat leak along the wires. "Bowstring" thermocouples, which are junctions formed at the middle of thermocouple wires flattened into ribbon, can be pressured against moving and stationary objects for approximate surface temperature measurements.

Accuracy can be maintained by using selected and calibrated thermocouples, backed up by a program of replacement as the elements deteriorate from stress and corrosion. Calibration methods are discussed in Section XIV. When thermocouples are exposed to temperatures no greater than the maximum temperatures listed in Table 3.III, they can be expected to maintain calibration for at least 100 hr at heat. It is good practice to calibrate thermocouples before and after critical experiments.

Local fabrication of thermocouples from wires, insulated pairs, and metallic-sheathed pairs is the easy, economical, and fast way to fit thermocouples into experimental apparatus. Any convenient welding, soldering, or clamping technique will form the junction. The junction should be made as small as is practical for the wire size, so that it will be isothermal in use. Kinks and sharp bends should be avoided because they can introduce nonhomogeneity in the conductors.

Homogeneity is tested as follows: Connect the measuring junction and the end of the positive lead to a galvanometer or null detector and then pass a flame, a small heater, or a piece of dry ice slowly along the wire. No emf should appear. Repeat the test with the negative lead connected. If an emf greater than the desired limit of error appears, discard the thermocouple. Similar test for homogeneity should precede recalibration of a used thermocouple.

Average temperatures may be measured by connecting n thermocouples in series, using a separate reference junction for each thermocouple (Fig. 3.9d). All reference junctions must be at the same temperature. Divide measured emf by n before converting to temperature. n Thermocouples in parallel will also measure average temperature with all thermocouples connected to the same reference junction (Fig. 3.9e). Parallel thermocouples should be equal in resistance for correct averaging. When thermocouples of various lengths are connected in parallel, equal resistance in each thermocouple can be approximated by padding each thermocouple with a series resistor. If the thermocouples are grounded, padding resistors of equal resistance must be added to *each* thermocouple leg.

F. THERMOPILES

Thermopiles, a group of thermocouples connected in series, are used in situations where the signal from a single thermocouple is not large enough to make a satisfactory measurement (Fig. 3.9d). They are also useful for making average temperature measurements.

These sensors take many forms and are usually designed for a specific purpose. Generally fine-gauge wires or ribbons are used for the thermoelectric elements. The material is selected for high thermoelectric power in conjunction with satisfactory mechanical properties. The measuring junctions can be bonded together thermally in a small area, or they may be spread out to measure the average temperature of a large body. They must, however, be electrically insulated from each other. The same applies to the reference junctions. Care must be used to keep heat transfer between the two sets of junctions to a minimum by thermally connecting them to the two areas where the temperature difference is being measured.

Recently, thin-film construction has been successfully used in radiation thermometry and other thermopile applications. More junctions can be deposited in a given area, with very small sensor heat capacity, making a fast responding sensor.

Typical applications are: Radiation thermometry (Section VIII.A), calorimetry, heat transfer, infrared gas analysis, differential thermal analysis, and other areas where small differences in temperature must be measured.

Thermopiles are the sensors of choice for many situations in calorimetry. Details will be found in Chapter 6. Sensitivity to temperature differences less than 0.001°C is easily obtained. For example, with 10 Type-T thermocouples in series, output will be about 400 μV/°C. Therefore, 0.4 μV sensitivity in the potentiometer and null detector is needed to achieve 0.001° readout of temperature difference. Microdegree sensitivity can be obtained by using either more thermocouples in series or a more sensitive readout instrument. Reference and measuring junctions may be located to measure temperature rise above ambient as in a bomb calorimeter, or temperature difference between chambers as in a differential calorimeter. When the measuring junctions measure calorimeter temperature and the reference junctions measure jacket temperature, the output can be fed to a temperature controller to maintain zero temperature difference in adiabatic calorimeters. Large-scale calorimeters may not have uniform temperatures. Distributing the reference and measuring junctions will average the temperature difference readings, improving accuracy. In general, wire or strip thermocouples are suitable for calorimeters designed for 10 ml or larger samples. Film thermocouples may be used for smaller-scale apparatus.

Heat-transfer applications include measurement of thermal conductivity of plate and rod samples of conductors and insulators, and performance measurements of heat exchangers. In all of these applications, the sensitivity advantage of a series of thermocouples is apparent. The ability to spread the junctions to obtain average temperature differences improves accuracy in experiments using large samples or complex apparatus. When thermal guarding is needed, as in measuring thermal conductivity of plate and sheet samples, a thermopile with junctions distributed around the edge of the sample and the edge heater provides means for controlling the edge heater at the sample temperature. Local hot spots in heat exchangers can be identified if the thermocouples are arranged with switching to permit measurement of individual TC temperatures.

In differential thermal analysis, thermopiles increase the signal from the small temperature difference between the reference and test samples. Very fine wires must be used to obtain reasonable speed of response and to minimize heat leak errors. Film thermopiles, if available, can provide better performance.

Output of the thermopiles is a function of the measured temperature, generally increasing as the temperature increases. Hence, the thermopiles must be calibrated over a range of temperature, and readings must be corrected for the actual temperature achieved in the experiment. Heat leaks must be minimized, and accounted for when significant. Readout instruments may be affected by electrical interference. Filtering, shielding, and guarding, as suggested for other thermoelectric thermometers are recommended.

In infrared gas analysis, thermopiles function as radiation sensors with output in reasonably constant relation to energy level over a wide range of wavelengths. In nondispersive analyzers, the measuring junction receives energy through the unknown gas and the reference junction receives energy through the standard sample.

VII. ELECTRICAL RESISTANCE THERMOMETERS

A. METAL RESISTANCE THERMOMETERS

1. Principle of Operation

Free electrons carry current through a perfect metallic crystal in proportion, by Ohm's law, to the applied voltage and the conductance of the metal. In theory, the flow of electrons at absolute zero would be unimpeded, since the internal electric field would be periodic in space corresponding to the periodicity of the lattice. In real life, superconductivity does not work quite that way. The thermal vibration of the lattice at higher temperatures, and the increase in the mean-free path of the electrons with temperature, increase the resistance. Thus, we have a temperature coefficient of resistance for pure metals as the basis for resistance thermometers. Imperfect lattice structures and foreign atoms as impurities or as alloys scatter the electrons to increase the resistance, and to decrease the temperature coefficient of resistance. Evanohm, constantan, and manganin—alloys used in the manufacture of high-quality resistors for electrical measuring instruments—have only a few parts per million temperature coefficient at ordinary ambient temperatures. Platinum, copper, nickel, and tungsten are the metals normally used in resistance thermometry. Temperature-resistance ratios for these metals are shown in Fig. 3.12.

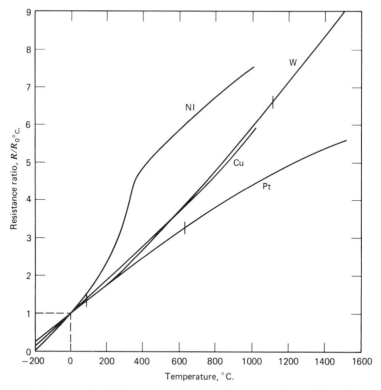

Fig. 3.12. Resistance ratio $R/R_{0°C}$ vs. temperature of some metals used in resistance thermometry. The bar on each curve indicates the usual upper limit of use. The resistivities at 0°C. are as follows: Cu 1.55, W 4.89, Ni 6.14, Pt 9.81 μohm-cm. (Courtesy of NBS.)

Resistors, and the generation of heat by passage of electric current through them, were included in Joule's experiments that established the mechanical and electrical equivalent of heat. Ohm's law and Wheatstone's bridge circuit for measuring resistance permitted observation of the effect of temperature on the resistance of metals. Development of telegraph systems, following S. F. B. Morse's 1837 invention, and development of electric generators and motors following Picinotti's 1860 invention of the dynamo encouraged substantial research on the characteristics of electrical conductors.

Winding several centimeters of fine-gauge insulated wire onto a small bobbin was the easy part of developing resistance thermometers. Unfortunately, the resistance of the lead wires between the sensors and the terminals of a Wheatstone bridge (Fig. 3.13) can be large, in comparison with the change of sensor resistance with temperature. Consider Fig. 3.13a. If X is 10Ω, and increases 0.04 Ω/°C, and a and b are each only 1 meter of #18 copper wire, the lead wire resistance will add about 1°C to the apparent temperature. William Siemens (73) devised in 1871 the bridge arrangement

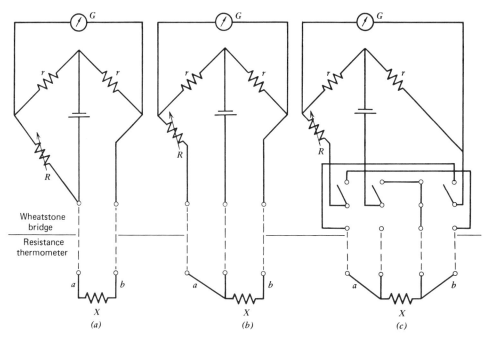

Fig. 3.13. Schematic drawings of three Wheatstone bridges. In each case the bridge is balanced by adjusting the rheostat, R, until the galvanometer, G, reads zero. Then at balance the following relations apply: (a) Two-lead thermometer, $R = X + a + b$. (b) Three-lead thermometer, $R = X + b - a \simeq X$. (c) Four-lead thermometer and reversing commutator, $R_1 = X + b - a$; $R_2 = X + a - b$; $X = (R_1 + R_2)/2$.

and three-lead thermometer of Fig. 3.13b. This compensates for lead resistance, provided that a and b are equal in resistance. In practice, equal lengths of copper wire of the same gauge are nearly equal in resistance. Thermometers and bridges of this basic design are the most common arrangements for resistance thermometry today. Precision of 0.1°C is typical for these arrangements. Mueller (52) in 1916 developed the four-lead thermometer and reversing commutator connection to the bridge shown in Fig. 3.13c. Interchanging the a and b leads and averaging the readings cancels out lead resistance even if the resistances are unequal, and also cancels out parasitic thermal emf in the thermometer and its connections, provided that the reversal is done quickly enough that there is no change in lead resistance or parasitic emf.

2. Available Resistance Thermometers

In 1888, Callendar (20) described platinum resistance thermometers he had developed for precise temperature and temperature difference measurements in calorimetry. These were platinum wires, loosely wound on notched mica frames to provide support without introducing strain from differential thermal expansion. The sensors were annealed in air at high temperature by passing electric current through the coil. The elements, about 5 cm long, were sealed with dry air in glass or metal tubes about 50 cm long. Gold leads were used inside the tube, and copper extension leads were used outside the tube for connection to a Wheatstone bridge. Lead resistance was compensated for by a loop of gold and copper wire, equal to the lead resistance, added to the adjustable standard resistance in a bridge such as Fig 3.8a. Platinum resistance at 0°C was about 25.5, providing a resistance change of about 10 Ω over the fundamental interval 0–100°C. Callendar's platinum thermometer, improved by Mueller (52), who substituted current and potential leads for the loop compensator, and by Meyers (49) who devised the helical helix winding and the helium fill for a more compact and faster-responding sensor, is the platinum resistance thermometer (PRT) used today for interpolating IPTS 68.

Precision thermometers for use below 90 K are enclosed in platinum capsules. These may be used at temperatures up to about 300°C. For temperatures above 500°C, to about 635°C, specially selected mica is needed for the supporting frame on the PRTs. Thermometers with $R_0 = 100$ are available to provide a gain in sensitivity of readout, at the expense of a more fragile element. "Bird cage" construction, (12,30) using a larger gauge platinum wire supported on an aluminum oxide frame in a fused quartz tube extends the useful range to about 1100°C. R_0 for these thermometers is about 0.25 Ω: a substantial loss in sensitivity of readout. Stability is also lower.

Computation of temperature from the polynomial interpolating equations is complex (63). Use of the precision thermometers is simplified by presentation of calibration report data by NBS and other calibration laboratories as computer printouts of R_t/R_0 ratios, usually in 1° increments, over the calibrated range of a particular thermometer.

PRTs are expensive, reflecting the high cost of the skilled handcraft manufacturing and calibration processes. Hence, they are usually reserved for use as calibration standards and for experimental and analytical work requiring the highest precision.

Industrial-grade resistance temperature detectors, RTDs, at a fraction of PRT cost, sacrifice the precision of strain-free construction. The economy of quantity production, and the ruggedness of elements bonded to suitable bobbins suggest use of RTDs for much of the work in an analytical laboratory. Platinum windings, with R_0 ranging from 10 Ω to 1000 Ω, and with a variety of mounting arrangements and protection tube materials, are available (68). Copper windings, with R_0 ranging from 10 Ω to 100 Ω, and with a variety of mounting arrangements and protection tube materials, are available, and suitable for temperatures from -100°C to 150°C. Nickel windings are suitable for the range -40°C to 300°C. Tungsten windings, theoretically useful to temperatures of 1000°C, are not readily available in the marketplace. The manufacturers provide conversion tables for their standard forms of industrial grade thermometers. Limits of error are usually 0.5°C or 0.1%

of reading. With proper calibration and correction for errors, one can achieve about 0.1°C accuracy. Using a bridge with 0.001% accuracy, temperature change can be measured with precision approaching 0.001°C over small ranges.

RTDs using deposited metal films have advantages of ruggedness and low cost. They are available in two types—thick film and thin film.

Thick-film elements are made by printing a special metallic paste onto a suitable substrate. The unit is then fired to drive off the vehicle and "cure" the metal. The resulting deposit is a relatively pure metal film adhering to the substrate. This technique has been used for many years to apply metallic decorations to china and glass. Research on the pastes that produce stable, adherent metallic film of proper temperature coefficient made this type of construction practicable.

Thin-film thermometers are made by typical thin film depositing techniques such as evaporation, sputtering, etc. It is much more difficult to control the deposit of a thin film to produce a stable, adherent element having the proper temperature coefficient. Once the technique has been mastered, a very small and rugged sensor can be produced.

Film thermometers are typically made of relatively pure platinum to match existing calibration curves, deposited on a sapphire or other pure aluminum oxide substrate. Alumina is used because its coefficient of thermal expansion is a close match to that of platinum, minimizing resistance change from strains in the film as temperature cycles.

"Stick-on" thermometers are grids of fine wire bonded within insulative plastic encapsulations.

3. Readout Instruments for Resistance Thermometers.

The Mueller bridge (Fig. 3.14a) is the instrument most commonly used for measuring resistance of platinum resistance thermometers. A range of 100 Ω is required to measure 25 Ω thermometers at 650°C. 400 Ω is required to measure 100 Ω thermometers. Decade steps of 1 $\mu\Omega$ are required to read the triple point of water with precision of 0.00001 K. The eight-decade Mueller bridge and a galvanometer or electronic null detector sensitive to a nanovolt will accomplish this. This is the present state of the art precision in both temperature and resistance measurements. Six-decade Mueller bridges, with less resistance range and 10 $\mu\Omega$ minimum resistance steps are less expensive and also useful for many laboratory purposes. These bridges read out in ohms, R_t/R_0 must be calculated, and converted to temperature. Mueller bridges may be used also with RTDs, thus omitting the reversal of leads. This is technical overkill for industrial-type thermometers, but it provides the needed sensitivity for use of such thermometers in precise measurement of temperature changes.

Similar bridges, powered by ac at about 1 KHz have comparable precision, eliminate thermoelectric effects, and take advantage of the high sen-

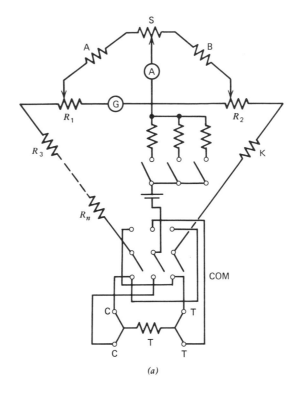

(a)

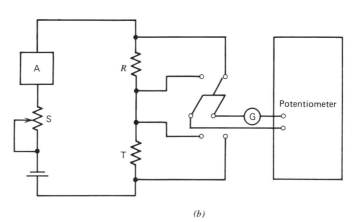

(b)

Fig. 3.14. (a) Mueller bridge method. $R_1, R_2 \ldots R_n$. Five to eight decades of resistance. $A = B$. Ratio resistors. T. Four-lead resistance thermometer. A. Meter to measure thermometer current. G. Null detector. COM. S.P.D.T. commutator for lead reversal. S. Trimmer for ratio arms. K. Bridge zero adjustment. At balance $R_T = R_1 + R_2 + \ldots R_n$. (b) Potentiometer method. A. Meter to measure thermometer current. S. Adjustable resistor which controls thermometer current. T. Four-lead resistance thermometer. R. Four-terminal standard resistor. G. Galvanometer. (c) Siemens three lead industrial thermometer bridge, manual or self-balancing. G. Null detector. T. Three-lead thermometer. $R_1 = R_2$. Bridge ratio. S. Adjustable standard resistance. Lead resistance B must equal lead resistance A; then, at balance, $R_S + B = R_T + A$ or $R_S = R_T$.

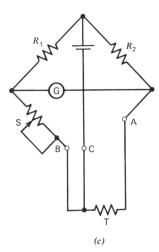

(c)   Fig. 3.14 *Continued*

sitivity of ac null detectors (37). A modified Kelvin bridge (72) provides the very wide range required to accommodate the low resistance bird-cage thermometers as well as the common 25 and 100 platinum thermometers.

The potentiometer method, Fig. 3.14b, compares the voltage drop across a four-terminal thermometer to the drop across a four-terminal standard resistor when both are connected in series and carrying a small current. Voltage range is from about 10 mV to about 400 mV, depending upon the thermometer used and the temperature range to be covered. Voltage ratio readings convert to R_t/R_0 values, for computation of temperature. Accuracy and sensitivity are less than that attainable by the bridge methods.

The Siemens bridge (Fig. 3.14c) and appropriate three-lead thermometers are used in most industrial forms of resistance thermometer systems. Manually balanced analog instruments may use decade or slide wire types of adjustable resistors, R, for balancing the bridge. Reading may be made in ohms, to be converted to temperature by reference to conversion tables, where accuracy of 0.1°C and precision of 0.01°C are required. Direct reading bridges, calibrated for the standard curve of a particular RTD, are typically accurate to 0.25% of span, and precise to 0.1% of span. Servo-balanced analog instruments are available as automatic indicators, chart recorders, and automatic controllers. Selector switches are available, permitting a single instrument to measure temperature of several RTDs. Multiple point recorders, with built-in selector switches synchronized with print wheel displays, are available.

The Siemens bridge may also be arranged with a fixed resistor at R, adjusted to the midpoint of the temperature span of the thermometer. A constant current such as 5 mA powers the bridge circuit. A high-resistance d'Arsonval millivoltmeter or a potentiometer measures the voltage across the galvanometer terminals. This will be zero when temperature of the RTD is at the midpoint of the temperature span. It will increase in a positive direction with increase of temperature, and will change in a negative direction as temperature falls. If a, b, and r are equal in resistance, voltage change will be essentially linear with resistance change in the RTD over a substantial temperature range. Conversion to temperature is calculated from the bridge characteristics and the characteristics of the sensor.

Digital voltmeter readout devices are equally adaptable for use with resistance thermometers as they are with thermocouples (Section VI.B). Measuring circuit is similar to Fig. 3.9 f. A constant current is supplied to the winding by an electronically regulated power supply. The voltage drop across the winding is proportional to its resistance. It is measured by a linearized digital voltmeter calibrated directly in temperature.

3. Application Notes and Caveats

The precise platinum thermometers, and precise bridges and potentiometers for reading them, are usually reserved for use as temperature standards and for the direct measurement of temperature or temperature change requiring better than 0.01°C precision. The ratio, R_t/R_0, is the important characteristic of the thermometer. Hence, R_0 should be checked at the beginning and the end of an important experiment, *using the bridge or potentiometer that will be used in the experimental measurements*. This will eliminate the effect of small errors in the resistance standards used in the experiment. Checking ratio of the resistance decades or potentiometer decades according to the procedures recommended by the manufacturer will insure that the ratio values measured are accurate within reasonable limits.

R_0 of a thermometer will drift over a period of time. Shock, vibration, temperature cycling, and general aging contribute to the drift, which usually tends to increase R_0. This will not usually affect the calibration ratio until the R_0 value has changed by the equivalent of 0.01°C or more. Sudden change greater than 0.01°C probably indicates damage caused by mechanical shock or rough handling. Before recalibrating, try soaking the sensor at the maximum temperature limit specified by the manufacturer for 1–2 hr. This treatment will frequently reanneal the winding sufficiently to bring the R_0 back near its original value.

The general precautions to insure that the sensor measures the temperature of interest, as described in the sections on mercury thermometers and thermocouples, also apply to RTDs and PRTs. Response time constants of 10–60 sec in stirred water baths are typical. Experiments have to be designed for slow and sustained changes if the full capability of these sensors is to be realized. Joule heat, T^2R, is part of the heat balance that must be accounted for in calorimetry. Note also that Joule heat raises the temperature of the element slightly. This is accounted for in the calibration, *when the thermometer is used with the same current that was used in calibration*.

Heat leaks through the thermometer mountings and lead wires can cause erroneous temperature readings. Leaks can also upset the apparent heat balance in critical experiments. Thermal barriers, immersion depths, heat sinks, and other practices to prevent or control heat leaks must be considered in designing experiments.

"Stick on" types of RTDs, useful for measuring surface temperatures, must be carefully bonded to the surface for good heat transfer. Heat leaks down the leads can usually be minimized by using fine wire leads bonded to the surface under test for several centimeters.

Four-lead thermometers, measured with a potentiometer or Mueller bridge, may be located at any distance, even kilometers, away from the readout instruments. Electrical interference must be minimized by using shielded twisted cables for the connections, and by using null detectors with at least 120 dB common-mode rejection. Power leads must not share a conduit with thermometer leads.

Three-lead RTDs may also be located at substantial distances from the readout instruments, with the same precautions about electrical interference. In addition, care must be taken to assure equality of resistance in the a and b leads. Commercial cable is reasonably uniform. Resistance measurement of the installed cable will usually permit selecting two of three conductors equal within a few milliohms. Adding a small loop of fine-gauge copper wire to a conductor to equalize resistance will have negligible effect on accuracy of measurement.

Commercial annealed copper magnet wire is usually pure and strain free. A copper winding on a heating element, for example, can therefore serve double duty as a heater and as a thermometer. Resistance change is linear over the range 0–200°C. Measure resistance at a known temperature for reference, and at the temperature attained in the experiment. The formula for approximate conversion is

$$t_x = \frac{R_{tx}}{R_{tr}} (234.5 + t_r) - 234.5 \qquad (14)$$

B. SEMICONDUCTOR THERMOMETERS

1. Description of Available Sensors

In semiconductors, the valence electrons form localized chemical bonds which must be overcome so that the electrons may carry current. This provides a gap in the energy states over which the electrons must be excited to conduct a current. In thermometry, the energy needed to surmount the gap is thermal energy. Conductance increases with temperature according to the equation

$$G = k\, e^{-E_g/2\sigma T} \qquad (15)$$

where k is Boltzmann's constant and E_g is the energy gap.

Thus, semiconductors usually have negative temperature coefficients, since the number of excited electrons increases with increase of temperature. Resistance is high, reflecting the small number of current carriers. Temperature coefficients are typically an order of magnitude greater than for metals, when the energy gap exceeds the mean available thermal energy. Materials with a very weak bond may have a positive temperature coefficient, similar to the coefficients of metals.

Thermistors are the most common form of semiconductor sensors. These are mixtures of metallic oxides, fused into beads or plates, with electrical connections. The useful temperature range is from about $-190°C$ to $400°C$, depending upon the chemical composition and the technique of fabrication. The sensors available range in size from about 1 mm to about 5 mm in diameter. Resistance is high, several thousand ohms at room temperature. Thus, a simple Wheatstone bridge of moderate accuracy will suffice for readout, and lead resistance has negligible effect. The resistance temperature relation is approximately

$$\log R = A + B/T \tag{16}$$

where A and B are constants unique to a particular construction.

Molded carbon resistors, the commercial radio resistors in 0.1–1 W sizes, measured with ordinary Wheatstone bridges, are suitable for use as sensors in the cyrogenic region. If the resistor measures 50 Ω at room temperature, it will measure about 10,000 at 10°K. The empirical equation for the low temperature region is

$$\log R + K/\log R = A + B/T \tag{17}$$

Constants A, B, and K are determined by calibration against a platinum thermometer or a ^{4}He vapor pressure thermometer in a helium cryostat (18).

Carbon-glass resistors are constructed from carbon-impregnated glass as strain-free encapsulated units, similar in size and function to molded carbon resistors. Temperature-resistance relationship is similar to the relationship in the molded carbon devices. The strain-free construction results in better stability after thermal cycling and permits their use over a wide temperature range. After selection for stability by cycling individual units over the range from room temperature to liquid He temperature about 10 times, these units can be depended upon for use as temperature sensors for precision of about 0.01 K in the region 1–50 K (25).

A wafer of Ge, doped with AsP or Ga, exhibits a substantial change in the R/T relationship over the range 0.5–50 K. Developed by Kunzler (44), GeRTs are secondary standards for the temperature range below 20 K, where sensitivity of PRTs and TCs is falling off. Commercial units are four-terminal resistors, cut from strain free wafers, encapsulated, and tested for stability by about 10 thermal cycles between room temperature and liquid He temperature (25). Measurement is made with a Kelvin or Mueller bridge. Calibration is stable to 0.005 K; precision of 0.001 K or better can be attained. NBS (59) and other laboratories can provide calibration data in tabular conversion format.

2. Readout Instruments for Semiconductor Thermometers

A conventional decade Wheatstone bridge, with range 1 Ω to 10 MΩ will read resistance of thermistors and carbon resistors. Conversion to temperature requires reference to calibration formulae, tables, or charts. A four-decade bridge will provide accuracy of about 0.2°C and precision of about 0.01°C in resistance equivalent. A five-decade bridge will provide about 10 times improvement. Digital electronic bridges and ohmmeters have similar capability, their advantage being faster readout and digital display.

Direct reading instruments, also available, are less accurate and less precise. They are limited to thermistors of particular characteristic curves. The modified bridge circuits use various network tricks to approximate a linear temperature scale. Trimmers are included to compensate for differences in the A and B constants and the R_0 for individual thermistors with the same nominal calibration. Electronically balanced chart recorders, unbalanced bridges with d'Arsonval meter display, and digital electronic instruments are available. In digital instruments, the linearization may be done in the analog domain, or with a microprocessor in the digital domain.

Similar choices of Kelvin bridge instruments are available for Ge thermometers.

3. Application Notes and Caveats

Like all contact thermometers, semiconductor sensors must be located in the process where the temperature of interest will be accurately transferred to the sensor. The small-size units are very fast, time constants under 0.1 sec. Larger units are similar in response time to mercury and resistance thermometers. Protecting tubes and wells add time lag and reduce heat transfer, as with other contact temperature sensors.

In differential temperature measurements, the sensors must be selected to have calibration and time constant characteristics matched as closely as possible. Mountings must be arranged to have substantially equal heat transfer, compensation for heat leaks, etc. Since the small thermistors are usually chosen for differential measurements, Joule heating must be kept as low as possible by circuit design in the readout instrument. Joule heat must be accounted for in the heat balance for small-scale experiments.

Semiconductors are mechanically rugged and capable of withstanding shock and vibration in reasonable amounts. Calibration will drift as a result of temperature cycling, contamination, and aging. Frequent calibration is advisable.

When semiconductor thermometers are used in the cryogenic region, self-heating of the sensor element by the measurement current can be especially critical. Since the sensing elements are small, the scale of cryogenic experiments is usually small, and sensor resistance may be thousands of ohms, I^2R can become a serious problem. At temperatures above about 5 K, 1 μW is not likely to introduce appreciable error. At lower temperatures, below 1 K using GeRTs for ecample, 1 nW can introduce substantial errors. If the sensor is calibrated *and used* with the same current at the same temperature, self-heating errors can be reduced *if* there is good heat transfer from the sensor to the working fluid of the experiment, *and* the heat added by temperature measurement is accounted for (5).

Heat leaks *into* the cryogenic experiment, along sensor leads for example, can cause substantial errors in measurement. Leaks may distort experimental results by introducing heat that cannot be accounted for in the heat balance. Or they may introduce heat to the sensor faster than it can be absorbed by the experiment, producing a falsely elevated apparent temperature. Judicious tie of leads to properly cooled heat sinks and use of small size sensor wiring can minimize these errors.

Cryogenic experiments in the presence of strong magnetic fields require special attention. The magnetoresistance effect of the field is a function of B, T, and the sensor used. The magnitude of the error, expressed as $\Delta T/T$ may be a large percentage of the apparent reading. In general, the error increases rapidly as temperatures are reduced below 20 K, and as magnetic fields are increased above 2 Tesla. Sample and Rubin (69) have provided a survey of the effects noted for various sensors, and comments on the techniques useful in making accurate low-temperature measurements in the presence of magnetic fields.

Cryogenic experiments are unusually susceptible to errors from radio frequency interference. Since thermal energy levels are very low, rf interference at levels less than 1 pW can be significant. The effect can be heat added to the experiment, distorting results, or rf interference with measuring circuits causing false or ambiguous measurements. Connections to power lines should include filters to block rf interference from these sources. Faraday screening may be needed to provide an experimental area substantially free of radiant fields emitted by computing and communications apparatus or high-frequency energy sources. Radiation from a pocket calculator or radio operated close to an experiment can cause detectable interference.

At temperatures below 5 K, encapsulated sensors filled with He are no longer useful. The sensors for this and lower temperatures should be directly immersed in the working fluid, or mounted in perforated capsules which the fluid can enter directly. Heat transfer from the fluid to the sensor is subject to increased thermal impedance of the conducting path, and to thermal impedance of the covering of the sensing element (5).

VIII. RADIATION THERMOMETERS

In Section III.B, the laws of electromagnetic radiation, developed by Wein, Planck, Stefan and Boltzmann are described. These are the basis of radiation thermometry.

Fig. 3.3 plots the spectral intensity of radiation at a specific wavelength as a function of temperature. If the entire amount of energy enveloped at each temperature is measured, the signal will be related to temperature as specified by the Stefan-Boltzmann equation:

$$W = E\sigma\, T^4 \tag{18}$$

Devices providing this measurement are called total radiation thermometers.

If we select the energy at a specific wavelength; e.g., draw a vertical line on Fig. 3.3 at any position, the length of the line from 0 to where it intersects the curve for each temperature, is governed by the Planck distribution equation.

A measurement of the energy in a specific spectral band is thus a function of temperature. The optical pyrometer and other instruments that measure temperature in this manner are called spectrally selective radiation thermometers.

A. TOTAL RADIATION THERMOMETERS

1. Available Total Radiation Thermometers

By definition, a total radiation thermometer is one that responds equally to thermal radiation of all wavelengths. Commercial instruments suitable for service in rugged industrial and laboratory environments fall slightly short of the defined capability in that they cannot be sensitive to all wavelengths emitted from a body. Reduction in sensitivity to some wavelengths is imposed by the selective absorption of the transparent member needed to protect the inside of the unit from contamination. Practical total radiation thermometers are therefore, in fact, instruments that are sensitive to a very broad band of thermal radiation. The difference from "total" imposes no restriction on performance, but it does alter the calibration relationship from the fourth power Stefan-Boltzmann law.

Commercial units comprise three basic parts or assemblies: a thermal radiation-sensitive element, an optical system to focus the energy on the sensor, and a protective housing or case.

The radiation sensor must be rugged, stable, and equally sensitive to energy at all wavelengths. The most used sensor is the thermopile. the individual thermocouples are arranged so that their measuring junctions are located in the radiation path defined by the optical system. The reference junctions are located outside of the radiation path, but are thermally connected to the case of the instrument. The measuring junctions are covered with a black coating that absorbs all wavelengths equally, providing a radiation receiver. The thermocouple materials can be fine wires or strips that stand alone or are metallic films laid on an insulative substrate. Speed and sensitivity improve with low-mass sensitive elements and substrates. Time constants may vary from a few milliseconds, to 10 sec or more, depending upon the design and the service intended.

To measure temperature properly, the radiation thermometer signal must be independent of the temperature environment surrounding it. Mounting the reference junctions outside of the radiation path provides a first-order compensation effect. Materials that make the most sensitive and stable thermopiles, however, have thermal characteristics that produce substantial ambient temperature coefficients. These can be compensated for by using either a positive coefficient electrical shunt, or its own built-in thermal shunt. There is also the room temperature or reference-junction effect to be accounted for. For the high temperature ranges (where the pyrometer is not sensitive to measuring room temperature targets), the reference-junction effect is negligible. For low temperature ranges where the sensor is measuring target temperatures near the housing temperature, compensation for variation in reference-junction temperature must be provided. This compensation is usually achieved by inserting an appropriate biasing voltage in series with one of the leads. The level and polarity of the bias voltage is controlled by an additional temperature sensor, such as, a thermistor built into the housing.

Two other types of broad-band radiation sensors are sometimes used. They are resistance-varying elements, generally called bolometers, and the more recently developed pyroelectric sensors.

Bolometers can be made from either metallic elements or from semiconductor thermistors. To obtain acceptable time constants, film-type metallic elements are generally deposited on a suitably strong insulative substrate. Thermistor materials are also deposited on similar substrates. The greatest mass of the composite member is in the substrate so that it governs the time constant when the sensor is used in the dc mode. When faster response is required, the radiation is optically chopped to take advantage of the shorter time constant of the deposited film.

Elements are used in matched pairs to compensate for ambient temperature variations. The two are connected in a bridge circuit. One element, blackened to absorb radiation, is located at the focal point of the received radiation. The other element is protected from radiation. Resistance ratio of the elements is a function of temperature.

Pyroelectric devices are photon detectors that have substantially more sensitivity to thermal radiation than bolometers or thermopiles. A thin flake of a pyroelectric single crystal is sandwiched between two electrodes. As the temperature of the crystal changes, the lattice spacings are altered, producing a change in spontaneous electrical polarization existing below the Curie point.

When the electrodes are connected into an external circuit, a current is generated that is proportional to the rate of change of crystal temperature. These high-impedance, capacitive-type detectors are usually packaged with a low-noise field effect transistor to provide an output impedence that is compatible with typical preamplifiers. Being capacitive, they do not operate in the steady-state mode and only are usable when the incoming energy is optically chopped.

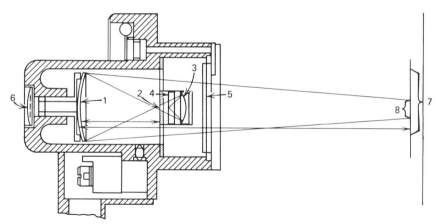

Fig. 3.15. Typical total radiation thermometer. 1. Primary Mirror. 2. Target Limiting Aperature. 3. Secondary mirror. 4. Thermopile. 5. Protecting Window. 6. Viewing lens. 7. Target. 8. Target diameter limited by stop 2. (Courtesy of Leeds & Northrup Co.)

Pyroelectric sensors operate near room temperature. Their upper temperature limit depends upon the Curie point of the crystal, where they become inoperative from depolarization. This limit is usually between 70°C and 100°C, depending on the crystal material. Triglycine sulfate is frequently used, although other compounds that exhibit pyroelectric effects, and are suitable for making this type of sensor.

Fig. 3.15 shows a typical total radiation thermometer.

The optical system has three functions: (1) It selects the target area and directs the energy to the sensor. (2) It provides a means for aiming the instrument and viewing the limits of the specific target area producing the signal. (3) It provides optical attenuation for adjusting the level of the signal as required for the predetermined temperature range. This feature makes these instruments very versatile. By blocking out more and more energy to adapt to increasingly higher temperature ranges, the radiation level reaching the sensor is held within safe limits. In addition, a trimming device is usually included to adjust each unit intended for a particular temperature range to the same signal level so that they are interchangeable.

The protective housing maintains optical alignment and shields the sensor and the optical parts from dirt and contaminating materials. It also eliminates air currents that would create signal ambiguity by upsetting the thermal balance of the sensor. The case should be sealed, requiring the use of a transparent member to transmit the energy to the sensor. This member can be either a plain window if reflecting optics are used, or it can be a lens if refractive optics are preferred. Since there are no materials that transmit all wavelengths equally, selection is made on the basis of maximum transmission in the temperature range required, balanced by the cost factor.

The case also provides light baffling to trap stray energy from outside the defined target to prevent it from affecting the signal. For general information on radiation thermometers see reference 35.

2. Radiation Thermometer Readout Devices

Radiation thermometers generally produce dc signals less than 100 mV. Since they are sensitive to nearly all wavelengths emitted by the target, the calibration curve of emf versus temperature has nearly a fourth-power relationship. The power law, however, is not quite constant because of the long wavelength cut-off of the protective window or lens. This cut-off results in a greater attenuation of signal at the lower portion of the temperature range compared with the high end, causing a varying departure from the theoretical fourth power.

Analog recording display is the most common readout for radiation thermometers, since trends are easily followed and the recorder can be equipped for automatic control as needed. The electronic potentiometer, similar to those used with thermocouples and described in Section VI.B is widely used. The recorder is usually calibrated directly in temperature, and the scale is nonlinear to match the exponential calibration curve of the sensor. The potentiometer circuit can be modified when necessary to provide a linear scale display. The range of temperature displayed is limited to the useful range of the sensor. Millivoltmeters and manually balanced potentiometers have limited value for radiation thermometers because of the low sensor output and the exponential calibration curve.

Digital voltmeters are frequently used to display the temperature. Linearizing is essential for useful digital readout. This may be accomplished in an analog signal processor ahead of the digital conversion, or in a digital microprocessor after conversion. A digital readout instrument may include an analog output to drive a chart recorder if a record is required.

With any type of readout, adjustment for the departure of the target from black-body emissivity is required. The signal from a real body will usually be less than the black-body signal for which the sensor was originally calibrated. In effect, the sensitivity of the readout instrument must be increased to compensate for the emissivity factor. This adjustment is made by a gain control in the circuitry of the readout instrument.

3. Application Notes and Caveats

Radiation thermometer signals are modified by parameters other than temperature. The main one is the emissivity of the surface being measured. This is the term ϵ in equation 18. Calibrations supplied by the pyrometer maker are referenced to a black body, so temperature measurement of real bodies requires knowledge of the quality of the emitting surface.

Other factors affecting the signal are radiation-absorbing media, such as smoke and gases, between the target and the receiver. Emissivity data for very broad radiation spectra are not generally available. It has been found to be most practical to lump all of the deficiency of energy available at the sensor into one quantity and calibrate the system for the specific conditions of use. This procedure requires measuring the temperature with another device, and then adjusting the readout to display the proper temperature under that condition. A reliable contact thermocouple or an optical pyrometer with proper corrections applied, is suitable for this purpose. Once the temperature is known, the radiation thermometer will repeat the reading when that temperature condition recurs. Total-radiation thermometers are, therefore, considered to be temperature-duplicating devices, and not absolute temperature-measuring instruments. The data supplied by them allows product quality control to be held within satisfactory limits. To maintain proper control, periodic ''on site'' checks should be made. When a trend requiring frequent readout readjustment is required, it is desirable to re-check the sensing head and readjust it. (Refer to Section XIV on Calibration.)

The following applies in installation of total radiation thermometers. The target must be large enough to fill completely the field required by the sensing head from the available observation distance. Less than adequate target will result in errors and inconsistent performance.

The sighting path must be free from radiation-absorbing obstructions. Absorbing gases such as CO_2, water vapor, smoke, etc., will cause errors. It is good practice to purge the sighting path with a flow of clean air, although care should be used not to make the flow strong enough to cool the target.

Make certain sufficient cooling is available to maintain the sensor head temperature below its safe upper limit.

B. SPECTRALLY SELECTIVE RADIATION THERMOMETERS

1. Theory of Brightness-Temperature Relationship

Temperature can be measured by measuring the relative spectral intensity of radiation at a specific wavelength. Instruments operating from this principle are called spectrally selective radiation thermometers. They have several definite advantages over total-radiation thermometers. They are very much less affected by emissivity variations when sighting upon objects in cooler surroundings, and by radiation-absorbing interference. Spectral selection has a definite advantage when measuring materials, such as glass, that have very different emitting characteristics at different wavelengths.

The reason for the smaller effect of emissivity in spectrally selective pyrometry lies in the power law, which may be explained as follows:

Over a small temperature range, we may express the luminosity L as a constant k multiplied by T to power n,

$$L = KT^n \tag{19}$$

Differentiating,

$$dL = nKT^{n-1}dT \tag{20}$$

Dividing the second equation by the first, we obtain:

$$\frac{dL}{L} = n\,\frac{dT}{T} \text{ or } \frac{dT}{T} = \frac{1}{n}\cdot\frac{dL}{L} \tag{21}$$

If we make an error in measuring L, the error in T will be only i/n times as great. Refering to equation 5 we find that

$$n = \frac{14388}{\lambda T} - 1 \tag{22}$$

Taking the optical pyrometer as an example where λ is approximately 0.65, n is approximately 22,000 divided by the temperature in Kelvins. At a temperature of 1100°C (1373 K), the optical pyrometer would have a power law of approximately 16, as compared with 4 for the total-radiation thermometer. The error caused by an error in the knowledge of the emissivity factor would affect the temperature reading of the optical only one-fourth as much.

2. Monochromatic Optical Pyrometers

The most important and familiar spectrally selective instrument is the optical pyrometer. It is the interpolating instrument for IPTS 68 above the gold freezing point. Modern optical pyrometers depend upon the Morse invention, patented in 1902 (50), of the disappearing filament principle. As the current through the filament of an incandescent lamp increases, the filament temperature increases approximately as the square of the current. The filament glows visibly at temperatures above 700°C, red at first, and then orange, yellow, etc., as current is increased. When this filament is placed in the plane of the image of an incandescent object in a telescope or microscope, the current through the lamp can be adjusted until the filament brightness matches the image brightness; the filament image disappears into the bright image of the object. The band selected for the optical pyrometer is the visible red band. Since the human eye judges the balance point, the band must be in the visible spectrum. Red is the first energy to become visible as the temperature rises, making the starting point of the temperature range as low as possible. The human eye selectivity can thus be used to restrict the pyrometer's band on the long wavelength side. To restrict the band on the short wavelength side, a red filter glass opaque to other colors is used. The resultant band of sensitivity is usually centered at about 65 μm and is about 4 nm wide. The temperature span starts at 760°C and there is no upper limit.

The essential optics—the lamp for observing the test object and the control box for adjusting the lamp current—are shown in Fig. 3.4. The object is sighted, the filament current is adjusted until its image disappears into the image of the object, then the current is read. Absorbing filters reduce the apparent source temperature when it is higher than 1250°C, the maximum lamp temperature allowed to maintain stability.

For a more detailed description of the disappearing filament optical pyrometer, refer to Kostkowsky and Lee (43,42) and Lovejoy (48).

While the waveband of the pyrometer (the band between the longwave
cut-off of the human eye, and the shortwave cut-off of the red glass) is
considered to be "monochromatic," it must have substantial width to provide adequate sensitivity for a disappearance within acceptable narrow temperature limits.

When the pyrometer is fitted with additional absorbing screens to extend
its range upward, the shade of color match between the lower-temperature
filament and the higher-temperature target changes. This is because the
spectral distribution of the energy within the accepted band becomes increasingly weighted toward the shorter wavelength side. It is, therefore,
desirable to use an absorbing screen that is not neutral, but has a lower
transmission for the shorter wavelengths in the band than in the longer ones.
In other words, the special screen alters the distribution of radiation from a
high-temperature target in such a way as to make it appear through the
telescope identical in color with the lower-temperature filament image. By
adding sufficient screens, we can extend the range of the pyrometer upward
as far as we desire.

The optical pyrometer is a very versatile tool for making precision high-
temperature measurements in the laboratory. Since it can be made portable,
it is also well suited for making on-site checks on industrial processes. It is,
therefore, the work-horse standard of radiation thermometry. Being dependent upon human operation, however, it does not produce a record of the
data nor fit into automatic control schemes.

The optical pyrometer has been automated by substituting a multiplier-
phototube for the human eye, and electronically adjusting the lamp current
to obtain a brightness balance. One scheme is shown in Fig. 3.16. Automatic
pyrometers of this type are about 10 times as sensitive to temperature

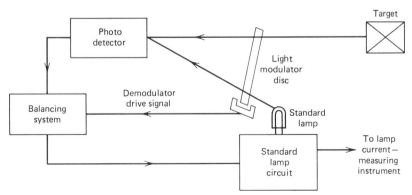

Fig. 3.16. Automatic optical pyrometer—alternate image type. (Courtesy of Leeds & Northrup Co.)

changes (about 0.1°C) at the freezing point of gold. The time required for the instrument to make a balance is about 1 sec, governed mostly by the time constant of the lamp filament. Precision calibration work that was previously handled by manually operated pyrometers is now performed by automatic instruments.

3. Color-Ratio or Two-Color Pyrometers

If the intensity of radiant energy in two spectral bands of wavelength is measured, and the ratio Q of these values is calculated, the relationship of this ratio and temperature T will be given by the equation:

$$Q = \frac{\epsilon_{\lambda_1} \lambda_2^5}{\epsilon_{\lambda_2} \lambda_1^5} \exp - \frac{C_2}{T} \left(\frac{1}{\lambda_2} - \frac{1}{\lambda_1} \right) \tag{23}$$

This equation is derived from Wein's approximation equation.

If the target is a "gray body" (one that has constant emissivity with respect to wave-length, i.e., $\epsilon_{\lambda_1} = \epsilon_{\lambda_2}$, the relationship between Q and T will be independent of emissivity.

Whether or not the instrument can be used for emissivity independent measurements, depends upon the material being observed, and the wavelengths selected. There are a number of materials that can be accurately measured with this system, but the user is cautioned to investigate the data available concerning the material in question before applying the pyrometer.

A second feature of two-color pyrometers is the ability to make accurate measurements through attenuating smoke, vapors, or other obscuring media that have equal absorption coefficients in the two selected spectral bands.

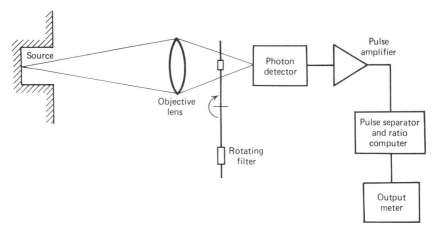

Fig. 3.17. Color ratio pyrometer.

This ability allows accurate readings to be made in areas where total or monochromatic instruments are useless due to continual variation of optical interference. Ruffino (66) discusses two-color pyrometry in detail.

It is desirable to select the wavelength of the two bands as close to each other as instrument sensitivity will permit. Specific instruments have spectral bands ranging from the middle infrared to the short-wavelength side of the visible spectrum. One very effective choice is 0.85, and 1 μm. These bands lie within the high sensitivity area of a silicon diode sensor which provides excellent stable readings for this service.

Temperature readout for two-color pyrometers requires the computing of the temperature from the two readings. This is done by electronic circuitry supplied as an integral part of the instrument. An analog or a digital display is usually incorporated (Fig. 3.17). If a record is also needed, the system provides and analog voltage output which can be displayed on a potentiometer recorder.

4. Spectrally Selective Radiation Thermometers with Photon Sensors

For less precise measurement than that supplied by the optical pyrometer, devices using silicon diode sensors in the dc mode are available. They are similar to total-radiation thermometers, except that the thermopile is replaced with the silicon diode. The spectral sensitivity of the sensor lies between 1.1 μm and 0.5 μm, giving an effective wavelength of about 0.9 μm, which is slightly longer than for the optical pyrometer. If spectral sensitivity closer to that of the optical is desired, additional filters can be added but with a reduction in sensitivity.

Temperature ranges for silicon pyrometers start at about 500°C. Some very sensitive types are sometimes used below this limit, but since they are also sensitive to visible radiation, care must be used to screen out all ambient light. There is no upper limit of target temperature, since these instruments can be optically attenuated in the same way as total-radiation thermometers.

The millisecond time constant of the silicon cell is much faster than potentiometer recorders or digital voltmeters so that the speed of response of the temperature system is determined by the readout. In cases where very fast temperature measurements are required, oscilloscopes may be appropriate.

Radiation thermometers using other solid-state sensors are also available. These instruments are necessarily more complicated, less rugged, and less precise than silicon cell types, because the sensors are less stable and generally have very high sensitivity to ambient temperature changes. There is usually an internal calibrating system built into the instrument to provide means for frequent recalibration to take care of sensor drift. The sensor temperature is usually controlled within very close limits to eliminate ambient temperature effects.

The list of sensors that can be used for these instruments is somewhat limited due to the impracticality of cryogenic cooling for this service. Therefore, only sensors with high sensitivity in the room temperature area can be used. In addition to sensitivity, sensor selection is also based on the spectral sensitivity band required for the particular measurement requirements.

Lead sulfide is frequently used between 1–2.5 μm, while indium antimonide provides sensing capability from 2.5 μm to approximately 5 μm. Additional optical filtering may be used to further restrict the band if available sensitivity will permit.

Readout systems for these thermometers, like the two-color pyrometer, have very specific requirements and are thus provided by electronic systems built into the assembly. They are usually equipped with an analog output signal that can be used to drive a standard potentiometer recorder or provide input to a computer.

5. Application Notes and Caveats

Manufacturers of optical pyrometers and other spectrally selective radiation thermometers provide comprehensive application information in their instruction books. These include detailed information on target size, observation distances, and focusing arrangements, as well as, information on emissivity corrections, ambient temperature effects, etc. These instructions should be followed precisely for best results. Corrections for readings from targets having $\epsilon < 1$ can be found in reference 60.

C. FLUOROPTIC®* THERMOMETRY

The change in intensity of fluorescent light emitted from certain phosphors is a function of temperature. This permits the phosphors to become temperature-measuring devices. The method of operation is described by the developers of commercial devices using the principle as follows (76a):

"When certain chemical compounds (phosphors) are excited by ultraviolet light, they emit light (fluoresce). The intensity of fluorescence is dependent on phosphor temperature. However, the fluorescent intensity also depends on the intensity of excitation. For certain phosphors, the fluorescent emissions at different wavelengths have different dependences, and yet their intensities have a linear dependence on intensity of excitation. Under these conditions, the ratio of emissions is independent of excitation and yet dependent on temperature.

The use of ratios not only removes excitation dependence, but also reduces the effects of variations in optical coupling within the probe assembly and variations in fluorescence from probe to probe. With the rare earth phosphors, which exhibit very sharp line emission spectra, the isolation of

* Registered trademark of Luxtron Corp.

the emissions different temperature dependences becomes especially simple and facilitates the fluoroptic principle''.

The sensing probe is an optical fiber, typically 0.7 mm in diameter and from 2–15 m long. The sensing end is coated with a thin film of the temperature-sensing phosphor. The other end terminates in an optical connector which couples it to an instrument package containing an optical head, a microprocessor, and associated electronic circuitry. Excitation of the phosphor and the return transmission are both completed through the same optical fiber. The fiber is coated with a thin layer of high-temperature-resisting, chemically inert plastic.

The probe is thus electrically nonconductive and its small physical size minimizes thermal conduction error, and maximizes speed of response. Temperature range of the thermometer is −50°C to +200°C. Accuracy and sensitivity are about 1°C and 0.1°C respectively.

Cost, compared with other thermometer systems, is high, probably restricting its use to areas that are not well handled by other techniques.

Applications where the method might be used are ones where high-strength electrical fields exist and the temperature area is obscured from other optical systems of measurement.

IX. QUARTZ CRYSTAL THERMOMETERS

The temperature coefficient of resonant frequency in quartz crystals has been developed by Bell Telephone Laboratories and Hewlett-Packard Company (13) for temperature measurement over the range −70–300°C. The quartz crystal, mounted in a probe about 7 mm in diameter and 30 mm long, drives the oscillator at a frequency proportional to the temperature. A count of the frequency over a suitable short time interval provides direct display of temperature with short term precision of better than 0.01°C.

The linear response of the sensor makes it attractive for temperature-difference measurements. This device is no longer available on the commercial market.

X. ACOUSTIC THERMOMETERS

The velocity of sound in a perfect gas relates to absolute temperature by the equation:

$$V^2 = \frac{\alpha R}{M} T \tag{24}$$

Practical realization of this principle for temperature measurement requires a sound-generating transducer, a sound-receiving transducer, electric circuitry to measure delay time in receiving the generated wave, and a known

distance between the two transducers. Temperature measurement is reduced to a measurement of time. The gas itself is the thermometer element,
so that rapid temperature changes, free from heat transfer, conduction, and
other errors, can be followed. The measurement gives the average temperature over the sound path.

The National Bureau of Standards (58) has developed an acoustic thermometer of high precision for calibration of germanium resistance thermometers over the range 4–20 K. Other laboratories have used acoustic
thermometers for special purposes, but no commercial acoustic thermometer is now available.

The velocotiy of sound in a metal rod can measure the temperature of the
rod. Theoretical velocity depends upon the ratio of Young's Modulus of
elasticity to density. Young's Modulus E decreases logarithmically with
increase of temperature. Transducers and associated circuitry to measure
time for the sound to traverse a measured length of rod would be similar to
the apparatus used for acoustic gas thermometers. Calibration would be
empirical. No commercial apparatus is presently available.

XI. LIQUID-CRYSTAL THERMOMETERS

In 1888 Reinitzer (62) discovered that cholesteryl benzoate exhibited a
new and peculiar property. As the compound was heated, the crystals
melted sharply at 63.1°C, but the melt was opaque. As more heat was applied, the opacity disappeared suddenly at 81.4°C transforming into a true
isotropic liquid. Investigation revealed that while the compound was fluid to
a greater or lesser extent between these temperatures, it also exhibited
anisotropic properties when viewed in thin section between crossed
polaroids. This suggested that the material was partly solid and partly liquid
in that temperature range. To describe the apparent anomaly, Lehmann (46)
used the terms "Fliessende Krystalle" or "Flüssige Krystalle," i.e. flowing
crystals. Thus, the present-day term "liquid crystal" was derived.

This phenomenon is being used in specific cases to indicate temperature.
The change from the opaque state to transparency has been used to construct some novelty temperature-measuring devices, but appears to have
little usage in the scientific area.

Cholesteric liquid thermography introduced by Fergason (32) is capable
of indicating human skin temperature to fractional degree precision.
Changes of color of thin layers of various cholesteral compounds applied to
the skin indicate temperature gradients by changing color similar to the
response provided by infrared thermography. The latter, however, requires
the use of expensive and complicated infrared scanning equipment. Indication of the presence of malignant tumors is a potential application.

More information on liquid crystals may be found in references 34 and 19.

XII. INDICATORS FOR SPECIFIC TEMPERATURES

Josiah Wedgwood, the potter, observed, and used, the softening of cones of ceramic materials as controls for firing of kilns before 1800. Seger (71) established a series of standard cones for the range of temperatures 600–2000°C commonly used in the firing of ceramic ware. A series of cones with different softening points, cemented to a slanted block, is placed in the furnace. A slumped cone indicates that the temperature has reached the softening point of that cone. This indicates the maximum temperature attained during that run, within about 50°C.

Paints, crayons, and pellets compounded of proprietary materials extend the range of these melting point indicators down to ambient temperatures, with accuracy of perhaps 10°C. Paints with temperature-sensitive pigments serve a similar purpose. These "tattle tale" devices are useful for "one-shot" estimates of temperature. Accuracy of actual results is influenced by the rate of heating and other factors (15).

XIII. TEMPERATURE MEASUREMENT IN MIXED ENERGY SITUATIONS

By definition, temperature measures the thermodynamic energy associated with thermal agitation of molecules. An equilibrium condition is implied. The test object must have sufficient density to attain equilibrium with a contact sensor, or to provide sufficient radiation to activate a radiation sensor. These conditions are usually met in the analytical laboratory.

In a flowing stream of fluid, the kinetic energy due to the flow rate adds to the kinetic energy due to thermal agitation of the molecules. The "temperature" measured by a sensor inserted into the stream will include the mean kinetic energy of the fluid in the stream. At stream velocities below 100 m/sec, the "temperature" added by stream velocity will be trivial. The effect in high-velocity streams can be reduced by mounting the sensor in a stagnation-type probe. Alternatively, the velocity of the stream can be measured, and the rise in temperature due to stream velocity can be calculated.

Flames, plasmas, and arcs present a more complex situation. Thermodynamic energy, energy of fluid velocity, and electronic energy in the electrons and ions are all present. If an object in the stream can be sighted through a window, an optical pyrometer sighted on the object will read a "temperature" reflecting the mean kinetic energy of the stream. Spectroscopic observation of the emitted radiation will permit simultaneous measurement of composition and temperature. Procedures are complex. At temperatures above about 5000°C, thermal energy overcomes the strength of molecular bonds and the solid state ceases to exist. Most of the molecules have been stripped. The plasma is composed mainly of ions and free electrons. High-energy physics displaces thermodynamics. The electron-volt competes with the thermodynamic temperature scale as a measure of energy. The approximate conversion is 1 eV = 10,000 K. Techniques for measurement in this difficult area are beyond the scope of this chapter. A tutorial survey will be found in references 29, 40, and 21.

XIV. CALIBRATION OF TEMPERATURE SENSORS

A. COMPARISON WITH A CALIBRATED STANDARD

Most calibrations are made by comparing the readings of the sensor under test with a calibrated standard at several points over the range of interest. A table of corrections to the scale readings at the test points permits interpolation at the intermediate values. In general, the standard thermometer should be of better quality than the unit under test, and its errors should be known to at least the accuracy and precision desired in the calibration. Some error will usually be added in making the intercomparison.

Well-stirred baths, operated at constant temperature, deep enough for needed immersion of the sensor, are most commonly used. Liquid may be a suitable organic fluid for the range ambient down to $-100°C$, water for the range 1–95°C, oils up to about 250°C, and molten tin or molten salts up to about 500°C.

At temperatures above 500°C, thermocouples are compared in a furnace, typically 60 cm deep and 10–15 cm diameter; fitted with an equalizing block of copper or bronze. Deep holes in the block provide adequate immersion. In addition to equalizing the temperature, the block forms a radiation shield to prevent excess heating from the walls of the furnace. Alternatively, a group of thermocouples may all be welded together, surrounded with a metal tube for a radiation shield, and heated for test in a tubular furnace. The latter method requires cutting off the junctions and rewelding them after the calibration has been completed.

Optical and radiation pyrometers may be compared by sighting them into the cavity of a furnace through a small viewing port, simulating a black-body situation. Optical pyrometers may be compared by sighting on a ribbon filament tungsten lamp held at constant temperature by regulating the current, or by sighting on the inside frosted surface of an ordinary tungsten lamp bulb. In each of these cases, the sighting path must be protected from stray visible radiation.

Detailed procedures for comparison calibrations will be found in the following references: for liquid-in-glass thermometers, NBS Monograph 150 (79); for installed thermocouples in place, ANSIMC96.1 (6); for thermocouples, ASTM E 220-72 (8) and NBS Circular 590 (65); for all sensors, NBS Special Publication 250, Calibration and Test Services (53).

B. CALIBRATION AT FIXED-POINT TEMPERATURES

The defining points for IPTS 68, and the secondary reference points, tabulated in Section IV, may all be used for either primary or comparison calibrations. The procedures used by NBS and other primary calibration laboratories are outlined in the text of IPTS 68 (14), and detailed in an NBS Special Publication (53). Basic principles are described briefly below.

The ice point, 0°C, 273.15 K, is easy to achieve with accuracy of 0.01°C or better, in any laboratory. Shaved ice, preferably from distilled water although the clear outer portions of commercially frozen block ice are substantially pure, is packed into a Dewar flask to a depth of about 15 cm. The ice will slowly melt to form a firm slush. Thermometers, thermocouple reference junctions, and other sensors will attain the temperature of the slush, provided that they are inserted into the slush, and not into the water at the bottom of the flask. Bottom water temperature may rise up to 4°C. Excess water must be removed periodically, and shaved ice added, to maintain a firm slush. Ice-point checking is the most convenient way to assure dependability and to correct for zero drift of mercury and resistance thermometers.

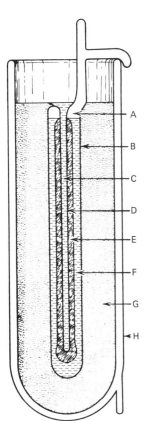

Fig. 3.18. Triple-point cell. A. Water vapor. B. Pyrex cell. C. Water from ice bath. D. Thermometer well. E. Ice mantle. F. Air-free water. G. Flaked ice and water. H. Insulated container. (Courtesy NBS.)

Triple points are more precise, but tricky to establish. A triple-point cell for water is shown in Fig. 3.18. The cell is mounted in a pack of shaved ice in a large Dewar flask. It is cooled from within until a thick layer of ice is frozen around the well. Warming from within, after insertion of the test thermometer, will produce an interface of ice and water, close to the surface of the well. The triple-point temperature will be attained, within about 0.0002°C, after a few hours. Temperature will remain constant for a month or more as the ice mantle slowly melts. Triple-point cells for other fluids are operated in a similar manner.

Boiling points require manometry to make pressure corrections to the readings, and rather intricate plumbing and heating arrangements. A hypsometer for steam-point measurement is shown schematically in Fig. 3.19. The open refluxing system, and barometric correction, will be reliable to a few millidegrees. N_2, O_2, and A, boiling points are realized with a small bulb containing a few milliliters of liquid, immersed in commercial liquid air. The

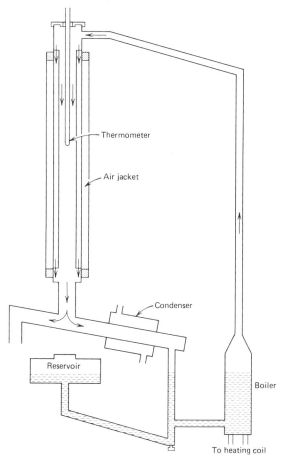

Fig. 3.19. Schematic diagram of team point apparatus hypsometer. (Courtesy NBS.)

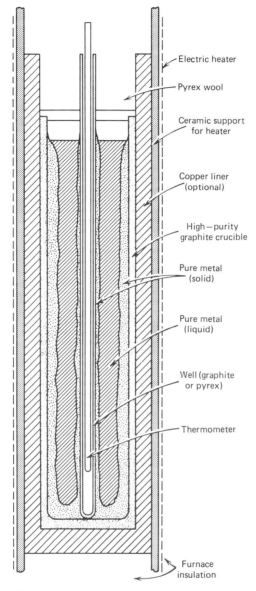

Electric heater

Pyrex wool

Ceramic support
for heater

Copper liner
(optional)

High—purity
graphite crucible

Pure metal
(solid)

Pure metal
(liquid)

Well (graphite
or pyrex)

Thermometer

Furnace
insulation

Fig. 3.20. Schematic diagram of apparatus for freezing points of metal such as zinc. Outer shell of furnace is not shown. (Courtesy NBS.)

thermometer is immersed in the boiling fluid. The nuisance, and the procedural pitfalls, of precise manometry and barometric corrections are usually avoided by using the boiling-point apparatus as a constant temperature bath for comparing test thermometers with a more precise standard.

Metal freezing-point measurements require a crucible, usually of graphite, containing the metal and a well for the sensors under test. Fig. 3.20 shows a suitable arrangement, to be inserted in a furnace with the required operating temperature. A typical crucible is about 60 cm deep and about 12 cm in diameter.

Oxidation of the metal sample, and dissolving of oxygen in the sample are inhibited by covering the metal surface with powdered graphite. The crucible is closed with a graphite coverplate, pierced for insertion of the thermometer. The furnace is heated to a temperature several degrees above the freezing point. After the metal is completely melted; heat is turned off. A cold metal rod is inserted into the test well to initiate freezing, then removed and replaced with the thermometer under test. Temperature readings are made at intervals of several seconds and recorded. Supercooling will be observed and followed by a rapid rise to the true freezing point. Temperature will be constant until heat of fusion is released, then fall slowly after the metal has completely frozen. Furnace cooling should be slow enough to sustain a plateau at freezing temperature for several minutes. If the furnace is inadequately insulated, the plateau time can be extended by reducing the heat about 20% below that required for fusion instead of cutting it off completely.

Automatic control of furnace temperature is a desirable convenience. The supercooling and the shape of the plateau at the true freezing point then may be displayed graphically. Final measurement is made with a manually balanced or digital instrument of suitable accuracy.

A melting-point furnace, charged with silver, gold, or copper, can be used as a black-body furnace for calibrating radiation and optical pyrometers. The pyrometer is sighted through the hole in the graphite cover plate into the thermometer well.

A quick melting-point calibration of thermocouples, accurate to a few tenths of a degree, is also useful. A small piece of gold or other reference wire is wrapped around the tip of a thermocouple. On slowly heating through the melting point, a brief plateau in the rising output emf will be observed as heat of fusion is absorbed at the melting point. The method is also useful for determining melting points of metals and alloys. The thermocouple tip will be contaminated. The tip must be cut off and the thermocouple rewelded after each such determination.

C. CALIBRATION SERVICES

NBS provides calibration services for thermometers, pyrometers, and readout instruments of high quality. NBS also offers standard metals for sale for many of the freezing points, and a few other standard materials useful in maintaining temperature calibrations. NBS circular 250 (53) describes the tests and materials available, and the prices for these services. Many instrument manufacturers and commercial testing laboratories offer calibration services, traceable to NBS standards.

XV. AUTOMATIC TEMPERATURE CONTROL

Most thermal methods of chemical analysis require some attention to the temperature at which the analytical reaction occurs. Controlled experimental conditions may, of course, be measured and adjusted by the analyst, but this consumes time and attention that can be better spent in more productive effort. An automatic temperature controller can be expected to do a better control job than an operator, in addition to eliminating the need for continuous personal attention. Analytical presentation of available control actions, and the process conditions dictating the use of particular control systems, is beyond the scope of this section. Instead, it is intended to provide an overview of available temperature controllers to aid in choosing appropriate equipment to meet experimental needs at reasonable cost.

Table 3.IV lists the sensors and auxiliaries available for automatic temperature control, in approximate order of increasing cost. Costs of sensors and auxiliaries range from less than $100 for bimetal contacting thermostats to more than $2000 for the most sensitive and sophisticated controllers equipped with time-temperature programming features. Cost of control actuators ranges from a few dollars for an electromechanical relay or solenoid valve to hundreds of dollars for the continuously variable electronic devices for electric heating and the continuously variable valves that are adjusted pneumatically or electrically. Brief discussion of the various control actions follows:

A. TWO POSITION CONTROL

Two-position, commonly called on-off control, maintains heat until the sensor reaches the set point, then turns it off. Process temperature continues to rise, overshooting the control point, until stored heat in the heating element transfers to the process. Heat loss to ambient conditions cools the process. When the set point is reached, heat is turned on again. Temperature continues to fall, undershooting the control point, until sufficient heat head is built up in the heating element to start raising the process temperature.

The most numerous on-off controllers are the thermostatic switches, usually bimetal expansion elements operating electric switches at the set point, used to control heating and air-conditioning systems, ovens, and electrical appliances. A common variant uses expanding fluid as the sensor, fluid pressure in a capillary extension operates the switch. Another variant uses expanding fluid as the sensor, but the fluid pressure operates a valve at the

TABLE 3.IV
Selection Guide for Temperature Control

Control sensor	Usual application	Display	Actuating means
Bimetal Thermostat (lowest cost)	On-off control in 0–300°C range; dead band 0.1 to 1% of span; locate at process; set point usually adjustable	None	Electric contract or pneumatic pilot valve
Mercury in glass thermostat (low cost)	On-off control in −30 to 300°C range; dead band 0.02 to 0.5°C, depending on range and quality of thermometer; locate at process; fixed set point	Emergent thermometer stem	Electric contact
Expanding fluid (low to medium cost)	Use for any form of control in −50 to 500°C range; locate control and display instrument within 30 meters of process; dead band 0.2 to 0.5% of meter span; also available for pneumatic or electronic transmission to control instruments; adjustable set point	Usually indicator or recorder	Electric contact, or pneumatic pressure or electric current depending on control form used
Electrical resistance thermometer-millivoltmeter (medium cost)	Use for on-off or proportioning control in range −100 to 500°C; locate display and control within 100 meters of process; dead band 0.2 to 0.5% of meter span; adjustable set point	Indicator or recorder	Electric contacts or electric current
Electrical resistance thermometer-balanced bridge (high cost)	Use for any form of control in range −100 to 600°C; dead band 0.1% of meter span; high output and short meter spans provide excellent sensitivity for control of temperature difference; set point adjustable	Usually indicator or recorder; "blind" controllers display only deviations from set point	Electric contacts or electric current or pneumatic pressure or slidewire position
Thermocouple-millivoltmeter (medium cost)	Use for on-off or proportioning control in range −100 to +1500°C, depending on choice of thermocouple; locate instrument within 100 meters of sensor; set point adjustable	Indicator or recorder	Electric contacts or electric current
Thermocouple-potentiometer (high cost)	Use for any control form in range −200 to 1500°C, depending upon thermocouple choice; dead band 0.1% of instrument span; locate instrument within 700 meters of sensor; use thermocouples in series to increase sensitivity for short temperature spans or control of small temperature difference	Usually indicator or recorder; "blind" controllers display only deviation from set point	Electric contacts or electric current or slidewire position or pneumatic pressure

TABLE 3.IV (*Continued*)

Control sensor	Usual application	Display	Actuating means
Radiation thermometer potentiometer (highest cost)	Use for any control form in range 300–3000°C, depending on sensor selected; locate instrument within 700 meters of sensor, dead band will be 0.5% of span near low end, reducing to 0.1% of span near high end of instrument scale; account for emissivity of target and absorption in sighting path	Usually indicator or recorder; "blind" controllers display only deviation from set point	Electric contacts or electric current or slidewire position or pneumatic pressure
Temperature difference (high cost)	Use thermistors, or resistance thermometers or thermcouples as sensors, with balanced bridge or potentiometer controller	Usually recorder	Electric contacts or electric current or slidewire position or pneumatic pressure
Time-temperature programs (added cost to basic controller)	Available as extra-cost features and auxiliary apparatus for all of the above arrangements except bimetal and mercury in glass thermostats	Recorder	Electric contacts or electric current or slidewire position or pneumatic pressure
Remote transmitter (added cost to basic controller)	Temperature transmitters convert sensor output into pressure or electric current signal for remote recorder or controller; used in control systems for large-scale processes and sometimes in laboratory situations where safety requirements demand it	Record or indication at receiving instrument	Usually electric current or pneumatic pressure

set point. A calibrated dial, linked to the trigger mechanism that operates the switch or valve, determines the set point. The simplest forms of these thermostats can be set within 5% of the dial span, and there will be an uncertainty dead band of about 2% of dial span in the actual temperature at which control action occurs. At higher cost, simple thermostats can be made more sensitive to provide setability within 1%, and dead band with 0.1% of a short range dial span. There is no display of actual temperature.

Mercury thermometers are converted into thermostats by sealing contacts into the capillary stem, and using the mercury column to close an electric circuit. Precision of contact closure, dead band, may be 0.1°C or even 0.01°C in the best constructed units. They are useful and inexpensive for controlling baths and ovens at preselected temperatures. Operating temperature is not adjustable. Actual temperature is displayed when the emergent stem can be mounted in view.

Any of the filled-system thermometers described in Section V.B can be provided with control contacts or pilot valves actuated by the Bourdon tube in the remote indicating or recording gauge. Control-point setters are easily adjustable. Temperature is displayed. Contact closure or pilot valve operation will be precise within 0.2–0.5% of the temperature span of the instrument.

Millivoltmeter display instruments for electric and radiation thermometers may be fitted with contacts to close control circuits. Set point is easily adjustable. Contact closure will be precise within 0.2–0.5% of the temperature span, depending upon the quality of the indicating meter.

Electronic potentiometers and bridges displaying output of electric and radiation thermometers provide the highest precision available in on-off controllers. Typical dead band is 0.1% of span. The narrow temperature spans, made possible by electronic instruments, permit millidegree dead band under very carefully controlled experimental conditions. Servo-balanced instruments, the most common form, display the temperature or temperature difference directly.

On-off control arrangements do an adequate job, at low cost, for relatively stable processes where upsets are small, and slow recovery after upset is permissible. A laboratory oven or constant temperature bath may be required only to maintain a set temperature over a long period of time. Material to be conditioned has small heat capacity compared with the vessel. It is stored in the vessel for a long time, and small temperature variations in the specimen are not significant. A slow heating rate and good insulation to provide slow cooling will minimize overshoot and undershoot of the set point. For example, a thermostat with a dead band of 0.5°C, applied to an oven that heats at 10°/hr with heat on and cools at the same rate with heat off, will probably maintain temperature at the set point within 1°C. Ware placed in the oven, after sufficient soaking time, will equalize with the average oven temperature. Control can be improved, without using a more sophisticated control system, by such tricks of the trade as adjusting the heating rate so that full "on" gives a very slow rise and "off" is merely a reduction of heat by several per cent to give a very slow cooling rate. A muffle, or baffle, between the heater and the controlled space, with the temperature sensor outside the muffle, will substantially reduce temperature variations in the controlled space. These, and other matters are discussed in Chapter 19, Temperature-Controlled Baths.

Dynamic experiments involving large or rapid changes in temperature, substantial exothermic or endothermic reactions, or substantial thermal upsets from any cause, should use one of the more sophisticated control forms described below. Fig. 3.21 shows a graphic comparison of the response of the different available control actions to changes in temperature, and to upsets in heat demand.

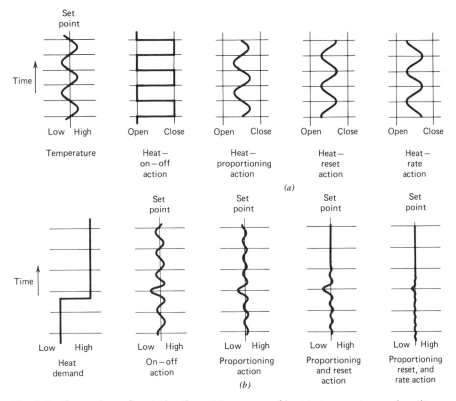

Fig. 3.21. Comparison of control actions: (a) response of heat to temperature cycles, (b) response of temperature to increased heat demand.

B. PROPORTIONING CONTROL

Proportioning control action provides a continuous linear relationship between the error signal, deviation of process temperature from the set point, and the addition of heat to the process. The range of process temperature required to change the heating rate from the minimum to the maximum of the controlled amount is called the proportioning band. It is usually dimensioned as per cent of instrument span. In practical proportioning control applications, the heating power will be gradually reduced to about half its maximum value as the set point is reached, and it will reduce toward zero as the process temperature moves above the set point. The throttling action of the proportioning controller will usually permit faster heating with less overshoot and undershoot cycling about the control point than can be obtained with on-off control. Larger upsets can be handled with a given requirement for precision of control results.

Proportioning control action is available in many forms. Pneumatic proportioning control requires a smoothly operating pilot valve that will stroke full scale, usually providing output air pressure over the range 3–15 psig, for a given percentage of span change in the measured temperature. An adjustment in the linkage among the temperature indicator, the control point setter, and the pilot valve stem permits adjusting the proportional band over limits such as 1–50% of the span of the instrument.

Electric proportioning control may operate through timed impulses with the duration of heat-on time in each impulse proportioned to the deviation from the set point error signal. Impulses may be timed thermally, mechanically, or electronically. Actual switching of the power on and off may be done with electromechanical relays or a solid-state switch. The impulses may also operate an electric valve to supply fuel to the process in duration-timed "slugs."

Electric proportioning control may control electric current to a process through a silicon-controlled rectifier or a magnetic amplifier which electronically proportions the load current to the error signal.

Electric proportioning control may drive an electric motorized valve to a position within its travel in proportion to the error signal. A position signal from the valve drive unit and the error signal from the control instrument are matched, usually in a bridge-type position control circuit, to maintain proportionality.

In all forms of electric proportioning control, the proportioning band is adjustable by means of a slidewire, usually marked in percent of span.

Proportional control, when properly tuned, can provide excellent control of processes subject to upsets, when recovery time from an upset can be relatively long and when lag in detecting or responding to process upset is not large. The actual control achieved will tend to run higher than the set point when the load is light, and to run lower than the set point when the load is heavy. Additional sophistication can overcome this limitation.

C. PROPORTIONAL PLUS RESET CONTROL

Proportional plus reset control provides control action proportional to the linear sum of the error signal plus the time integral of the error signal. Action may be pneumatic, in which case the integrating action is provided by a storage vessel in the pneumatic system, or it may be electronic, in which case the integrating action is provided by electronic or thermal storage. Reset action is dimensioned as "repeats per minute" of the error signal, adjustable to trim the process response.

Proportional plus reset control eliminates the offset between the set point and the process temperature after an upset has stabilized. The two actions may add or subtract, depending upon the size and direction of the upset requiring action and the amount and direction of recent error signals. Proportional plus reset action will stabilize a process faster after an upset than simple proportioning control. Process lag and detection lag are not directly accounted for.

D. RATE CONTROL

Rate control action provides an output proportional to the rate of change of the error signal—the velocity of the temperature change. This provides a way to account for lags in the thermal response of the process to changes in heat input and in the rate of response of the detector to process change, provided that these are lags associated with the heat transfer and storage capacity in the system. Rate action can be added to proportional control alone or to proportional control with reset. Rate action is available for pneumatic and electronic control systems. Rate adjustment is calibrated in rate time, over a range from about a second to several minutes, to provide for a wide range in process lag characteristics.

The effect of rate action is to reduce heat substantially when the process temperature is rising rapidly and to increase the heat substantially when the process temperature is falling rapidly. The rate action adds algebraically to the other control actions in the system to provide for rapid stabilization of process temperature after an upset.

E. COMPLEX CONTROL SYSTEMS

Many control instruments can be obtained as program controllers, where the set point can be automatically changed according to a predetermined time and temperature schedule. Motor-driven control point setters can change the set point at a constant rate, according to the motor speed. Cams driven by timing motors can be coupled to the control point setter to provide complex functions.

Electronic programmable controllers are displacing the established electromechanical types and electropneumatic types. Microprocessors and other forms of integrated circuits can provide timing, relay tree, and set-point changes, as well as various degrees of sophistication in control action. Solid-state electronic controllers, properly engineered for the application, may be more sensitive and more reliable than the older forms of programmable controllers.

Complex action is available when more than one input signal is used as a basis for control. Ratio control regulates the temperature as a function of some other measured quantity. Cascade control is used to adjust the set point of a flow controller, for example, according to a temperature measurement. In multivariable control systems, feed-forward signals can be judiciously used to signal and control upsets in fast-responding loops as a means of minimizing upsets in the more complex reactions of the process.

Control of heat added, to maintain temperature above ambient, has been described above. Control of coolant is similar. Chilled fluids are regulated by control valves. Compressed refrigerants are regulated by expansion valves. Refrigerating machines are regulated either by cycling on and off or by speed control at the compressor. Thermoelectric coolers are regulated by controlling the direct current.

F. APPLICATION NOTES AND CAVEATS

Control temperature sometimes crosses ambient temperature, as in a bath or oven maintained at 20°C or 25°C when ambient temperature is uncontrolled. Eliminate the cross-ambient problem, as automobile manufacturers do in climate control for air conditioning systems, by cooling the bath below the expected low ambient and adding heat to maintain controlled temperature. A trickle of cold water through a cooling coil may suffice. More cooling capacity may require thermoelectric or mechanical refrigeration. Alternatively, add heat to raise temperature above expected high ambient and regulate coolant. Unusual requirements for cross-ambient control require complex systems. Consult a manufacturer of control systems.

Thermocouples may break in hostile environments. According to Murphy's law, these failures will ocur at times of catastrophic inconvenience. Failures result from such conditions as high temperatures, thermal expansion, temperature cycling, aging, and corrosion. Break protection, available for most forms of thermocouple controllers, will fail safe by turning off the process or transferring control to a back-up instrument. Thermocouple break protection and back-up control should be provided if control or thermocouple failure will cause problems during unattended control operation.

ACKNOWLEDGMENT

Some of the narrative and figures used in Sections II, V, VI, VIII, XII, and XV, of this chapter were previously published in *"Temperature Measurement and Control,"* pages 599–642 Vol. 3, Snell-Hilton *Encyclopedia of Industrial Chemical Analysis*, John Wiley and Sons, Inc., New York, 1966, Vol. 3, pp. 599–642.

REFERENCES

1. AIP et al., *Temperature, Its Measurement and Control In Science and Industry*, Vol. I, Reinhold, New York, 1940.

2. AIP et al., *Temperature Its Measurement and Control In Science and Industry*, Vol. II, Reinhold, New York, 1954.

3. AIP et al., *Temperature, Its Measurement and Control In Science and Industry*, Vol. III. Reinhold, New York, 1962.

4. AIP et al., *Temperature, Its Measurement and Control In Science and Industry*, Vol. IV, Instrument Society of America, Research Triangle Park, N.C.

4a. AIP et al., *Temperature, Its Measurement and Control in Science and Industry*, Vol. V, Instrument Society of America, Research Triangle Park, N.C., 1982.

5. Anderson, A. C., *Instrumentation for Germanium Resistance Thermometry at Ultra Low Temperatures*, ISA, Proceedings, ISA-78, paper 78–802.

6. ANSI, *Temperature Measurement Thermocouples* MC96-1, Instrument Society of America, Research Triangle Park, N.C., 1982.

7. ASM, *Temperature Measurement*, American Society for Metals, Cleveland.

8. *Calibration of Thermocouples by Comparison Techniques.* E-220-72, ASTM, Philadelphia, 1974.

9. *Theory and Properties of Thermocouple Elements*, 492, ASTM, Philadelphia, 1974.

10. *Manual on the Use of Thermocouples in Temperature Measurement*, 470–72. ASTM, Philadelphia, 1975.

11. Baker, H. D., E. A. Ryder, and R. H. Baker, *Temperature Measurement in Engineering*, Wiley, New York, 1961.

12. Barber, C. R., and W. W. Blanke, *J. Sci. Inst.* **38**, 17 (1961).

13. Benjaminson, A. *Hewlett-Packard Journal*, **16**, no. 7, Mar. (1965).

14. BIPM, *Metrologia* **17**, July 17 (1976).

15. Bole, G. A., "Pyrometric Cones," in *Temperature, Its Measurement and Control in Science and Industry*, Vol. I, Reinhold, New York, 1940, p. 988.

16. Bolton, H. C. *Evolution of the Thermometer, 1592–1743*, The Chemical Publishing Co., Easton Pa., 1900.

17. Boyle, R. *The Laws of Gases* in *Works*, Peter Shaw, Ed. Vols. I and II, London, 1725.

18. Brickwedde F. G. et al., *NBS J. Res., A Physics & Chem.*, No. 1, p. 64a (1960).

19. Brown, G. H., G. J. Dienes, and M. M. Labes, *Liquid Crystals* Proceedings of the International Conference on Liquid Crystals at Kent State Univ., Aug., 1965.

19a. Burley, N. A., G. W. Burns, R. L. Powell, and M. C. Scroger, *The Nicrosil versus Nisil Thermocouple: Properties and Thermoelectric Data*, Monograph 161, NBS, Washington, 1978.

20. Callendar, H. L. *Phil. Trans. Roy. Soc. London* **178**, 61–220 (1888).

21. Campbell I. E. and E. M. Sherwood, Eds. *High Temperature Materials and Technology*, Wiley New York, 1967.

22. Camper, R. A., and J. E. Zimmerman, *Noise Thermometers With The Josephson Effect*, J.A.P. Vol. 42, #1, P 32, 1971.

23. Carnot, N. L. R. S., *Reflections on the Motive Power of Heat*, (French) Paris, 1824; Transl. by R. H. Thurston, Wiley, New York, 1890.

24. Celsius, A., Vetenck Akad. Hendl, (Swedish), Stockholm, 1742.

25. Clark, Jr., C. F., et al, *Stability of Cryogenic Temperature Sensing Elements: Germanium and Carbon Glass*, ISA, Proceedings, ISA-78, paper 78–804.

26. Coxon, W. F., *Temperature Measurement and Control* MacMillan, New York, 1960.

27. Dauphinee, T. M., *Can. J. Phys.* **33**, 255 (1955).

28. Day A. L. and R. B. Sossman *High Temperature Gas Thermometry*, Monograph 157, Carnegie Inst. of Washington, 1911.

29. Dickerman, P. J., *Optical Spectrometric Measurement of High Temperatures*, University of Chicago Press, Chicago, 1961.

30. Evans, J. P., and G. W. Burns, "Study of Stability of High Temperature Resistance Thermometers," in Temperature, Its Measurement and Control in Science and Industry, Vol. III, Reinhold, New York, Pt. I, pp. 313–318 1962.

31. Fahrenheit D. G., Five Papers, *Phil. Trans.* **XXX** and **XXXIII** (1724).

32. Fergusson, J. L., Sci. Am. **211**, 77 (1964).

33. Gay-Lussac, L. J. *Attempt to Determine Changes in Temperature Which Gases Experience*, (French) Mmoires d'Arcueil, I-1807, transl. by J. S. Ames, ed. in *Free Expansion of Gases*, Scientific Memoires, Harper, New York, 1898.

34. Gray, G. W., *Molecular Structure of the Properties of Liquid Crystals*, Academic Press, London, 1962.

35. Harrison, T. R., *Radiation Pyrometry and its Underlying Principles of Heat Transfer*, Wiley, New York, 1960.

36. Hill J. J., and A. P. Miller, *Proc. I.E.E. (London)*, **110**, (2), 453 (1963).

37. Hudson, R. P., EPT-76: *A Provisional Temperature Scale for the Region 0.5K to 30K*, ISA, Proceedings, ISA-78, paper 78–801.

38. Hudson, R. P., H. Marshak, R. J. Soulen, Jr., D. Button, *J. of Low Temp. Phys.* **20**, July (1975).

39. Joule J. P. and W. Thomson (Lord Kelvin), *Phil. Trans.* **CXLIV**, 321 (1854).

40. Kettani, M. A., M. F. Hoyeaux, *Plasma Engineering*, Wiley, New York, 1973.

41. Kirchhoff G., *Emission and Absorption*, 1861, (German), in *Klassiker #100 der Exakten Wissenchaften*, W. Ostwald, ed. Barth, Leipzig, 1896.

42. Kostkowski, H. J., *High Temperature Technology*, McGraw-Hill, New York, 1960.

43. Kostkowski, H. J. and R. D. Lee, *Theory and Methods of Optical Pyrometry*, Washington, NBS, Monograph 41, 1962.

44. Kunzler J. E. et al., "A Germanium Resistance Thermometer for Cryogenic Work."

45. Lamb, A. H. *Modern Thermometry*, Weston Engineering Notes, No. 22, Feb. 1947, Schlumberger-Weston Co., Newark, N.J.

46. Lehmann, O. *Ann. Physic,* **25**, 852; **27**, 1099; Ber. **41**, 3774 (1908).

47. LeChatelier, H. and O. Boudonard, *High Temperature Measurements* (French), Paris, Curre and Nand, 1900, transl. by G K. Burgess, Wiley, New York, 1901.

48. Lovejoy, D. R. *Can. J. Phys.,* **36**, 1397 (1958).

49. Meyers, C. H., *NBS J. Res.,* **9**, 807.

50. Morse, E. F., U. S. Patent 696878; U. S. Patent 696916, 1902.

51. Mortlock, A. J., *J. Sci. Inst.*, **35**, 283 (1958).

52. Mueller, E. F., *N.B.S. Bull.*, **13**, 547 (1916).

53. NBS, *Calibration and Test Services*, NBS Cir. 250, Washington, 1972.

54. NBS, *Units of Weights and Measures (U. S. Customary and Metric) Definitions and Tables of Equivalents*, NBS Misc. Pub. M.P. 233, 1960.

55. News report, *C&EN*, August 7, p. 21, (1978).

56. Ohte, A. and H. Iwaoka, *A Precision Nuclear Quadrupole Resonance Thermometer*, IEEE Trans. on Inst. & Meas., Dec., 1976.

57. Planck, M., *Treatise on the Theory of Radiation*, (German), Leipzig, Barth, 1906; transl. by H. L. Brose, MacMillan, London, 1932.

58. Plumb, H. et al., "An Acoustic Thermometer Calibration Standard," in *Units of Weights and Measures (U.S. Customary and Metric) Definitions and Tables of Equivalents*, NBS, part 1, p. 129, 1960.

59. Plumb, H. H., and L. M. Besley, *Stabilities of Germanium Thermometers at 20K and Below*. ISA, Research Triangle Park, NC. Proceedings, ISA-78, paper 78-803.

60. Poland, D. E., J. W. Green, and J. L. Margrave, *Corrected Optical Pyrometer Readings*, NBS Monograph 30, 1961.

61. Powell, R. L., W. J. Hall, C. H. Hyink, Jr., L. L. Sparks, G. W. Burns, M. G. Scroger, and H. H. Plumb, *Thermocouple Reference Tables based on IPTS-68*, Washington NBS Monograph 125, 1974.

62. Reinitzer, F., *Monatsh* **9**, 421 (1888).

63. Riddle, J. L., G. T. Furukawa, H. H. Plumb, *Platinum Resistance Thermometry*, NBS Monograph 126, 1973.

64. Roantree, R. R., *Guide for Use of International System of Units, S. I.*, NBS, F.I.P.S., 34, 1975.

65. Roeser, W. F., and S. T. Lonberger, *Methods of Testing Thermocouples and Thermocouple Materials*, NBS Cir 590, 1958.

66. Ruffino, G. *Increasing Precision in Two-colour Pyrometry* Inst. Phys. Conf. Ser. No. 26, Chapt. 5, 1975.

67. SAMA, *Thermocouple Thermometers*, Standard RC7-b. SAMA, PMC Section, Inc., New York.

68. SAMA, *Resistance Thermometers*, Standard RC5-b. New York, SAMA, PMC Section, Inc., 1966.

69. Sample, H. H., and L. G. Rubin, *Instrumentation and Methods for Low Temperature Measurements in High Magnetic Fields*, ISA. Proceedings, ISA-76, Paper 76–608.

70. Seebeck, T. J., *Fundamentals of Thermoelectricity*, Pogg. Amer. 6, p. 133, 1826.

71. Seger, H. A., *Collected Writings*, 2, 577, The Chemical Publishing Co., Easton, Pa., 1902.

72. Shirk, W. H., and R. F. Pistol, *A New High Accuracy Extended Range Resistance Thermometer Bridge*, ISA Tech. Paper 68–620, 1968.

73. Siemens, W., *Bakerian Lecture* (Royal Society), Proc. Roy. Soc., 1871.

74. Slaughter, J. I., and J. L. Margrave, *Temperature Measurement* U. S. Dept of Comm., Washington, D. C., Office of Tech. Services Report AD298142, (1963). (Also, Chap 24, p 717, Ref. 69.)

75. Sullivan, D. B., L. B. Holdeman, and R. J.Soulen, Jr., *The Future of Superconducting Instruments in Thermometry*, Proceedings of the A.I.P. Conference, Future Trends in Superconductive Electronics, Charlottesville, Va., 1978.

76. Thomson W. (Lord Kelvin) and J. P. Joule, Phil. Mag., Oct. (1848).

76a. U. S. Patents, 4,075,493 and 4,215,275.

77. Wein, M., *Ann. Physik*, **58,** 662 (1896).

78. White, W. P., *Phys. Rev.*, **31,** 562 (1960).

79. Wise, J. A., *Liquid in Glass Thermometry*, N. B. S. Monograph 150, 1976.

Chapter 4

CONSTANT-TEMPERATURE BATHS

By Lee D. Hansen, *Thermochemical Institute,*
Brigham Young University, Provo, Utah
and Roger M. Hart, *Hart Scientific, Provo, Utah*

Contents

I. INTRODUCTION

Constant-temperature baths are used commonly in all fields of chemistry. Because physical and thermodynamic properties of materials, values of thermodynamic quantities for reactions, and reaction rates and mechanisms are all temperature dependent, meaningful measurements of such parameters can only be made under conditions in which the temperature can be specified. This is usually done by performing the experiment in a vessel immersed in a constant-temperature bath. Frequently, the constant-temperature bath is so simple that we do not recognize it as such. For example, ice-water mixtures and steam baths are in common use in large undergraduate laboratories. Even when an experiment is supposedly not done in a bath it is, in fact, carried out with the room air as a bath fluid.

For the purposes of discussion, constant-temperature baths may be divided into six categories: (1) baths in which the temperature control is obtained through use of a phase change, (2) low-precision baths where the temperature control band is greater than 0.01 K, (3) high-precision baths where the control band is less than 0.01 K, (4) baths for extreme temperatures, (5) air baths, and (6) temperature-controlled metal blocks with a well in them for the reaction vessel. Each of these categories has its own unique advantages and disadvantages. The real problem is to match the bath best to the needs of the experiment to be carried out in it. Also, it should be apparent that any given bath may be in more than one of the above categories.

The simplest possible type of constant temperature bath is the type in which the temperature is controlled by a phase change. For a solid–liquid bath, no controller, temperature sensor, or external source of cooling or heating is needed while the bath is operating. External heating or cooling must be provided, however, for periodic "recharging" of the bath. For a liquid–gas bath, only very roughly controlled external sources of heat and/or cooling must be provided to maintain the liquid at its boiling point or to condense the vapor, respectively, if desired.

Aside from their simplicity, such phase-change baths can provide both very precise and very accurate temperature control. In this connection, it should be noted that such baths are the standards by which thermometers are calibrated to the absolute temperature scale. The use of ice-water and water-steam baths as fixed reference points in the early development of thermometry is obvious to anyone who is familiar with a centigrade (or Celsius) thermometer.

TABLE 4.I
Two-Phase Refrigerating Bath Fluids

Fluid	Temperature (°C)
Water-ice, 1 atm	0
NaCl-ice (33 g salt/100 g ice)	−21
CaCl₂-ice (100 g salt/81 g ice)	−40
Chloroform slush	−64
Dry ice-acetone or isopropanol	−78
Toluene slush	−95
Pentane slush	−130
Isopentane slush	−160
N₂ liquid-gas, 1 atm	−196
N₂ liquid-gas, pumped	−220
He liquid-gas, 1 atm	−269

The disadvantages to such phase change baths are (1) their limited capacity to absorb or release heat unless provided with external sources of heating and/or cooling, (2) the presence of large temperature gradients if the bath is not properly constructed or is used for rapid dissipation of large quantities of heat, (3) the temperature cannot be easily varied, and (4) both the precision and accuracy of the temperature control depend on the purity of the material used in the bath. Table 4.I gives some commonly used two-phase bath fluids for use at and below 0°C where such baths are most commonly used.

Low-precision liquid baths are the most common type of constant temperature bath found in routine laboratory use. If it is to be used at or above ambient temperature, the bath usually consists of a large container filled with water, a stirrer, an electrical heater connected to a controller, and a coil of copper tubing with coolant flowing through it. If the bath is to be used below ambient, the controller is usually used to control the coolant flow, the heater is eliminated and the room air is used as the source of heat. Such baths are frequently based on tradition rather than on proper design principles, and their success is often achieved more through luck than through planning. A poorly designed water bath of this type may only be controlling the temperature within a few degrees, regardless of the sensitivity of the controller, amount of money invested, or claims of the manufacturer. And, unless the experimenter properly tests the bath, he may remain ignorant of the problem. Properly designed liquid baths for operation around ambient temperatures can be constructed or purchased for only a few hundred dollars and should control with a precision of $\pm 0.01 - \pm 0.05$ K. This kind of control is usually sufficient for the accuracy to which most measurements, e.g., equilibrium constants, rate constants, spectrophotometric data, are made. The great majority of commercially available baths fall into this category.

High-precision liquid baths are available that will control with a precision as small as a few microdegrees. These baths must be designed as a system including the environment in which they are to be operated. The components must be carefully chosen to function as a system and the price will go up an order of magnitude for every two decades the precision is reduced. If a control precision of 0.001 K or less is required, it will usually be less expensive to purchase a bath system than for the individual experimenter to build his own.

Baths to be used below about -80°C (cryogenic baths) are also a highly specialized design problem, largely because of materials selection. This is illustrated by the difficulty in selecting a bath fluid, since none of the common liquids remain so below about -80°C. For this reason, most cryogenic baths either use a gas for a working fluid or a liquid–gas phase change with a variable pressure controller.

Likewise, baths to be operated above about 250°C are difficult to design because of materials selection. The selection of a working fluid is again a problem since common liquids either boil or decompose below 250°C. Thus, either a gas, a fluidized solid, or a metal block must be used for the thermostatted material. Another difficulty is that nearly all glues, adhesives, and organic electrical insulation materials melt or decompose between 250°C and 500°C.

Air baths have a major advantage in their simplicity. They are used very commonly to control the temperature of electronic components, cell cultures, and large pieces of apparatus. They range in size from a few cm³ to very large walk-in rooms, and in sophistication from a cardboard box with a light bulb for a heater to very elaborate stainless steel and chrome units. Their major advantage lies in not having to provide any special enclosure for experiments placed in them.

The major disadvantages of air baths stem from the low heat capacity and low viscosity of the gaseous working fluid. Because of these properties, heat transfer and mixing of the fluid are slow. Temperature gradients of several degrees can easily exist within an air bath. Also they are slow to respond to any temperature upset because heat transfer to the sensor is slow. Anyone who has carefully observed the performance of the refrigerator or freezer in his home is aware of these problems. Also, any object placed in an air bath will usually take several hours to reach the temperature of the bath. Thus, the temperature control in an air bath is usually not better than one or two degrees.

The low heat capacity of the air in an air bath can be used to advantage, however, in temperature-programmed baths such as ovens on gas chromatographs. In these applications, the temperature of the bath can be changed quite rapidly with only a moderate heater power, the objects being controlled are of low thermal mass so they respond reasonably, and neither very accurate nor very precise control is necessary.

Thermostatted metallic blocks, usually copper or aluminum, can also be used to avoid the problems of containing a liquid. They have the advantage of a high thermal mass and can be used over a wide temperature range. The disadvantages are the existence of temperature gradients, the difficulty of transferring heat to and from the other components of the thermostat and the experiment containment vessel, and their slow recovery from a temperature upset.

Metallic-block thermostats have found wide use in microcalorimeter designs in recent years where temperature stability of less than a microdegree has been claimed (12). Metallic blocks are also useful for high-temperature applications.

II. SPECIFICATION OF CONSTANT-TEMPERATURE BATH CHARACTERISTICS

"How well does a particular constant-temperature bath control the temperature?" "At what temperature?" "With respect to what kind of sen-

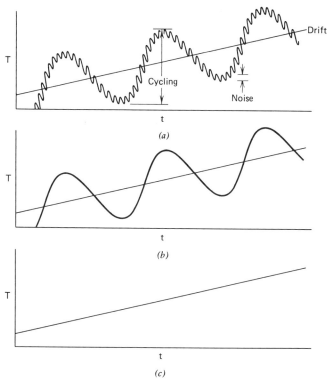

Fig. 4.1. Types of thermal noise in a constant-temperature bath. (*a*) Actual temperature of the bath fluid at a point in the bath. (*b*) Temperature as measured with a sensor with a time constant greater than the reciprocal of the noise frequency. (*c*) Temperature as measured with a sensor with a time constant greater than the reciprocal of the cycling frequency.

sor?''. In this section, we shall be concerned with the definition, calculation, and measurement of terms that can be used to describe fully the important characteristics of a constant temperature bath.

"Gradient" is used to describe the long-term average temperature differences that exist between different parts of the bath. "Noise" is used to refer to short-term random fluctuations in the temperature of the bath fluid. In contrast, "cycling" is used to denote a regular pattern (usually sinusoidal) of short-term temperature fluctuations. "Drift" connotes a long-term change in the average bath temperature. These four parameters must all be specified to describe the temperature control in a bath. An additional parameter, the "dead band," is also required to specify bath operation if its magnitude is comparable to the other four. The "dead band" is twice the temperature offset from the control temperature which is required to activate the controller. The meanings of some of these parameters are illustrated in Fig. 4.1. Also illustrated in Fig. 4.1 is the fact that the sensor used to determine these parameters must have a time constant shorter than the signal being measured. This may seem obvious to the reader, but values of these parameters have often been reported based only on short-term temperature observations made with a mercury in glass thermometer which has such a long time constant that drift is probably the only measurable parameter. Further, unless such observations are carried out over a period of several hours, drift may be entirely missed or confused with a very low-frequency cycling.

The "settling time" is the time required for the bath to come back into control after a small (<0.1 K) temperature upset occurs. The term "time constant" is not properly used in this context because a constant-temperature bath is a controlled system and may have several time constants, each of which may contribute to the settling time. Only when a system has one time constant that is much larger than any other can the term "time constant" be used to describe the time behavior of a bath after a temperature upset. In its strictest sense the "time constant" of such a system is equal to k in equation 1,

$$\Delta T = Ae^{-kt} \tag{1}$$

where ΔT is the difference between the temperature at time t and the control temperature and A is a constant.

The "slew rate" is the maximum rate at which a bath temperature change can occur and is an important parameter in applications where large temperature upsets occur and a rapid return to control is important.

III. THE BATH AND CONTROLLER AS A SYSTEM

A. THE STEADY-STATE DESCRIPTION OF THE SYSTEM

Although a controlled bath maintains a constant temperature, it is not an equilibrium system, but rather a dynamic system that is usually in a steady-state condition with respect to heat flow. The thermostat is actually maintained at constant temperature by balancing the rates of heat outputs and inputs to the bath. Because fluctuations in the temperature of the surroundings of the bath system will affect the rate of heat flow into or out of the bath, they must also be considered in designing the bath. Because any controlled steady-state system undergoes a transient response of some kind when the steady state changes to a different condition, and, since control is not maintained during the transient response time, it should be obvious that precise control of a thermostat with regard to cycling depends (1) on precise control of the heat flows in and out of the bath and (2) on keeping the heat flow rates as constant as possible.

Consider for example a bath with an on-off heater that has a power output which is large compared with the rate of heat loss from the bath. Such a system violates the second control principle above since the rate of heat flow into the bath will be far from constant. When the heater turns on the bath

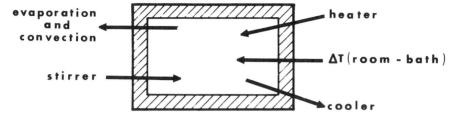

Fig. 4.2. Heat fluxes between a constant-temperature bath and the surroundings.

will very quickly become hotter than the control temperature. The temperature sensor will, however, lag behind and hence the bath will be much too hot when the heater turns off and passive cooling begins. During the cooling transient period, the bath is not under control. This cycle will be repeated each time the heater comes on. Note that the control precision is determined by the cycling and depends only on the gain and time constants of the sensor, controller, and heater.

An example of a bath that violates both of the control principles above is an uninsulated bath in a location where an on-off air conditioning or heating system for the room impinges directly on the bath. The rate of heat loss from the bath will change when the room conditioning system turns on. The controller must then compensate for this change by increasing or decreasing power to the heater. Because the sensor, controller, and bath will all have finite time constants, there will be some time during which the bath temperature will deviate from the true control temperature. Since the room can have a time constant near that of the bath, the room air conditioning system can cause regular fluctuations in the bath temperature. In measuring this effect, care must be taken to insure that it is actually bath temperature that is changing and not external circuitry that is being affected by room temperature changes.

The heat flows which must be considered are shown schematically in Fig. 4.2. Of course, not all baths will have all of them and some of the heat flows will be insignificant under certain circumstances. A consideration of the heat flows in Fig. 4.2 leads to further considerations in designing a thermostat.

The rate of heat exchange caused by evaporation or condensation of the bath fluid will depend on the vapor pressure and heat of vaporization of the fluid, and on the rate of transfer of the saturated vapor phase out of the space above the liquid. For example, a tightly sealed lid is a necessity on a water bath which is to be controlled better than ± 0.001 K. Fluctuations in air currents over an open water bath can cause large changes in the rate of heat loss from the bath. A constant-speed stirrer should be used since the heat input from this source is frequently the largest. A 0.05 horsepower motor, such as is often used to stir general-purpose laboratory baths, puts in approximately 30 watts of heat continuously. Unless the surroundings are sufficiently well controlled and several degrees removed from the bath temperature, using heat transfer through the walls for the cooler or heater is not a good idea. It is generally best to make the heat transfer rate through the bath walls and top as small as possible by thoroughly insulating the bath. This relieves one of the necessity of tightly controlling the surroundings as well as eliminating transients caused by sudden temperature changes in the surroundings.

Either the cooler or the heater should be operated at a constant power level while the other is controlled. The cooler and heater should never both be controlled since this only leads to excessive cycling. The heater is usually the best to control because it will respond faster and is more convenient. The one which is to be controlled should supply about 50% of the heat flow in that direction. For example, assume a system in which evaporation, convection, and heat transfer directly to the surroundings are negligible; the stirrer input is 30 watts; and the cooler is to be operated at a constant rate. For such a system, the cooler should provide about 60 watts of cooling to remove the heat from the stirrer and to provide the 50% control range for the heater. A 60-watt controlled heater then needs to be used. The principles involved are that the heater and cooler should be as small as is practical and the heater power should be on about 50% of the time. If rapid heat up or cool down is required, auxiliary heaters or coolers can be used independently of the control system of the bath.

It should be noted that the cooler may be controlled while the heater is operated at a constant rate. There are disadvantages to this procedure, however. Coolers are usually more difficult to control because they require electromechanical equipment while an electrical heater requires only electrical control. Peltier coolers are of course as easily controlled as a heater, although their response time is slower and their capacity is low.

B. PLACEMENT OF BATH COMPONENTS

Placement of the heater, cooler, stirrer, and sensor in the bath is critical for precise temperature control. The principle that governs this aspect of bath design is that heat can only flow if there is a temperature gradient. Thus, if the heater is put in one end of a bath and the cooler in the other end, there must exist a temperature gradient between them. The way to avoid such gradients is then obvious—place the heater and cooler at the same location. A combined cooler-heater in which a cooling fluid flows through the heater coil has in fact been patented (15). The above general principle applies to all heat inputs and outputs and not just to the cooler and heater. Thus, all heat inputs and outputs should be clustered at one location in the bath.

The relative locations of elements within the cluster with respect to fluid flow are also important. This becomes apparent from a consideration of what happens if the fluid flows over a constant power cooler, then over a controlled heater, and then over the sensor. For the system to function, the cooler must lower the temperature below the control temperature, the controlled heater brings the temperature of the fluid back to the control temperature. Reversing the locations of the heater and cooler will usually result in cycling since the sensor sees temperature fluctuations caused by the cooler in a shorter time than those caused by the heater. Placing the sensor too far downstream from the heater can also cause cycling, because the sensor will then appear to have a long time constant. The general idea then is to place all of the constant heat inputs and outputs upstream of the controlled heat input or output and the controlled element upstream of the sensor. Placing the object to be thermostatted downstream of, but very close to, the sensor then guarantees that there will be a flow of constant temperature fluid surrounding the object.

It is well to keep in mind that the only point in the bath that is actually controlled is the fluid surrounding the sensor. For this reason the sensor should be placed upstream of a thermally neutral object to be thermostatted, but downstream from an object that is generating or absorbing heat. The temperature of an object that is generating heat will be above the control temperature if the sensor is placed upstream. If the sensor is downstream, then the surface of the object will be maintained at the correct temperature, even though the temperature of the rest of the bath would be below the control temperature. This is a further illustration of the fact that heat can only be transferred if there is a temperature gradient.

One major disadvantage exists to placing the sensor downstream of the object to be thermostatted. This disadvantage is that a major change in the average bath temperature will occur each time an object not at bath temperature is inserted into the bath. Since the fluid flowing off the object will be at a temperature different from the control temperature, a downstream sensor will see this as an indication that the entire bath is away from the control temperature. As a result, the entire bath temperature will be changed and, depending on bath response time, a long time may be required for the system to restabilize.

C. ILLUSTRATION OF PRINCIPLES: CONSTANT-TEMPERATURE ROOMS AND FLUID FLOW LOOPS

Two examples of temperature control problems that are usually done improperly, i.e., constant-temperature rooms and the control of constant-temperature jackets by fluid flow loops, will be used to illustrate the application of the principles in the previous two sections.

Constant-temperature rooms are frequently designed so that only heating or cooling is available at a given time. The controller is usually a simple on-off type. (A typical American house with central heating is a good example of the usual method). The room is usually thought of as a static system in which heating or cooling must be supplied to balance the activities in the room. Because the time constants of the sensor, the room, and the heating or

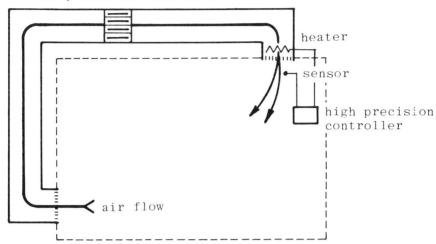

Fig. 4.3. Schematic diagram of a constant-temperature room. More than one air inlet and outlet will usually be required to obtain mixing of the air in the room in actual practice.

cooling unit are all likely to be quite long, this design leads to both temperature cycling and temperature gradients of several degrees. The sensor generally gets the blame for being too insensitive, and a more sensitive one is installed with no improvement noted.

Viewing the room as a dynamic system of heat flows (air flows) quickly leads to the conclusion that it is not the air in the room that must be controlled, but the air flowing into the room. The air flow must be constant and large enough to change the air in the room every few minutes. The temperature of the air flowing into the room can be accurately controlled by first cooling (or heating) it to one or two degrees below the control temperature with a roughly controlled cooler (or heater) and then reheating with a small, controlled electric heater. The air flows and relative placement of various elements are shown schematically in Fig. 4.3. Using these techniques, a 400 ft^2 laboratory has been controlled to ± 0.3 K (6). The room must be well insulated, of course, so that large uncontrolled heat fluxes do not occur through the walls, floor, or ceiling and careful attention must be paid to mixing of the air in the room.

The control of temperature jackets remote to the bath by fluid flow loops are used when it would be very inconvenient to place the object in the bath itself. A good example is the thermostatting of spectrophotometer cells. The design of such a remote jacket presents special problems however.

In considering such a remote jacket, it is well to keep in mind that it is the jacket and contents that must be kept at the correct temperature and not the bath feeding the fluid flow loop. This suggests that the sensor should be

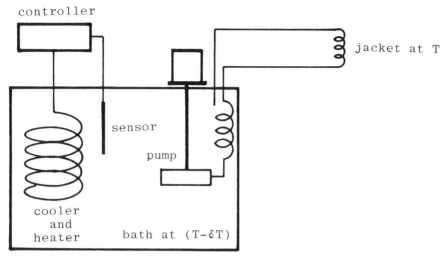

Fig. 4.4. Design for temperature control of a remote jacket by a fluid flow loop.

placed in the jacket and not in the bath. Also, the controlled heater (or cooler) should be placed in the flow loop just upstream of the sensor and should be of very low power. The bath in this case should be controlled at a temperature slightly below the desired temperature in the loop. Although this approach is the correct one in agreement with the design principles outlined above, it is not easy to implement for two reasons. First, most sensors are flow sensitive, i.e., changes in the flow rate over the sensor will cause noise in the signal to the controller, and flow lines will always contain instabilities in the flow at the rates that must be used. Second, it is difficult to obtain the proper relationship between the time constants of the heater (or cooler), sensor, and flowing fluid.

A remote constant-temperature jacket can be controlled equally well if proper attention is paid to the various sources of heat exchange in the fluid in the lines. Such a system is shown in Fig. 4.4. Use of a submersible, centrifugal pump and placing a few coils of the pump exit tubing in the main bath will insure that the fluid leaves the bath at the proper temperature. Positive displacement pumps and pumps mounted outside the bath will heat the fluid significantly above bath temperature and should be used only in the return line if they are used at all. Fluid flow rates in the flow loop should be as large as possible (i.e., greater than 5 l/min for most applications). To obtain large enough flow rates without causing a significant pressure drop (which would result in heating of the fluid), no constriction in the line should be less than 0.25 in and the flow lines should be a minimum of 0.375 in, with 0.5 in being better. The fluid flow lines should also be as short as possible and well insulated. Such a flow loop will control a remote jacket as well as the bath is controlled, but the actual control temperature at the remote jacket usually will be slightly different. As will be shown later, a much better way of solving this problem is to build a very small bath in place of the jacket. Such small self-contained temperature-regulating jackets are commercially available for spectrophotometer cells (7).

D. THE EFFECT OF BATH SIZE, FLUID, AND STIRRING ON TEMPERATURE CONTROL

It has generally been assumed that, at least up to a point, the bigger a bath is the better it can be controlled. This is an invalid assumption. The thermal inhomogeneity, δT, in a stirred fluid bath depends on the power being dissipated through the bath, P, the eddy current lifetime in the fluid, τ, the heat capacity per unit volume of the fluid, C, and the volume of fluid in the bath, V, as shown in proportionality 2, which reiterates the importance of keeping

$$\delta T \propto P\tau/CV \qquad (2)$$

the total power input to the bath as small as possible. In the context of this section, we shall be mainly concerned with power input from stirring.

Because it is difficult to assign a numerical value to τ, an accurate a priori calculation of δT is not usually feasible. The relationship can be used to predict qualitatively how the temperature noise in a bath will vary with bath size, fluid, power inputs, and mixing efficiency. The prediction is not straightforward, however, because the variables in proportionality 2 cannot be varied independently. For example, doubling the volume of the bath while maintaining P constant would at first appear to halve the value of δT, but doubling the volume without doubling the energy input for stirring will result in an increase in τ, probably by more than a factor of 2, and thus δT will probably increase instead of decreasing. As the size of a bath approaches about 100 liters, it becomes more and more difficult to maintain τ small while using only one stirrer. Placing more stirrers in the bath at different locations will decrease τ, but will cause gradients to exist in the bath since the extra stirrer(s) will be an additional source of power input. As the size of a bath drops below about 1 liter, it becomes possible to get very efficient mixing (small τ) with very low power input from the stirrer. The value of δT thus decreases markedly as the volume decreases for very small baths.

The thermal inhomogeneities described by proportionality 2 define the ultimate lower limit that can be realistically achieved in controlling a bath. Figure 4.5 gives some results that have been obtained on 25°C water baths in the authors' laboratory in attempts to determine what this theoretical limit might be. The absolute values of the results depend to some extent on bath shape and might be improved slightly in the future by better design of the stirrer, but the data in Fig. 4.5 do clearly show how bath size affects temperature regulation.

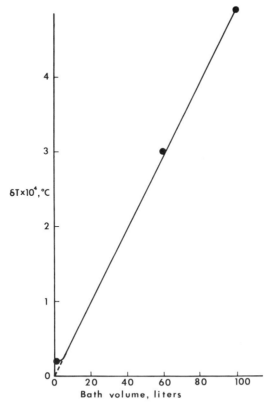

Fig. 4.5. Thermal noise in a constant-temperature bath as a function of bath size. δT is the minimum achievable peak-to-peak short-term noise with a sensor response time of 10 sec.

The properties of the fluid used in the bath will affect three of the parameters in proportionality 2. The effect of the heat capacity is obvious from the equation, which suggests that a fluid with a lower heat capacity would improve bath control. The problem with this approach is that most fluids with low heat capacities also have poor mixing or heat transfer properties, and thus the P term increases by a larger factor than C is decreased. We have found by experience that substituting silicone oil or all-weather type motor oil for water in a bath degrades the control around room temperature, but temperature control steadily improves as the temperature rises to about 80°C, at which point the control is again the same as in the room temperature water bath. These effects were attributable largely to changes in the viscosity of the oil which decreased τ faster than P was increased to maintain the higher temperature.

Another factor related to stirring and bath control is bath shape. A rectangular bath, for example, may have dead spots in the corners. These blobs of fluid will occasionally break loose from their corners and cause large random fluctuations in the bath temperature as they are mixed in. To avoid this problem, a bath shape with slightly rounded corners should be chosen. Cylindrical baths where the length is long compared with the diameter should also be avoided since lengthwise mixing is difficult and temperature stratification will result. Baffles may also be necessary to prevent stratification near the walls of the bath. A handful of paper punches or glitter dropped into the bath is useful for visualizing flow patterns, eddy time, and dead spots.

IV. SELECTION OF BATH COMPONENTS

The following series of questions need to be answered in order to define the bath characteristics for a given application.

1. Temperature control?
2. Temperature operating range?
3. Absolute resettability?
4. Bath size?

The answers to these four fundamental questions will largely determine the components that will be required and the cost of the bath system.

The temperature control is probably the most important parameter to be set. Generally, it is also very easy to determine this parameter since it will usually be known beforehand what effects temperature fluctuations will have on the experiment to be performed in the bath. The length of the experiment in time must also be considered because it will determine whether drift and long-term cycling will cause significant effects. High-frequency thermal noise is usually of little consequence because it will be filtered out by the experiment container unless the container has a very high thermal conductivity. Placing some insulation such as a glass or plastic wall between the thermostatted object and the bath fluid is in fact a good way of reducing the transmission of short-term noise from the bath if the longer equilibration time then required can be tolerated.

If peak-to-peak fluctuations of ± 0.1 K can be tolerated, almost any set of components can be used to construct such a liquid bath and it can be an inexpensive do-it-yourself project. Air baths with about ± 1 K control can be just as easily constructed.

To improve the control to ± 0.01 K, a controller and sensor with at least this much capability must be chosen, although a simple on-off controller may still be used if the sensor and heater have a response time of less than about 5 sec. This, together with proper placement of the bath elements, the use of an insulated container, and sufficient stirring, will usually guarantee success to this level of control. Simple phase-change-type baths can be used to near this level. Many brands of relatively inexpensive baths are commercially available to fill a need in this range, although construction is still simple and depending on the skill of the worker some money may be saved by building rather than purchasing. Cosmetic appearance and reliability usually suffer, however.

Bath control to ±0.001 K requires much more careful attention to detail and a more sophisticated approach in the controller. A controller with at least proportional gain and zero dead band must be used. The controller itself may be subject to drift or to fluctuations in the set point caused by room temperature effects. The bath surroundings may have to be temperature controlled. The top of the bath must be covered to prevent evaporative and convective heat losses. Bath cooling (or heating, whichever is held constant) must be held within a narrow range. This may require a second rough controller on the cooling source. Placement of components and stirring efficiency will be critical. The number of commercially available systems decreases markedly at this point.

To improve control to less than ±0.001 K, in addition to the above, a type-1 (proportional, derivative, and integral) controller must be used and the cooler and heater must be tightly coupled so that no heat actually flows through any part of the bath.

The temperature operating range will determine the materials that can be used to construct the cooler, heater, sensor, and the bath itself. In addition, this will narrow the choice of bath fluids that are compatible with the temperature range and construction materials. The maximum temperature can also be used to establish the insulation that will be necessary to keep the power input requirements within reason. It is well to keep in mind that a given bath system should not be expected to operate well over more than about a 75 K range, since this much of a change in temperature will alter the properties of the components enough to make them incompatible in several ways. For example, if a thermistor or other nonlinear element is used as the sensor, a different one will have to be used to maintain the same sensitivity. Also, the viscosity of the bath fluid will usually change so much that a new stirrer configuration may be required. Chemical reactivity of materials will increase as the temperature rises and some parts may be rapidly corroded at one temperature and not at another. Physical properties will also change and structural parts may become soft or brittle.

The minimum or the maximum operating temperature will also partially determine whether or not the bath is a do-it-yourself project. Unless a simple phase-change bath is suitable, the construction of a bath for temperatures above about 150°C or below about −20°C is a difficult undertaking and will almost certainly be more costly than purchasing a bath.

V. TEMPERATURE-SENSING ELEMENTS

The temperature-sensing element is the source of information for the controller. The properties of this element have a great deal to do with the satisfactory control of the bath. The effect of a long-time constant has already been discussed. The time constant can also be too short. If there are short time variations in the bath, such as might be detected by a bead thermistor, it may be advantageous to integrate out this noise by deliberately increasing the time constant. The time constant can be increased by adding heat capacity or thermal insulation to the temperature-sensing element itself or by adding a resistance-capacitance network in an external circuit. This element must also have sufficient sensitivity; it must detect changes in the bath temperature that are smaller than the permissible variation. If the bath temperature must be held constant for a long time, the set point (the reference point for the controller) must not change with time. If the absolute temperature response of the sensing element changes with time, the bath temperature will also drift. If the sensing element lacks sensitivity or reacts too slowly, consider a better one, although sometimes the control elements can be made to compensate for these shortcomings.

The type of sensing element must be compatible with the type of control elements selected. The more subtle types of control require more information than is available from a simple on-off sensing element. Resistance thermometer bridges, thermistor bridges, and thermocouple circuits, in addition to "too hot" and "too cold" give the information "how much," which can be used to provide closer control. Since signals from such elements are continuous, they can be easily differentiated and integrated by the controller to find how fast the controlled temperature is approaching the desired temperature or how the time-average temperature differs from the desired temperature. This information may then be used by the control elements.

A. THERMAL EXPANSION SENSING ELEMENTS

A number of sensing elements are available commercially that depend on thermal expansion to make or break a contact. The mercury thermoregulator is of this type. These regulators operate very much like a mercury-in-glass thermometer. When heated, mercury in a bulb expands into a fine capillary. When the mercury column is long enough, it makes contact with a wire fixed in the capillary. The sensitivity increases with the amount of mercury and with smaller capillary diameter. To avoid a long time constant, the bulb is made long and of rather small diameter to get a large surface in contact with the bath. The operating temperature is determined by adjusting the amount of mercury in the bulb or the position of the wire contact. Commercial mercury thermoregulators can be obtained which are sensitive to 0.01 K. The contacts and the mercury are hermetically sealed to avoid corrosion, and a reservoir allows for adjusting the control point. These regulators will operate satisfactorily for years if the contacts are not overloaded. For this reason, it is usually necessary to use the regulator to drive a power relay or to switch a triac. The expansion of liquids other than mercury can also be used, (8) however a short length of mercury in a capillary is still retained to make the electrical contacts.

Bimetallic regulators utilize the difference in thermal expansion of two metallic strips fused back-to-back. One end of the composite strip is held rigidly and the other is fitted with an electric contact. On heating, one side expands more than the other, causing the strip to bend so that the contact is made (or broken). These regulators are usually somewhat less sensitive than the mercury regulators, but they are rugged and easy to set to different control temperatures. Their contacts can carry rather large currents, often eliminating the need for a relay. The contacts of the more sensitive bimetallic regulators however will handle little power and may require an auxiliary relay, like the mercury regulator. These more sensitive types offer the convenience of easy change of the set point with only a small loss of sensitivity. Bimetallic regulators are useful as safety devices when more sensitive sensing elements are used for control. They can be set to turn off all power in case the control system fails and thus prevent the bath from reaching a temperature dangerously above the set point. Bath control from ± 10 K to ± 0.05 K is practical with these regulators.

Another application of differential expansion is a thermal switch that employs two coaxial tubes of different metals fastened together at one end with an electrical contact at the other. These are also useful as safety devices, but the temperature of operation is set in manufacture and is not adjustable in the laboratory. A variety of these thermal switches is used in home appliances and in automobiles, so they are readily available.

The expansion of gases may also be used to operate a thermostat (8). Helium thermometers are commonly used for control of cryogenic baths. The control temperature is set by adjusting the mercury level in a manometer. Other gases can also be used.

B. RESISTANCE THERMOMETERS

The increase with temperature of the resistance of pure metals is a property that can be used in the measurement and control of temperature. The sensing resistance, T in Fig. 4.6, is placed in a Wheatstone bridge circuit with three resistors, R_{ref}, R_1, and R_2. A signal voltage will develop at the output when the ratio of T/R_{ref} is not equal to R_1/R_2. The current may be either alternating or direct depending on the requirements of the controller. The sensitivity can be increased by increasing the resistances and/or the current. Useful signals corresponding to 0.0001 K or less are obtainable. For control purposes, resistance thermometers have the disadvantage that variations in lead resistances must be taken into account. Leads from the resistors to the branch points of the bridge can be kept short or the resistances themselves made large to reduce the relative effect of the leads. To obtain a large resistance of pure metal, however, one must use long wires of small diameter. The leads can be made of an alloy such as manganin whose resistance changes very little around room temperature. The problem of leads can also be solved by winding the entire bridge on a single probe. This

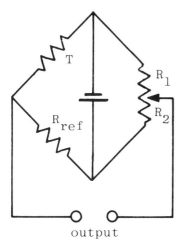

output

Fig. 4.6. Wheatstone bridge circuit for use with resistance thermometers and thermistors.

technique sacrifices the advantage of easy change of the control temperature, which is varied by changing the ratio of R_1/R_2.

Platinum, nickel, tungsten, molybdenum, and copper have been used for resistance thermometers. Thermometers of fine copper wire varnished on copper bodies are highly reproducible, and give good thermal contact and short time constants. Still better reproducibility may be obtained from a strain-free platinum winding at the cost of increased time constant. Such platinum thermometers are used as temperature standards. In any bridge used for control, one should be careful that the reference resistors, R_{ref}, R_1, and R_2 (Fig. 4.6) do not change with time.

C. THERMISTORS AND OTHER ELECTRONIC DEVICES

Thermistors are semiconductor devices that have found wide application to temperature control and measurement problems. Like wire resistance thermometers, thermistors are usually used in Wheatstone bridge circuits. Their resistance decreases as they are heated and the per cent change per degree is about 10 times greater than for metals. They are available in a wide variety of sizes, shapes, and resistances. Even a very small thermistor may have a resistance so large that the effect of leads can be negligible. Because of their small size, thermistor sensors can have a very short time constant. Thermistors will maintain a calibration on absolute temperature of about 0.0025 K/year if they have been properly manufactured and are treated properly (17).

Other electronic devices can also be used as temperature sensors, although none of them have yet achieved the general acceptance of thermistors. The properties of transistors, capacitors, oscillators, and integrated circuits can all be made very sensitive to temperature changes.

D. THERMOCOUPLES AND THERMOPILES

The use of thermocouples to measure temperature differences is a well-established technique. If one of the junctions is at a known temperature, as in an ice bath, the temperature at the other junction can be calibrated as a function of the observed emf. This emf can be compared with a reference emf to obtain a dc signal proportional to the difference between the observed and the desired temperature. This signal can be used to regulate the observed temperature. By changing the reference, the set point is easily changed. Using multiple thermocouples connected in series, the sensitivity in microvolts per degree is increased in proportion to the number of couples since the noise only increases in proportion to the square root of the number of couples.

Thermocouples for bath control require that the ice bath or other reference, and hence the reference voltage, remain constant within the desired tolerance. Ice baths are easily made reproducible to ±0.002 K, and with more trouble, constant to 0.0001 K (16). Commercially available water triple-point cells are reproducible over long periods and constant to ±0.0004 K (1,10), so that the reference junction need not limit the accuracy of control with thermocouples.

Because of their low cost and ease of replacement, thermocouples are more frequently used for high-temperature work than other sensing elements. The small signal from thermocouples requires a good amplifier for sensitive control. Several papers have reported apparatus using thermocouples as sensing elements operating in various ranges from −196°C to 1000°C. (2,5,9,11,14).

E. VAPOR PRESSURE

Another useful property that can be used for temperature control is the vapor pressure of a boiling liquid. A vapor–liquid bath can be very useful because large quantities of heat can be transferred by a boiling and condensing liquid. The temperature of the vapor–liquid interface is fixed when the pressure is controlled. This property is used to control the baths used for resistance thermometer calibrations. Stimson (1a) describes water and sulfur boilers used for this purpose in conjunction with a very precise manometer. The same principle is utilized to control liquid helium baths, varying the controlled pressure to change the bath temperature (3). Many refrigeration units use the vapor pressure of the refrigerant as a means of temperature sensing and control.

VI. HEATERS, COOLERS, AND POWER RELAYS

Both a source of heating and of cooling must be provided for control of a constant temperature bath. For baths operating above ambient, the surroundings are frequently used as the source of cooling and a heater is used as the control element. For baths below ambient, the surroundings are sometimes used as the source of heat and the cooler is regulated.

By far the most common source of heat for baths is an electric heater. Electric power, either as alternating current or rectified to direct current, can be conveniently controlled in a variety of ways to obtain a large range of power levels. The simplest "valve" for regulating electric current is a switch. The bimetallic regulator is built with switch contacts adequate to handle power for most baths. Relays, which have equivalent contacts, can be operated by current from a controller or through a mercury regulator, although this practice is good only for very sensitive relays. For on-off or two-level control, relays are inexpensive and adequate for many applications. When operated within their ratings, the coils last indefinitely, but the switching contacts eventually go bad.

Alternating current is often regulated by means of triacs. These have the advantage over relays in that the power level can be adjusted in small increments by shifting the phase of the control signal. Often they are used merely as on-off switching elements for two-level control. Larger amounts of alternating current can be regulated smoothly by saturable core reactors. Reactors are rugged and are available for handling various power levels from around 100 watts to many kilowatts. Power transistors can also be used to switch reasonable amounts of power. They are especially applicable to switching dc power at lower voltages and higher currents than tubes. Heaters for use with transistors can therefore be of lower resistance for the same power.

Electric bath heaters should have short time constants so as to avoid too much storage of heat and too long a time before an increase in power is actually transmitted to the bath. To obtain small time constants, the electrical insulation and the material for protection from the bath fluid should be of low heat capacity, high thermal conductivity, and no thicker than necessary. The thermal resistance to the bath can also be decreased by making the area of contact between the heater and bath as large as practicable. A good practice is to wind the heater on a tube through which a stirrer drives the bath liquid. Air spaces between the heater wire and the bath should be avoided wherever possible. Even such poor thermal conductors as silicone grease or epoxy resin are a great improvement over air. With nonconducting bath fluids and a glass or plastic container, the heater can be a coil of resistance wire suspended directly in the bath. Such a heater has a very small time constant and can be used with most liquids if a low-voltage, low-resistance, high-current heater is used. A heater that uses a coil of stainless steel tubing directly in contact with water as the resistive element and which is driven by an 18-V ac current has been patented (15). This particular design further uses the stainless steel tubing as a cooler by flowing chilled water or refrigerant through a silicone rubber tube inside the stainless steel tube. In an arrangement where the heater is in direct contact with the bath fluid, it is usually necessary to insure that the sensor is well insulated electrically and not affected by stray currents. An isolation transformer in the heater circuit is useful in this respect.

Ordinary incandescent light bulbs can also be used effectively as heaters. They do not need to be immersed directly into the bath fluid, but can be suspended above the fluid or even outside the bath container if a window is provided. The heat transfer from the light bulb to the bath takes place by infrared radiation and thus a light bulb may have a very short time constant. The wattage of the bulb should be used as a guide to the power it will produce rather than the resistance of the filament, since the latter changes greatly as the filament heats up.

Gas is a convenient source of heat in the laboratory, but not much used for bath heating. A solenoid valve, opened or closed by electric current, can be used to regulate gas flow to a bath heater. At first, it appears that a gas heater has a negligible time constant, but the flame is analogous to the wire in an electric heater and is usually insulated from the bath. The time constant must accordingly allow for the flow of heat from the flame to the bath through a wall or baffle.

Another source of heating is the laboratory hot water supply. Water flow can also be regulated by means of a solenoid valve. Similarly, chilled water can be used as a source of cooling. The simplest arrangement is simply a coil of metal tubing through which the coolant or hot water flows. For precise control, both the flow rate and the temperature of the coolant or hot water should be kept constant. Tap water will often suffice for cooling, but its temperature usually fluctuates from daytime to nighttime and from winter to summer so that it is not satisfactory for very precise bath control. For precise control, a refrigerated bath controlled to ± 0.1 K and a circulating pump can be used to provide coolant flow to the cooler in the main bath. Such systems, including the pump, are commercially available in a wide range of sizes.

For bath temperatures below about 5°C or above about 80°C a fluid other than water must be used to flow through the cooler. Solutions of automotive antifreeze (ethylene glycol) or alcohol are useful down to about -30°C. Below this one must resort to an organic liquid. Antifreeze solutions are usable up to about 100°C. Above that, silicone oils, fluorocarbons, mineral oil, and xylenes are the most commonly used fluids.

An elegant way of providing both heating and cooling is by the use of a Peltier cooler. A temperature gradient is generated across these thermoelectric devices when an electrical current flows through them. The temperature gradient and hence the rate at which they will transport heat is proportional to the electrical current. The direction of heat transport depends on the direction of the electrical current, so they can be switched from cooling to heating by reversing the current. Peltier coolers have a limited power output and are somewhat expensive so their application is limited to small baths (<1 liter). Also, they are subject to slowly decaying hysteresis effects when the electrical current is changed. This effect makes them unsuitable for temperature control to better than about ± 0.01 K.

VII. CONTROLLERS AND TYPES OF CONTROL

The term "controller" in this discussion applies to the part of the system which, on the basis of information from the sensing element, increases or decreases the power to the bath heater or cooler. The controller decides what action is to be taken on the information and is the "brain" of the system. It can be omitted entirely for some purposes, and, when the sensing element is a bimetallic regulator, it usually is. For thermocouples and resistance bridges, the signals must usually be amplified to increase sensitivity, and the amplifier is usually also included in the controller.

For the purposes of this discussion, the various types of control functions will be termed two-level, proportional, reset, and derivative control. The first of these is probably the simplest to use and understand and certainly the most common, although it is difficult to treat mathematically because of the discontinuous signal and the corresponding power in the heater. A number of operations can be performed with simple two-level controls which are more easily understood if the same operations are understood for the case of continuous signal and power supply.

A. TWO-LEVEL CONTROL

The most common mode of control is usually based on thermal expansion-sensing elements, especially the mercury thermoregulator. Such a controller alternately switches from a high power level, which will heat the bath under all anticipated conditions, to a lower power level, which will allow the bath to cool under all anticipated conditions. When the low-power level is zero, the control is of the on-off type. With two-level control, some heat is supplied while the bath is cooling, so that it cools more slowly. The temperature of the sensing element will be closer to that of the bath for slow heating or cooling rates. The bath will overheat or undercool less if the two power levels are adjusted as closely as possible to the actual average power level. The limitation on this approach to improvement of the control is the variation in ambient conditions. For example, if the low power level is just adequate when room temperature is one degree below the desired bath temperature, the system will go out of control when the difference exceeds one degree.

For laboratories having little temperature variation, it is possible to adjust the lower levels in the heaters if they have been designed to allow for large variations in ambient conditions. If the bath has two heaters, both power levels may be adjusted. In baths having only one heater, the power circuit can be arranged to supply two different voltages to the heater. A circuit such

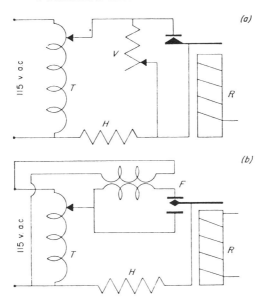

Fig. 4.7. Circuits for two-level control. (*a*) The relay R bypasses the current from a variable transformer T around a series resistor V to increase the power in the heater H. (*b*) The relay adds the voltage of a filament transformer F to increase the power.

as that in Fig. 4.7a may be used. The variable transformer *T* is adjusted to give a power level just exceeding the maximum requirements and the variable resistance *V* to give a level just below the minimum. Many simple variations of this circuit are possible. For example, to avoid wasting power in an external resistor, the double throw relay with an auxiliary transformer *F* (Fig. 4.7b) can be used. The voltage applied to the heater is increased or decreased by the voltage of the transformer, which might practically be one of the various low-voltage filament transformers available from electronics suppliers.

Two-level control is adequate in a great many applications but inherent in it is the alternating rise and fall of the bath temperature, which is termed hunting. If this effect is too large, modifications or other types of control are necessary. Under the best of conditions, two level control may be used to obtain control in a bath to about ± 0.001 K. Under more typical laboratory conditions, control to about ± 0.01 K can be expected.

B. PROPORTIONAL CONTROL

Resistance bridges and thermocouples provide more information than thermal expansion elements. In addition to "too hot" and "too cold," they indicate "how much;" that is, the signal is to a good approximation proportional to the difference between the observed temperature and the desired temperature. Signals from such sensing elements can be used to decrease the fluctuations in bath temperature due to changes in room temperature, power line voltage, and other causes. A controller that changes the power in direct proportion to the signal is called a proportional or type-zero controller. Its function is expressed algebraically by equation 3

$$\dot{Q} = \dot{Q}_0 + P(\theta_0 - \theta) \tag{3}$$

where $\dot{Q}$ is the total power, $\dot{Q}_0$ is the constant power from a heater connected directly to a supply voltage, θ_0 is the desired temperature or set point, and θ is the actual temperature. The proportional constant P is a factor that can be selected on the controller.

For example, suppose the bath loses 10 w for each degree above room temperature. If room temperature is 25°C and the bath is set for $\theta_0 = 50°C$, $\dot{Q}$ must be 250 w. Obviously, $\dot{Q}_0$ cannot exceed the smallest expected power, so let us take it to be 200 w, which would allow for an increase in laboratory temperature to 30°C. According to the equation, the deviation of the bath temperature from the set point can be as small as desired by selecting a sufficiently large value of P. Actually, the time constants of the control system set an upper limit on P. When P is large, a small increase in θ increases $\dot{Q}$ to full power. At this point, proportional control degenerates to two-level control, and the bath temperature oscillates continuously.

Suppose the upper limit for P for a particular system is 5000 w/K. To obtain the extra 50 w to maintain the bath at 50°C, $\theta_0 - \theta$ must be 0.01 K, but the variation in the bath temperature for a one-degree change in room temperature to supply the additional 10 w is only 10/5000 K. The proportional control also compensates for changes in the line voltage. If the voltage changes so that the directly connected power $\dot{Q}_0$ increases 10 w, θ will increase only 10/5000 K.

The bath control is affected by insulating the bath. Doubling the insulation, assuming no evaporation, will reduce the power requirement to 5 w/K. If the same proportional factor can be maintained, a change of 1 K in room temperature will alter the bath by 5/5000 = 0.001 K, an improvement equal to the improvement in insulation.

The advantage of proportional control over two-level control is that the bath temperature can be held steady when the heat loss to the surroundings and power sources are steady. With two-level controls under the same circumstances, the heating-cooling cycle cannot be avoided. Proportional control, however, is at the mercy of slow variations in the bath environment and in the power supply. A change in the environment temperature causes a corresponding, but smaller, change in the bath temperature which is called droop, sag, or off-set error. In many systems, these small variations can be tolerated. When slow variations are too large, one way to reduce them is by automatic reset.

C. AUTOMATIC RESET

In the preceding section, it was seen that a change in room temperature causes a much smaller change in the bath with proportional control. If an attendant observed the change in the bath temperature, he could restore the temperature to that preceding the change by manually altering the constant power $\dot{Q}_0$. It might be said that the attendant had reset the steady power. Adjustment of $\dot{Q}_0$ by automatic devices is called automatic reset, integral, or type-1 control.

The proportional control is adjusted to take care of the short-term variations and the automatic reset is adjusted to allow for slower changes in room temperature or supply voltages. In this case, the power is given by

$$\dot{Q} = \dot{Q}_0 + P(\theta_0 - \theta) + R \int_0^t (\theta_0 - \theta)\,dt \qquad (4)$$

where R is the reset time constant selected on the controller, t is time, and the other symbols are the same as in equation 3. A change in $\dot{Q}$, the power required, is immediately supplied by the proportional term at the expense of a deviation of the temperature from the set point. The reset term then adds to the power over a period of time until $\theta_0 = \theta$. Deviations from the set point can thus be brought back to zero even for permanent changes in the power requirement.

A bath controlled with a proportional-reset controller should achieve a temperature stability of ± 0.001 K or perhaps slightly better.

D. DERIVATIVE CONTROL

A third type of control function is derivative or rate control, which is sometimes used with proportional-reset control to dampen the instabilities inherent in type-1 controllers. Such a circuit is also referred to as an anticipation or lead network when it is used to prevent overshooting the set-point. Derivative control changes the power in proportion to the rate of change of the temperature. When used with proportional and reset it gives a controller output according to the equation.

$$\dot{Q}_0 = \dot{Q} + P(\theta_0 - \theta) + R \int_0^t (\theta_0 - \theta)\,dt + D\frac{d\theta}{dt} \qquad (5)$$

where D is the derivative time constant selected on the controller and the other symbols are the same as in equations 3 and 4. Derivative control can be used to increase power when the bath is cooling rapidly so as to slow down this cooling rate. In this way, it resembles a viscous force in mechanical systems having a damping effect on the "velocity" of the temperature. It is one of the features of three-mode controllers available commercially. Such controllers have been used to control large baths to ± 0.0002 K for long periods of time (13). Small baths have been controlled to within a few microdegrees with this type of controller (4).

There are at present so many brands of relatively inexpensive temperature controllers available that it is seldom economically practical to build one unless a simple on-off control is all that is desired. In selecting a general purpose controller, three features should be present to make the controller compatible with a variety of uses. The controller should have controls to set the response time of the electronics, the gain of the amplifier, and a switch so that sensors with either positive or negative temperature coefficients can be used.

VIII. CIRCULATING FLUIDS

The purpose of a constant-temperature bath is to maintain constant the temperature of the objects placed in it. The bath must therefore be able to transfer heat from the heater to objects placed in it or from the objects to the cooler. It is possible to place objects in a metal box whose walls are at constant temperature and depend mainly on heat transfer by metallic conduction. This method is effective for small enclosures where small amounts of heat are involved, but conduction of larger amounts of heat over greater distances tends to set up intolerable temperature differences. Usually a bath is filled with a fluid that circulates past the objects whose temperature is to be controlled.

The bath fluids usually wet the surfaces of objects placed in them and a film of the fluid adheres to the surfaces, forming a static layer through which heat must flow by conduction. The thickness of this layer increases with the viscosity of the fluid and decreases with the velocity of the fluid past the surface. To give good thermal contact between the thermostatted object and the bath, the fluid should have a high thermal conductivity, a low viscosity, and be stirred vigorously. For best circulation, the thermostatted objects should be placed so as to allow the fluid to flow freely about them.

Other desirable properties of the fluid are low cost, easy availability, and that it not create a housekeeping problem or fire hazard. Water is often a very satisfactory bath fluid. Its useful temperature range is from the ice point up to about 80°C where evaporation begins to transfer heat too rapidly from the bath. Air is another commonly used fluid, especially in low-precision applications such as ovens and refrigerators.

Other bath fluids that have been used include organic liquids, silicone oils, fused salts, and fluidized beds of small solid particles (8). The use of any of these pose various hazards and problems for which the user must be prepared. Many of these fluids are toxic, flammable, and may react violently with water or other common materials that may be accidently spilled into them. Some of them are corrosive and severely limit the choice of materials that can be used in contact with the fluid. Substances that are solid at ordinary temperatures and expand on melting should be melted from the top to avoid breaking the bath container. Those that expand on solidifying may break the bath container if allowed to cool below their melting point because of a loss of power or failure of a controller component.

IX. BATH CONTAINERS

Four factors must be kept in mind in selecting or designing a container for liquid baths; material, heat transfer, geometry, and the requirements of the experiment to be done in the bath. The last of these obviously must be considered first. For example, if the experiment must be visible when in the bath, a glass container or at least one with a glass window is about the only choice available. The volume occupied by the experiment will also likely determine the bath size. The geometry of the bath should be such that it will help promote stirring and mixing of the bath fluid, i.e., there should be no stagnant spots. The heat transfer from the bath to its surroundings should be minimized for the reasons given earlier. This means that the bath should be insulated and liquid baths should be provided with a lid to minimize evaporative losses. If a lid is not practicable, the liquid can be covered with a layer of a less volatile fluid or a layer of floating solid particles such as ping-pong balls, styrofoam pieces, etc.

X. STIRRERS

The stirrer should be so designed that the rate of mixing is maximized while the energy input is minimized. For the usual type of stirrer, which is a propeller driven by an electric motor, the important parameters are the propeller diameter, number of blades, blade pitch, and the motor speed. These parameters are all interdependent, and the choice of these parameters will also depend on the bath volume and the physical properties of the bath fluid. In general, all of the stirrer parameters should be as large as feasible. The upper limit is set by the boundary condition where turbulent flow around the stirrer begins. At that point, the efficiency of the stirrer (i.e., energy/mixing rate) will decrease markedly. Sometimes stirrer efficiency can be improved by placing more than one propeller on a shaft or by using two or more small stirrers in place of one large one.

Other means of stirring that have been used commonly are various types of circulating pumps and blowing a gas through the fluid. Circulating pumps tend to create flow in the bath instead of the mixing that is desired unless special precautions are taken in designing the bath container. Blowing a gas through the fluid generally creates problems by increasing the heat transfer between the bath and its surroundings. It does work rather nicely for very small baths if very tight temperature control is not necessary.

XI. A FINAL ILLUSTRATIVE EXAMPLE: THE DESIGN OF A GENERAL-PURPOSE LABORATORY WATER BATH FOR CONTROL TO ± 0.002 K

Since a convenient, insulated, and chemically inert container is available in the form of the polystyrene ice chests commonly available in sizes ranging from about 45 to 65 liters, we can begin with the bath container. The thermal conductivity of the walls (a 2–4 cm air gap or plastic foam layer) of such an ice chest is about 2×10^{-4} watt cm^{-2} K^{-1} and the surface area is about 1×10^4 cm^2. Therefore the heat transfer rate will be about 2 watts/K. Assuming that we wish to operate the bath at 25°C and that the laboratory temperature

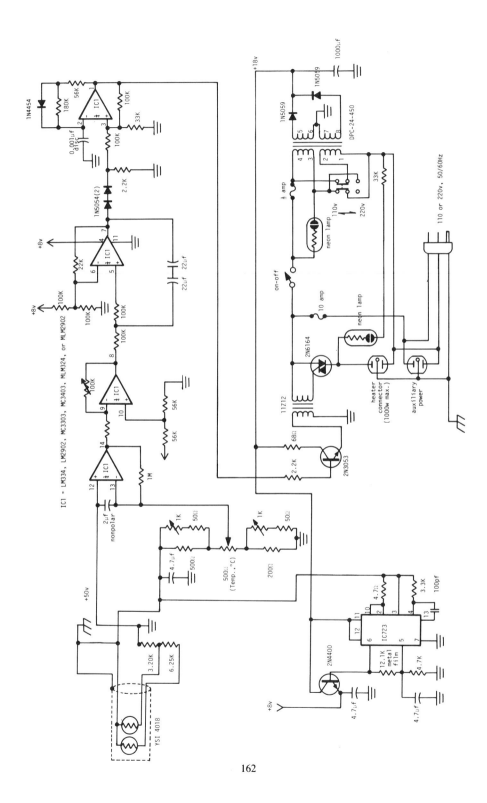

162

may fluctuate between 18°C and 32°C, a heater and a cooler with a power of at least 14 watts will be required to compensate for the heat lost or gained through the walls. Assuming a 0.05-horsepower motor is used to drive the stirrer means that another 30 watts of heat will be generated in the bath from this source. Thus a cooler with a power of 100 watts will be required. If the cooling source is chilled water at 5°C flowing through a coil of copper tubing, the flow rate will need to be greater than 1.2 ml/sec. The heater should have a minimum of about 100 watts of power. To permit operation up to 60°C, a 200-watt heater may be used with only a slight degradation of control.

Adding a controller, sensor, and power relay to the heater completes the system. The controller can be a simple on-off design if the response time of the sensor is not more than a few seconds. Such controllers and sensors which also incorporate the heater relay are readily available commercially. If it is desired to construct one, however, a fairly simple circuit diagram is given in Fig. 4.8.

The heater, stirrer, and cooler should all be placed in close proximity in one end of the bath. The sensor should be placed either in the middle of the bath or near the apparatus whose temperature is to be controlled.

The system described above including the controller can be constructed for less than \$200 (1979 dollars) and should easily achieve control better than ±0.01 K.

To recapitulate the steps in designing a constant temperature bath:

1. Determine the size of the bath.

2. Determine the operating temperature range.

3. Determine the ambient temperature and range.

4. Select the materials for the container and bath fluid.

5. Select the stirrer.

6. Calculate the rates of all heat inputs to and losses from the bath at the extreme conditions.

7. Using the information from 6, select the heater and cooler.

8. Determine the control precision required and select a controller, sensor, and power relay.

9. Carefully plan the locations of the heater, cooler, and sensor in the bath.

REFERENCES

1. American Institute of Physics, *Temperature, Its Measurement and Control in Science and Industry,* Reinhold, New York, Vol. I, 1941; Vol. II, Ed., H. C. Wolfe, 1955; Vol. III, Ed., C. M. Herzfeld, 1962.

1a. American Institute of Physics, *Temperature, Its Measurement and Control in Science and Industry,* Vol. II, Chapter 9.

Fig. 4.8. Circuit diagram for a simple proportional temperature controller sensitive to 0.002 K. (Based on a Tronac Model PTC-20. Compliments of Tronac, Inc.)

2. Benson, G. W., and G. C. Benson, *Rev. Sci. Instrum.*, **24**, 1070 (1953).

3. Cataland, G., M. H. Edlow, and H. H. Plumb, *Rev. Sci. Instrum.*, **32**, 980 (1961).

4. Christensen, J. J., J. W. Gardner, D. J. Eatough, R. M. Izatt, P. J. Watts, and R. M. Hart, *Rev. Sci. Instrum.*, **44**, 481 (1973).

5. Darken, L. S., *Rev. Sci. Instrum.*, **20**, 323 (1949).

6. Eatough, D. J., Thermochemical Institute, Brigham Young University, Provo, Utah 84602, unpublished data.

7. Perkin Elmer, Oak Brook, Ill. 60521, Bulletin ADS-100, B47, 1979.

8. Griffiths, R., *Thermostats and Temperature-Regulating Instruments,* Charles Griffin and Co., Ltd., London, 1951.

9. Hoell, P. C., *Rev. Sci. Instrum.*, **29**, 1120 (1958).

10. Plumb, H. H., Ed., Instrument Soc. of America, *Temperature, Its Measurement and Control in Science and Industry,* 5th Symposium, Washington, DC, Vol. IV and V, 1973.

11. Primak, W., *Rev. Sci. Instrum.*, **27**, 877 (1956).

12. Prosen, E. J., "Design and Construction of the NBS Clinical Microcalorimeter", NBSIR 73–179, April 1973.

13. Tronac, Inc., Orem, Utah 84057, Catalog SF-1.

14. Vajda, J., and D. B. Hart, *Rev. Sci. Instrum.*, **24**, 354 (1953).

15. Watts, P. J., and M. D. Crawford, U. S. Patent No. 3680630 Issued to Tronac Inc., 1 Aug. 1972.

16. White, W. P., *J. Am. Chem. Soc.*, **56**, 20 (1934).

17. Wood, S. D., B. W. Mangum, J. J. Filliben, and S. B. Tillett, *J. Res. NBS*, **83**, 247 (1978).

CALORIMETRY

By Robert L. Montgomery, *Department of Chemistry, Rice University, Houston, Texas*
AND
John L. Margrave, *Department of Chemistry, Rice University, Houston, Texas*

Contents

I. INTRODUCTION

The use of calorimetry as an analytical tool appears to be on the increase in spite of a substantial overall decrease of research activity in the field of calorimetry, which has been especially severe in the United States.* Regular symposia on analytical calorimetry have been held by the American Chemical Society (221), the number of different calorimeters that are commercially available is much greater than it was a dozen years ago, and it is not unusual for the designers of new calorimeters to patent them. A number of American Society for Testing and Materials (ASTM) specifications relate to calorimetry (5).

It is actually easier today than ever before to apply calorimetry to analysis (a) because many of the calorimeters can be purchased instead of being designed and built by the user, (b) because of the development of more standard materials and reactions for testing calorimeters, and (c) because of the wide adoption of computers for data collections and data processing. The standards somewhat reduce the pitfalls involved in the use of calorimetry as a tool by persons who are not experts in calorimetry per se, and computational difficulties have also been reduced.

Probably, the main problems that will result from the decreases in calorimetric research are not yet being recognized. They are (a) a continuing shortage of basic thermodynamic data on materials of new or renewed industrial interest; (b) a loss of the expertise needed to recognize and solve special problems which occur in calorimetry, as specialists in this field retire or devote themselves to noncalorimetric work; and (c) a diminished supply of new ideas for calorimetric designs and applications.

We shall define "analysis" broadly to include not only the determination of the chemical compositions of materials, but also the determination of properties, where the need for knowledge of the property is the main reason for requiring an analysis. An example of the former is the determination of the amount of carbon dioxide in a gas stream through its heat of reaction with lithium hydroxide (299,300). A sensitivity in the picomole range is claimed, making this one of the more sensitive analytical techniques. An example of the latter use, the determination of a property, is the direct calorimetric measurement of the "heating value" of a fuel. Determination of the chemical composition followed by calculation of a heating value would be many times more difficult and expensive.

A very important application of calorimetry is the determination of basic thermodynamic data for substances. This is not, strictly speaking, an analytical application. However, results based on the data are used in analytical chemistry, as well as in every other branch of chemistry, including theoretical studies. In applied research, development, and engineering, these data are used to predict optimal conditions for maximizing yields of desired products and in predicting thermal quantities for the design of heating, cooling,

---

* A statement from the U.S. Calorimetry Conference (115) documents substantial decreases in the funding of, and the number of persons involved in, calorimetry at a number of U.S. research institutions. The statement also notes that the number of papers on calorimetry abstracted by *Chemical Abstracts* has decreased, and that the fraction of these papers that is of U.S. origin has diminished greatly.

and heat-exchange systems. Because of the importance of basic thermodynamic data, some attention will be given to it in this chapter.

We shall define "calorimetry" rather narrowly, excluding for the most part those aspects that are covered in other chapters. The discussion in this chapter will concern primarily "classical" calorimetry, in which the heat to be measured is confined as well as possible to a small region, the calorimeter proper, and determined through its effects in that region. The "thermal fluxmeter" type of calorimeter, exemplified by the Tian-Calvet microcalorimeter, in which the heat is rapidly conducted away from the calorimeter proper through thermoelectric devices that measure the heat flow as a function of time, will be discussed briefly. These devices have been made more sensitive qualitatively than classical calorimeters, but not as quantitatively precise or accurate. They are best applied to very small heat effects, very small samples, or very slow processes. Heat-capacity calorimetry of the steady heating or cooling type will receive brief mention. Although it has some advantages over the intermittent-heating type, extremely slow rates of temperature change are required in many cases to avoid spurious effects caused by lack of equilibrium in the sample; occasionally it becomes absolutely necessary to use intermittent heating.

Differential scanning calorimetry is covered in another chapter, as is thermometric titration.

It should be pointed out that calorimetric techniques that are useful for analytical purposes may be useless for the determination of basic thermodynamic properties, due to the systematic errors present. These are not always apparent to the casual user of calorimetry as a tool. Misunderstanding of the true errors present in measurements can cause a tremendous understatement in the uncertainty assigned to a property measurement and may result in the publication of seriously inaccurate values. In general, the greatest caution is advised in reporting calorimetric measurements of basic thermodynamic properties. This caution is not needed often with purely analytical applications, because there are usually ways to check the results against other methods until the technique has been proven.

The various categories of calorimetry overlap somewhat. Therefore, discussions of particular types of calorimetry may not always be found under the expected headings below. The uses of calorimetry in nonchemical application, such as radioactive power determination and measurement of electromagnetic energy, will not be covered here.

II. PRINCIPLES AND THEORY OF THE METHOD

The discipline of thermodynamics is treated in another chapter. Detailed discussion is also given in textbooks on chemical thermodynamics (156,235) and, from a somewhat different point of view, in numerous textbooks covering its engineering applications.

In review and summary, thermodynamics applies to all systems. The central and distinguishing concept is that of entropy. The entropy is a function of the state (i.e., pressure, temperature, volume, composition, magnetic field, etc.) of a system. In modern thought, it is related to the probability of occurrence of that state as compared with other possible states. It can also be considered as a measure of the degree of disorder. It is directly connected with the heat and temperature, and can often be derived from calorimetric measurements. The second law of thermodynamics requires an increase of the total entropy of the universe in every spontaneous process.

From the laws of thermodynamics, an astonishing number of interrelationships between heat, temperature, and other variables determining the state of a system can be derived. For reasons of convenience, most chemical applications refer to temperature and composition as the controlled variables, with the volume allowed to change as it will, at constant pressure, and with all other conditions held constant.* A number of convenient thermochemical state functions of a system have been defined: C_p°, the heat capacity at constant pressure; H°, the enthalpy (sometimes called "heat content"); S°, the entropy; and G°, the Gibbs free energy at constant pressure. The superscript, $^\circ$, in each case, when used, indicates that the material is in its standard state.

For a chemical reaction at constant temperature and pressure, the symbol, Δ, preceding one of these functions indicates the change of that function during the reaction. When the values of the functions are known, Δ can be calculated by subtracting the value for the reactants from the value for the products. The heat absorbed during a reaction at constant pressure is given by $\Delta H = \Sigma H_{products} - \Sigma H_{reactants}$. The constant-presure free energy (energy derivable from the reaction at 100% efficiency) is given by $\Delta G = [\Sigma G_{products} - \Sigma G_{reactants}]$. A useful relationship is that $G = H - TS$, where T is the absolute temperature. The properly defined equilibrium constant K for the reaction is given by $K = \exp(-\Delta G/RT)$ where R is the universal gas constant. Thus, all of the information about the heat, free energy, and equilibrium constant of a reaction is given by a knowledge of any two of the three quantities G, H, and S for the materials involved at temperature T.

At the present development stage of theoretical chemistry, precise and accurate thermodynamic functions can be calculated only for ideal gases. For all other materials, experimental measurements are needed to determine these functions, and the experiments will usually be calorimetric. Even for gases, the absolute values of H cannot be calculated, so experiments must determine appropriate values of ΔH.

---

* Relatively little calorimetric work has been done for chemical purposes where other system variables are used intentionally to determine changes of thermodynamic properties. Results from the many experiments with materials sealed in containers at constant volume are usually converted to constant-pressure values for usage. The experiments on heat capacities with varying magnetic fields are of greater significance in the field of physics.

The variations of H and S with temperature can be determined by measurements of C_p over the pertinent temperature ranges. If the measurements are extended down to near the absolute zero of temperature, absolute values of S (but not of H) can be derived from them. Our knowledge of the thermodynamic functions can then be completed by measurement of the appropriate ΔH or ΔG at one temperature, and this can be a calorimetric measurement of ΔH.

Modern tabulations of thermochemical data usually give the enthalpy of formation, ΔH_f°, and free energy of formation, ΔG_f°, of each compound from the elements, with the compounds and the elements in their standard states at a standard temperature (formerly 18°C, now 25°C), as well as the standard entropy of the compound at that temperature.* The most complete tables give also the entropy, enthalpy, and free energy as functions of temperature, generally in the form of S_T^0, $(H_T^0 - H_0^0)/T$ and $(G_T^0 - H_0^0)/T$. The application of these tables to chemical calculations has been discussed previously (156,169) with special reference to the convenience of the free energy function $(G_T^0 - H_T^0)/T$. For analytical purposes, it is usually H that is required, since the reactions chosen generally go to completion, and the extent of reaction is usually determined through measurement of the heat produced.

Another use of calorimetric data is to reveal relationships among the thermodynamic properties of different compounds, permitting reliable estimation of unknown values and aiding theoretical studies. Some of the compilations listed here contain estimated values.

III. CALORIMETRIC MEASUREMENTS AND TECHNIQUES

Specific devices will be discussed later in greater detail, but the basic ideas for quantitative measurements were developed by experimentalists of the 19th and early 20th centuries. Work of a precision and accuracy nearly equal to that achievable today has been in progress since about 1930. Modern improvements have been primarily in extension of the technique to new compounds and processes, in miniaturization of apparatus to study compounds available only in small quantities, and in improving the speed and

---

* The first self-consistent table of chemical thermodynamic properties appeared in the International Critical Tables. It was brought up to date a few years later by Bichowsky and Rossini (19) and again with Circular 500 (238), which covered the data through 1948, except for organic compounds with more than two carbon atoms per molecule. This work is continuing with the publication of the TN-270 series (292). Cox and Pilcher provided tables for organic and metal organic compounds (56). A committee of the ICSU (126) is selecting data for certain important materials. The foregoing are primarily devoted to information at the standard temperatures. Compilations which also give functions at various temperatures include the work of Stull and Sinke (269) on the elements, Stull, Westrum, and Sinke (270) on organic compounds, the JANAF tables (128), and the API Project 44 Tables (322). There are quite a number of other compilations and literature and data-bank-searching services, particularly in specialized fields.

convenience of the measurements through automation and electronic computation. Completely new calorimeters are seldom reported (278), although ingenious and creative variations are often described.

The standard of energy for modern calorimetry is the absolute joule, with all measurements being ultimately traceable to a comparison of the measured heat with the heat produced by dissipation of electrical power in a resistance. Whenever a calorimeter does not have provision for a direct electrical calibration, it is calibrated by the heat of a process that has been compared previously with electrical heat.

The use of the heat of combustion of benzoic acid in oxygen, with material certified by a national standardizing laboratory, and the heat of reaction of hydrogen and oxygen in a flame have been generally accepted as satisfactory for the calibration of calorimeters for the most precise and accurate work. No other standards have yet achieved such good acceptance. Some of the standard materials are recommended only for testing calorimeters, to insure that the instruments do not give results very far from those observed with the best equipment and techniques. The usual reasons for not using them as actual calibration standards are that existing calorimeters already have electrical calibration apparatus, and that the agreement among different laboratories on measurements of the same materials are not quite as good as expected, considering the potential accuracies of the measurements. Information on the current status of new standards and test substances is given in the *Bulletin of Thermodynamics and Thermochemistry* (77). The U.S. Calorimetry Conference is very active in the field of test and standard substances (85). A number of these substances, accompanied by explanatory certificates, are available from the U.S. National Bureau of Standards (290). Standards approved by the IUPAC are described in their journal (125).

The prospective user of calorimetry has three choices: (1) commercially available calorimeters can be purchased and used with or without modification; (2) the literature can be reviewed for a suitable design, which can be constructed, with or without modification, being careful not to infringe on patent or other rights held by other parties; (3) a new design can be attempted. The first of the alternatives is obviously the safest from the standpoint of avoiding unexpected difficulties that can be expensive and time-consuming. New designs should be used only when the available equipment is not adequate for the experiments to be performed. Electrical calibration equipment is very desirable, except where calibration with the heat of combustion of benzoic acid or a hydrogen-oxygen flame is possible, although accurate calibration of water-solution calorimeters with a standard material may soon be proved (188,189) (see Addendum). Wherever possible, a standard or test reaction should be used to test the calorimeter.

The design of calorimeters is a special field involving considerations of structural design; thermal insulation; selection of materials for strength, heat conductivity, and chemical resistance, often over a wide range of temperatures; temperature measurement and control technology; electronics, often involving measurements of very low voltages and of very high precision and accuracy; and others. Vacuum technology and provision for adequate and reproducible stirring are often involved, as is proper provision for automation and electronic computation. It is not uncommon for the first design to fail to operate properly, and unrecognized systematic errors may occur.

The prospective designer of a calorimeter should consult the classical works on the subject (37,143,236,240,271,281,308,314) as well as some of the more modern reviews (176,254,291,316). These references will also be helpful to users of existing calorimeters. He should then make a careful study of the literature on the specific type of calorimeter to be designed. He should become familiar with related subjects such as temperature measurement. Finally, he should consider the following new developments: (a) the availability of new materials, such as high-temperature, corrosion-resistant alloys; (b) the commercial availability of a wide variety of electronic and electromechanical devices which can be used as "building blocks" in the system (these include, for example, digital voltmeters of high accuracy); (c) the commercial availability of precise thermometers, some capable of directly producing a digital output readily adapted to automatic or semiautomatic systems; (d) the commercial availability of small, relatively inexpensive computers for automatic control and computations; and (e) any other developments that might be applied to the particular type of calorimetry proposed. The problems involved are greatly reduced if the highest accuracy is not necessary, and especially if the conditions of temperature, pressure, etc., are not extreme. Attention will be given below to very simple apparatus for approximate results.

A. HEATS OF REACTION

1. Combustion Calorimetry

The development of combustion calorimetry has been spurred primarily by the need to measure the heats of combustion of fuels and by the need to acquire precise and accurate values of the heats of formation of compounds. Since the measurements can be made with very high precision, the requirement for taking differences between heats of combustion to obtain heats of formation does not cause serious errors. Combustion calorimetry is the preferred method for determining the heats of formation of many compounds. An ASTM specification (5) for combustion calorimetry calls for about 0.1% precision, and precisions of 0.01% or better are not uncommon in scientific work. The field was reviewed by Huber and Holley (120).

a. OXYGEN BOMB CALORIMETRY

The oxygen bomb combustion method has been used to determine the heats of formation of many substances and is in routine use for determining the heats of combustion of liquid and solid fuels. A variety of commercial oxygen bomb calorimeters are available.

The usual calorimeter of this type has a heavy-walled reaction chamber (bomb), filled with oxygen gas under pressure and containing the material to be burned. The bomb is surrounded by a stirred liquid (or a block of solid material in the aneroid type) to prevent excessively high temperatures from being maintained in the bomb over long periods and to keep the temperature in a range where accurate corrections are possible. The stirred liquid (or block) is in turn surrounded by a shield maintained at a controlled temperature, which may be constant or adjusted to be equal to the temperature of the calorimeter proper (bomb plus stirred liquid, or block, surrounding it). The rise in temperature of the calorimeter proper, with appropriate corrections, is proportional to the heat generated in the combustion reaction, which is initiated by electrical ignition. The rise in temperature produced by the combustion of benzoic acid, certified by a national standardizing laboratory, is usually used for calibration. Combustion bomb calorimetry was originated by Berthèlot in 1869.

Many ingenious variations of technique have been applied in calorimetric bomb combustion measurements to guarantee complete combustion of the material to a reproducible final state. Some of these are described in the reviews and general references on the technique listed below. It was discovered early that the addition of a small amount of water to the bomb before the combustion of organic materials improved reproducibility and simplified the Washburn correction (236,308) to the standard states. Lack of reproducibility in measurements on materials such as organic sulfur and halogen compounds was caused by variations in the concentration of the final solution present in different parts of the bomb. Moving-bomb techniques introduced by Popoff and Schirokich (220) and developed at institutions including the University of Lund (257,276), the U.S. Bureau of Mines (119), and the University of Wisconsin (197) eliminated this problem by stirring the final solution. Calorimeters with unusual means of providing the stirring motion include a spherical bomb which is rotated by unwinding a cord wrapped around it (209) and a bomb which vibrates (141).

For detailed information on oxygen bomb calorimetry, the reader is referred to general descriptions and instructions for the measurements (5,191, 236,291), to reviews of the subject (108,114,120,226,291), and to the original literature.

A typical oxygen combustion bomb is shown in Fig. 5.1. Like most bombs in use today, it still resembles very closely the bomb described by Berthelot (18) in 1885. It consists of a heavy-walled body, with an internal volume about 350 ml, onto which a cap can be screwed. The lid can be pressed firmly against the body of the bomb by a number of set screws in the cap; a rubber or Teflon gasket gives a tight seal. Two valves are fitted into the lid so that the bomb can be flushed and filled with oxygen (usually to about 30 atm pressure) and the products of the combustion swept out and analyzed. The bomb is either constructed from corrosion-resistant alloys or lined with platinum and has platinum internal fittings.

The sample [a solid pressed into a pellet or a liquid in a thin-walled glass ampul (228,282), or a bag made from polyester film (159,236)] is placed in a platinum crucible. It is ignited through a fuse wire connected between the insulated electrode and the sample support rod; the wire, usually platinum, is either fused or momentarily heated by an electrical current, igniting a cotton thread which in turn ignites the sample.* A small platinum cup, not shown, contains 1 ml. of water to ensure saturation of the oxygen.

The constant-temperature-environment calorimeter assembly designed by Dickenson (59) has been widely used. The bomb is immersed in a metal can containing about 3000 ml of water, which can be stirred. This is completely surrounded by an isothermal jacket, the temperature of which is kept constant to within a few thousandths of a degree. The change in the temperature of the can on combustion of the sample (usually between one and two degrees) is measured to within at least 0.001°C. The heat exchange between can and jacket is quite large, but, provided that heat transfer by convection is kept small and the thermal head does not exceed 3°C, it follows Newton's law and can be accurately calculated from the temperature drift of the calorimeter before and after the combustion (236).

The many factors that must be considered in obtaining precise and accurate results have been discussed in the literature cited above. The purity of the sample is very important. Since one can measure the heat produced to 0.01%, the errors produced by impurities should be insignificant compared with 0.01%. When impurities cannot be removed, corrections should be applied for them. Fortunately, the presence of small amounts of isomeric materials in organic compounds seldom produces a significant error. The products of combustion should preferably be analyzed quantitatively to determine whether the combustion was complete and to determine the amount of reaction which has taken place. Because of the time and experimental difficulty involved in the analysis of the combustion products, this analysis is seldom carried out, except for the most precise work. The products of combustion are instead (for organic compounds) tested for carbon monoxide

---

* Platinum, iron, and Chromel are the fuses recognized by the U.S. National Bureau of Standards (for the calibration of calorimeters by the heat of combustion of standard benzoic acid). Fusion of the wire can ignite a sample, even if enclosed in a glass ampule. When a platinum fuse is merely heated enough to ignite a cotton thread, an accessory material with a known heat of combustion will usually be necessary to generate enough heat to ignite a material sealed inside glass. Filter paper and hydrocarbon oil have been used as accessory materials. The oil may also be helpful in lowering the overall rate of combustion of those materials that tend to burn too rapidly, causing incomplete reaction. With proper positioning of the platinum fuse, the cotton thread technique permits use of the same platinum wire for many combustion experiments. The Bureau of Standards certificate does not recognize the cotton thread technique for calibration purposes, presumably because of the effect of the additional combustible material on the energy of combustion of the benzoic acid. It would appear, however, that the error from this effect, with only a few milligrams of cotton thread, can be made negligible even in work of fairly high precision, and experimenters at Rice University have sometimes used cotton fuses to ignite the benzoic acid.

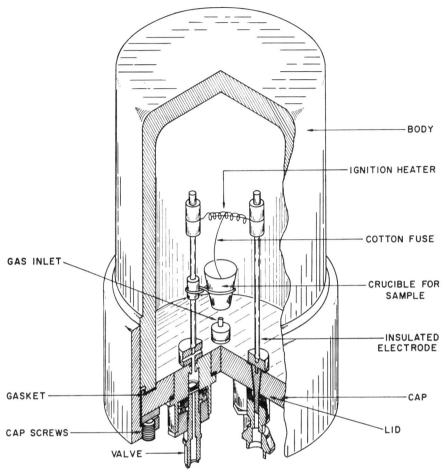

Fig. 5.1. Typical combusion bomb (Parr Instrument Co., Type 1001).

and examined for soot. These tests are simple, rapid, and sensitive. The amount of reaction is then determined from the mass of sample burned, and this puts even greater reliance on its purity (e.g., water must be absent).

Appropriate corrections must be applied for the exchange of heat between calorimeter and surroundings, including the heat of stirring, evaporation of water, etc. Corrections must be applied for the energy of ignition and for any cotton thread and accessory materials used. Then the Washburn correction (236,308) is applied to convert the energy to the standard energy, with all reactants and products in their standard states. Without this correction, the energy depends upon the size of the bomb, the amount of sample burned, the oxygen pressure, the amount of water present, etc., and the results are not comparable with results from other techniques. Finally, the result is usually converted to constant pressure conditions. Lack of understanding of all these factors in the first 50 years or so of combustion calorimetry has caused the older work to be of limited utility.

The Washburn correction is generally small compared with the total energy of combustion, and is often neglected in work for engineering and industrial purposes. Since the correction is not small compared with the differences between heats of combustion, which are used for calculating energies of formation and reaction, such uncorrected work is not usually useful in science. Electronic computing has greatly reduced the labor involved in computing corrections (108), and it is now practicable to apply them wherever the necessary information is available.

It is recommended that engineering and industrial work, where possible, be carried out with attention to the possibilities it may have for other uses, and that all the information needed for the corrections be presented in published reports. This would require very little increase in the time and publication space needed, and could help prevent unnecessary duplication of this work. This would be especially useful where the equipment employed is inherently capable of high precision (such as equipment originally constructed for scientific use) and when unusual compounds are studied. The enhanced usefulness of the results would eventually be reflected in improvements in technology.

In engineering work aimed at such problems as the determination of heats of hydrogenation from heats of combustion, all the corrections are significant, and should be applied. In these cases, the desired quantity is determined from differences between heats of combustion.

A rotating-bomb oxygen combustion calorimeter is shown in Fig. 5.2. and described by Hubbard, Katz, and Waddington (119). The bomb, of conventional design, sits in a ball-race which is connected by a miter gear to a drive wire. Rotation of the drive shaft by a motor outside the isothermal jacket gives the bomb an end-over-end motion. At the same time, it walks around a gear fitted circumferentially that turns it about its longitudinal axis. During the initial part of the combustion, the bomb is kept in the inverted position, where the bomb solution protects the valves. When the rotation is started, electrodes (not shown) connecting the firing circuit spring out of the way, and the crucible, held in a special gimbal, falls into the bomb solution and is washed. A solenoid is used to stop the rotation of the bomb at a predetermined point.

There has been no agreement on a standardizing compound for bomb calorimetry other than benzoic acid.* When the compound studied contains

---

* However, there has been extensive work on the development of such standards since the first edition of this treatise. See, for instance, the activities of the Calorimetry Conference listed in the Bulletin of Thermodynamics and Thermochemistry (77), the publications of national standardizing laboratories (2,3,7,131,163,260), and others (4,61,118,140). It must be emphasized that only one compound should be the ultimate standard. The difficulties caused by dual standards in electrical units have been described (237) and serve as an indication of the potential problems.

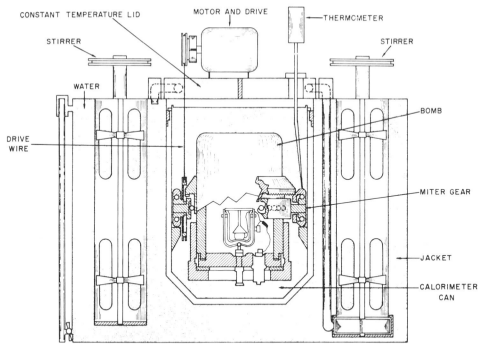

Fig. 5.2. Rotating bomb calorimeter (W. N. Hubbard, C. Katz, and G. Waddington, *J. Phys. Chem.*, **58**, 112 (1954); copyright (1967) by the American Chemical Society. Reprinted by permission of the copyright owner).

sulfur, the conditions and products of the calibration and combustion experiments will be different. Consequently, systematic errors will no longer tend to cancel out. In addition, if the heat of formation is required, the thermodynamic state of the products, which is a complex mixture of gases and solution, must be known. This usually cannot be calculated accurately and is established by doing a third combustion with a mixture of orthorhombic sulfur and hydrocarbon oil chosen so that the products are identical to those from the organosulfur compound. The two combustions can be written in a simplified form as

$$C_aH_bO_cS_d + \left\{a + \frac{b}{4} - \frac{c}{2} + \frac{3d}{2}\right\} O_2(g) + \left\{nd + d - \frac{b}{2}\right\} H_2O \text{ (liq)} =$$

$$aCO_2(g) + d(H_2SO_4 \cdot nH_2O) \quad (1)$$

and

$$aCH_2 \text{ (liq)} + dS \text{ (Rh)} + \left\{\frac{3a}{2} + \frac{3d}{2}\right\} O_2(g) + (nd + d - a) H_2O(\text{liq}) =$$

$$aCO_2(g) + d(H_2SO_4 \cdot nH_2O) \quad (2)$$

The difference between the heats of reactions 1 and 2 is the heat of the reaction

$$C_aH_bO_cS_d + \left\{\frac{b}{4} - \frac{a}{2} - \frac{c}{2}\right\} O_2(g) = dS\,(Rh) + aCH_2(liq) + \left\{\frac{b}{2} - a\right\}$$

$$H_2O(liq) \quad (3)$$

If the heat of formation of the hydrocarbon oil (expressed as CH_2) is known, the heat of formation of the sulfur compound can be calculated from reaction 3.* A detailed account of the procedure has been given by Waddington, Sunner, and Hubbard (236).

The good agreement, to within 0.03%, between results obtained in different laboratories using slightly different procedures, for compounds such as 1-pentanethiol and thianthrene, has shown that the method is of high precision; there is also good agreement with values obtained by equilibrium and solution techniques (277). This method has been successfully used to determine the heats of combustion of compounds containing many other elements.

b. Fluorine Bomb Calorimetry

Many inorganic compounds, such as the metallic hydrides, borides and silicides, either do not burn completely in oxygen (at least not without specially developed techniques using auxiliary materials), or give combustion products that are slightly different in each combustion experiment. The latter situation makes it impossible to define exactly what the reaction was, unless separate experiments by another technique are performed with the end products, to determine their heat of conversion to a well-defined final product. Even if the combustion products are the same in every experiment, they may not be compounds for which the heats of formation are known, so separate experiments may still be required.

Fluorine combustion bomb calorimetry was developed for determining the heats of formation of many compounds that either give very poor results or cannot be burned in oxygen combustion calorimeters. The precision of measurement of fluorine combustion reactions is often 0.01% or better, indi-

---

* In calorimetry, it is a good idea to write the reactions in a form such as that of equations 1, 2, and 3, showing quantitatively all materials participating in the reaction (including solvents, if present) and all pertinent conditions that are not constant. This helps to avoid errors in adding or subtracting the heats of reaction to determine the heat of another reaction. For instance, if the amount of water in the bomb in the second experiment were not exactly as given by equation 2, so that the final solution became $d(H_2SO_4, mH_2O)$, one could immediately see the need for adding a correction (the heat of reaction of one final solution with water to give the other final solution). Accurate values of the heats of dilution of sulfuric acid, and many other quantities needed for corrections in calorimetry, can be found in tabulations of thermochemical data.

Reports need not necessarily use equations in this form, but when space permits their inclusion, they facilitate the reader's understanding.

cating that most problems involved in measuring the heat of the reactions have been solved.

With materials which burn to well-defined products, the overall uncertainty is small. For instance, the heat of combustion of silicon in oxygen (253) was used to derive -209.33 ± 0.25 kcal $\cdot$ mol^{-1} as the standard enthalpy (heat at constant pressure) of formation of quartz.* An improved oxygen combustion technique used by Good (87) involved mixing vinylidene fluoride polymer with silicon, and using a rotating bomb with aqueous hydrofluoric acid in it. The result, combined with the heat of solution of quartz in hydrofluoric acid, gave the standard enthalpy of formation as -217.5 ± 0.5 kcal $\cdot$ mol^{-1}, showing that the ordinary oxygen combustion technique gave incorrect results. Fluorine combustion calorimetry of silicon and quartz by Wise and coworkers (318) gave -217.75 ± 0.34 kcal $\cdot$ mol^{-1}, in excellent agreement with the specially developed oxygen bomb technique. Benzoic acid again is the usual calibration material, the bomb being filled with oxygen for the calibration.

It is critical in fluorine combustion calorimetry to determine the state and composition of the final products. In those cases where these are not well-defined pure compounds, the overall uncertainty in the heat of combustion may be 0.1% (ten times the precision uncertainty). The products can be the subject of other experiments, as was mentioned above, or they can be analyzed and the final state defined from the results of the analysis.

A simple constant volume reaction cell made of Pyrex glass and immersed in a conventional isothermal calorimeter, and similar to that used previously to determine the heats of chlorination of metals has been used by Gross and co-workers (95). It consisted of two bulbs separated by an internal break-off seal; one bulb contained fluorine at several atmospheres pressure, the other the sample under vacuum. When the seal was broken, using a magnetic plunger, the fluorine filled the reaction vessel and reacted with the sample. This method is very convenient for compounds that react spontaneously with fluorine but cannot be used with high pressures of fluorine because of the risk of breakage of the glass apparatus and a consequent explosive reaction of fluorine with the calorimetric fluid. It has been used to determine the heats of formation of sulfur hexafluorides, titanium tetrafluoride (96), and boron trifluoride (97), by burning the elements in excess fluorine.

The apparatus developed by Hubbard and the calorimetry group at the Argonne National Laboratory is similar to that for oxygen bomb calorimetry but built with fluorine-resistant materials. The bomb is of nickel or monel, with Teflon gaskets in the valves and the main cap seal (see Fig. 5.1) of aluminum, gold, or Teflon. The apparatus used to purify the fluorine (to 99%

---

* The calorie used in this chapter is the thermochemical calorie, defined as exactly 4.184 absolute joules, and used in most scientific calorimetric work. The "steam table calorie" is slightly different in size. New data in the calorimetric literature are being reported in kilojoules mole^{-1} and new tabulations are also being converted to this unit.

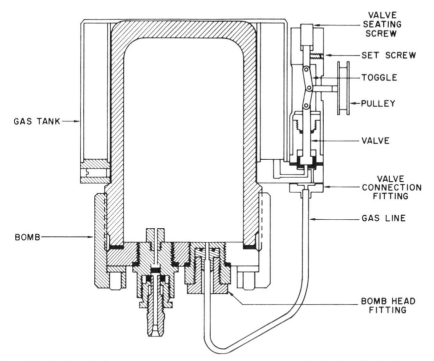

Fig. 5.3. Combustion bomb for spontaneously reactive materials (R. L. Nuttall, S. Wise, and W. N. Hubbard, *Rev. Sci. Instr.*, **32**, 1402 (1961)).

purity) by fractional distillation at liquid nitrogen temperatures and to charge and discharge the bomb have been described (99,119). One of the most formidable problems has been to obtain complete combustion of the sample without reaction of the sample support. Ductile metals were used as sheets, suspended by a wire from a nickel frame. Powders were supported on dishes made from a stable fluoride (calcium fluoride or the fluoride formed in the combustion), or, in some cases, nickel.

Compounds that did not react spontaneously with fluorine were mounted in the bomb, which was filled with fluorine at 3–25 atm pressure, and the sample ignited at the appropriate time by passing an electric current through a fuse wire. Compounds that react spontaneously have been studied with a two-chambered device (201), which could be used in a rotating bomb calorimeter. A nickel tank was fitted concentrically around the bomb in place of the usual rotating gear mechanism (see Fig. 5.3) and was connected by a tube to an inlet valve of the bomb. The bomb containing the sample was evacuated, the tank filled with fluorine, and the apparatus placed in the calorimeter. The fluorine could be introduced into the bomb when required by opening a valve in the fluorine tank with the bomb rotating drive fitted to the calorimeter. This method was used to determine the heats of formation of silicon tetrafluoride and silica (318). A two-chambered bomb that could be adapted for use with fluorine has been described by Ivanov and Tumbakov (127). The two compartments were separated by a metallic foil which could be broken. Kybett and Margrave (148) have developed a two-compartment system in which reactive samples are isolated from the fluorine by a titanium diaphragm and have applied this apparatus for studies of the fluorination of TiO_2 (rutile, anatase, and brookite) and of boron hydrides. Two-compartment systems were also described by Settle, Greenberg, and Hubbard (249) and by Klyuev et al. (139), Bisbee et al. (21), Bosquet et al. (28), and Schroder and Sieben (246). Fluorine combustion calorimetry has been reviewed by Barberi et al. (13) and by Gross et al. (98).

c. Observation Bombs

A great deal of useful information about prereaction of the sample and unstable combustion processes can be readily obtained by visual examination of trial combustions with the same conditions as the calorimetric determinations. The glass combustion bomb shown in Fig. 5.4 is made (200) from a 3-in. diam. Pyrex "pipe spacer" with 0.5-in. thick walls clamped between stainless steel end pieces. The cap, which is equipped with two valve ports and an insulated electrode, can be screwed down against an O-ring seal. A version with metal parts made of nickel and with Teflon seals has also been used with fluorine. Other bombs with windows have been designed by Tiehl and Roth (284), Pavore and Holley (205), Peters et al. (210,211), Gurevich et al. (104), and Armstrong (8).

d. Safety Practices for Oxygen and Fluorine Bomb Experiments

An essential precaution in all bomb calorimetry, for the safety of persons in the laboratory, is a test for the absence of leaks. In the usual oxygen bomb system with a water can, the bomb is observed carefully for the formation of bubbles after it is placed in the water and before the lid is placed on the can. Because of the high pressures temporarily reached in the bomb, a jet of flame could erupt from the bomb at a leak. The Parr instrument company recommends that, when Teflon gaskets are used, the bomb lid should be fully tightened twice; once initially, and once more after waiting 10 min for any creep of the Teflon under pressure to have taken place. Special safety procedures are required when fluorine is used.

e. Flame Calorimetry

Flame calorimetry has changed very little in recent years; in fact, the precision and accuracy obtained by Rossini in 1931 have not been surpassed. Flame calorimeters are used to determine the heats of combustion of substances that are gaseous or have high vapor pressures at room temperature. The reaction vessel of the highly accurate apparatus developed by Rossini (233) to determine the heat of combustion of hydrogen in oxygen to form water is shown in Fig. 5.5. It was made from Pyrex glass, except for the silica burner tube. The inlet tubes led the gases, at controlled rates, into the

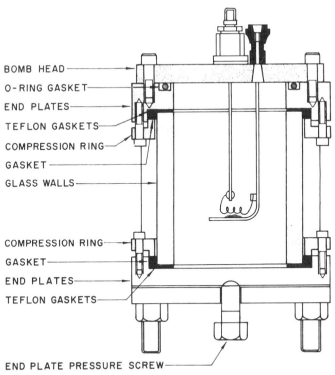

BOMB HEAD
O-RING GASKET
END PLATES
TEFLON GASKETS
COMPRESSION RING
GASKET
GLASS WALLS

COMPRESSION RING
GASKET
END PLATES
TEFLON GASKETS

END PLATE PRESSURE SCREW

Fig. 5.4. Glass combustion bomb (R. L. Nuttall, M. A. Frisch, and W. N. Hubbard, *Rev. Sci. Instr.*, **31**, 461 (1960).

reaction chamber, where they were burned. The leads for the spark circuit used to ignite the flame were also passed through these tubes. Most of the water formed collected in the condensing chamber. The excess gas, plus the small amount of water vapor which saturated it, was led out through the cooling of heat exchanger tubes and analyzed. The reaction vessel was completely immersed in a calorimeter containing water and surrounded by a constant-temperature jacket. Both calorimeter and jacket were similar in design to those used in bomb calorimetry.

At the start of an experiment, the reaction vessel was flushed out with the excess gas, oxygen or hydrogen. At a given moment the other gas, previously led to exhaust, was switched into the burner tube, and the spark circuit closed for a fixed length of time. The flame, about 5 mm long, burned quietly. After a sufficient amount of combustion (2–4° rise in temperature) had taken place the gases were again led to the exhaust. The rise in temperature of the calorimeter was determined as in bomb calorimetry while the amount of reaction was determined from the mass of water formed.

A slightly different reaction vessel was used to determine the heats of combustion of some hydrocarbons (234). A cleaner combustion was ob-

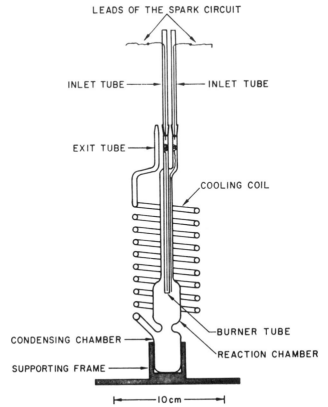

Fig. 5.5. Flame calorimeter (F. D. Rossini, "Experimental Thermochemistry" Interscience Publishers, Inc., New York, 1956, Vol. 1, page 61).

tained by injecting some oxygen into the hydrocarbon gas just before it reached the burner. In later designs (217,225) the proportion of oxygen could be more easily varied. The proportion of oxygen is quite critical since, with low concentrations the flame is sooty and unstable, whereas with high concentrations there is precombustion and carbon is deposited inside the burner. In the calorimeter used by Pilcher and co-workers (217), the organic vapor was carried into the reaction vessel in a stream of argon.

Flame calorimeters can be calibrated electrically (233), but it is preferable to use a standard reaction.* The Commission of Thermochemistry (IUPAC)

---

* A striking advantage of flame calorimetry is that the temperature–time relationship during electrical calibration can be made nearly identical to that during a measurement of a flame reaction. This is illustrated by Fig. 4 in Chapter 4 of reference 236. Thus, the usual heat-exchange correction required in calorimetry is reduced to an extremely small correction for the difference in heat exchange between two nearly identical situations. In most calorimetry, the temperature rise as a function of time is approximately linear during electrical calibration, while it is approximately exponential during the measurements of reactions. Therefore calibration with a standard reaction is preferred for flame calorimeters because it avoids any possible systematic differences between laboratories in the accurate comparison of electrical energy with the basic standards.

in reports issued in 1934 and 1936 (123,124) accepted the reaction of hydrogen and oxygen to form water as the standard for reactions in a flame at constant pressure, using the data reported by Rossini (233). The heat due to the ignition spark is determined by a blank experiment in which the flame is extinguished after it has started burning evenly (about 45–60 sec), so that the heat of combustion is in effect determined from that portion of the combustion when the flame is in a steady state. If the entering gases are not at the same temperature as the calorimeter, a correction can be calculated from the volumes and heat capacities of the gases. There is also a correction to allow for the fact that some water leaves the calorimeter as vapor. It has been calculated that the error due to incomplete transfer of heat from the gases to the cooling coils is less than 1 in 60,000. In flame calorimetry, the heat of combustion in the gaseous state at a constant pressure of 1 atm is measured directly and no Washburn corrections, which can be a source of error, have to be calculated. The combustion is not isothermal, as the products leave the calorimeter at a steadily increasing temperature, but to a good approximation the result can be identified with the heat of the isothermal reaction occurring at the mean temperature of the experiment.

The need to obtain accurate values for the heats of formation of hydrogen fluoride and carbon tetrafluoride, which are fundamental to the thermochemistry of compounds containing fluorine, has led to the development of flame calorimeters using fluorine as the oxidant. Early workers used small concentrations of fluorine (239,295,296). This limited corrosion of the apparatus but increased the relative importance of side reactions. Improvements in techniques and materials suitable for handling fluorine have made it possible to construct flame calorimeters which can be used to determine the heats of combustion of gases in fluorine to within 0.3%.

The combustion chamber used by Jessup et al. (130) to determine the heat of formation of carbon tetrafluoride is shown in Fig. 5.6. It was constructed of copper, with Teflon gaskets. The reactants ignited spontaneously so they could not be premixed. The fluorine was brought through B to an annular opening in the base of the chamber. Methane entered at A and burned at the burner tip F. The products and excess fluorine passed through a cooling tube. To prevent deposition of carbon on the burner, the flame was lifted slightly by passing helium through C to an annular opening surrounding F. The amount of reaction was determined from the mass of sample and by analyzing the product hydrogen fluoride. The rest of the apparatus, which has been described by Armstrong (6), was similar to that used in oxygen flame calorimetry. This apparatus has also been used to determine the heat of formation of hydrogen fluoride from the combustion of ammonia in fluorine (9).

2. Solution Calorimetry

Most solution calorimeters are relatively simple in design and operation. Thermal and chemical equilibrium are generally achieved through stirring,

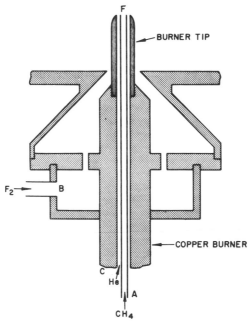

Fig. 5.6. Burner of flow colorimeter for $F_2 + CH$ reaction (R. S. Jessup, R. E. McCoskey and R. A. Nelson, *J. Am. Chem. Soc.*, **77**, 244 (1953) copyright (1967) by the American Chemical Society, reprinted by permission of the copyright owner).

and the proper design of the stirrer for effective stirring with a constant heat produced per unit time is one of the important design considerations.

Formerly, every experimenter who wished to perform experiments in solution calorimetry had to construct his own calorimeter, and so there were a great number of different designs. As was mentioned in the previous edition of this chapter, it was very difficult to decide on the reliability of the measurements, because of the many different design features, the impracticability of presenting every detail of the design in the published material, and the absence of an accepted test reaction for solution calorimeters. These problems should be less severe in the future, since solution calorimeters are now commercially available, and three test reactions have also been approved.

The history of the three approved test reactions is given in a recent report (188). One is the reaction of tris-(hydroxymethyl)-aminomethane ("tris") with hydrochloric acid solutions, an exothermic reaction. Another is the reaction of "tris" with sodium hydroxide solutions and the third is the reaction of potassium chloride with water. The latter two reactions are endothermic. The reactions involving "tris" have been certified by the U.S. National Bureau of Standards (223), which supplies certified samples of "tris." The reaction involving potassium chloride is accepted by the IUPAC

(113). The U.S. Bureau of Standards work with "tris" plus sodium hydroxide solution showed a poorer precision than with "tris" plus hydrochloric acid. Recent experiments also led to the suggestion that the procedure for using potassium chloride as a standard material should be modified (188).

None of the test reactions, however, has been accepted as a method of calibrating calorimeters, and electrical calibration is still the method of choice. The test reactions are recommended only for verifying that the results from a solution calorimeter are not too different from those obtained with other calorimeters and techniques generally considered to be accurate. It is hoped (188) that calibration with a standard reaction for water solution calorimeters will be practicable within a few years. The ratios of the heats of solution of a few salts to those of "tris" in HCl and KCl in H_2O have been measured with the expectation that these will be usable (189). Even though the test reactions have not been accepted for calibration, they already have important applications in checking new calorimeters for systematic errors, and thus improving the reliability of solution calorimetry (see Addendum).

The accuracy obtainable with modern solution calorimetric techniques is believed to be 0.1% or better. The precision is several times better than this in many cases. The agreement between different laboratories with specially selected reactions and very careful work can be very close but in many cases it is much poorer (224). This can usually be ascribed to the measurement of the amount and type of reactants (errors being due to impurities, side reactions, etc.) rather than the measurement of the heats. If all of the problems with purity, definition of the amount of reaction, side reactions, etc., could be eliminated, it is likely that the most accurate solution calorimeters would give results at least two or three times better than the nominal 0.1%.

The quantities measured in solution calorimetry are generally much smaller than those measured in combustion calorimetry. Thus, the absolute uncertainty (in $kcal \cdot mol^{-1}$ or $cal \cdot gm^{-1}$) in the heat of a process as determined from solution calorimetry with an uncertainty of 0.1% may be an order of magnitude less than the absolute uncertainty of combustion calorimetry with an uncertainty of 0.01%. For this reason, results obtained with a very simple solution apparatus may be useful, even if the uncertainty of measurement in percent is rather large.

One very important point in solution calorimetry is that the experiments can involve either exothermic or endothermic reactions, while the calibration is nearly always exothermic. (This problem does not occur with bomb or flame calorimetry.) Therefore the temperature-versus-time relationships in calibration and experiment may be quite different, and can result in substantial systematic errors. Establishment of accurate endothermic calibration methods would reduce this source of uncertainty (see Addendum).

Unknown systematic errors remain one of the greatest problems in solution calorimetry. Although the presently available test reactions can reduce the uncertainty from this problem, much work on test reactions remains to be done before the potential accuracy of solution calorimetry can be realized. In analytical work, the absolute accuracy may not be a problem where the precision is good.

For detailed information on solution calorimetry, the reader is referred to general descriptions and instructions for such measurements (271,291,314, 316) and to the original literature. The authors of this chapter believe that better results will be obtained for exothermic reactions if the final temperature of the calorimetric measurement is close to the temperature of the jacket, surrounding the calorimeter, when a constant-temperature jacket is used. This appears to be supported by recent theoretical work (51), but conflicts with some of the recommended procedures described in the older literature. Since solution calorimetry is still a developing field, one should critically study even the most recent work to obtain the highest accuracy.

Sunner and Wadso (279) critically investigated four solution calorimetric designs before arriving at their choice, as shown in Fig. 5.7. Two of these are similar to those used by a number of other investigators. Design *d* was the most satisfactory. It consisted of a thin-walled metal can suspended from an outer metal jacket by a thin-walled glass tube. The lid of the can was fitted with two tubes which held a thermistor and an electrical heater; an ampul-breaking pin was fastened to the base. A shaft mounted in Teflon-coated gaskets passed through the glass tube and held the combination stirrer and ampul holder. Heat transfer between the can and jacket was reduced by chromium plating all the reflecting surfaces and evacuating the space between them through a cone joint fitted to the jacket. The solute was contained in a thin-walled glass bulb fitted into a recess in the stirrer. The can was filled with solvent and the apparatus completely immersed in a water bath thermostatically regulated to within a few thousandths of a degree. The ampul was broken against the pin by pushing down on the stirrer shaft from outside the thermostat. This calorimeter had a very well-defined boundary with its surroundings, and reached thermal equilibrium in the short time of 2 min.

Calorimeter *a* was based on a glass Dewar vessel, with an additional glass shield. This was fitted with a long-necked ground-glass cap through which passed the leads for a thermistor, electrical heater, and the shafts for the stirrer and ampul holder. It was immersed in a water thermostat with only the upper part of the cap exposed. This design, in common with all designs based on glass Dewar vessels, had an ill-defined boundary with its surroundings and the long equilibration time of 60 min. It gave results reproducible to within 0.5%. This is adequate for many purposes, and because they are inexpensive and easy to construct, glass Dewar calorimeters have been extensively used. The extra glass shield in this design is not usually included. Simple designs using glass Dewar calorimeters published within the last dozen years or so are listed in references (192,229). A simple and inexpensive thermometer is described in reference 69.

A rather simple calorimeter that was very satisfactory for its purpose was used by Turner (288) to determine heats of hydrogenation in acetic acid

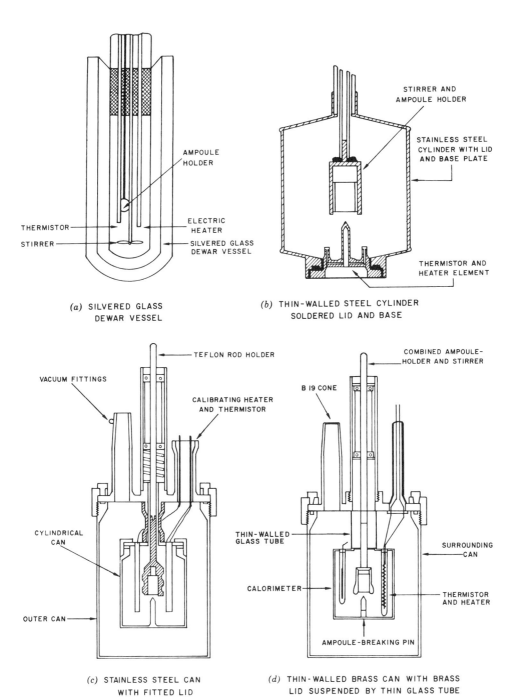

(a) SILVERED GLASS
DEWAR VESSEL

(b) THIN-WALLED STEEL CYLINDER
SOLDERED LID AND BASE

(c) STAINLESS STEEL CAN
WITH FITTED LID

(d) THIN-WALLED BRASS CAN WITH BRASS
LID SUSPENDED BY THIN GLASS TUBE

Fig. 5.7. Designs for vacuum-jacketed calometers (H. A. Skinner, J. M. Sturtevant and S. Sunner, "Experimental Thermochemistry," H. A. Skinner, Ed., Interscience Publishers, Inc., New York, 1962, Vol. 2).

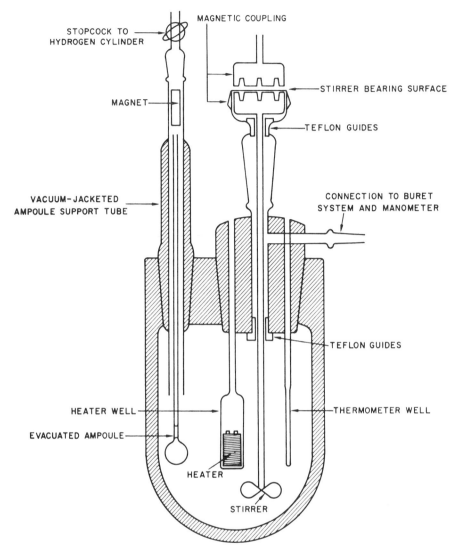

Fig. 5.8. Calorimeter for hydrogenation studies (R. B. Turner, W. B. Meador, and R. E. Winkler, *J. Am. Chem. Soc.*, **79**, 4116 (1957); copyright (1967) by the American Chemical Society; reprinted by permission of the copyright owner).

solution as shown in Fig. 5.8. It consisted of a Dewar of approximately 450-ml capacity with three ground glass joints at the top. Two of these held vacuum-jacketed glass tubes, one fitted with a stop-cock was used to flush the Dewar with hydrogen, the other (not shown) was stoppered at this point. The sample and the catalyst were in ampuls supported inside the tubes, and these could be broken against the side of the calorimeter by moving the magnets. The third joint was equipped with a stopper through which passed the leads for a thermistor, an electrical heater inside a glass well, and a stirrer that ran in Teflon bearings and was driven through a magnetic coupling. The calorimeter was immersed in a water thermostat to a point just above the hydrogen inlet tube.

In a typical experiment about 225 ml. of acetic acid was poured into the Dewar. The system was evacuated, flushed with hydrogen, and its contents heated to just above the thermostat temperature. The hydrogen pressure was kept constant and slightly (3 mm) above atmospheric pressure during the experiment which consisted of measuring the change in temperature when (*a*) the sample ampul was broken, allowing the sample to dissolve in the acetic acid; and (*b*) the catalyst ampul was broken and both sample and catalyst hydrogenated. Afterwards, the calorimeter was cooled to its initial temperature and the heat capacity of the calorimeter and its final contents determined electrically, duplicating as closely as possible the rate of change of temperature in the hydrogenation run. The heat of hydrogenation was the difference between the heat evolved during the hydrogenation of the sample and the catalyst, and that associated with the reduction of the catalyst. The latter was found in a separate experiment. The mass of sample hydrogenated was calculated from the volume of hydrogen absorbed, with an allowance for the hydrogen absorbed by the catalyst, the free space in the evacuated ampuls, and the partial pressure of acetic acid. The accuracy achieved was around 1%, which was adequate because heats of hydrogenation are relatively small (of the order of 25 kcal/mol of hydrogen). To obtain equal accuracy with combustion measurements (on a material with a heat of combustion near 1000 $kcal \cdot mol^{-1}$), the combustion measurements would have to be precise to within 0.02% (systematic errors that were the same in the combustion experiments on the original and hydrogenated materials would not affect the result). The combustion experiments have the advantage that the techniques necessary for the required precision are available and that the heats of hydrogenation need not be corrected for differences in the heats of solution in acetic acid between the original and hydrogenated materials.

Other hydrogenation calorimeters include those of Skinner et al. (70,71), the simple apparatus of Rogers et al. (230,232), and others (158,231).

A novel isothermal constant-flow-type of calorimeter has been used by Lacher and co-workers (149) to determine heats of halogenation and hydrohalogenation at elevated temperatures, 103–248°C. A Monel metal reaction chamber containing a catalyst was suspended from the metal lid of a Dewar vessel, which in turn was enclosed in a brass can and immersed in a "condensing vapor" thermostat. The Dewar was filled with a volatile liquid and also contained an electrical heater. The reactant gases, organic fluorides or chlorides and hydrogen, were passed at definite constant flow rates through heat exchangers in the thermostat and entered the catalyst chamber through concentric tubes. The product gases passed through a heat exchanger coiled around the catalyst chamber. The temperature difference between the Dewar and the thermostat was measured with a thermopile, and the heat of the reaction was balanced by the cooling produced by bubbling an inert gas such as nitrogen or hydrogen, brought to the thermostat temperature, through the liquid in the Dewar.

In an experiment, a high rate of flow of hydrogen through the volatile liquid in the Dewar was set and maintained constant. The cooling produced by this was counterbalanced by passing an electric current through the heater. When isothermal conditions had been produced, the flow of reactants was started and the electrical heat input reduced to compensate for the heat of reaction. The difference between the two energy readings gave the rate of heat evolution due to the reaction. The amount of reaction was determined by analyzing the products. The main advantage of the isothermal system was that, as the temperature did not change, catalysts of high adsorptive capacity, such as palladium on charcoal, could be used. Since large quantities of reactants were adsorbed, they had a long mean life in the reaction chamber and rather slow reactions could be brought to completion. The calorimeter was believed to give results reliable to within 1%.

A rocking-bomb calorimeter for measuring heats of solution has been described by Gunn (100). It is useful for vigorous reactions, especially when the products are gases, since the reaction takes place inside a sealed, pressure-tight vessel. Thermal equilibrium between gas and liquid phases is readily achieved, and the products can be recovered quantitatively. The body of the bomb, Fig. 5.9 was made of ⅛-in-thick copper, gold-plated inside and out. It was fitted with a resistance thermometer, an electrical heater, and a special hammer device. The bomb was held by nylon loops in a submarine mounted inside a thermostat. The submarine could be evacuated, and rocked through an arc of up to 240°. The usual procedure was to put the glass bulb containing the sample in place, assemble the bomb, and half fill it with water or solution. If necessary, the solution could be degassed by evacuating the bomb. The bomb was mounted in the submarine, which was closed and evacuated. The calorimeter was continually oscillated through an arc of 150°. This provided efficient stirring of the solution with a low heat of stirring. When the stirring rotation was increased to about 240°, the extra rotation released the hammer which swung around and smashed the sample bulb. Test runs indicated that the combined effect of the energy of the moving hammer and the fracture of the bulb was negligible and that the heat generated by the stirring action was the same whether the angle of oscillation was 150° or 240°. This is believed to be one of the most accurate solution calorimeters. In addition to its uses for special purposes where the type of construction makes it especially useful, it has been used to compare the heat of combustion of standard benzoic acid with the calorimeter's electrical calibration and thus with heats of solution. It has also been used for comparison purposes with the next calorimeter described below.

Prosen and Kilday (224) give an unusually detailed description of their solution calorimeter, shown in Figs. 5.10 and 5.11. Its precision is ± 0.01% under optimum conditions. It is of the adiabatic type (i.e., its jacket tempera-

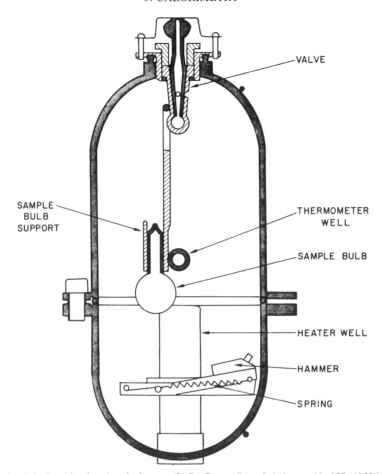

Fig. 5.9. Rocking bomb calorimeter (S. R. Gunn, *Rev. Sci. Instr.*, **29**, 377 (1958)).

ture is controlled to remain equal to that of the calorimeter proper), and is capable of higher precision than existing isoperibol (i.e., constant-jacket temperature) devices when used for the measurement of very slow reactions.

The holder (not shown) for the sample to be dissolved is a special device, made mostly of platinum, which contains a cylinder open at each end. The ends of the cylinder are covered by flat plates (which are part of a framework) and sealed by O-rings. To initiate the reaction, the cylinder is allowed to drop away from the flat plates. It comes to rest at the bottom of the framework, in a position such that the stirred liquid flows through the cylinder and between the plates.

For endothermic reactions, electrical energy is added to the calorimeter to maintain a rising temperature overall (223). This helps to maintain near-

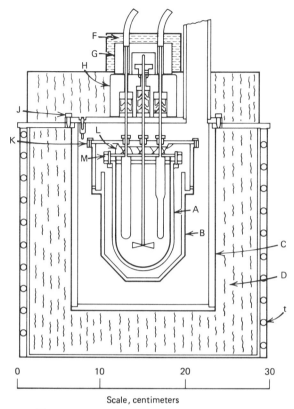

Fig. 5.10. Adiabatic solution calorimeter (224).

optimum conditions and to avoid systematic errors such as those mentioned above.

Prosen and Kilday (224) chose the reactions between aqueous sulfuric acid ($H_2SO_4 \cdot 8H_2O$) and aqueous sodium hydroxide solution at two different concentrations for a comparison between this calorimeter and the one by Gunn (100) described immediately above. The results from the two laboratories (102,224) agreed within 0.01%. Because of the great differences between the calorimeters, their systematic errors should be different, and the agreement implies a high accuracy in both measurements.

The measurements with $H_2SO_4 \cdot 8H_2O$ and aqueous sodium hydroxide are too difficult to be used as a routine test or calibration technique. The agreement between the two calorimeters was much poorer for the reaction between "tris" and aqueous HCl. The results (101,102,223) suggest that the "tris" + HCl(aq) reaction is not as well-defined as is desirable, but it remains the best practical test reaction for solution calorimetry in general. Development of the reaction of potassium chloride with water as a test or calibration reaction has been hampered by limited interest in endothermic reactions (100) (see Addendum).

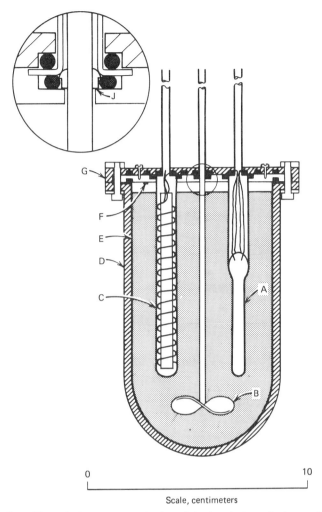

Fig. 5.11. The calorimeter proper in the adiabatic solution calorimeter (224).

Zinc oxide is a standard material for the determination by solution calorimetry of the heat of hydration of cement, an important quantity in the design of large concrete structures (199). Microcalorimeters (see the discussion below in Section III.A.3) are often used for the measurements because of the slowness of the reactions. A number of papers on calorimeters for this measurement have been published recently (1,55,57,73,179,180,183,208, 321).

3. Microcalorimetry

Microcalorimetry is a general term for the measurement of heat in very small quantities, at very slow rates, or from very small samples. Microcalorimetry is quite an active field, with biochemical applications providing much of the impetus toward its development and use. Several types of microcalorimeters are commercially available.

Analysis by calorimetry in the case of biochemical materials is particularly feasible, because the lack of specificity in the calorimetric technique is compensated by the specificity of the reactions (for instance, enzyme reactions). Because of the limited quantities of many biochemical materials available for laboratory study, biochemical calorimetry usually involves some microcalorimetric technique.

As was mentioned above, microcalorimetry is applicable to a wider range of reactions and conditions than "classical" techniques of calorimetry, but is less precise and accurate. Many of the older methods of microcalorimetry were discussed by Swietoslawski (281) and by Calvet and Prat (37). More recent reviews, some specifically for biochemical uses are in references (17, 24,30,31,38,49,72,86,89,90,206,259,272,302–6,319). Other uses of microcalorimetry include kinetic studies, detection and characterization of bacterial growth, testing of electrical power cells for critical applications (107), hydration of cement (see above under Section III.A.2) and other heats of wetting, heats of adsorption, heats of recovery and recrystallization of metals, etc. The number of potential applications is enormous.

Tian (283) developed a microcalorimeter in which the heat flux between the reaction cell and its surroundings was determined. The heat liberated in the cell was compensated by use of the Peltier effect, so that the change in the temperature of the cell was very small. Calvet and co-workers have developed the Tian calorimeter into a high-precision instrument, and they have used it to determine small energy changes due to many different phenomena (38,236,254). Figure 5.12 is a diagram of one of their calorimeters. The experimental cell sat in a socket consisting of 144–1000 small silver plates placed as close as possible to each other while still electrically insulated, and covered with thin mica to insulate them electrically, but not thermally, from the cell. A thermocouple was welded to the back of each plate. The other junctions of the thermocouples were attached to the walls of the thermostatted jacket in which the cell was situated. One set of thermocouples was used to determine the heat flux between the cell and jacket, the other was connected to a battery and was used to compensate, by the Peltier effect, any heat produced in the cell. Absorption of heat was compensated with an electrical heater.

The cell was long and narrow (2 cm^2 cross-section by 8 cm long). This gave quite good thermal equilibration without stirring. If stirring had been used, it would often have generated as much heat as the effect being measured. This shape also ensured that nearly all the heat flow between the cell and its surroundings passed through the thermocouples, which could not be placed over the lid of the cell.

A differential arrangement was used to provide the extremely steady thermostat that was needed for accurate measurements. Two identical mi-

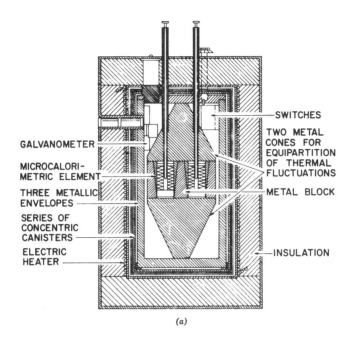

(a)

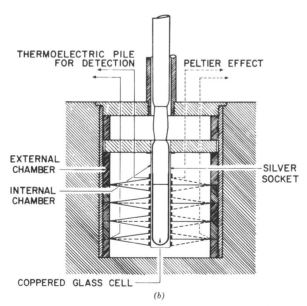

(b)

Fig. 5.12. Calvet microcalorimeter (E. Calvet, *Experimental Thermochemistry,* H. A. Skinner, Ed., Vol II, and F. D. Rossini, Ed., Vol. I. Interscience Publishers, Inc., New York, 1962 and 1956).

crocalorimetric elements were placed in a massive, double cone-shaped jacket that had been carefully designed to keep the temperatures of their boundaries identical, although not constant, to better than 1×10^{-6}°C over very long periods of time. Temperature differences between the external and internal boundaries of the cell as small as 1×10^{-6}°C could be determined, and energy changes of 0.001 cal/hr^{-1} measured.

The Tian-Calvet microcalorimeter has been extensively used to determine heats of adsorption (35). If the rate of adsorption was slowed, to give time for diffusion over the whole available surface and prevent premature formation of multimolecular layers on those parts of the surface more directly accessible to the vapor, the thermogram (heat versus time record of the experiment) showed a sharp discontinuity corresponding to the completion of the formation of a monomolecular film on the surface of the adsorbent. The specific surface of the adsorbent was derived from the weight of the adsorbent and the weight of the adsorbed material at the break in the thermogram (39). It was found that the solution of alumina gel in soda solutions takes place in two stages, a rapid exothermic solution of amorphous alumina followed by a slow endothermic solution of crystalline alumina (40). The two stages were readily distinguished from the thermogram of the solution process, enabling the degree of crystallinity in a gel sample to be readily determined. Many different cells have been constructed for the Tian-Calvet microcalorimeter, adapting it to many different types of measurements. An interesting development in this type of thermal fluxmeter microcalorimeter is a simplified means of constructing the thermopiles (17) by electroplating a coil or using a stack of disks (164).

A twin adiabatic calorimeter was built by Buzzell and Sturtevant to measure slow thermal processes in solution (33). Two identical systems were equipped with electrical heaters and resistance thermometers. The thermometers were in adjacent arms of a Wheatstone bridge. The output of the bridge, which was proportional to the temperature difference between the calorimeters, was amplified and fed back to control the voltage to one of the heaters. In addition, the output, which was accurately proportional to the total electrical energy feedback, was integrated and recorded on a strip chart recorder. In this way, the energy resulting from a reaction proceeding in one of the calorimeters was duplicated by electrical heating, and the temperatures of the two calorimeters held closely equal. This apparatus gave satisfactory kinetic and thermal data for reactions having initial rates of heat changes as small as 10^{-4} cal/min^{-1} · ml^{-1} of reaction solution, provided the half-time was between 2 min and several hours.

Microbomb calorimeters have been developed that can be used to determine heats of combustion to within a few hundredths of 1% with only milligram quantities of material. Although not as accurate as the conventional apparatus, they are of great value in studying compounds that are only available in small quantities. In early microbomb calorimeters, such as that of Roth and co-workers (241), the accuracy was poor because of incomplete combustion. McEwan and Anderson (177) developed a microbomb in which the crucible was made from very thin platinum to permit a large local temperature rise, and an elevated lid of thin platinum was placed over the crucible. They obtained a precision of 0.2% with samples of approximately 40 mg of benzoic acid. A number of other designs, in which more or less unusual measures were taken to solve the particular problems of this technique, have been reported. These include the semimicro combustion calorimeter of Mackle and O'Hare (160), Charlu and Kleppa (46), and of Muller and Schuller for 10^{-5}-gram samples (193). Among others is the design of Mansson (165,166) from which results with a standard deviation less than 0.05% are reported on 0.01-gram samples. Electrical calibration agreed with standard benzoic acid measurements within the uncertainty of 0.03%.

Combustion microbombs are commercially available. Important factors to keep in mind are: (1) the very small samples must be weighed with an appropriate balance and must be handled with more care than samples for larger bombs; and (2) the particular microbomb selected should have a precision and accuracy adequate to the application.

New microflow calorimeters have been reported, including those of Stoesser and Gill (264), Picker et al. (92,213–5), and others (68,88,184). (Also, see Section III.B.7.)

An interesting series of papers by Wadsö et al. (66,80,142,147,185,280, 307) illustrates the development of a microcalorimeter and its adaption to several very different types of measurement by inserting different cells. This is a field in which the prospective user can profitably exercise his ingenuity in adapting such calorimeters for a specific purpose.

Many theoretical and experimental studies have been made concerning the best ways for calibrating, using, and correcting data from microcalorimeters. Only a few of these, in addition to the reviews mentioned above, can be listed here (29,36,50,105,293). The interested reader should consult the original literature in his particular field of interest. Thus, the recently described microcalorimeters with unusual features include twin calorimeters on the electrodes of a cell (132), a device for photochemical reactions (16), a calorimeter with a vapor effusion cell (242), a quadruple microcalorimeter (207) designed for analytical use, a microcalorimeter made of alumina for high-temperature reactions (137), and a miniaturized low-temperature heat capacity apparatus (273).

B. ENTHALPY INCREMENTS AND SPECIFIC HEATS

1. Definitions

The specific heat, or heat capacity per unit mass, is the amount of heat that must be supplied to a unit mass of material to raise its temperature by an infinitesimal amount, divided by that increase in temperature. When specific heat is nearly constant over a range of temperature, it can be defined as the amount of heat supplied per unit mass per unit resultant temperature change. The specific heat will vary depending upon how the other conditions defining the state of the material, such as pressure and volume, are allowed to vary during the heating. The term "heat capacity" as used in this chapter is understood to be that per unit mass.

Two types of heat capacity are in common use:

1. The heat capacity at constant volume, C_v. This is equal to the temperature derivative of the internal energy and may be written $(\partial E/\partial T)_v = C_v$.

2. The heat capacity at constant pressure, C_p. This is equal to the temperature derivative of the enthalpy (heat content), and may be written $(\partial H/\partial T)_p = C_p$. The constant-pressure heat capacity is the one most directly applicable to deriving the frequently used thermodynamic functions for the compilations cited earlier.

Since $H = E + PV$ by definition, where P is pressure and V is volume, C_p and C_v are nearly equal for most condensed phases near 25°C and atmospheric pressure. The work done on these phases by a moderate pressure increase is very small compared to the overall heat capacity. For gases, there is a substantial difference between C_p and C_v.

A third type of heat capacity, C_s, is sometimes used. It is the heat capacity of a condensed phase under its own saturated vapor pressure, the subscript s representing "saturated." Calorimetric measurements of heat capacity are often made on samples sealed into a calorimeter at constant volume, with a vapor space to allow for expansion of the condensed phase during heating. Providing that the vapor space is not too large and the vapor pressure not too great, C_s is nearly equal to C_p at a low and constant pressure.

There are exact relationships between C_s, C_v, and C_p, for which the reader is referred to Chapter 2 on thermodynamics and to texts and expositions of thermodynamics.

The heat capacity becomes extremely large near temperatures where "first-order" transformations, such as melting and certain crystal structure changes, occur. These transformations are considered to take place abruptly at one temperature, where the heat capacity is undefined. Therefore, we must speak of the enthalpy of transformation, ΔH_{trans}, the heat absorbed in the transformation at constant pressure (or of the corresponding heat quantity for variable pressure conditions). The measurement of this enthalpy of transformation is performed in essentially the same way as the measurements of C_p. Since it is impossible to perform an experiment in which the temperature rise is infinitesimal, or in which the rate of heating is infinitely slow, all measurements of C_p are actually measurements of average heat capacity over a temperature interval. The average C_p, by definition, is the enthalpy increment $H_{T_2} - H_{T_1}$ divided by the temperature change. So in actual measurements, the heat capacities are derived from enthalpy increments.

The total enthalpy increment for a material between two temperatures T_1 and T_2 is given by:

$$H_{T_2} - H_{T_1} = \int_{T_1}^{T_2} Cp\,dT + \sum \Delta H_{trans}$$

where the ΔH_{trans} represents the enthalpies of any "first-order" transformations between T_1 and T_2.

The entropy increment $S_{T_2} - S_{T_1}$ is given by a similar expression in which C_p is replaced by C_p/T and ΔH is replaced by $\Delta H/T$. Since the entropy often approaches zero (or a known quantity) as the temperature approaches absolute zero, measurements of the absolute entropy can be made by calorimetry.

2. Uses

There are obvious valuable results to be obtained from the measurement of heat capacities, so far as completing thermodynamic property value tables is concerned. Additionally, even very rough measurements may provide results useful in industry. Obviously, the heat input or extraction required to raise, lower, or maintain the temperature of a system can be calculated if its mass and heat capacity are known.

Many of the transformations seen in heat capacity calorimetry are associated with important properties. These transformations, whether of the "first order" (such as melting) or higher order (such as "glass transformation"), can be spotted and their temperatures and enthalpy changes identified. Even a rough and approximate measurement may be adequate for certain purposes, such as determining the approximate glass transformation temperature of a plastic material. The reader should again be cautioned that such useful results may have very little to do with the basic thermodynamic quantities applying to the same transformation, and such basic values should be published only when the experiments are appropriately low in systematic errors.

The purity of a material can be determined from the ΔT required to melt successive portions of the material. This is possible because the melting point is not exactly constant except for the idealized case of a perfectly pure material. The method has certain limitations, often not important in practice. Primarily, it is assumed that the impurities are soluble in the melted material but do not form solid solutions with the frozen material. This technique is mentioned below in Section IV.

Heat-capacity measurements have been of great value in the study of structural changes in materials as a function of temperature. Conversely, in developing a proposed analytical method, the change of heat capacity can sometimes be predicted from a corresponding structural change known to occur in the material. Modern statistical thermodynamic theory predicts the entropy changes that will result from structural changes. Examples of structural changes that affect heat capacity are disordering in alloys, uncoiling in macromolecules, melting, transformation to a liquid crystal and other changes of crystal structure, glass transformation, etc.

3. Adiabatic Heat-Capacity Calorimetry

In the adiabatic method, the sample is isolated as well as possible from heat exchange with its environment, and the remaining heat exchange is made as reproducible as possible. Measured amounts of heat are then supplied to the sample and the resulting temperature changes are measured. The heat supplied is usually electrical, but may be from other sources such as lasers (154).

The principles are exactly the same, regardless of temperature, but there are important differences in design between low-temperature and high-temperature calorimeters. One of the most obvious is the choice of construction materials, and the choice of such materials and design with them is one of the more difficult problems in heat-capacity calorimetry. It becomes more difficult as the range of temperatures become larger and the maximum temperature higher.

Another important difference is the effect of heat exchange due to radiation. At very low temperatures, it is negligible compared with the effect of conduction; at very high temperatures, it is the main means of heat exchange.

The sample being studied is usually sealed into a container (the calorimeter proper), even if it needs no protection from the conditions encountered in the measurement. The heat capacity of the sample is then determined as a difference between the heat capacity of the filled container and that of the empty container. This permits the cancellation of many systematic errors in the measurements. All errors that are the same in the two measurements (empty and filled calorimeter) will cancel out. In the most accurate work, a great deal of attention must be given to ensuring that the heat exchange of the surroundings with the filled calorimeter is the same as that with the empty calorimeter. At the higher temperatures where radiation is important, the variation of temperature distribution on the outside of the calorimeter is important, and this leads to the provision of numerous radiation shields around the calorimeter. Attempts have been made to create "hybrid"-type calorimeters capable of accurate operation from about 4 K to temperatures well over 100°C by adding a single radiation shield to a low-temperature-type calorimeter (171,187). In general, different heat-capacity calorimeters are used for very low temperatures (below 30 K), for low temperatures (about 4 K to 50°C), for high temperatures (about 50°C to 1000°C), and for very high temperatures (over 1000°C), when the best accuracy is needed.

Adiabatic heat-capacity calorimeters may be heated intermittently (the usual procedure in accurate work), giving average heat capacities over small ranges of temperature, or heated continuously, giving a curve of heat capac-

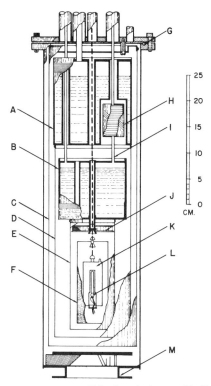

Fig. 5.13. Cryostat for low-temperature adiabatic calorimetry (E. E. Westrum, Jr., *J. Chem. Ed.*, **39**, 433, 1962)). 1, Liquid nitrogen tank, *B*, liquid helium tank; *C*, *D*, *E*, radiation shields; *F*, adiabatic shield; *G*, O-ring gasket-sealing brass vacuum jacket; *H*, effluent helium vapor exchanger (economizer); *I* helium exit tube; *J*, ring for adjusting temperature of leads; *K*, calorimeter assembly; *L*, platinum-resistance thermometer; *M*, connection to vacuum diffusion pump.

ity versus temperature. Continuous heating has some advantages, especially for analytical work where the speed of analysis is important. However, it is essential for accurate work with continuous heating that the rate of heating be slow compared with thermal equilibration in the sample, and desirable that two different heating rates be compared (261). The accuracy requirements for a particular technique may be such that a single rapid heating rate can be used.

Adiabatic heat-capacity calorimeters for the low-temperature range and for the very high-temperature range will be described here. A modern version of the Nernst adiabatic vacuum cryostat (310) may be used to carry out calorimetric experiments under adiabatic conditions (Fig. 5.13). The complete apparatus is enclosed in a brass vacuum jacket sealed with an O-ring gasket, G. The calorimeter is suspended in isolation by a braided silk thread and surrounded by an adiabatic shield, F. The calorimeter and shield may be brought into direct thermal contact with a refrigerant tank, B, when cooling is necessary to obtain the desired operating temperature, and isolated again by lowering them. Two chromium-plated copper tanks, A and B, are used as thermal sinks. The radiation shields, C, D, and E, are also of this material, which not only conserves the refrigerants but generates zones of uniform and progressively lower temperature.

If a temperature above 90 K is needed, liquid nitrogen is used in both tanks A and B; for temperatures down to 50 K, the lower tank is evacuated and the nitrogen solidified; and when temperatures between 4 and 50 K are desired, liquid helium is used in the lower tank. By pumping on the liquid helium, an operating temperature approaching 1 K may be obtained. Evaporation of ^{4}He and ^{3}He will extend the working range to approximately 0.9 and 0.3 K, respectively, and adiabatic demagnetization may be utilized to yield even lower temperatures, although very little calorimetry has been done in the latter case (212). The upper limit of the working temperature may also be extended to 600 K by suitable modification of the apparatus.

A high vacuum, produced by a diffusion pump connected at M, exists inside the apparatus, eliminating heat transfer by gas conduction, and this facilitates both thermal isolation and the achievement of adiabatic conditions. Thermal conduction along the electric leads is minimized by anchoring them to the refrigerant tanks, and to an "economizer", H. This acts as a heat exchanger by utilizing the cold effluent helium gas, I, to absorb the heat conducted down the leads, thus conserving the liquid helium. The temperature of the lead bundle is tempered by a ring, J, and also adjusted to the temperature of the calorimeter by the adiabatic shield which is kept to within ± 0.002 K of the calorimeter temperature. Copper-constantan thermocouples monitor the differences in temperature between the calorimeter and shield and between shield and ring, and actuate three separate channels of recording electronic circuitry provided with proportional, rate, and reset control actions for the reestablishment of adiabaticity. Low-temperature calorimetry has been reviewed by Westrum (311).

In a book edited by Skinner (254), Kubasckewski and Hultgren have described an automatically controlled high temperature adiabatic calorimeter used by Backhurst (12) to study titanium and various steels up to a temperature of 1600°C (Fig. 5.14). If the sample is solid a molybdenum crucible is used and for molten metals it is made of alumina. The thermal capacity of the crucible must be relatively small compared with that of the sample to achieve accurate results, and for this reason a large amount of sample was used (2 kg). The thermocouples, A and C fixed to the outside of the crucible and B and D fixed to the inside of the outer enclosure, were of the 5% Rh-Pt/20% Rh-Pt type. One from each sheath was connected to one coil of a double-coil mirror galvanometer, so that equal currents in each coil acted against each other and produced no deflection. The reflected beam was made to operate an electropneumatic controller, which adjusted the power supplied to the tungsten heaters so that the temperature of the enclosure remained nearly equal to that of the crucible. The emf of the second ther-

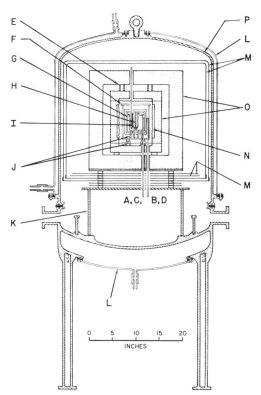

Fig. 5.14. Adiabatic calorimeter for measurements up to 1600°C. (O. Kubaschewski and R. Hultgren, *Experimental Thermochemistry*, Vol. II, H. A. Skinner, Ed., Interscience Publishers, Inc., New York, 1962, Chapter 16). *A, B, C, D*, thermocouples; *E*, ceramic supports; *F*, alumina pot., *G*, rhodium-platinum radiation shields; *H*, heater inside alumina tube; *I*, crucible; *J*, alumina cone supports; *K*, four-legged steel table; *L*, water jacket; *M*, nickel radiation shields; *N*, heater grid of tungsten wire; *O*, firebrick enclosures; *P*, steel bell.

mocouple in each sheath was indicated as a trace on a chart, and fine adjustments to the power input were made by hand to keep the mean deviation as near to zero as possible. A temperature–time trace of the specimen thermocouple was made simultaneously, and the specific heats were calculated using the gradient of this curve and the power input to the sample heater. Precautions were taken to ensure that the temperature of the sample crucible and enclosure was uniform.

Most reaction calorimeters belong to the "solution calorimeter" category. However, it is also appropriate to mention them here, since high-temperature reaction calorimeters have many similarities to high-temperature heat-capacity devices in design and in the materials studied. A high-temperature adiabatic reaction calorimeter was described by Kubaschewski and Walter (145,146). This field is relatively inactive at present; for instance, O.J. Kleppa (138) commented that there were no liquid tin solution calorimeters in operation in the United States at the time (1977). Recently reported high-temperature heat-capacity calorimeters with unusual features include one for reactive materials (93), one with electron-beam heating (262), and one of modest accuracy for rapid work (261). The subject of adiabatic heat-capacity calorimetry has been reviewed by Westrum (312). High-temperature reaction calorimetry was reviewed by Leach (153), combustion and reaction calorimetry by Head (109), calorimetry of solids by Chandrasekharaiah (45), general reaction calorimetry by Franke (74), and calorimetry of phase and order transformation by Westrum (313). Metallurgical applications were reviewed by Kubaschewski and Slough (144). An interesting new high-temperature solid–solid reaction calorimeter is that of Capelli, Ferro, and Borsese (42); designs are given by Neckel and Nowotny (194) and Dienstbach and Blachnik (60). See also Section III.B.5.e. A proposal for a microgravity high-temperature reaction calorimeter for reactive materials has been advanced (106).

No thermal insulation system is capable of preventing significant heat transfer in a heat-capacity experiment except for a few special techniques, generally used at extremely low or high temperatures where other techniques are impracticable, in which the heating is extremely fast (see Section III.B.6). For most experiments, then, the surroundings of the calorimeter proper will have to be maintained at the same temperature as the calorimeter. Proper design of the controlled temperature shields around the calorimeter and of their control systems is one of the main problems of adiabatic calorimetry.

The simplest shield control method is, of course, manual control of heaters and/or cooling devices. This technique can give satisfactory results in some cases, but is wasteful of skilled labor. Automatic control is capable of providing a closer and more reproducible matching of temperatures. A large number of devices for such control are sold commercially. The mathematics of the design of such control systems is analogous to that involved in servomechanisms and can be handled with the mathematical tools of modern electrical engineering. The design is usually based on systems found satisfactory in similar calorimeters, so that a detailed mathematical analysis is not necessary. An understanding of the principles of control is desirable, however, for speeding the final adjustment of instrument settings for best performance, and possibly for making adjustments in the physical positioning of the temperature sensors and heaters.

The design problems of shield control are considerably simplified if continuous heating is used. Any device that is to be used for accurate measurements on a wide variety of materials, however, should be capable of an intermittent-heating operation for those cases where thermal equilibrium in the sample is inherently a slow process. Intermittent heating is also desirable for the determination of purity through the variation of temperature as the amount melted varies. The amount melted can be determined from the amount of heat supplied.

The recent availability of ^{3}He–^{4}He dilution refrigeration for conveniently achieving low temperatures to about 0.3 K has led to a number of designs and transient techniques for the range 0.3–30 K. These are discussed in Section III.B.6.

4. Differential Calorimetry

Differential calorimetry is discussed at this point because heat-capacity determination is a common application. Differential microcalorimeters were discussed in Section III.A.3.

Just as in the differential microcalorimetric techniques, differential calorimetry in general determines the difference between two heat effects in two different cells of the same calorimetric apparatus. Its advantage is that the overall accuracy of measurement required for a given accuracy in the difference is much less than for separate experiments with the two processes.

Differential calorimetry for heat capacities usually involves an adiabatic device surrounding one cell with the sample to be studied and another cell with a sample of known heat capacity. The power to the two cells is adjusted so that their temperatures are both equal to the shield temperature.

Differential scanning calorimetry (DSC) has become a very popular form of this technique. The term "scanning" refers to the continuous heating program. Commercial instruments are available for DSC, which is described in detail in another chapter.

5. Isothermal Drop Calorimetry

In "drop" calorimetry, the sample is heated to the desired temperature in a furnace and then dropped into a calorimeter of known heat capacity at a temperature near room temperature. The resulting heat effect measured in the calorimeter gives the average heat capacity (or enthalpy increment) of the sample between the calorimeter temperature and the original sample temperature. This is sometimes called the "method of mixtures," since the thermal energy in the sample is "mixed" with that of the calorimeter.

Drop calorimetry is somewhat simpler to perform than adiabatic calorimetry and is more readily adaptable to extremely high temperatures where adiabatic calorimetry becomes impractical. Its main disadvantages are: (a) the heat capacities are found from the differences between large enthalpy increments, requiring a higher precision of measurement for the same precision in the final heat capacities; and (b) the sample may not come to equilibrium as it cools very rapidly through a large temperature difference.

Isothermal drop calorimetry refers to dropping the sample into a calorimeter containing a material at its melting (or occasionally, boiling) point so that the temperature remains constant. The amount of heat supplied by the dropped sample is determined by the change in volume due to additional melting or vaporization. The calorimeter may also be a large copper block or other material that will change only slightly in temperature when the sample is dropped into it. In this case, the calorimetric measurement is similar to that in a solution calorimeter. In drop calorimetry, some of the systematic errors can again be cancelled by using a container with the sample (as in adiabatic calorimetry), or by proper procedures in using a standard material for calibration of the equipment.

The furnace used to heat the sample is an important part of the drop calorimeter. It must be capable of reaching the desired temperature and holding it. The region of the furnace where the sample is located must be uniform in temperature within narrow limits. The heat from the furnace must not disturb the calorimeter. Design of the special furnaces needed for drop calorimetry is discussed in the literature (23,41,136,170). Electric power is usually the most convenient and tractable means of heating; other methods that have been used include solar heat (82), electron beams (76), etc.

The drop mechanism should operate as reproducibly as possible to permit proper cancellation of errors, and the heat transferred to the calorimeter while it is open should be minimized. Various methods have been used to achieve these goals, such as automatic shutters which open for a precise length of time to permit passage of the sample and the use of "lifting" of the sample rather than dropping, with the sample pulled by a fine wire (75,94).

Calibration of drop calorimeters with a standard sample has been described in detail (78). It is advisable to check the calorimeter with such a sample even if the calorimeter is calibrated electrically and the measurements carried out with the intention of cancelling systematic errors.

The measurement of temperature requires a special effort, especially when working at very high temperatures. The exact scale used should be specified. (See Chapter on temperature measurement). Since the temperature can sometimes be measured with much greater precision than that with which the absolute thermodynamic temperature scale is known, meaningful comparison of temperatures between different laboratories requires great care. This is especially true for adiabatic heat capacity calorimetry at melting and transformation temperatures.

Standards for heat capacity calorimetry have been made available by the Bureau of Standards (290) and the Calorimetry Conference (85). The nominal accuracy of heat-capacity calorimetry in the best temperature ranges is about 0.1%. As in solution calorimetry, there still remains much work to be done in establishing the most accurate values for the standards, but they are already known well enough for all but the most accurate work.

The proper application of temperature corrections for heat exchange, etc., in calorimetric work, especially drop calorimetry, has been the subject of a number of recent papers in which improved methods of calculation are sought. However, it would be desirable to remain with the currently recommended methods until sufficient experience has been obtained with the newer methods to warrant new recommendations.

a. Ice Calorimetry

Ice calorimetry, in which heat effects are determined by the amount of ice that they are capable of melting, is certainly one of the oldest calorimetric techniques (22,152). Bunsen (32) described the first of the modern, practical ice calorimeters.

The calorimetric sample is dropped into a well in the apparatus. Around the well is a coating of ice in equilibrium with liquid water. An inert gas is flushed through the system to prevent condensation. As the heat is absorbed from the sample into the ice, the ice melts and the volume of the water plus ice decreases, drawing mercury through a capillary tube. The volume change is determined from the motion of the mercury along the tube. The method is simple, sensitive, and accurate, but must be performed with meticulous attention to experimental details to prevent errors. For instance, the water must be scrupulously purified, and the coating of ice must be frozen properly and allowed to establish equilibrium with the liquid water, etc.

Ice calorimetry was used by the Bureau of Standards in determining the heat capacity of standard Al_2O_3 for drop calorimetry (78).

b. Diphenyl Ether Calorimetry

Diphenyl ether is another convenient and popular material for isothermal drop calorimeters. Its volume change on melting is about three times as great as that of ice for the same amount of heat. Its melting point is a secondary fixed point on the International Temperature Scale of 1968 and is conveniently close to the standard temperature of 25°C. The material is stable, its thermodynamic properties near the melting point are known accurately (79), and no serious disadvantages of its use have been reported.

A description of the diphenyl ether calorimeter of Hultgren et al. (121), will serve to illustrate the principles of both ice and diphenyl ether calorimetry. The calorimeter is shown in Fig. 5.15. The water-cooled gate is opened to allow the sample to drop into the copper tube. The sample passes the copper baffles inside the tube and is caught in a copper screen basket. The baffles prevent the sample from radiating heat out of the tube and retard the rise of hot air from the tube. The screen basket prevents the sample from producing a hot spot on the tube. The constant temperature bath is maintained about 0.2°C above the melting point of the diphenyl ether so that in the absence of external heat there will be a very slow and constant melting rate of the mantle of frozen diphenyl ether. As the diphenyl ether melts, its volume increases, and mercury is displaced into the capillary tube. The volume displaced, which is corrected for the steady heat input from the constant temperature bath, gives the amount of heat received from the sample. Procedures for diphenyl ether calorimetry are discussed by Davies and Pritchard (58).

Another material used for this type of calorimeter is liquid argon (263). In

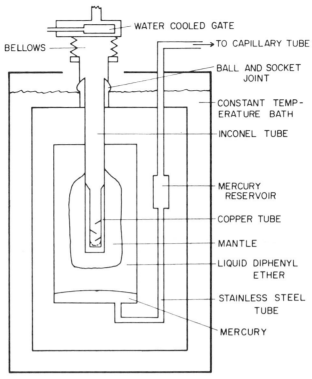

Fig. 5.15. Diphenyl ether calorimeter (R. Hultgren, P. Newcombe, R. L. Orr, and L. Warner, *The Physics and Chemistry of Metallic Solutions and Intermetallic Compounds,* Vol, I. Her Majesty's Stationery Office, London, 1959).

this case, the volume change is that due to vaporization rather than melting, so there are a few additional complexities introduced into the measurements.

c. COPPER-BLOCK CALORIMETRY

The copper-block calorimeter is not completely isothermal, but rises slightly in temperature when the sample is dropped into the tapering receiving well in the block. The average heat capacity of the sample is determined from this rise in temperature. Southard (258) described a calorimeter that has served as the prototype for copper-block calorimeters. The typical copper-block calorimeter of Margrave and Grimley (91) was made in three sections: a receiving well, an outer jacket, and the main body of the calorimeter. The well has a gold-plated exterior and a rhodium-plated interior. A 10Ω heater made from silk-covered manganin was bifilarly wound round the well for the electrical calibration. A Maier-type resistance thermometer (162), protected by a chromium- and nickel-plated jacket, was wound in a recess on the outside of the calorimeter block to provide a measure of its rise in temperature. The calorimeter connection to the furnace includes a water-cooled gate which is operated from the outside by a shaft consisting of two steel sections joined by a nylon center section to reduce the thermal leakage along the shaft. A double-junction copper-constantan thermocouple is mounted in the gate to check the temperature of the resistance thermometer. All coils are covered with several coatings of Glyptal varnish. A copper coil is connected to the lower part of the outer jacket and argon gas introduced to provide an inert atmosphere inside the calorimeter. The top section of the calorimeter has three tubes attached: (1) one tube is for the shaft to the calorimeter-closure device: (2) a central tube connects through the water-cooled gate to the calorimeter; and (3) one tube is for the electrical leads. The two sections of the jacket were sealed with a neoprene gasket.

The calorimeter is immersed in a well-stirred oil bath, the temperature of which is held constant to within $\pm 0.001°C$. A thyratron relay in conjunction with a mercury regulator controls the bath temperature. The furnace consists of a 1-in inside diameter Alundum tube around which is wound Pt-10% Rh wire cemented in position with Alundum cement. Outside this is 4.5 in of zirconia insulation, which in turn is surrounded by an annular casting of Johns-Manville Firecrete. The two end plates are made from transite and the furnace is controlled by regulated voltage supply ($\pm 1\%$) and a Variac. A Pt versus Pt-10% Rh thermocouple, placed near the sample, is used to measure the temperature of the sample with the help of a White double potentiometer. A drop tube was fitted on the top of the furnace, and a spring-loaded, solenoid-activated trigger contained inside the drop tube was used to control the fall of the sample. The calorimeter was calibrated electrically and checked periodically by measuring the heat content of Al_2O_3.

d. LEVITATION CALORIMETRY

A somewhat similar device is being used for "levitation calorimetry" (27,48,286). In levitation calorimetry, which permits very high-temperature enthalpy measurements on reactive materials, the sample is heated by high-frequency induction heating in such a way that the sample is suspended by electrical effects without any mechanical support, i.e., without any container, eliminating problems of reactions with container materials. When the power to the levitator-heater is turned off, the sample falls into the calorimeter. The temperature of the sample is measured by optical pyrometry, and that of the calorimeter block by a quartz crystal oscillator thermometer, which is simpler to use than metal-resistance thermometers.

The temperature can be read from the digital display of the thermometer, recorded manually, and punched on cards for computing processing. The thermometer automatically displays new temperature readings at precisely equal time intervals. An automatic system for temperature-time data can contain an interface to convert the digital thermometer output to serial AS-CII format and either a teletype equipped with paper tape or a computer terminal equipped with a magnetic tape recorder. After the experiment, the paper or magnetic tape recording is transmitted directly to the computer over a telephone line, eliminating the possibility of copying and keypunch errors.

Such an automatic system may be shared between several different calorimeters. If each calorimeter has its own thermometer, including display, it is not even necessary to schedule experiments for times when the automatic system is available, since the temperatures can be recorded manually. At Rice University, an automatic temperature-time data recording system is shared between levitation, combustion (64), and solution (188) calorimeters.

e. THE TIN CALORIMETER

The tin, or liquid-metal-solution, calorimeter combines elements of a solution calorimeter (since a solution process takes place) and a drop calorimeter (since the material to be studied is often dropped into the tin from a furnace at a definite temperature). As was mentioned earlier, this is a relatively inactive field at present. The technique is extremely well adapted to metallurgical studies, including determinations of heat capacities, heats of mixing and alloying, and heat effects of cold working, recovery, and recrystallization.

Tin is the usual working material because it is a good solvent for many metals, the heats of solution of metals in tin are of the same order of magnitude as heats of alloying, the melting point is not extremely high, and the maximum temperature of operation is reasonably high provided that the tin is protected from oxidation by an inert atmosphere in the calorimeter. The constant temperature bath around the tin can be $LiNO_3$-KNO_3-$NaNO_3$ mixture (117). Just as in other drop calorimetry, the "inverse drop" method can be used, meaning that the sample is dropped from a temperature lower than that of the calorimeter. This term is sometimes used to indicate that a cold sample is lifted into a hot calorimeter.

Low-melting alloys are sometimes used instead of tin to permit operation at lower temperatures. This reduces some of the problems of operation at temperatures well above room temperature and still permits useful data to be obtained. Recent literature on liquid metal solution calorimetry includes devices for high-temperature operation (14,63,219) to about 1200°C as well as others (103,181,251,252) for operation as low as 70°C.

6. Transient Techniques

There are high-temperature limits on conventional calorimetry set by such factors as the limits on measuring techniques, the difficulty of finding suitable construction and container materials, and the extremely rapid and unavoidable heat transfer due to radiation. At very low temperatures there are problems due to the difficulty of providing thermal contact with a heat sink (for cooling) and then isolating the sample from that material (for measurement). Heat capacities become so small at very low temperatures that a slight mechanical vibration or heat conduction down a fine wire can cause substantial warming. Transient techniques are used to make the duration of the experiments extremely short, so that these problems do not prevent measurements. The uncertainties, of course, are likely to be larger than in conventional calorimetry.

High-temperature heat pulse methods were described by several authors and have been reviewed in detail by Beckett and Cezairliyan in a book edited by McCullough and Scott (176). Samples may be heated (if electrically conducting) by pulses of current from batteries or capacitors. It may be necessary to start at a temperature substantially above room temperature. Other methods include a shuttered arc-image furnace (222) and a light flash (135,195,196), which may be used for nonconductors. The energy supplied electrically can be measured electronically. When the measurement is too rapid to permit temperature measurements, the temperature must be deduced from the resistance of the sample, which is derived from the electrical measurements.

A heat-pulse method for metals was developed by Worthing (320) and modified by Avramescu (10). Systems for 1300–1600°C have been designed by Cezairliyan (44) and by Strittmater and Danielson (267). In the latter, the sample was a loop of wire 7 in long and 0.005 in in diam. The resistance-temperature relation was determined with the wire at controlled temperatures in a furnace. The wire was then placed in a vacuum and a 0.1-sec, 9-volt pulse applied. The loss of heat by conduction and radiation over this short period was neglected, and the heat capacity calculated from the instantaneous power and temperature values measured electronically. Results were within 10% of literature calorimetric values. Considering the difficulties of measurement at such high temperatures, a 10% discrepancy cannot be regarded as serious. Strittmater, Pearson, and Danielson (268) concluded that some variation of the method could become the best method at very high temperatures. Cezairliyan (44) improved and extended this method in studies of the refractory metals. Shaner and Gathers et al. have applied this technique for studies of liquid metals up to 5000 K (323).

Since the total energy in a charged capacitor can be determined by electrical measurements, the only additional measurement needed to obtain the average heat capacity of a sample is the maximum temperature reached when a capacitor is discharged through it. This method is suitable for small solid and liquid samples up to 5000 K. Electrical measurements of heat capacity with transient techniques can be repeated and the results averaged electronically, so long as the temperature is not high enough to alter the samples permanently.

In low temperature ranges, heating can be electrical and cooling can be by a constant conduction path between the sample and a heat sink. This method avoids mechanical disturbances of the sample and resultant unknown heating effects. This method has become popular for studies of very small samples at temperatures of 0.3 K to a few degrees K.

The thermal behavior of a sample under repetitive electrical heating resembles electrical behavior of ac circuits. ac techniques are sometimes used, the principles being the same as when dc pulses are employed, and the mathematics being similar to those for ac electrical circuits. Recent equipment and methods for low-temperature transient techniques are described in references 11,52,83,167,168,221,247,248,274,275,285,294.

7. Flow Calorimetry

Flow calorimetry was first used to determine enthalpies and heat capacities of fluids and has become popular for the determination of enthalpies of reaction and even for kinetic studies. Kinetic work may involve the "stopped-flow" technique, in which the reaction is started and then the flow is suddenly terminated and the progress of the reaction is followed by the change in temperature. Stopped-flow techniques are applicable to faster reactions than are continuous-flow techniques. Some flow calorimetric systems are small enough to be described as flow microcalorimeters. Flow calorimetric systems are commercially available.

Flow calorimeters may be applied as analytical tools for appropriate reactions, and may be operated continuously as monitors of processes. See the description of a flow reaction calorimeter (149) under the Section III.A.2.

The early method of Callendar and Barnes (34) for heat capacities of liquids, and its variations for determining heat capacities of gases and vapors, consists of passing the gas or vapor at a constant predetermined rate over an electrical heater in the calorimeter, which is supplied with a constant power input, and measuring the rise in temperature of the gas stream by means of two resistance thermometers located before and after the heater, once a steady-state condition has been established.

If the gas flow rate is maintained at γ g/sec, the power input to the heater is W watts, the rise in temperature is Δt, and the temperature correction due to Joule-Thomson cooling is δt, then the mean heat capacity of the substance $C_{p(obs)}$ of molecular weight M, in cal deg^{-1} mol^{-1} is given by

$$C_{p(obs)} = MW/4.184\,(\Delta t + \delta t)\,\gamma$$

As the measurements are conducted under strictly steady-state conditions, the heat capacity of the calorimeter does not enter into the calculation. The time for the attainment of steady-state conditions could be greatly reduced by using a minimum volume of material of low heat capacity in the construction of the calorimeter. Heat exchange between the calorimeter and its surroundings and between the heater and the measuring thermometers, resulting from conduction of the material of the calorimeter and from radiation, are primary sources of error in flow calorimetry. These heat losses, however, can be reduced by increasing the gas flow rate. In actual practice, one determines the heat capacity at a number of different flow rates at each temperature and pressure and extrapolates to infinite flow rates.

The possibilities for internal heat exchange are vastly minimized by incorporating an insulating material around the gas path between the first thermometer and heater, by evacuating the jacket space around the calorimeter and by using properly spaced radiation shields between the thermometers and the heater. Because of the pressure drop through the calorimeter and the resultant Joule-Thomson cooling, blank experiments must be undertaken in which one measures the decrease in temperature, δt, for the gas flowing through the calorimeter when no power is applied to the heater.

Accurate control of flow rates is very important to obtain reproducible values for Δt. Substances that can be easily condensed pose no problem, as the liquid can be boiled under constant pressure by a constant input of electrical energy with the boiler under adiabatic conditions. Suitable types of boilers for this purpose, along with complete designs for apparatus and the methods of operation, are described by Pitzer (218), Waddington (301), Montgomery (186), McCullough (175), and Bennewitz and Rossner (15). In the case of gases, highly sensitive diaphragm valves have been used by Osborne (203) and Masi and Petkof (172,173) to achieve accurate flow control. Scheele and Heuse (243–245) have used a mercury-regulating pump for this purpose.

A simple flow calorimeter that could be constructed from readily available laboratory materials giving heat capacities within ±3% of accepted values has been described by Pierce (216). Lacher et al. (150) used flow calorimetry extensively in determining the heats of chlorination, hydrogenation, etc., of organic compounds.

Flow calorimetry is used in determining enthalpies of fluids for engineering design work. A number of papers on calorimetric equipment for determination of fluid enthalpies have been published recently. One is for measuring gas streams in natural gas processing (317). A review of flow calorimetry is given by Counsell (54).

Stoesser and Gill (264) developed the first satisfactory flow microcalorimeter (272) and the field has seen rapid development. The differential flow microcalorimeter described by Picker et al. (92,213–5) is shown schematically in Figure 5.16. This device was designed primarily for rapid determination of the heats of mixing and dilution of liquids using small samples. The two thermostated liquids A and B are mixed and passed through a countercurrent heat exchanger, where the heat of mixing is absorbed by the exchanger fluid. The mixed liquid AB then flows back through a reference heat exchanger. The "modulator" C delivers equal pulses of heat exchange fluid alternately to the two heat exchangers. The thermometer d is thus exposed alternately to heat exchanger fluid from each of the heat

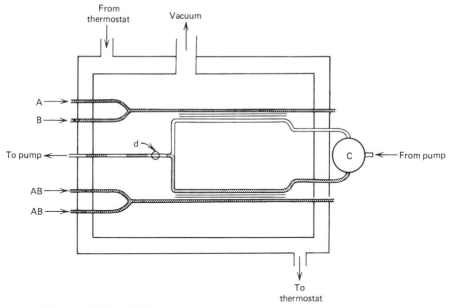

Fig. 5.16. Differential flow microcalorimeter (Adapted from Fig. 6 or Ref. 213).

exchangers. With this arrangement, it is not necessary to provide matched thermometers. The difference in temperature between the heat exchanger fluid from the two heat exchangers is used to calculate the heat of mixing. The pump for liquids A and B is designed to permit rapid variation and recording of the relative flow rates of the two liquids. Thus, a range of compositions can be investigated in a short time. A similar but simpler calorimeter is described by Goodwin and Newsham (88). Another rapid-scanning flow microcalorimeter is described by Falk and Sunner (68).

8. Other Techniques

There are many facets of calorimetry that have not been mentioned above and others about which very limited information was given. Some of these have little or no analytical application at the present time. Others are in developmental stages where it would be premature to predict what the accepted procedures will be in the future. A number of the most important fields are covered by other separate chapters in this Treatise.

There has been a considerable amount of theoretical work on systematic errors and means of avoiding them in design, operation, and calculations. Details of automated calorimetric systems and computer programs for calculations have been published.

IV. SOME APPLICATIONS OF CALORIMETRY IN ANALYTICAL CHEMISTRY

General references on analytical applications of calorimetry include (5,49, 86,133,206,221,259,306). The general references on calorimetry and microcalorimetry listed in the sections above also contain information pertinent to analytical uses. A few analytical applications will be described in detail.

A quantitative distinction can be made between amorphous and crystalline samples by calorimetric determinations of the heats of a reaction process involving them. The thermodynamic reasoning is based on the fact that an amorphous material is more disordered than the crystalline, so the amorphous has a higher entropy. Heat must be absorbed to increase entropy in reversible processes, so the amorphous form is expected to have a higher enthalpy (heat content). Therefore, more heat will be evolved (or less absorbed) in a reaction involving the amorphous form than in the same reaction involving the crystalline form.

The determination of the amount of crystallinity in a sample is obviously important in many commercial processes. An example of this determination was given above in Section III.A.3.

Neugebauer and Margrave (198) determined the enthalpies of samples of amorphous carbon relative to graphite by combustion calorimetry, and were thus able to complete their determination of the heat formation of tetrafluoroethylene. Lewis, Frisch, and Margrave (155) similarly determined the enthalpies of several pyrolytic graphites. These enthalpies are directly related to the relative thermodynamic stabilities of the materials.

The determination of purity by calorimetry has been discussed above. White (315) and Glasgow, Streiff, and Rossini (81) discussed identification of substances and determination of purity by analysis of cooling curves. More recent reviews pertinent to this subject are given by Smit (255,256) and Streiff (266). West (309) has described an interesting situation in the case of sulfur, a material where the molecular composition of the melt is inherently different from that of the solid, even in the absence of impurities.

In establishing new techniques of analysis with calorimetry, caution should be used concerning possible errors due to reaction rates much slower than the time scale of the measurements. This is especially likely to occur in the case of gas–solid reactions. In most cases, the analysis can be carried out with little difficulty, once the proper procedure has been established. Most of the problems are due to incorrect interpretations in terms of fundamental thermodynamic properties.

In the partial list of applications below, the intent is to indicate the general range of applications and possible applications. It is not intended as a survey of proven methods; in fact, some of the methods may not be practicable in the long run. Not all references on a given technique are necessarily listed, and one should also consult the general references listed above.

TABLE 1
Calorimetric Analyses and Applications

Air pollutants (acid particles)	(62)
Aldehydes, total amount, by oximation	(298)
Antibiotics, sensitivity of bacteria to	(20)
Antibodies/antigens	(49)
Bacteria (detection of growth; identification of)	(26,30)
Blood (general)	(306)
Cancer, screening of potential chemotherapeutic agents	(129)
Carbon dioxide, total	(299,300)
in CO/CO_2 mixtures	(151)
in blood	(289)
Catalysts, activity of	(89,90,287)
Cells, electrical, quality of (internal leakage)	(107)
Cellulose industry, analysis of pumping liquors in	(122)
Cement, heat of hydration of	(1,5,55,57,73,179,180,183,199,208, 321)
Cholinesterase, assay of	(184,202)
Clothing, effectiveness of, against cold wind	(250)
Clothing, heats of wetting of fibers and textiles	(47, 161)
Combustible gases, amount in a gas stream	(67)
Concrete Structures—see Cement	
Crystalline/Amorphous	(40,155,198)
Enzymes/substrates	(25,53,174,184,190,206,207,227)
Fermentation, industrial, monitoring of	(65)
Ferrous ion	(206,207)
Fuels, heating values of	(5) See Section III.A.1
Gases in metals	(182)
Glucose	(84,178)
Heavy Metals, total	(206)
Hyaluronic acid	(206)
Hydrogen peroxide (by peroxidase)	(206,207)
Hydroxyl groups, total number in a molecule, (polyoxyalkenes and polyesters) by acetic anhydride	(134)
Insulation, aging of	(204)
Ketones, total amount, by oximation	(297)
Laundering, power in	(157)
Metals—see Heavy Metals or Gases in Metals	
Particle size, by heat of solution (may also be determined by heat of adsorption)	(116)
Phosphate, organic (dimethyl dichlorovinyl phosphate)	(30)
Polymerization,	(254)
Process control, polyethylene	(43)
Process control, acrylonitrile + styrene	(110,111)
Process control, homogeneous copolymerization	(112)
Purity, by melting/freezing behavior	(81,255,256,266,315)
Surface area—same techniques as for particle size	
Structural changes	(See Section III.B.2)
Unsaturation of organic compounds, by hydrogenation	(See Section III.A.2)

ACKNOWLEDGMENT

This chapter reflects calorimetric experiences of the authors while at Rice University, the University of Wisconsin, Oklahoma State University, and the U.S. Bureau of Mines, where calorimetric research has been supported by the U.S. Atomic Energy Commission, the U.S. Army Research Office, the Petroleum Research Fund of the American Chemical Society, the National Science Foundation, and the Robert A. Welch Foundation. We also appreciate the cooperation of our many colleagues for permission to use figures and diagrams from their publications.

It is inevitable that some important and pertinent work will be overlooked in any review of a subject as broad as calorimetry. We hope that the references cited and the brief descriptions of calorimetric techniques will help the reader find the information he needs.

To keep up-to-date in calorimetry and thermochemistry, the interested scientist should utilize the computerized special topics surveys now available through the Chemical Abstracts Service, the Chemical Society, and other technical societies.

REFERENCES

1. Adams, L. D., *Cem. Concr. Res.*, **6**, 293 (1976).

2. Aleksandrov, Yu. I., V. P. Demidov, G. A. Novikov, and V. F. Yushkevich, *Tr. Metrol. Inst. SSSR*, **155**, 41 (1974).

3. Aleksandrov, Yu. I., B. N. Oleinik, B. R. Psavko, and G. R. Usvyatseva, *Tr. Metrol. Inst. SSSR*, **4** (1971).

4. Aleksandrov, Yu. I., V. P. Varganov, V. B. Yushkevich, G. A. Novikov, V. P. Demidov, and A. K. Ivanov, *Izmer. Tekh. 1974*, **37**.

5. American Society for Testing and Materials. Standards.
   (a) C186. Method of Test for Heat of Hydration of Hydraulic Cement.
   (b) D240. Test for Heat of Combustion of Liquid Hydrocarbon Fuels.
   (c) D407. Definition of the Terms Gross Calorific Value and Net Calorific Value of Solid and Liquid Fuels.
   (d) D1826. Test for Calorific Value of Gases in Natural Gas Range by Continuous Recording Calorimeter.
   (e) D2015. Test for Gross Calorific Value of Solid Fuel by the Adiabatic Bomb Calorimeter.
   (f) D2382. Test for Heat of Combustion of Hydrocarbon Fuels by the Bomb Calorimeter (High Precision Method).
   (g) D2766. Test for Specific Heat of Liquids and Solids.
   (h) D3286. Test for Gross Calorific Value of Solid Fuel by the Isothermal-Jacket Bomb Calorimeter.

6. Armstrong, G. T., in H. A. Skinner, Ed., *Experimental Thermochemistry*, Vol. 2, Interscience, New York, 1962.

7. Armstrong, G. T., *Colloq. Int. Cent. Nat. Rech. Sci.*, **201**, 77 (1971).

8. Armstrong, G. T., *5th All-Union Conference on Calorimetry*, Moscow, U.S.S.R., 1971.

9. Armstrong, G. T., and R. S. Jessup, *J. Res. Natl. Bur. Standards*, **A 64**, 49 (1960).

10. Avramescu, A., *Z. Tech. Physik*, **20**, 213 (1939).

11. Bachman, F., F. J. DiSalvo, Jr., T. H. Geballe, R. L. Greene, R. E. Howard, C. N. King, et. at., *Rev. Sci. Instrum.*, **43**, 205 (1972).

12. Backhurst, I., *J. Iron Steel Inst.* (London), **189**, 124 (1958).

13. Barberi, P., J. Carre, and P. Rigny, *J. Fluorine Chem.*, **7**, 511 (1976).

14. Barboian, R., D. Laing, and S. N. Flengas, *Can. J. Chem.*, **45**, 382 (1967).

15. Bennewitz, K., and W. Rossner, *Z. Physik. Chem.*, **B39**, 126 (1938).

16. Benzinger, T. H., and R. L. Kiesow, U.S. Pat. 3,245,758 (1966).

17. Benzinger, T. H., and C. Kitzinger, in *Temperature, Its Measurement and Control in Science and Industry,* Vol. 3, Reinhold, New York, 1963.

18. Berthelot, and Vieille, *Ann. Chim. Phys.,* **6**, 546 (1885).

19. Bichowsky, F. R., and F. D. Rossini, *The Thermochemistry of the Chemical Substances,* Reinhold, New York, 1936.

20. Binford, J. S., L. F. Binford, and P. Adler, *Biochemistry* **10**, 4136 (1971).

21. Bisbee, W. R., J. V. Hamilton, R. Rushworth, T. J. Houser, and J. M. Gerhauser, *Advan. Chem. Ser.* **54**, 215 (1966).

22. Black, J., unpublished work, 1760; see S. Kopperl and J. Parascandola, *J. Chem. Educ.* **48**, 237 (1971).

23. Bockris, J. O'M., J. L. White, and J. D. McKenzie, *Physico-Chemical Measurements at High Temperatures,* Academic, New York, 1959.

24. Biovinet, P., *Eur. Biophys. Congr. Proc. 1st,* **4**, 277 (1971).

25. Bolen, D. W., and J. L. Slightom, Calorimetry Conference, Argonne, Illinois, (1976). Unpublished.

26. Boling, F. A., G. C. Blanchard, and W. J. Russel, *Nature* **241**, 472 (1973).

27. Bonnell, D. W., Thesis, Rice University, Houston, Texas, 1972.

28. Bosquet, J., J. Carre, M. Kollmannsberger, and P. Barberi, *J. Chim. Phys.,* **72**, 280 (1975).

29. Brie, C., J. L. Petit, and P. C. Gravelle, *Rev. Gen. Therm.,* **11**, 315 (1972).

30. Brown, H. D., Ed., *Biochemical Microcalorimetry,* Academic, New York, 1968.

31. Brown, H. D., *J. Agric. Food Chem.,* **19**, 669 (1971).

32. Bunsen, R., *Pogg. Ann. 1870,* 141.

33. Buzzell, A., and J. M. Sturtevant, *J. Am. Chem. Soc.,* **73**, 2454 (1951).

34. Callendar, H. L., and H. T. Barnes, *Trans. Roy. Soc.* (London), **A199**, 55 (1902).

35. Calvet, E., *J. Chim. Phys.,* **35**, 69 (1938).

36. Calvet, E., *Colloq. Int. Centre Nat. Rech. Sci.* (Paris), No. 156, 95 (1967).

37. Calvet, E., and H. Prat, *Microcalorimetrie,* Masson, Paris, 1956.

38. Calvet, E., and H. Prat, *Recents Progres en Microcalorimetrie,* Dunod, Paris, 1958.

39. Calvet, E., H. Tibon, and J. Chapuis-Seite, *Bull. Soc. Chim. France,* **23**, 1939 (1956).

40. Calvet, E., H. Thibon, A. Maillard, and P. Boivinet, *Bull. Soc. Chim. France,* **17**, 1308 (1950).

41. Cambell, I. E., Ed., *High Temperature Technology,* Wiley, New York, 1967.

42. Capelli, R., R. Ferro, and A. Borsese, *Thermochim. Acta,* **10**, 13 (1974).

43. Carter, D. E., U.S. Pat. 3,521,479 (1970).

44. Cezairliyan, A., *Advances in Thermophysical Properties at Extreme Temperatures and Pressure,* American Society of Mechanical Engineers, New York, 1965.

45. Chandrasekharaiah, M. S., *Solid State Chem. 1974,* 793.

46. Charlu, T. V., and O. J. Kleppa, *J. Chem. Thermodynamics* **5**, 325 (1973).

47. Charuel, R., and P. Traynard, *J. Chim. Phys.,* **52**, 441 (1955).

48. Chaudhuri, A. K., D. W. Bonnell, L. A. Ford, and J. L. Margrave, *High Temp. Sci.,* **2**, 203 (1970).

49. Chignell, C. F., and T. H. Benzinger, *Meth. Pharmacol.,* **2**, 465 (1972).

50. Churney, K. L., G. T. Armstrong, and E. D. West, *Nat. Bur. Stand. Spec. Publ. No. 338,* p. 23 (1973).

51. Churney, K. L., E. D. West, and G. T. Armstrong, *Nat. Bur. Stand. Report NBSIR-73-184,* 1973.

52. Collan, H. K., T. Heikkila, M. Crusius, and J. R. Pickett, *Cryogenics,* **10,** 389 (1970).

53. Cooney, C. L., J. C. Weaver, S. R. Tannenbaum, D. V. Faller, A. Shields, and M. Jahnke, *Enzyme Eng. (Pap. Res. Rep. Eng. Found. Conf.) 2nd,* 411 (1973).

54. Counsell, J. F., *Chem. Thermodyn.,* **1,** 204 (1973).

55. Courtault, B. and P. Longuet, *Ind. Chim. Belge 1967,* Spec. 2, Pt. 2.

56. Cox, J. D., and G. Pilcher, *Thermochemistry of Organic and Organometallic Compounds,* Academic, New York, 1970.

57. Danielsson, U., *Svenska Forskningsinst. Cem. Betong Kgl. Tek. Hogsk. Stockholm,* Handl. No. 38, (1966).

58. Davies, J. V., and H. O. Pritchard, *J. Chem. Thermodynamics,* **4,** 9 (1972).

59. Dickenson, H. C., *Bull. Natl. Bur. Stand.,* **11,** 189 (1914).

60. Dienstbach, F., and R. Blachnik, *Z. anorg. allgem. Chem.,* **412,** 97 (1975).

61. Ducros, M., R. Levy, and G. Meliava, *Bull. Soc. Chim. France 1969,* 1387.

62. Eatough, D. J., T. E. Jensen, R. M. Izatt, and L. D. Hansen, U.S. Calorimetry Conference, Argonne, Illinois 1966. Unpublished.

63. Elliott, J. F., M. G. Benz, and R. N. Dokken, *Steelmaking: Chipman Conf., Proc. Dedham, Mass,* p. 60, (1962).

64. Engel, P. S., R. L. Montgomery, M. Mansson, R. A. Leckonby, J. L. Foyt, and F. D. Rossini, *J. Chem. Thermodynamics,* **10,** 205 (1978).

65. Eriksson, R., and T. Holme, *Biotechnol. Bioeng. Symp. No. 4,* p. 581, (1973).

66. Eriksson, R., and I. Wadso, *Proc. 1st Eur. Biophys. Conf.,* p. 319 (1971).

67. Eyraud, C., French Pat. 1,407,736 (1965).

68. Falk, B. and S. Sunner, *J. Chem. Thermodynamics,* **5,** 553 (1973).

69. Ferguson, A. M., and L. F. Phillips, *J. Chem. Educ.,* **50,** 684 (1973).

70. Flitcroft, T., and H. A. Skinner, *Trans. Faraday Soc.,* **54,** 47 (1958).

71. Flitcroft, T., H. A. Skinner, and M. C. Whiting, *Trans. Faraday Soc.,* **53,** 784 (1957).

72. Forrest, W. W., in *Methods in Microbiology,* vol. 6B, Academic, New York, 1972, p. 285.

73. Forrester, J. A., *Chem. Technol.,* **1,** 95 (1970).

74. Franke, E. K., *Method Chim.,* **B1,** 641 (1974).

75. Frederickson, D. R., R. D. Barnes, M. G. Chasanov, R. L. Nuttall, R. Kleb, and W. N. Hubbard, *High Temp. Sci.,* **1,** 373 (1969).

76. Frederickson, D. R., R. Kleb, R. L. Nuttall, and W. N. Hubbard, *Rev. Sci. Instrum.,* **40,** 1022 (1969).

77. Freeman, R. D., ed., *Bulletin of Thermodynamics and Thermochemistry,* Pub. at the Oklahoma State University, Stillwater, Oklahoma. Name changed to *Bulletin of Chemical Thermodynamics* with volume 21 (1978).

78. Furukawa, G. T., T. B. Douglas, R. E. McCoskey, and D. C. Ginnings, *J. Res. Natl. Bur. Stand.,* **57,** 67 (1956).

79. Furukawa, G. T., D. C. Ginnings, R. E. McCosky, and R. A. Nelson, *J. Res. Natl. Bur. Stand.,* **55,** 195 (1955).

80. Gill, S. J., N. F. Nichols, and I. Wadso, *J. Chem. Thermodynamics,* **7,** 175 (1975).

81. Glasgow, A. R., A. J. Streiff, and F. D. Rossini, *J. Res. Natl. Bur. Stand.,* **35,** 355 (1945).

82. Glasser, J. E., *J. Solar Energy,* **2,** 7 (1958).

83. Gobrecht, K. H., and M. Saint Paul, *Proc. Int. Cryog. Eng. Conf., 3rd,* p. 235 (1970).

84. Goldberg, R. N., Calorimetry Conference, Argonne, Illinois 1976. Unpublished.

85. Goldberg, R. N., committee chairman, 1977, U.S. Calorimetry Conference Committee on Standards.

86. Goldberg, R. N., and G. T. Armstrong, *Med. Instrum.*, **8**, 30 (1974).

87. Good, W. D., *J. Phys. Chem.*, **66**, 380 (1962).

88. Goodwin, S. R., and D. M. T. Newsham, *J. Chem. Thermodynamics*, **3**, 325 (1971).

89. Gravelle, P. C., *Advan. Catal.*, **22**, 191 (1972).

90. Gravelle, P. C., *Catal., Proc. Int. Congr., 5th*, **1**, 65 (1972).

91. Grimley, R. T., and J. L. Margrave, *J. Phys. Chem.*, **62**, 1436 (1958).

92. Grolier, J. P. E., G. C. Benson, and P. Picker, *J. Chem. Thermodynamics*, **7**, 89 (1975).

93. Gronvold, F., *Acta. Chem. Scand.*, **21**, 1695 (1967).

94. Gronvold, F., *Acta Chem. Scand.*, **26**, 2216 (1972).

95. Gross, P., C. Hayman, and D. L. Levi, *Trans. Faraday Soc.*, **51**, 626 (1955).

96. Gross, P., C. Hayman, and D. L. Levi, *Intern. Congr. Pure Appl. Chem.*, *17th*, Munich (1959).

97. Gross, P., C. Hayman, and D. L. Levi, *Fulmer Res. Inst. Rept. R146/4/23* (1960).

98. Gross, P., C. Hayman, and M. C. Stuart, *Proc. Brit. Ceram. Soc.*, **8**, 39 (1967).

99. Greenberg, E., J. L. Settle, H. M. Feder, and W. N. Hubbard, *J. Phys. Chem.*, **65**, 1168 (1961).

100. Gunn, S. R., *Rev. Sci. Instrum.*, **29**, 377 (1958).

101. Gunn, S. R., *J. Phys. Chem.*, **69**, 2902 (1965).

102. Gunn, S. R., *J. Chem. Thermodynamics*, **2**, 535 (1970).

103. Gupta, B. K., and A. K. Jena, *Trans. Indian Inst. Metals*, **26**, 44 (1973).

104. Gurevich, Ya. A., V. S. Pervov, V. Ya. Leonidov, and I. M. Ievleva, *Zavod. Lab.*, **35**, 1510 (1969).

105. Gutenbaum, J., E. Utzig, J. Wisniewski, and W. Zielenkiewicz, *Conf. Int. Thermodyn. Chim. (C.R.) 4th*, **9**, 144 (1975).

106. Haessner, F., W. Hemminger, and H. L. Lukas, *American Institute of Aeronautics and Astronautics*, Paper No. 74-666 (1974).

107. Hart, R. M., L. D. Hansen, and J. H. Christensen, 1976 Calorimetry Conference, Argonne, Illinois, Unpublished.

108. Hawtin, P., S. A. Gardner, and R. A. Huber, *Atomic Energy Research Establishment, Britain, Report. No. AERE-R 5249* (1966).

109. Head, A. J., *Chem. Thermodynamics*, **1**, 95 (1973).

110. Hendy, B. N., U. S. Pat. 3,740,194 (1973).

111. Hendy, B. N., *Adv. Chem. Ser.*, **142**, 115 (1975) (*Copolym., Polyblends, Compos., Symp.*, 1974).

112. Hendy, B.N., *Chemtech*, **6**, 38, 69 (1976).

113. Herington, E. G., and J. D. Cox, *Pure Appl. Chem.*, **40**, 399 (1972).

114. Holley, C. E., Jr., *Pure Appl. Chem.*, **8**, 131 (1964).

115. Holley, C. E., Jr. committee chairman. U.S. Calorimetry Conference Com. for a Statement on the Funding of Calorimetric Research. 1976–77. Statement adopted at Sherbrooke, Quebec, Canada, July, 1977; in Ref. 77 for the year 1978, Vol. 21, p. 476. *High Temp. Sci.*, **9**, 215 (1977).

116. Hondros, E., *Colloq. Int. Cent. Nat. Rech. Sci.*, **201**, 357 (1971).

117. Howlett, B. W., J. S. Leach, L. B. Ticknor, and M. B. Bever, *Rev. Sci. Intrum.* **33**, 619 (1963).

118. Hu, A. T., G. C. Sinke, M. Mansson, and B. Ringner, *J. Chem. Thermodynamics*, **4**, 283 (1972).

119. Hubbard, W. N., C. Katz, and G. Waddington, *J. Phys. Chem.*, **58**, 142 (1954).

120. Huber, E. J., Jr. and C. E. Holley, Jr. *Tech. Metals Res.*, **4**, 243 (1970).

121. Hultgren, R., P. Newcombe, R. L. Orr, and L. Warner, *Phys. Chem. Metallic Solution and Intermetallic Compounds,* Natl. Phys. Lab., Teddington, England, *Symp. No. 9* (1959).

122. Hultman, B., A. Dalborg, and L. Uhlin, *Sven. Papperstidn.*, **78**, 471 (1975).

123. International Union of Pure and Applied Chemistry. *First Report of the Standing Commission for Thermochemistry* (1934).

124. Appendix to Reference 123 (1936).

125. International Union of Pure and Applied Chemistry, *Pure and Applied Chemistry* (periodical), Butterworths, London.

126. International Council of Scientific Unions, Committee on Data for Science and Technology (CODATA), *Key Values for Thermodynamics,* published intermittently in Committee Bulletins and in the *J. Chem. Thermodynamics.*

127. Ivanov, M. I., and V. A. Tumbakov, *Zhur. Fiz. Khim.*, **33**, 224 (1959).

128. JANAF Thermochemical Tables. Dow Chemical Co., Midland, Michigan.

129. Jensen, T. E., Calorimetry Conference, Argonne, Illinois, 1976. Unpublished.

130. Jessup, R. S., R. E. McCoskey, and R. A. Nelson, *J. Am. Chem. Soc.*, **77**, 244 (1953).

131. Johnson, W. H., *J. Res. Natl. Bur. Stand.*, **79A**, 561 (1975).

132. Joncich, M. J., and H. F. Holmes, *Proc. Australian Conf. Electrochem. 1st, Sydney, Hobart, Australia,* p. 138 (1963).

133. Jordan, J., J. K. Grime, D. H. Waugh, C. D. Miller, H. M. Cullis, and D. Lohr, *Anal. Chem.*, **48**, A 427 (1976).

134. Kaduji, I. I., and K. H. Rees, *Analyst* (London), **99**, 435 (1974).

135. Kay, J. G., N. A. Kuebler, and L. S. Nelson, *Nature*, **194**, 671 (1962).

136. Kingery, W. D., *Property Measurement at High Temperatures,* Wiley, New York, 1959.

137. Kleppa, O. J., *U.S. Army Research Office Report AD-766967* (1973).

138. Kleppa, O. J., U.S. Calorimetry Conference, Sherbrooke, Quebec, Canada, 1977. Unpublished.

139. Klyuev, L. I., V. Ya. Leonidov, O. M. Gaisinskaya, and V. S. Pervov, *Zh. Fiz. Khim.*, **48**, 212 (1974).

140. Kniebas, D. V., *Am. Gas. Assoc., Oper. Sect. Proc.*, p. D70 (1972).

141. Kolesov, V. P., L. S. Ivanov, S. P. Alekhin, and S. M. Skuratov, *Zh. Fiz. Khim.*, **44**, 2956 (1970).

142. Konicek, J., J. Suurkuusk, and I. Wadso, *Chem. Scr.*, **1**, 217 (1971).

143. Kubaschewski, O., E. L. Evans, and C. B. Alcock, *Metallurgical Thermochemistry,* 4th Edition, Pergamon Press, Oxford, 1967.

144. Kubaschewski, O., and W. Slough, *Prog. Mater. Sci.*, **14**, 1 (1969).

145. Kubaschewski, O., and H. Villa, *Z. Elektrochem.*, **53**, 32 (1949).

146. Kubaschewski, O., and A. Walter, *Z. Elektrochem.*, **45**, 630 (1939).

147. Kusano, K., B. Nelander, and I. Wadso, *Chem. Scr.*, **1**, 211 (1971).

148. Kybett, B. D., and J. L. Margrave, *Rev. Sci. Instrum.*, **37**, 675 (1966).

149. Lacher, J. R., J. J. McKinley, C. M. Snow, L. Michel, G. Nelson, and J. D. Park, *J. Am. Chem. Soc.*, **71**, 1334 (1949).

150. Lacher, L. R., and J. D. Park, *Am. Chem. Soc., Div. Petrol. Chem. Reprint 3, No. 4B,* 51 (1958).

151. Lanneau, K. P., U.S. Pat. 3,560,160 (1971).

152. Lavoisier, H. L., and P. S. de la Place, *Mem. Acad. Sci. 1780* (3), 355 (1784); see G. T. Armstrong, *J. Chem. Educ.*, **41**, 297 (1964).

153. Leach, J. S. L., *Tech. Metals Res.*, **4**, 197 (1970).

154. Lee, K. N., R. Bachman, T. H. Geballe, and J. P. Maita, *Phys. Rev*, **B2**, 4680 (1970).

155. Lewis, D. C., M. A. Frisch, and J. L. Margrave, *Carbon*, **2**, 431 (1965).

156. Lewis, G. N., and R. Randall, *Thermodynamics*, revised by L. Brewer and K. S. Pitzer, 2nd ed., McGraw-Hill, New York, 1961.

157. Loeb, L., and R. O. Shuck, *J. Am. Oil. Chem. Soc.*, **48**, 25 (1971).

158. Lopes, M. T. R., M. F. T. Nunes, M. H. F. S. Florecio, M. M. G. Mota, and M. T. N. Fermandez, *Rev. Port. Quim.*, **15**, 129 (1975).

159. Mackle, H. and R. G. Mayrick, *Pure Appl. Chem.*, **2**, 25 (1961).

160. Mackle, H., and P. A. G. O'Hare, *Trans. Faraday Soc.*, **59**, 2693 (1963).

161. Maggs, F. A. P., and P. H. Schwabe, *J. Sci. Instrum.*, **39**, 364 (1962).

162. Maier, C. G., *J. Phys. Chem.*, **34**, 2860 (1930).

163. Makhnina, H. P., V. D. Mikina, B. N. Oleinik, and B. R. Psavko, *Tr. Metrol. Inst. SSSR*, **129**, 83 (1971).

164. Mananikov, B. P., and V. N. Pankratov, *Zh. Fiz. Khim.*, **46**, 787 (1972).

165. Mansson, M. *J. Chem. Thermodynamics*, **5**, 721 (1973).

166. Mansson, M. *J. Chem. Thermodynamics*, **6**, 1018 (1974).

167. Manuel, P., H. Niedoba, and J. J. Veyssie, *Rev. Phys. Appl.*, **7**, 107 (1972).

168. Manuel, P., and J. J. Veyssie, *Rev. Gen. Therm.*, **12**, 337 (1973).

169. Margrave, J. L. *J. Chem. Educ*, **32**, 520 (1955).

170. Margrave, J. L. and R. G. Bautista, in H. B. Jonassen and A. Weissberger, Eds., *Technique of Inorganic Chemistry*, Vol. 4, Interscience, New York, 1966.

171. Martin, D. L., and Snowdon, R. L., *Can. J. Phys.*, **44**, 1449 (1966).

172. Masi, J. F., H. W. Flieger, and J. S. Lickland, *J. Res. Natl. Bur. Stand.*, **52**, 275 (1954).

173. Masi, J. F., and B. Petkoff, *J. Res. Natl. Bur. Stand.*, **48**, 179 (1952).

174. Mattiasson, B., B. Danielsson, and K. Mosbach, *Anal. Lett.*, **9**, 217 (1976).

175. McCullough, J. P., et al., *J. Am. Chem. Soc.*, **76**, 4791 (1954).

176. McCullough, J. P., and D. W. Scott, Eds., *Experimental Thermodynamics, Vol. 1, Calorimetry of Non-reacting Systems*, Butterworths, London, 1968.

177. McEwan, W. S., and C. M. Anderson, *Rev. Sci. Instrum.*, **26**, 280 (1955).

178. McGlothlin, C. D., and J. Jordan, *Anal. Chem.*, **47**, 786 (1975).

179. Mchedlov-Petrosyan, O. P., A. Sen, and A. V. Usherov-Marshak, *Silikattechnik*, **20**, 229 (1969).

180. Mchedlov-Petrosyan, O. P., A. V. Usherov-Marshak, and A. M. Urshenko, *Silikattechnik*, **25**, 261 (1974).

181. Misra, S., H. P. Singh, and P. U. Nayak, *Indian J. Technol.*, **6**, 254 (1968).

182. Mogutnov, B. N., and P. A. Gomozov, *Metody Opred. Issled. Sostoyaniya Gazov Met., Vses. Konf., 3rd*, **1**, 119 (1973).

183. Monfore, G. F., and B. Ost, *J. Portland Cement Assoc. Res. Development Lab.*, **8**, 13 (1966).

184. Monk, P., and I. Petterson, *Process Biochem. 1969*, 63.

185. Monk, P., and I. Wadso, *Acta Chem. Scand.*, **22**, 1842 (1968).

186. Montgomery, J. B., and T. DeVries, *J. Am. Chem. Soc.*, **64**, 2372 (1942).

187. Montgomery, R. L., *Science*, **184**, 562 (1974).

188. Montgomery, R. L., R. A. Melaugh, C. C. Lau, G. H. Meier, H. H. Chan, and F. D. Rossini, *J. Chem. Thermodynamics*, **9**, 915 (1977).

189. Montgomery, R. L., R. A. Melaugh, C. C. Lau, G. H. Meier, R. T. Grow, and F. D. Rossini, *J. Chem. Eng. Data*, **23**, 245 (1978).

190. Mosbach, K., B. Danielsson, A. Borgerud, and M. Scott, *Biochim. Biophys. Acta*, **403**, 256 (1975).

191. Mott, R. A., and W. C. Thomas, *Fuel* (London), **33**, 448 (1954).

192. Mountford, G. A., and P. A. H. Wyatt, *Bol. Soc. Chilena Quim.*, **13**, 47 (1963).

193. Muller, W., and A. Schuller, *Ber. Bunsenges. Phys. Chem.*, **75**, 79 (1971).

194. Neckel, A., and H. Nowotny, *Int. Leichtmetalltagung, 5th*, p. 72 (1968).

195. Nelson, L. S., and N. A. Kuebler, *J. Chem. Phys.*, **37**, 47 (1962).

196. Nelson, L. S., and N. A. Kuebler, *J. Chem. Phys.*, **39**, 1055 (1963).

197. Neugebauer, C. A., and J. L. Margrave, *J. Phys. Chem.*, **61**, 1429 (1957).

198. Neugebauer, C. A., and J. L. Margrave, *J. Phys. Chem.*, **60**, 1318 (1956).

199. Newman, E. S. *J. Res. Natl. Bur. Stand.*, **A66**, 381 (1962).

200. Nuttall, R. L., M. A. Frisch, and W. N. Hubbard, *Rev. Sci. Instrum.*, **32**, 1402 (1961).

201. Nuttall, R. L., C. Katz, and W. N. Hubbard, *Rev. Sci. Instrum.*, **31**, 461 (1960).

202. O'Farrell, H. K., S. K. Chattopadhay, and H. D. Brown, U. S. Calorimetry Conference, Sherbrooke, Quebec, Canada 1977. Unpublished.

203. Osborne, N. S., H. F. Stimson, and T. S. Slign, *Natl. Bur. Stand. Sci. Papers*, **20**, 119 (1924–1926).

204. Palonieme, P., *Institute of Electrical and Electronic Engineers. Trans. Elec. Insul.*, **7**, 126 (1972).

205. Pavore, D., and Holley, C. E., Jr., unpublished.

206. Pennington, S. N., *Rev. Anal. Chem.*, **1**, 113 (1972).

207. Pennington, S. N. and H. D. Brown, *Chem. Instrum.*, **2**, 167 (1969).

208. Pepper, L., "Precision of Quartz Crystal and Mercury Differential Thermometers in Heat-of-Hydration Test." *U.S. Clearinghouse Fed. Sci. Tech. Inform.*, AD-692794 (1969).

209. Peters, H., and J. Malzahn, *Monatsber. Deut. Akad. Wiss. Berlin*, **13**, 893 (1971).

210. Peters, H., E. Tappe, and M. Urbanczik, *Monatsber. Deut. Akad. Wiss. Berlin*, **8**, 720 (1966).

211. Peters, H., E. Tappe, and M. Urbanczik, *Monatsber, Deut. Akad. Wiss. Berlin*, **9**, 28 (1967).

212. Phillips, N. E., *Phys. Rev.*, **114**, 679 (1959).

213. Picker, P., *Can. Res. Develop., 1974* (1), 11.

214. Picker, P., C. Jolicoeur, and J. E. Desnoyers, *J. Chem. Thermodynamics*, **1**, 469 (1969).

215. Picker, P., P. A. Leduc, P. R. Philip, and J. E. Desnoyers, *J. Chem. Thermodynamics*, **3**, 631 (1971).

216. Pierce, P. E., *J. Chem. Educ.*, **39**, 338 (1962).

217. Pilcher, G., H. A. Skinner, A. S. Pell, and A. E. Pope, *Trans. Faraday Soc.*, **59**, 316 (1963).

218. Pitzer, K. S., *J. Am. Chem. Soc.*, **63**, 2443 (1941).

219. Pool, M. J., and J. R. Guadagno, *U. S. Air Force Report AFML-TR-66-352* (1966).

220. Popoff, M. M., and P. K. Schirokich, *Z. Phys. Chem.*, **A167**, 183 (1933).

221. Porter, R. S., and J. F. Johnson, Eds., *Analytical Calorimetry*, American Chemical Society, Plenum, New York.
    (a) Volume 1, 1968

   (b) Volume 2, 1970
   (c) Volume 3, 1974
   (d) Volume 4, 1977

222. Prophet, H., and D. R. Stull, *J. Chem. Eng. Data*, **8**, 78 (1963).

223. Prosen, E. J., and M. V. Kilday, *J. Res. Natl. Bur. Stand.*, **A77**, 581 (1973).

224. Prosen, E. J., and M. V. Kilday, *J. Res. Natl. Bur. Stand.*, **A77**, 179 (1973).

225. Prosen, E. J., V. Maron, and F. D. Rossini, *J. Res. Natl. Bur. Stand.*, **42**, 269 (1949).

226. Pugh, B., *Fuel Calorimetry*, Butterworths, London, 1966.

227. Rehak, N. N., J. Everse, N. O. Kaplan, and R. L. Berger, *Anal. Biochem.*, **70**, 381, (1976).

228. Richards, T. W., and R. H. Jesse, *J. Am. Chem. Soc.*, **32**, 268 (1910).

229. Rivin, O. V., I. V. Basina, and G. Z. Khaidarov, *Probl. Teploenerg. Prikl. Teplofiz.*, **4**, 249 (1967).

230. Rogers, D. W., and F. J. McLafferty, *Tetrahedron*, **27**, 3765 (1971).

231. Rogers, D. W., P. M. Papadimetriou, and N. A. Siddiqui, *Mikrochim. Acta*, **2**, 389 (1975).

232. Rogers, D. W., and R. J. Sasiela, *Anal. Biochem*, **56**, 460 (1973).

233. Rossini, F. D., *J. Res. Natl. Bur. Stand.*, **6**, 1 (1931).

234. Rossini, F. D., *J. Res. Natl. Bur. Stand.*, **15**, 357 (1935).

235. Rossini, F. D., *Chemical Thermodynamics*, Wiley, New York, 1950.

236. Rossini, F. D., Ed., *Experimental Thermochemistry*, vol. 1, Interscience, New York, 1956.

237. Rossini, F. D., *Fundamental Measures and Constants for Science and Technology*, CRC Press, Cleveland, Ohio, 1974.

238. Rossini, F. D., D. D. Wagman, W. H. Evans, S. Levine, and I. Jaffe, *Selected Values of Chemical Thermodynamic Properties, U.S. Natl. Bur. Stand. Circular 500* (1950).

239. Roth, O. and W. Menzel, *Z. anorg. allgem. Chem.*, **198**, 375 (1931).

240. Roth, W. A., and F. Becker, *Kalorimetrische Methoden zur Bestimmung chemischer Reaktionswarmen*, F. Viewig. Braunschweig, 1956.

241. Roth, W. A., H. Ginsberg, and R. Lasse, *Z. Elektrochem.*, **30**, 417 (1924).

242. Sabbah, R., R. Chastel, and M. Lafitte, *Thermochim. Acta*, **5**, 117 (1972).

243. Scheel, K., and W. Heuse, *Ann. Physik*, **37**, 79 (1912).

244. Scheel, K., and W. Heuse, *Ann. Physik*, **40**, 473 (1913).

245. Scheel, K., and W. Heuse, *Ann. Physik*, **9**, 586 (1919).

246. Schroder, J., and F. J. Sieben, *Chem. Ber.*, **103**, 76 (1970).

247. Schutz, R. J., *Rev. Sci. Instrum.*, **45**, 548 (1974).

248. Sellers, G. J., and A. C. Anderson, *Rev. Sci. Instrum.*, **45**, 1256 (1974).

249. Settle, J. L., E. Greenberg, and W. N. Hubbard, *Rev. Sci. Instrum.*, **38**, 1805 (1967).

250. Sezin, L. N., and M. I. Sukharev, *Izv. Vysshikh Uchebn. Zavedenii, Tekhnol. Legkoi Prom 1966*, 35.

251. Singh, H. P., *Scr. Met.*, **6**, 519 (1972).

252. Singh, H. P., *Banaras Met.*, **5**, 6 (1973).

253. Sinke, G. C., *Thermodynamic Properties of Combustion Products*, Dow Chemical Company, Midland, Michigan, 1959.

254. Skinner, H. A., Ed., *Experimental Thermochemistry, Vol. 2*, Interscience, New York, 1962.

255. Smit, W. M., *Rec. Trav. Chim.*, **75**, 1309 (1956).

256. Smit, W. M., *Anal. Chem. Acta*, **17**, 23 (1957).

257. Smit, L., and Sunner, S., *The Svedberg Memorial Edition*, Uppsala, Sweden, 1944, p. 352.

258. Southard, J. C., *J. Am. Chem. Soc.*, **63**, 3142 (1941).

259. Spink, E., and I. Wadso, *Meth. Biochem. Anal.*, **23**, (1975).

260. Splitstone, P. L., and W. H. Johnson, *J. Res. Natl. Bur. Stand.*, **78A**, 611 (1974).

261. Stansbury, E. E., and G. R. Brooks, *High Temp.-High Pressures*, **1**, 289 (1969).

262. Steffen, H., and H. Wollenberger, *Rev. Sci. Instrum.*, **44**, 937 (1973).

263. Stephens, H. P., *High Temp. Sci.*, **6**, 156 (1974).

264. Stoesser, P. R., and S. J. Gill, *Rev. Sci. Instrum.*, **38**, 422 (1967).

266. Streiff, A. J., *Ann. N.Y. Acad. Sci*, **137**, 375 (1966).

267. Strittmater, R. C. and G. C. Danielson, *U.S. Atomic Energy Commission Report ISC-666* (1955).

268. Strittmater, R. C., G. J. Pearson, and G. C. Danielson, *Proc. Iowa Acad. Sci.*, **64**, 466 (1957).

269. Stull, D. R., and G. C. Sinke, *Thermodynamic Properties of the Elements*, Advances in Chemistry Series, No. 18. American Chemical Society, Washington, D.C., 1956.

270. Stull, D. R., E. F. Westrum, Jr., and G. C. Sinke, *The Chemical Thermodynamics of Organic Compounds*, Wiley, New York, 1969.

271. Sturtevant, J., "Calorimetry," in *Techniques of Chemistry*, A. Weissberger and B. W. Rossiter, Eds., Wiley-Interscience, New York, 1971 (a revision and expansion of an earlier treatise).

272. Sturtevant, J. M., *Meth. Enzymol.*, **C26**, 227 (1972).

273. Sukhovei, K. S., V. F. Anishin, and I. E. Paukov, *Zh. Fiz. Khim.*, **48**, 1589 (1974).

274. Sullivan, P. F., and G. Seidel, *Ann. Acad. Sci. Fenn, Ser. AVI, No. 21*, 58 (1966).

275. Sullivan, P. F., and G. Seidel, *Phys. Rev.*, **173**, 679 (1968).

276. Sunner, S., Thesis, University of Lund, Sweden, Carl Bloms Boktrycheri, Lund, Sweden, 1949, Unpublished.

277. Sunner, S., *Acta Chem. Scand.*, **13**, 825 (1959).

278. Sunner, S., *Developments in Calorimetry over Thirty-Five Years 1945–1980*, Huffman Memorial Lecture, U.S. Calorimetry Conference, Gaithersburg, Maryland, 1970. Printed in the *Bulletin of Thermodynamics and Thermochemistry*.

279. Sunner, S., and I. Wadso, *Acta Chem. Scand.*, **13**, 97 (1959).

280. Suurkuusk, J., and I. Wadso, *J. Chem. Thermodynamics*, **6**, 667 (1974).

281. Swietoslawski, W., *Microcalorimetry*, Reinhold, New York, 1946.

282. Thompson, R., *J. Chem. Soc. 1953*, 1908.

283. Tian, A., *Bull. Soc. Chim. France*, **33**, 427 (1923).

284. Tiehl, I., and W. A. Roth, *Z. Elektrochem.*, **52**, 219 (1948).

285. Trainor, R. J., G. Knapp, M. B. Brodsky, G. J. Pokorny, and R. B. Snyder, *Rev. Sci. Instrum.*, **46**, 1368 (1975).

286. (a) Treverton, J. A., and J. L. Margrave, *Proc. Symp. Thermophys. Prop.*, 5th, 1970, p. 489; (b) *J. Phys. Chem.*, **75**, 3737 (1971).

287. Tsutsumi, K., H. Q. Koh, S. Hagiwara, and H. Takahashi, *Bull. Chem. Soc. Japan*, **38**, 3576 (1975).

288. Turner, R. B., W. R. Meador, and R. E. Winkler, *J. Am. Chem. Soc.*, **79**, 4116 (1957).

289. U.S. Dept. of Health, Education, and Welfare, U.S Patent Application 357,272 (1973); through *Chem. Abstr.* **81**, P116858n.

290. U.S. National Bureau of Standards, Institute for Materials Research, Office of Standard Reference Materials, catalog.

291. U.S. National Bureau of Standards, *Selected NBS Papers on Heat, U.S. Natl. Bur. Stds. Special Publication No. 366,* Vol. 6 (1970).

292. U.S. National Bureau of Standards. Technical Notes.
    (a) No. 270-1 Superseded by No. 270-3.
    (b) No. 270-2 Superseded by No. 270-3.
    (c) No. 270-3, by D. D. Wagman, W. H. Evans, V. B. Parker, I. Halow, S. M. Bailey, and R. H. Schumm. 1968.
    (d) No. 270-4, by D. D. Wagman, W. H. Evans, V. B. Parker, I. Halow, S. M. Bailey, and R. H. Schumm. 1969.
    (e) No. 270-5, by D. D. Wagman, W. H. Evans, V. B. Parker, I. Halow, S. M. Bailey, R. H. Schumm, K. L. Churney, 1971.
    (f) No. 270-6 by V. B. Parker, D. D. Wagman, and W. H. Evans, 1971.
    (g) No. 270-7 by R. H. Schumm, D. D. Wagman, S. Bailey, W. H. Evans, and V. B Parker, 1973.

293. Verhoff, F. H., *Anal. Chem.,* **43,** 183 (1971).

294. Viswanathan, P., and H. L. Luo, *Proc. Nucl. Phys. Solid State Phys. Symp. 17th,* C3615 (1972).

295. von Wartenburg, H., and O. Fitzner, *Z. anorg. allgem. Chem.,* **151,** 313 (1926).

296. von Wartenburg, H., and H. Schutsa, *Z. anorg. allgem. Chem.,* **206,** 65 (1932).

297. Vulterin, J., P. Straka, M. Stastny, and R. Volf, *Chem. Prum.,* **24,** 618 (1974).

298. Vulterin, J., P. Straka, M. Stastny, and R. Volf, *Cesk. Farm.,* **24,** 10 (1975).

299. Vurek, G. G., L. C. Stoner, and S. E. Pergram, *Proc. Annu. Conf. Eng. Med. Biol.,* **15,** 124 (1973).

300. Vurek, G. G., D. G. Warnock, and R. Corsey, *Anal. Chem.,* **47,** 765 (1975).

301. Waddington, G., S. S. Todd, and H. M. Huffman, *J. Am. Chem. Soc.,* **69,** 22, (1947).

302. Wadso, I., *Quart. Rev. Biophys.,* **3,** 383 (1970).

303. Wadso, I., in H. D. Brown, Ed., *Biochemical Microcalorimetry,* Academic, New York, 1968.

304. Wadso, I., *Protides Biol. Fluids. Proc. Colloq.,* **19,** 507 (1971).

305. Wadso, I., *Pure Appl. Chem.,* **38,** 529 (1974).

306. Wadso, I., in *New Techniques in Biophysics and Cell Biology,* R. H. Pain and B. J. Smith, Eds., Vol. 2, Wiley, New York, 1975.

307. Wadso, I., *Acta Chem. Scand.,* **22,** 927 (1968).

308. Washburn, E. W., *J. Res. Natl. Bur. Stand.,* **10,** 525 (1933).

309. West, E. D., *J. Am. Chem. Soc.,* **81,** 29 (1959).

310. Westrum, E. F., Jr., *J. Chem. Educ.,* **39,** 443, (1962).

311. Westrum, E. F., Jr., *Colloq. Int. Cent. Nat. Rech. Sci.,* **201,** 103 (1972).

312. Westrum, E. F., Jr., *Adv. High Temp. Chem.,* **1,** 239 (1967).

313. Westrum, E. F., Jr., *Pure Appl. Chem.,* **38,** 539 (1974).

314. White, W. P., *The Modern Calorimeter,* American Chemical Society Monograph No. 42, Chemical Catalogue Co., New York, 1928.

315. White, W. P., *J. Phys. Chem.,* **24,** 393 (1920).

316. Wilhoit, R. C., *J. Chem. Educ.,* **44,** A571 (1967); *J. Chem. Educ.,* **44,** A629 (1967); *J. Chem. Educ.,* **44,** A685 (1967); *J. Chem. Educ.,* **44,** A853 (1967).

317. Wilson, G. M., and S. T. Barton, *Proc. Annu. Conv. Natur. Gas Processors Assn.,* Tech. Pap. No. 46, 18 (1967).

318. Wise, C. J., H. M. Feder, W. N. Hubbard, and J. L. Margrave, *J. Phys. Chem.,* **67,** 815 (1963).

319. Woledge, R. C., *Pestic. Sci.*, **6**, 305 (1975).

320. Worthing, A. G., *Phys. Rev.*, **12**, 199 (1918).

321. Zielenkiewicz, W., and T. Kurek, *Przemysl Chem.*, **45**, 247 (1966).

322. Zwolinski, B. J., Selected Values of Properties of Hydrocarbons and Related Compounds, American Petroleum Institute Research Project 44, Texas A&M University, College Station, Texas. (Loose leaf sheets and booklets published at irregular intervals.)

323. Shaner, J. W., G. R. Gathers, and C. Minichino, *High Temp.-High Pressures*, **8**, 425 (1976).

ADDENDUM

This addendum contains a number of comments and references covering new developments up to early 1982. We have clearly omitted many other significant contributions to calorimetry, since time did not permit a complete survey and evaluation of the literature published since this chapter was written.

1. When this chapter was written, calibration of solution calorimeters by a standard reaction had not been widely accepted as a satisfactory procedure. Recently, however, the situation has been changing. The U.S. National Bureau of Standards has been preparing new data and recommendations for the use of potassium chloride as a calorimetric reference material, and the recommendations provide for calibration of water solution calorimeters with potassium chloride when electrical calibration equipment is not available. (M. V. Kilday, National Bureau of Standards, Private communication, March 1980).

2. Among the industrial applications of calorimetry described recently is the control of a combustion process. The heating value of fuel gas is determined in a calorimeter, and the flow rate of the gas is adjusted accordingly. This is claimed to be an improvement over control through measurement of the oxygen remaining after combustion. The calorimetric method is said to maintain steadier combustion conditions and to increase the efficiency of combustion by reducing the amount of excess air required. (*Chemical Processing*, March 1981, page 68).

3. A recent paper describes the indirect determination of the equations of state of gaseous compounds and mixtures, using calorimetry. The enthalpy changes during expansion of a gas are directly related to the temperature derivatives of the coefficients in the virial equation for the P-V-T relationship. Calorimetry was used to determine the effects of temperature on the second virial coefficients. [Lesnevskaya, L. S., Domracheva, T. J., and Nikiforova, M. B., *Zh. Fiz. Khim.*, **55**, 2436 (1981) through *Chem. Abstr.* **95**, 192571t (1981).]

4. Other recent publications on calorimetry include:
   a. Hemminger, W., "Grundlagen der Kalorimetrie," Verlag Chemie (1979).

b. Gokcen, N. A., Mrazek, R. V., and Pankratz, L. B., compilers, "Proceedings of the Workshop on Techniques for Measurement of Thermodynamic Properties," Albany, Oregon, Aug. 21–23, 1979, U.S. Bureau of Mines Information Circular Number 8853 (1981).

c. Recommendations regarding the data required to establish key thermochemical values [V. A. Medvedev and I. L. Khodakovskii, *Russian Chemical Reviews*, **48**, 1168 (1979)].

d. Recommendations about how to conduct and report calorimetric measurements on biological systems [CODATA Bulletin No. 44, Committee on Data for Science and Technology of the International Council of Scientific Unions, Paris, France (1981)].

e. Recommendations concerning the assignment and presentation of uncertainties in calorimetric data [International Union of Pure and Applied Chemistry, *J. Chem. Thermodynamics*, **13**, 603 (1981)].

f. Recommendations concerning symbols and terminology in chemical thermodynamics [J. D. Cox, *Pure Appl. Chem.* **51**, 393 (1979)].

EVOLVED GAS ANALYSIS

By H. G. Langer,
Dow Chemical USA,
Central Research New England Laboratory,
Wayland, Massachusetts

Contents

I. INTRODUCTION

A. HISTORY

The study of physical or chemical changes of materials, subjected to a temperature program, is defined as thermal analysis (438,440,441,442,460). Since the technique depends on precise temperature control and measurement, it developed when accurate and sensitive instruments became available. In addition, devices had to be modified to detect changes caused by varying the sample temperature or the amount of energy supplied. A large number of investigations involve chemical reactions, many with release of gaseous products. It soon became desirable to detect and identify volatile species, to measure the amounts evolved, and to understand the mechanisms by which gases are formed.

This part of thermal analysis, now known as evolved gas analysis (EGA), is the subject of this chapter. Even more than other thermoanalytical techniques, it depends on instrumentation. The importance of sufficiently accurate control of the atmosphere surrounding a sample was recognized early (539). Equipment was adapted to allow such control (16,17,612,623) and to regulate gas flow (673,674). The first "gas effluent analysis proper" was attributed (417) to work published in 1949 (56), that used stepwise heating. Several publications around 1960 described ways of detecting evolved gases (17,35,91,121,143,151,162,232,233,445,507,531,607,768). A mathematical treatment of EGA was reported in 1960, together with construction of an advanced apparatus for atmosphere control and temperature programming (232). Later, a commercial thermal evolution analyzer was developed (169,669).

With advances of modern instrumental designs, specialization developed, and gas analysis separated into various techniques based on "hardware" rather than purpose or results of the analytical procedure. At present, the most powerful techniques are gas chromatography and mass spectrometry, each with its own subdivisions. As part of thermal analysis, therefore, evolved gas analysis is no longer a simple method, but a combination of two or more, sometimes very sophisticated, analytical techniques.

By far the most significant effect was caused by the penetration of computer technology into analytical chemistry. In fact, today, data acquisition and storage capabilities of computers have outpaced human memory in utilizing the wealth of available data. Now, the limitation appears to be no longer in instruments, but in creating programs capable of reducing the data generated by a computer to a level capable of being handled by the human mind.

B. PURPOSE AND SCOPE

This chapter will discuss ways in which thermal analysis and gas-analytical techniques can be and have been combined to achieve results not obtainable with either technique alone. There are almost unlimited ways by which such combinations are possible. Consequently, this discussion cannot be exhaustive, but is intended to cover the principles and significant contributions. For detailed information on specific gas-analytical problems, the reader must be referred to the corresponding literature. In addition, a list of general references (G1–G64) should be useful as a starting point for extended studies.

Because of the advances and specialization of instrumental developments and applications, it should not be surprising that in the majority of reported experiments the investigators had to adapt, modify, even redesign or construct equipment, suitable for thermoanalytical studies. This in turn explains, why even today the reports on methodology and instrumentation appear to be more numerous than those on applications and results.

C. NOMENCLATURE

Simultaneous and independent developments caused considerable confusion with definitions and terminology. Efforts by the nomenclature committee of the International Confederation for Thermal Analysis (ICTA) (438,440,441,460) resulted in recommendations reported elsewhere in this volume. Every attempt is being made to abide by these recommendations. However, alternate terms have been in widespread use and therefore will have to be recognized here, provided they describe a significantly different approach. The two specific recommended terms for this chapter are "Evolved Gas Analysis" (EGA) and "Evolved Gas Detection" (EGD). To avoid further confusion and to recognize the usage of "qualitative analysis" for "detection," EGD will be used here as the subordinate term to EGA.

D. DEFINITION

As defined by ICTA (442), evolved gas analysis is "a technique in which the nature and/or amount of volatile products released by a substance is measured as a function of temperature whilst the substance is subjected to a controlled temperature programme." Even this definition is open to various interpretations. It will be used here with the greatest possible latitude, provided: (1) thermal reaction caused by a controlled or measured application of energy generated a gas (or vapor), (2) an observation or measurement is made on this gas, and, (3) some direct or indirect temperature measurement is made during the experiment.

Thus, gas evolution analysis as defined by Wendlandt (G58), effluence analysis by Garn (G25), and gas effluent analysis described by Lodding (G37) are all included. Neither the name of a method used to generate the gas, nor the purpose or result of any procedure, shall exclude it from consideration. The only exception is thermogravimetry (TG). Even though it essentially fits the definition above, it has become known as a technique in its own right and is being discussed in a separate chapter.

II. THEORY

A. BASIC LAWS

Theoretical considerations are restricted to gas formation and transport phenomena that affect the outcome of the analysis. Theories concerning specific gas analytical techniques that have been combined with thermal analysis are discussed in the specific sections. Summaries will be provided in the section on Practice.

The processes involved in EGA are heterogeneous and generally include the following steps:

1. gas formation (such as in a chemical reaction),
2. gas diffusion (through solid/liquid),
3. gas release (desorption) from the interface, and
4. transport from the sample area to a detector.

Lack of reproducibility in early experiments resulted from a lack of understanding of the effects of experimental conditions on these processes. It will be shown later, that several of these variables can be eliminated by proper design of the experiment.

Rogers and co-workers (607) presented a mathematical treatment for gas evolution from solids. They compared the theoretical curves with those observed with a thermal conductivity detector for the decomposition of two explosives in a stream of helium. Instead of a thermobalance, they used a simple pyrolysis block.

For nonvolatile compounds decomposing by a first-order rate law, the Arrhenius equation applies:

$$- \frac{dN}{dt} = ZN \exp\left(- \frac{E}{RT}\right) \tag{1}$$

where N is the number of moles of the sample.

For a constant rate of temperature increase:

$$r = \frac{dT}{dt} = \text{const.}$$

$$\text{and } p \doteq - \frac{dN}{dt}$$

Differentiation with respect to T yields:

$$\frac{d \ln p}{dT} = \frac{E}{RT^2} - \frac{Z}{r} \exp\left(- \frac{E}{RT}\right) \tag{2}$$

and after integration:

$$\ln\left(\frac{p}{p_0}\right) = \frac{E}{R}\left(\frac{1}{T_0} - \frac{1}{T}\right) - \frac{Z}{r}\int_{T_0}^{T} \exp\left(-\frac{E}{RT}\right)dT \tag{3}$$

where p_0 is the rate of decomposition at temperature T_0.

The integral of equation 3 is evaluated via asymptotic expansion of the exponential integral:

$$\int_{x}^{\infty} \frac{e^{-x}}{x}\,dx$$

to give:

$$\ln\left(\frac{p}{p_0}\right) = \frac{E}{R}\left(\frac{1}{T_0} - \frac{1}{T}\right) - \frac{ZR}{rE}\left[T^2 \exp\left(-\frac{E}{RT}\right) - T_0^2 \exp\left(-\frac{E}{RT_0}\right)\right] +$$

$$\frac{2ZR^2}{rE^2}\left[T^3 \exp\left(-\frac{E}{RT}\right) - T_0^3 \exp\left(-\frac{E}{RT_0}\right)\right] + \dots \tag{4}$$

Since RT/E is generally in the order of 10^{-2}, higher-order terms are negligible.

The relationship between the amount of gas released and the response of a thermal conductivity detector was discussed by Ingraham (304,307).

For dilute systems, the ideal gas law is valid:

$$PV = \frac{wRT}{M} \tag{5}$$

where P is the equilibrium pressure of w grams of gas liberated into volume V at temperature T. M is the molecular weight. In a flowing system, P becomes the partial pressure of the product gas in the carrier and V the total volume of product gas and carrier. For continuous flow, $V = vt$, where v is the flow and t the time of flow. Thus equation 5 becomes:

$$w\left(\frac{RT}{Mv}\right) = Pt \tag{6}$$

and differentiated,

$$\frac{dw}{dt} \sim P \tag{7}$$

The response of a thermal conductivity detector is proportional to the partial pressure of the sample gas surrounding it, and if the temperature increase is linear, time is proportional to temperature. Therefore, equation 7 describes the shape of the pyrolysis curve as recorded by the detector. The area under the curve is proportional to the amount of gas liberated, since $dw/dt = kf(t)$:

$$w = k\int f(t)\,dt \tag{8}$$

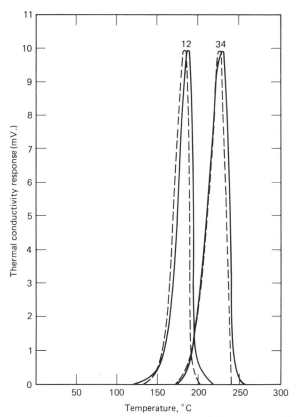

Fig. 6.1. Comparison of theoretical and experimental pyrolysis curves. 1. Theoretical curve for PETN. 2. Experimental curve for PETN, uncorrected for gas flow time lag. 3. Theoretical curve for RDX. 4. Experimental curve for RDX, uncorrected for gas flow time lag. (Ref. 607)

Rogers et al. (607) compared their experimental and theoretical curves (Fig. 6.1) for cyclotrimethylenetrinitramine (RDX) and pentaerythritol tetranitrate (PETN). For their theoretical curves, they used values of Z and E reported earlier (605,606).

At the maximum:

$$\frac{ZR}{rE} T_m^2 \exp\left(-\frac{E}{RT_m}\right) = 1 \tag{9}$$

At $p = \frac{1}{4} P_{max}$, equations 4 and 9, by neglecting the T terms, yield:

$$\frac{E}{R}\left(\frac{1}{T_{1/4}} - \frac{1}{T_m}\right) = 1 + \ln 4 - \frac{ZR}{rE} T_{1/4}^2 \exp\left(-\frac{E}{RT_{1/4}}\right) \tag{10}$$

This equation is also valid if every 4 is replaced by a different number. Therefore, if for instance T_m and $T_{1/4}$ are determined experimentally, E and Z can be calculated from equations 9 and 10.

For completely volatile compounds, Rogers et al. (607) obtained pyrolysis curves with a gradual rise and sharp drop-off at a temperature T_x. For saturation of the carrier gas with sample vapor, the rate of disappearance of the sample is:

$$- \frac{dN}{dt} = \frac{Pv}{RT} \tag{11}$$

Here, P is the vapor pressure and v the flow rate in ml/sec. The original moles of sample N_0, according to the Clausius-Clapeyron equation are:

$$N_0 = \frac{P_0 v \, \exp\left(\dfrac{\Delta H}{RT_0}\right)}{R_r} \int_0^{T_x} \frac{\exp\left(-\dfrac{\Delta H}{RT}\right)}{T} \, dT \tag{12}$$

By approximation of the integral above, the relation becomes:

$$P(T_x) \cong 3 \times 10^4 \left(N_0 \frac{r}{v}\right) \frac{\Delta H}{T_x} \tag{13}$$

where $P(T_x)$ is the vapor pressure in millimeters at the temperature T_x.

Rogers and co-workers (607) demonstrated the effect of the carrier gas, flow rate, heating rate, sample weight, and bridge voltage on pyrolysis curves of several samples. They also compared the maxima of pyrolysis curves with melting and boiling points and known dehydration temperatures and implied that the area under the curves can be used for quantitative analysis.

B. KINETICS

The effect of particle shape and size of reacting solids on kinetics has been discussed by Ingraham (307). If the reaction product is a solid with a porous open structure, such as in,

$$CaCO_3(s) \rightleftharpoons CaO(s) + CO_2(g)$$

then the solids have little or no influence on the rate of decomposition. On the other hand, as for the reduction of iron oxide by hydrogen or carbon monoxide,

$$Fe_2O_3(s) + 3\ H_2(g) \rightleftharpoons 2\ Fe(s) + 3\ H_2O(g)$$

$$Fe_2O_3(s) + 3\ CO(g) \rightleftharpoons 2\ Fe(s) + 3\ CO_2(g)$$

diffusion of product can be slow, even though the diffusion of reactant to the reaction interface may be rapid. In those cases, measured reaction rates reflect changes at the interface, which may be significant for ground or powdered materials. McKewan (465) reported a mathematical treatment of the equation:

$$rdf = kt \tag{14}$$

where r is a measure of particle size, such as the radius of a sphere or edge of a cube, d the density of the reactant, t the reaction time, and k the rate constant for weight loss per area unit of the reaction interface per unit of time. The fractional penetration of the interface f is related to the fractional weight loss α:

$$\alpha = 1 - \frac{[(1 - f)(a - f)(b - f)]}{ab} \tag{15}$$

Here, a and b are ratios of two particle dimensions to a third. For a cube, sphere, or cylinder with equal height and radius, equation 15 becomes:

$$\alpha = 1 - (1 - f)^3 \quad \text{and} \tag{16}$$

$$rd[1 - (1 - \alpha)^{1/3}] = kt \tag{17}$$

For a film or thin slab of sample, one particle dimension is very small compared with the other two, thus

$$rd\alpha = kt \tag{18}$$

with no significant change in geometry during the reaction and consequently no effect on the result. Ingraham and co-workers (303,306,722) extended this relationship from solid samples to dense pellets of uniform dimensions, prepared by compressing powdered materials. In this case, the space between grains is rapidly filled with gaseous decomposition product at the equilibrium decomposition pressure. This then effectively retards further reaction inside the compact, as if it were a uniform solid.

C. ACTIVATION ENERGY

Activation energies may be determined by a *series* of isothermal reactions at different temperatures (307). More important, however, is the method of using programmed, linearly increasing temperature in a *single* experiment.

The reaction rate for heterogeneous decompositions is the number of moles (n) of gas released from a unit of interfacial area per unit of time:

$$\frac{dn}{dt} = \frac{AkT}{\lambda^2 h} \exp\left(-\frac{\Delta H}{RT}\right) \exp\left(\frac{\Delta S}{R}\right) \tag{19}$$

where A is the interfacial area and λ is the distance necessary to remove the gas from the lattice.

For different sample sizes, α represents the fraction of reacted moles (n) at time (t) over the total number of moles (N), released after completed reaction

$$\alpha = \frac{n}{N}; \quad d\alpha = \frac{dn}{N} \tag{20}$$

For a linear temperature increase,

$$T = at + c \tag{21}$$

where a is the heating rate in degrees per minute, and c the starting temperature.

It follows, that

$$dT = adt, \tag{22}$$

and by substitution, the rate equation becomes:

$$\frac{d\alpha}{dt} = A\,(\lambda^2 aN)\,\frac{kT}{h}\,\exp\left(-\frac{\Delta H}{RT}\right)\exp\left(\frac{\Delta S}{R}\right) \tag{23}$$

and after combination of all temperature independent constants as k_0,

$$d\alpha = \left(\frac{k_0}{a}\right)T\exp\left(-\frac{\Delta H}{RT}\right)dT \tag{24}$$

For a linear heating rate, the area α under an evolved gas curve is an integral

$$\alpha = \frac{k_0}{a}\int T\exp\left(-\frac{\Delta H}{RT}\right)dT \tag{25}$$

between the limits of the temperature program. Obviously if the reaction is completed within this temperature range, the fraction α becomes 1.

To evaluate the integral, $1/T$ is set to X. Then $T = 1/X$ and $dT = -dX/X^2$. For $\Delta H/R = b$, the fraction becomes

$$\alpha = \frac{-k_0}{a}\int\left[\exp\left(-\frac{bX}{X^3}\right)\right]dx \tag{26}$$

By substitution, the integral can be expressed as a series for evaluation:

$$u = \frac{1}{X^3} \qquad\qquad dv = [\exp(-bx)]\,dx$$
$$\tag{27}$$
$$du = \frac{3\,dX}{X^4} \qquad\qquad v = -\exp\frac{(-bx)}{b},$$

thus,

$$\alpha = \left(\frac{k_1}{a}\right)T^3\exp\left(-\frac{\Delta H}{RT}\right)\left[1 - \frac{3RT}{\Delta H} + \frac{12R^2T^2}{\Delta H^2}\cdots\right] \tag{28}$$

and further

$$\ln\left(\frac{a\alpha}{T^3}\right) = -\frac{\Delta H}{RT} + \ln k_1 = \ln\left[1 - \frac{3RT}{\Delta H} + \frac{12R^2T^2}{\Delta H^2}\cdots\right] \tag{29}$$

Assuming,

$$\ln\left[1 - \frac{3RT}{\Delta H} + \frac{12R^2T^2}{\Delta H^2}\cdots\right] = \text{const.} \tag{30}$$

the plot of $a\alpha/T^3$ versus $1/T$ is a straight line, and the activation enthalpy ΔH is the slope of this line multiplied by $2.303\,R$. The error caused by eliminating the expression in equation 30 is small, since the term in brackets approaches zero for large ΔH and small T.

Ingraham (307) illustrated the method with precipitated calcium carbonate compressed into thin pellets. EGA curves were obtained at various heating rates and the areas under the curves determined by cutting out and weighing. Changes in geometry of the samples during each experiment were compensated for by normalizing the values of α, using a plot of α versus f. For each heating rate, $\log f$ was plotted against $1/T$. From these linear equations, values of $\log (af/T^3)$ were calculated in 20°C intervals. Plotted values of $\log (10^{10}\, af/T^3)$ showed a linear relationship for the decomposition range from 6% to 100%. The activation energy of 44.5 kcal/mole was in good agreement with values previously obtained by other methods.

D. OTHER CONSIDERATIONS

The foregoing deductions are valid only where setting limiting conditions is justified. This applies to mathematical simplifications and even more so to neglect of variables other than time and temperature. In many cases, however, either experimental elimination of complicating factors is not possible or the factors themselves are not understood. Alternative treatments have been proposed (198,296,358,575,639,702), and corrections and refinements are being offered, including the use of the area under a differential thermal analysis (DTA) exotherm for correlations with amounts of water produced in hydrogenolysis of coal (446).

Physical properties such as surface roughness (305) and particle size (540), as well as presence of additives (3,540), evaporation, desorption, and diffusion phenomena (2,3,159,422,581,640), retention volume (86,87), solid–solid interactions and nucleation (640), phase changes (21), and the reversibility of dehydration reactions (524,543,544,658) have been considered. A mathematical postulate was developed for *multiple* reactions during coal pyrolysis (108). The methods of Flynn and Wall (45) and Freemann and Carroll (198) have been expanded for EGA measurements by mass spectrometry (745), which will be discussed later. Since the mass spectrometer can monitor specific ion intensity, multiple reactions can be detected and evaluated, which is not possible by thermogravimetry. Another complication was observed when TG was used as an isothermal measurement for the mass of evolved gases. This was the downthrust of the sample container in a dynamic atmosphere (529,530) together with the effect of pressure and temperature. Mathematical treatments of kinetic problems include polynomial approximations (435,528), description of a thermoanalytical curve by a number of functions and consideration of the maximum of the nonisothermal transformation (252), and the incorporation of the heating rate (503).

In the first case (252), the author derived a "universal characteristic acti-

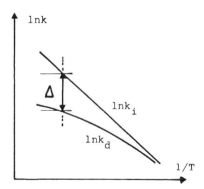

Fig. 6.2. Experimental determination of $\Delta = \ln k_i - \ln k_d$. (Ref. 503)

vation quantity ε_0," which makes a comparison of kinetic data for a number of chemical transformations more accurate:

$$E_1 \frac{f_1(dm)}{dmf_1'(dm)} = E_2 \frac{f_2(dm)}{dmf_2'(dm)} = E_3 \frac{f_3(dm)}{dmf_3'(dm)} = \varepsilon_0 \qquad (31)$$

where E's are the activation energies and m denotes the point of maximum rate of transformation. When the influence of the heating rate on reaction kinetics was considered (503), the general formula for nonisothermal conditions was:

$$g(x) = A_0 \exp(-\Delta) \exp\left(-\frac{Q}{RT}\right) t \qquad (32)$$

where Δ values are a function of ω, which is related to chemical bonds in the reacting solid:

$$\Delta = \ln k_i - \ln k_d = \frac{W_d - W_i}{RT} = \frac{\omega_d}{RT} \frac{1}{(1 + c_d T)^2} - \frac{\omega_i}{RT} \frac{1}{(1 + c_i T)^2}, \quad (33)$$

(i = isothermal; d = nonisothermal) and can be obtained experimentally (Fig. 6.2). For gypsum, calcium sulfate hemihydrate, and calcium oxalate monohydrate, the theory agreed well with the experimental data for heating rates below 20°C per hour.

"Rate of product formation" treatments (194,329) have been expanded with a computer model (19) for calculations on heating a mixture of independent solids. The rate of formation of product i from a solid j at a linear heating rate p, is:

$$R_{ji} = b_j A_{ji} \exp\left(\frac{-E_{ji}}{RT}\right) \sum_{i=1}^{i=y} (F_{ji}) \exp(-B) \qquad (34)$$

for the reaction order $n_j = 1$, or

$$R_{ji} = A_{ji} \exp\left(\frac{-E_{ji}}{RT}\right) \left[b_j \sum_{i=1}^{i=y} (F_{ji}) \right]^{1-n_j} + (n_j - 1)B \right]^{\frac{n_j}{1-n_j}} \qquad (35)$$

for $n_j \neq 1$.

$$B = \sum_{i=1}^{i=y} \left[\left(\frac{A_{ji}}{r} \right) \int_{T_0}^{T} \exp \left(\frac{-E_{ji}}{RT} \right) dT \right] \tag{36}$$

and b_j is a concentration proportionality constant, F_{ji} the amount of product i formed from solid j over the entire temperature range, and y the number of competitive reactions of a solid component. To use these equations for nonlinear heating programs, linear portions may be superimposed onto the real temperature curve.

Other algorithms and computer programs were reported (604) and applied to water desorption from silica gel (159) and zeolites (160) (in this case by resolving complex curves into the sum of a limited number of Gaussian curves) and to other thermal desorption experiments (422).

III. PRACTICE

EGA may be as simple as rubbing a substance between two fingers and detecting a product by its smell or it may be a combination of two, three, or more independent analytical methods. Each of these, as well as the various combinations, have specific advantages and limitations. Most of the individual processes may be combined, either as "coupled simultaneous techniques" or "discontinuous simultaneous techniques" (23). Minimum instrumental requirements include a heating device (such as the two fingers) and a detector (such as the nose). For combined methods, an interface is often required, due to either the construction of the individual instruments or their specific operating conditions. Thermal and analytical functions of EGA will be discussed separately in view of inherent properties of the samples under test. It should be pointed out here, however, that it is not always possible to combine the optimal heating process with the optimal analytical method. The reasons in most cases are instrument related. In fact, the choice of *one* function may determine the other or *define the type of results available*. For the analytical part, an attempt is being made to differentiate between detection, identification, and quantitative analysis of evolved gases. These are in most cases primary considerations for selecting a method, even though particular instruments may yield more than one of these types of data.

A. GAS PRODUCTION

A variety of techniques can produce gases or vapors by a thermal reaction. The oldest and most general form is pyrolysis. After technologically advanced instruments have been developed for precisely controlled heating of a test sample, today this term is used primarily for the most rudimentary isothermal procedures (396). These methods are treated in a separate chapter, including temperature measurements. Therefore, this discussion will be limited to the *principle* of gas evolution as it affects the analytical results and to those techniques that are no longer covered by the term "pyrolysis."

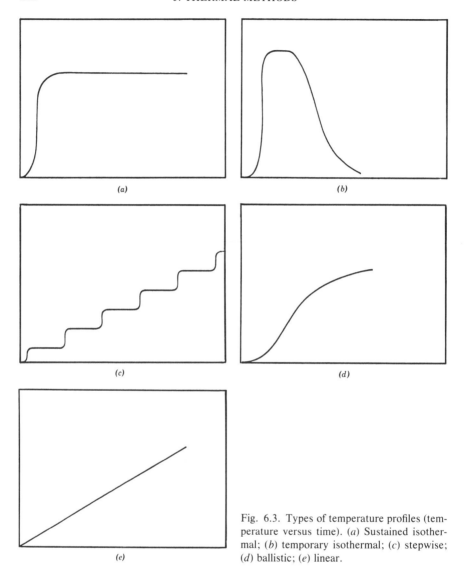

Fig. 6.3. Types of temperature profiles (temperature versus time). (a) Sustained isothermal; (b) temporary isothermal; (c) stepwise; (d) ballistic; (e) linear.

Gas evolution may be initiated by isothermal or programmed heating processes, resulting in temperature profiles schematically represented by Fig. 6.3.

For each mode, the amount and rate of evolved gas not only depends on sample-related factors described under "theory," but also on such parameters as temperature rise time, *actual* sample temperature over the duration of the experiment, and temperature changes of evolved gases between generation and analysis.

Let us consider the simplest case, where the sample size could be essen-

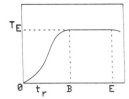

Fig. 6.4. Idealized temperature profile for isothermal reactions. T_E = Temperature of experiment; B = beginning of isothermal range; E = end of isothermal range; t_r = temperature rise time.

tially unlimited, only *one* reaction would occur at the temperature chosen, the products did not react further, and the temperature could be maintained and measured for extended periods of time. Under these considerations, a sample could be introduced into a preheated pyrolysis chamber, such as a tube furnace, (42,169,415,444,445,496,709) open or sealed in an ampule (625). The only concern might be the rise time of the sample temperature, which would depend on the sample size and geometry and the heat-transfer mechanism from the energy source to the sample (Fig. 6.4). Generally, the thermal mass of the furnace should be large compared with that of the sample, and the heat transfer should be rapid.

Unless a *different* reaction took place before the experimental temperature T_E was reached, the portion of the curve from 0 to B can be ignored, if the purpose of the experiment is identification of the evolved gas(es) or is a kinetic study. For *qualitative* analysis, knowledge of the *exact* sample temperature may not even be of great importance. On the other hand, in kinetic studies, for instance, maintaining and measuring the reaction temperature is essential.

Temperature can be measured directly or indirectly, or calculated via some energy relationship. Sometimes, precise measurement of the sample temperature is more difficult than exact control of the energy input. In this case, instruments must be able to operate under reproducible conditions to obtain comparable results (284).

In this form, isothermal pyrolysis is particularly useful for evolved gas detection (EGD), either in a static or dynamic flow system, or for monitoring specific gases. Examples include simple decompositions, e.g.,

$$CaCO_3 \longrightarrow CaO + CO_2,$$

or gas–solid interactions, where both the reactant and product gases can be measured (369). A typical case is the previously mentioned reduction of a metal oxide by hydrogen

$$H_2 + MeO \longrightarrow Me + H_2O$$

Variables for these investigations could be pressure, flow rate, physical properties of solids involved, and surface and catalytic phenomena (686).

Few experiments, however, fall into this category. Different reactions may take place before the test temperature is reached, reaction modes may change with temperature, the reaction temperature may be unknown, or it

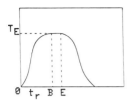

Fig. 6.5. Temperature profile for isothermal reactions at high temperatures. T_E = Temperature of experiment; B = beginning of isothermal range; E = end of isothermal range; t_r = temperature rise time.

may be *too high* to be maintained for extended time periods. In the latter case (Fig. 6.5), the energy source must be designed to attain relatively high temperatures quickly, and the sample is studied at the temperature maximum.

The rise time is critical because it is long compared with the flat portion (B–E) of the curve, within which measurements are made (397). This in turn requires extremely rapid temperature increase of the heating devices themselves, and their heat capacities become small compared with those of the samples. Consequently, the effects of strongly exothermic or endothermic reactions will be superimposed on the temperature profile expected from the heating device itself (398).

The method is also known as *flash pyrolysis*. Xenon or mercury lamps (352,515,634), electron beams (577), laser devices (177,276,295,428,429,430, 431,432) [sometimes in the presence of additives to increase the yield (694,699)], capacitor-discharged filament heaters, and Curie point pyrolyzers (214,366,398,549,550,668,767) are used. The arc-image furnace has also been evaluated in this connection (203). Inherently, temperature measurements are difficult, but most devices can be calibrated in terms of energy output for comparative studies.

The main applications for flash pyrolysis are volatilization of high boiling materials, release of trapped gases from high-melting solids, study of high-temperature reactions, and fast vaporization of thermally sensitive materials such as biological or polymeric samples.

Because of instrumental limitations, additional restrictions are placed on the samples. To assure rapid and *uniform* heating, they have to be cast into thin films or coated onto wires. This is particularly important for materials with low thermal conductivity. The smaller the sample size, the more efficient and reproducible will be the heating process. On the other hand, when only small amounts of gases are evolved, any dilution between the reacting sample and the detector should be minimized, while the sensitivity of the detector should be as high as possible. Finally, because of the high temperatures characteristic for this type of analysis, in most cases the evolved gases must be maintained and analyzed at or near the pyrolysis temperatures to avoid condensation or secondary reactions. Moving the detector as close as possible to the point of origin of the evolved gas avoids most of the problems. A good example for this arrangement is the so-called laser probe,

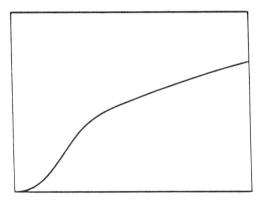

Fig. 6.6. Temperature profile for slow ballistic heating.

where a calibrated and focused laser beam hits a solid sample and the resultant plume is immediately analyzed spectrophotometrically.

Repeating the process depicted in Fig. 6.3a produces stepwise heating (Fig. 6.3c), which may be considered the simplest controlled program. It was first used by necessity (35), when instrumentation was not yet available to generate a linear temperature profile. Today stepwise heating is still applied, especially to study the kinetics of successive reactions. An example is the dehydration reaction, where different hydrates are stable at different temperatures. This will be more fully treated in Section IV.

Stepwise decomposition has been applied to polymers (170,286), described as thermal volatilization analysis (466,467), used on lunar rocks (242,245) and for adsorption and desorption studies (568,581,759).

Of the continuous nonisothermal methods, "ballistic heating" is represented by Fig. 6.3d. It can be carried out with a limited amount of control. The sample holder is generally a body of high heat capacity. Application of a fixed amount of energy, such as voltage applied to a heater coil, will lead to a relatively slow temperature increase (Fig. 6.6) until thermal equilibrium is reached. Amount of energy applied and thermal mass of the container influence the curve shape.

Alternatively, energy can be applied in increasing amounts, such as by a motor-driven variable transformer (214,347,349,350,730). This generates a more linear profile, (Fig. 6.7), limited by increasing heat loss of the sample container at higher temperatures. In either case, to obtain meaningful results, a fairly precise measurement of the sample temperature is necessary as a function of time. Reproducibility depends primarily on the control over the energy input.

Such devices are often found as parts of or attachments to gas-analytical instruments such as gas chromatographs or mass spectrometers, which will be described below (168,204,214,347,348,349,350,433,484,703).

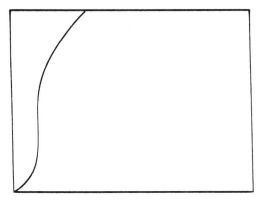

Fig. 6.7. Temperature profile for fast ballistic heating.

Recently a special tube furnace for high heating rates has been constructed, with a moving sample being exposed first to increasing and subsequently to decreasing temperatures (711), (Fig. 6.8).

Incorporation of feedback control, coupled with exact temperature recording and provisions for variation of heating rates, produces linear temperature profiles (Fig. 6.3 e), characteristic of the modern instruments developed for differential thermal analysis (DTA), differential scanning calorimetry (DSC) and thermogravimetry (TG). These instruments are described in separate chapters. In addition to increasing and decreasing heating rates linearly, commercially available programmers also provide for isothermal, cyclic, and stepwise modes above and below ambient temperatures.

Potentially any DTA, DSC, or TG apparatus with a closed gas-flow system qualifies as a gas generator for EGA. However, to maximize thermal sensitivity, the sample is often enclosed in a cell, and modifications may be necessary to collect and transfer evolved gases. One of the many solutions is shown in Fig. 6.9 a,b (376), where, by the right combination of heating rate

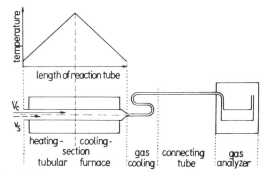

Fig. 6.8. Temperature-Programmed Tube Furnace. (Ref. 711)

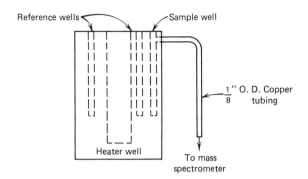

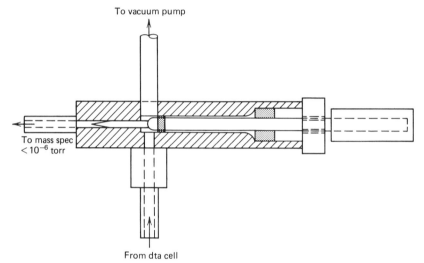

Fig. 6.9. Sampling from DTA block to mass spectrometer in "plug-flow" mode. (a) Modified DTA cell; (b) sampling valve. (Ref. 376)

and carrier gas flow, a "plug-flow" of the evolved gas has been achieved. This method also overcomes the second problem characteristic for DTA, DSC, and TG, namely sample dilution. Specific collection methods will be discussed later. A variety of interfaces have now become available as attachments to these instruments, most of them utilizing the full amount of carrier gas.

Finally, one has to be aware of the thermal effects of the sample itself, as described for the filament heating process. In both DTA and TG, the actual sample temperature may deviate considerably from the programmed linear profile during strongly endothermic and even more so during strongly exothermic reactions.

Particularly noteworthy is the work by Paulik and Paulik with the derivatograph, using a specially designed labyrinth sample holder to create "quasi-isothermal, quasi-isobaric" heating conditions (556,557,559,564,567) by a feedback mechanism, which allows for better separation of overlapping reactions.

Simultaneous DTA, DSC, or TG with EGA is a powerful method, especially for those gases and vapors that are not highly reactive and not easily condensed.

A few rather specific practices for evolved gas analysis are worth mentioning. Torsion effusion (608,712) is a useful technique to study kinetics of irreversible gas-producing reactions. It is based on the recoil effect of a gas escaping from a closed crucible through an orifice under molecular-flow condition. For eccentric location of orifices perpendicular to a suspending torsion wire, the angular deflection of the crucible has been mathematically related to the rate of reaction (120,712,721). Triggered evolution (721) was used for standardization by releasing a trapped gas at the melting point of a solid.

Combustion phenomena can be studied by subtractive oxygen monitoring (212) or analysis of oxidation products (28), and secondary gas-phase pyrolysis of evolved gases has been helpful to identify organic homologs (480).

Inorganic solids have been studied in the presence of reactive gases, monitored by a thermal conductivity detector (489).

It should further be pointed out, that isothermal reactions at ambient temperatures such as:

$$Zn + H_2SO_4 \longrightarrow H_2$$
$$CaCO_3 + HCl \longrightarrow CO_2$$

also fall into the category of evolved gas analysis.

Utilization of integral heating devices of gas chromatographs and mass spectrometers for evolved gas analysis will be described later.

B. GAS ANALYSIS

After a gas has been produced it must be removed from the reaction zone and either analyzed immediately by a direct method or temporarily stored. The latter may be accomplished by cold traps for condensable gases, simple gas bulbs (either evacuated or sealed by an inert liquid), gas burettes, or syringes (350,418). Absorption in liquids or adsorption on charcoal or other porous solids also are effective collection mechanisms (171).

In all these cases, quantitative determination is possible by either mass or volume, and the collected gases are available for *subsequent* separation and identification. While this is doubtless the most general and comprehensive procedure, it is also cumbersome and requires relatively large amounts of sample for sufficient accuracy.

By contrast, most of the modern gas analytical detectors are highly sensitive instruments, which therefore allow analyses on relatively small samples (263). However, most methods are either destructive or generally not designed to collect gases. For quantitative measurements, they often require

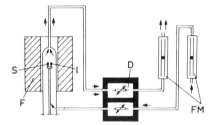

Fig. 6.10. Use of dual thermal conductivity detectors for evolved gas analysis. S, sample; F, furnace; I, reference; D, thermal conductivity detector; FM, flow meter. (Ref. 175)

calibration, and for identification they depend on unique and specific sample properties to provide unambiguous results. Very few instruments are capable of handling all gas analytical problems simultaneously and with equal sensitivity and accuracy. This often necessitates considerable prior knowledge of the reaction investigated or the use of separate methods for a *complete* analysis.

Detectors have been employed in EGA either directly or as part of instrumentation designed to maintain reproducible operating conditions, such as in gas chromatography (GC). Indeed, the development of gas detectors is an integral part of the advancement of GC.

1. Thermal Conductivity Detectors

The first device was described by Ray in 1954 (592). It is based on the cooling effect of a gas passing over a heated wire, which is a function of the different thermal conductivities of different gases. In a dual cell arrangement as shown in Fig. 6.10, the pure carrier gas is flowing through the reference cell. A different gas composition in the measuring cell will have a different cooling effect from that of the reference cell. The temperature difference of the two hot wires can be measured via a Wheatstone bridge circuit (Fig. 6.11), since the electrical conductivity of a metal wire is a function of the temperature. Because of its second highest thermal conductivity (Table 6.I), helium is mostly used as a carrier gas. A detailed review of the thermal conduction detector, which is also known as *Katharometer,* was published by Ingraham (307).

Early publications on EGA mention simple thermal conductivity detectors for both organic and inorganic samples (17,70,121,305,416). Later, the useful temperature range was extended above 400°C by employing model airplane glow plugs (184,607), instead of thermistors or filaments. Further improvements, automation (518,519), developments of commercial detectors (21,38,175,479,747), and applications in various fields were reported (32,164,232,234,312,368).

As an example of comparative detectability, results from EGD have been compared with those from DTG in an apparatus for simultaneous micro-DTA and EGD (755). For materials of boiling points ≤300°C, the sensitivity of a W/Re TCD has been greater than 10 times that observed by DTG.

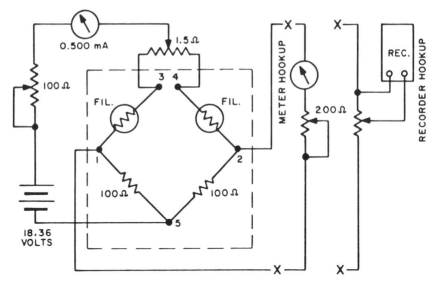

Fig. 6.11. Wheatstone bridge circuit for use with two thermal conductivity detectors. (Ref. 307)

A particular arrangement with thermal conductivity detectors has become known as the gas density balance. Its operating principle is illustrated in Fig. 6.12. Here the sample is injected into a split, balanced reference stream *after* the reference gas has passed the two detectors. Gravity of the sample gas will cause an imbalance of the flow rates in the two reference streams, thus creating a temperature difference between the two dectectors, which is

TABLE 6.I
Thermal Conductivities of Gases at 100°C[a]

Gas	$\lambda \times 10^5$
Acetone	3.96
Air	7.20
Ammonia	7.09
Argon	5.09
Benzene	4.14
Carbon dioxide	5.06
Carbon tetrachloride	2.05
Helium	39.85
Hydrogen	49.94
Methyl alcohol	5.16
Methyl chloride	3.84
Nitrogen	7.18
Nitrous oxide	5.06
Oxygen	7.43
Sulfur dioxide	3.65 (est.)
Water vapor	5.51

[a]Adapted from Ref. 307.

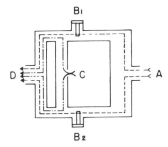

Fig. 6.12. Gas density detector. A, reference gas inlet; B_1, detector; B_2, detector; C, sample gas inlet; D, gas exit. (Ref. 417)

measured by the conventional bridge circuit. Thermistors are generally used as the sensing elements. An advantage of the gas density balance is the absence of contact between sample gas and sensing element, which makes it useful even for corrosive effluents.

The original gas density balance developed for gas-phase chromatography (451) was used by Liberty et al. (400) for molecular weight determinations. Therefore it is sometimes also called molecular weight chromatograph. A gas of known molecular weight serves as an internal standard, and a mathematical expression is derived for the molecular weight of an unknown gas, expressed by areas under the response curves of the detectors for both the reference gas and the unknown.

A theory developed by Nerheim (517) led to an improved gas density detector without a need for calibration. This design became available later as commercial models. Recently (346), a detailed theoretical analysis of the gas density balance has been reported in connection with the use of the so called mass chromatograph or molecular weight chromatograph (347,348,349,350, 387). This is an instrument that uses two GC systems and provides both molecular weights and GC retention times.

2. Ionization Detectors

Detectors based on ionization processes are many times more sensitive than thermal conductivity detectors. The argon detector was described in 1958 (424,425). It is based on ionization of the sample gas by collision with metastable argon atoms. Essentially at the same time, the flame ionization detector was developed; it utilizes the change of the electrical conductivity of a hydrogen flame by organic vapors (473). The first design consisted of two hypodermic needles serving as both a jet and an electrode. The other electrode was a wire grid above the flame. Specially designed circuitry provided much higher sensitivity than a TCD. The electron capture detector, which was introduced in 1960 (426), is based on different electron affinities of different molecular species, especially of those of organic compounds that are determined by functional groups.

Although ionization detectors can be used directly for EGA (66) either qualitatively (EGD) or quantitatively, their main use is in gas chromatography, which will be discussed below.

The choice of a detector depends on its inherent sensitivity and reliability, as well as on the physicochemical properties of the sample gases (417,671,744).

The thermal conductivity detector is the least specific, and therefore is applicable to both organic and inorganic materials. For oxidizable organic molecules, flame ionization detectors are highly sensitive, and the response is proportional to the number of carbon atoms. Methane, ethane, fluorocarbons, carbon monoxide or dioxide, and especially water vapor are *not* registered by the argon detector, whereas the electron capture detector is very sensitive to organic or inorganic halides, oxygen and oxygen-containing molecules such as peroxides, nitrates, anhydrides, and also organometallics.

Less widely used detectors (417,744) are based on magnetic susceptibility, ultrasonics, specific gravity, cross-section ionization, radio frequency, and flame photometry.

3. Gas Measurement

Quantitative analysis requires not only prior identification of the evolved species but often also separation of mixtures. Volume measurements under isobaric conditions are important for physicochemical investigations (92,108,580,733). The gas burette is still in use (98), but more recently a "new, Automatic Gas Volumeter" has been described (632), which allows the study of very slow or very fast (explosive) reactions. Its variable volume chamber is kept at atmospheric pressure by a servomotor, activated by an amplified signal from a pressure sensor. A "moving coil differentiator" provides high resolution with small sample sizes for slow processes.

While volume was the first quantity acquired in EGA, pressure is the most frequently measured one today. Since it is also an indiscriminate property, its value is in the area of isothermal measurements, studying mechanism, thermodynamics, and kinetics, and especially in determination of activation energies of known reactions. A variety of manometers and pressure gauges have been employed (264,458,461,462,538,541,626,688,732), including the silicon oil manometer (317), a quartz spoon gauge (318), differential (52) and multiplying manometers (443), ionization gauges (738), and transducers (19).

Of all pressure indicators, the Pirani gauge has found most widespread use in EGA (576). Like the Katharometer, it contains a hot wire as the indicator, whose temperature is changed with a pressure-dependent change in thermal conductivity of the surrounding gas (71,72,190,254,325,459,611). In general, Pirani gauges have been useful for pressures between 1 torr and 10^{-3} torr, with a Wheatstone bridge circuit or a sensitive thermocouple measuring the temperature of the wire (292). Taking advantage of convection of the gas around the wire, the measurable pressure range has been extended to cover about 0.0001 torr to about 1000 torr (283). Filament geometry of a tungsten wire has been related to sensitivity and response time (695). A thick filament provides high sensitivity but slow response. For both

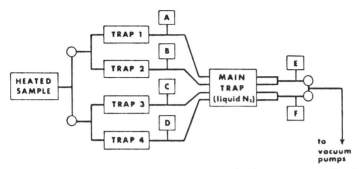

Fig. 6.13. Schematic for differential trapping. A,B,C,D, Pirani gauges; E,F, collection tubes; circles, stopcocks. (Ref. 468)

good sensitivity and fast response, a long thin filament has been recommended. Recently an ultralow noise Pirani gauge has been described that used gold-plated tungsten wires (532).

Pirani gauges are found in a variety of experimental setups. One method has been reported as thermobarogravimetry, measuring mass, pressure, and temperature (27,459). Another, called thermal volatilization analysis records pressure as a function of linearly increasing temperature or isothermally as a function of time (466,467,472). The gauge response has been reported to be similar to derivative thermogravimetry curves. By differential condensation in a series of cold traps with associated gauges, (Figs. 6.13, 6.14) (115,468,469,470,471,474), materials with considerably different volatilities have been studied.

For more detailed information of available commercial devices, the general references, encyclopedias (292), or manufacturer's literature should be consulted.

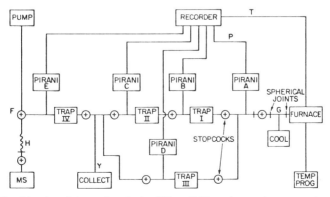

Fig. 6.14. Combination of thermal analysis, differential trapping, and mass spectrometry. (Ref. 115)

4. Chemical Detection Methods

Another form of quantitative measurements is based on chemical analysis (74). In its simplest form, an acidic or basic evolved gas is physically or chemically trapped, titrated with a base or an acid (238,263,628), or measured via the change in pH of an absorbent solution (495). The use of automatic titration has also been reported (341). A significant contribution to both methodology and instrumentation by Paulik, Paulik, and Erdey has become known as thermal gas analysis or thermo-gas-titrimetric method (339,450,477,553,554,555,560,561,562,565,566,610). The authors developed a multipurpose thermoanalytical instrument they named "Derivatograph" (554), which is described elsewhere in this chapter. It allows simultaneous measurements of changes in weight, enthalpy, and temperature, as well as the addition of various accessories including EGA devices. For thermal gas titrimetric analysis, the evolved gases are trapped in a liquid absorbent and changes of the potential in the liquid are measured by a pair of electrodes. The solution is automatically titrated, and the amount of titrant is recorded. In this fashion, acid-base titrations, redox reactions, and precipitation reactions have been carried out. Water analysis by the Karl-Fischer method has also been successful (450). In addition to selective absorbents employed by other investigators (104,362,368,418), the use of ion-specific electrodes (186,336) should also be mentioned.

In some cases, conductivity cells have been used successfully (265,337). Water has been determined by measuring the amount of oxygen and hydrogen evolved in a capillary with two coaxial helical platinum electrodes connected with a layer of P_2O_5 (195). In another example, an electrolytic hygrometer differentiated water from oxygen (291). Oxides of nitrogen could be detected by a temperature rise of a $V_2O_5 - Al_2O_3$ catalyst bed from the exothermic reaction with ammonia (308).

Very simple tests are often adequate for quick identification of an expected effluent. For instance, HCl can easily be detected with a piece of pH paper and a drop of silver nitrate solution at the exit port of a furnace tube (187).

5. Spectroscopy

Spectroscopic analysis can be based on either emission or absorption. For gas analysis, absorption methods are predominant. In its simplest form, analyses are called nondispersive if they measure the total amount of radiation absorbed by a sample placed between a radiation source and a detector.

This method is nondiscriminating, therefore useful only for general gas detection or for quantitative measurements of a single, known, evolved species. Ways of making spectroscopic methods discriminating are described in the following.

a. INFRARED ANALYSIS

Infrared absorption is more useful for gas analysis than any other optical method, both for qualitative and quantitative measurements (427,488). Primarily nondispersive methods have been applied to several specific gases (41,142,171,183,301,320,351,407,494). Absorption cells essentially consist of tube furnaces with IR transparent windows at both ends (94,763). A high-pressure cell has been patented for 500–50,000 psig and temperatures from 38°C to 260°C (718).

Various instruments produce selective information (317,347,349,427,709). This includes application of filters such as the sample gas itself. The Luft analyzer (427) is a double-beam instrument whose detector can be sensitized with several gases. It compares the pressure increase caused by infrared heating of a known amount of gas in a reference chamber with that due to the gas in a closed sample cell.

Other instruments can scan the infrared region with the aid of prisms, gratings, or band-pass filters. In combination with stationary or flow cells, identification of gaseous product becomes possible and quantitative measurements can be made for selective characteristic absorption bands.

The latest development is the interference spectrometer (427). It produces high signal-to-noise ratios, which are especially useful for evolved gas analysis, but it requires a computer and extensive electronic circuitry.

Interfaced commercial thermoanalytical and infrared instruments have recently become available (271,693), and the possibility for infrared emission spectroscopy has been pointed out (259).

b. OTHER SPECTROPHOTOMETRIC METHODS

Several other detection systems should be mentioned briefly, although their applications are limited to rather specific problems. They include, for instance, a selective spectral radiometer for multiple trace gas monitoring (742), a luminescence detector for oxygen (63), chemiluminescence detectors for oxides of nitrogen (6,453), laser-induced fluorescence analyzers (18) for continuous monitoring (73,421), and a polychromatic light-source instrument (609). For particulate matter, a number of photometric systems have been used (125,140,419,508). The effectiveness of X-ray fluorescence emission spectroscopy for gas analysis has been demonstrated for ethylbromide and methyliodide (297). Recently, a solid electrochemical cell has been patented for analysis of oxygen and unburnt fuel (624). The simple combination of a furnace and photoionization detector was suggested for EGA of organic materials (75).

6. Nonspectroscopic Detection

Included here are, for instance, commercial gas analyzers (217,666). An automatic gas-measuring device for high rates on a semimicro scale (215) uses a mercury drop in a PVC tube as the piston. Its movement is followed by electrodes in a high-frequency oscillating circuit. This method also yields rates of reactions. A patented apparatus for gas mixture analysis is based on electric or magnetic susceptibility without use of a reference gas (299).

Platinum resistance thermometers have been used specifically for hydrogen detection (568).

7. Combination Methods

Up to this point, methods were discussed that could be summarized as pyrolysis EGA. Operating conditions were mostly limited to control and measurement of the sample temperature, combined with direct detection of the evolved gas. At the detector, usually only *one* physical property could be measured.

Many analytical procedures, however, inherently can supply several pieces of information simultaneously on a single sample, provided proper detectors are available and can be combined in the right sequence. Such combinations are becoming more and more important.

In the introduction, the two basic thermoanalytical methods DTA/DSC and TG were mentioned. Coupling of either system with a gas analytical technique results in such a simultaneous method (442). It requires control of gas flow in the thermoanalytical instrument, an inlet system for the gas analytical device, and in most cases an interface. Thermoanalytical equipment is discussed elsewhere in this section, and for detailed studies of the method to be combined, the reader is referred to the general literature list (G60). Specific reports on thermoanalytical combination methods and instrumental DTA-TG modifications for EGA have also been written (30,40,193, 258,267,322,340,362,439,663,726,727,729).

The obvious use of weight-loss determination by TG as indirect EGA was mentioned earlier. Now, using high-pressure DTA in a flow-type apparatus together with differential pressure analysis, areas under DTA exotherm curves have been correlated with amounts of water produced in hydrogenolysis reactions (446).

Today, by far the two most important gas analytical methods are gas chromatography (GC) and mass spectrometry (MS). Like the thermoanalytical methods themselves, these two have also been combined as GC/MS. Each is a very powerful method, yet each is very different in terms of the results produced. Their role in combination with thermal analysis, specifically evolved gas analysis, is rapidly expanding.

a. Gas Chromatography

This is a *separation method* for mixtures of gases (263,671,744). It *is not* a primary *means of identification*. On the other hand, under reproducible instrumental conditions and by calibration with identified components of the test mixture, it is an accurate and sensitive quantitative tool.

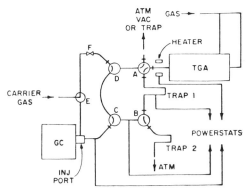

Fig. 6.15. Switching system for trapping evolved gases from TGA. A,B,C,D, stopcocks; E, switching valve; F, regulating valve. (Ref. 112)

Its main component is a column, filled with a nonvolatile liquid supported on a solid. In modern capillary columns, the solid support is the column wall itself. By differential adsorption and desorption on the liquid, gas mixtures may be separated as they are pushed through the column by an inert carrier gas at a constant flow rate.

Other components are the injector at the head of the column, a detector at the end, and a metering and recording device. Modern GC equipment allows heating of both the injector for immediate flash evaporation of nongaseous samples and of the column to increase mobility of the sample. By programmed temperature increase, additional separation may be achieved. Integration of the peak areas under the response curve of the detector yields the amounts of known effluents, identical with the direct determination of evolved gases discussed in the theoretical part.

Needless to say, there are conceptually and instrumentally almost indefinite modes to utilize GC for EGA (137). The earliest reports describe what is often called pyrolysis GC (5,68,85,143,255,261,392,396,497,542). In addition to the heating methods mentioned before in this chapter (276,366,398,549, 550,625), direct attachments for GC have been constructed, including a multiple sample attachment (119), a microreactor (516), a microdoser (4), and a microthermoanalyzer (479). Any reaction product trapped, as shown earlier, can be subsequently injected into a gas chromatograph for further separation (361). This may be done as the total evolved gas mixture, or already preseparated via differential trapping (112,114).

In combination with thermogravimetry, intermittent GC analysis has been carried out when indicated by the weight-loss curve (112,114). A switching valve directs the carrier gas flow through the TGA unit if desired (Fig. 6.15) (112), and the evolved gases are trapped for subsequent injection (Fig. 6.16) into the GC column. By incorporation of microvolume valves and solenoid switches, this method is now semiautomatic (Fig. 6.17) (113).

In another arrangement (233), the evolved gases from a DTA cell pass

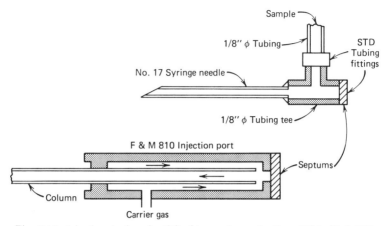

Fig. 6.16. Injection devices for GC of trapped samples from TGA. (Ref. 112)

through a thermal conductivity detector and then are split by a valve system for time-resolved gas chromatography (Fig. 6.18). Figure 6.19 is the schematic of a commercial instrument, which allows rapid backflushing of the trap for GC and also for attachment of other analyzers (552).

With a series of trapping tubes at low temperature, changed at short intervals, and flash-evaporated into a gas chromatograph, results have been obtained that were comparable to those of combined simultaneous methods (756).

Various coupling techniques have been reported for the Derivatograph (157,158,371), for TGA with GC and IR (136), with combustion processes

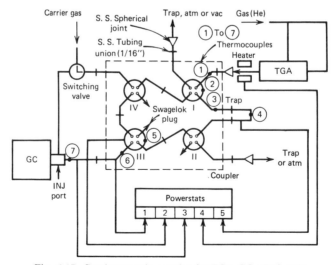

Fig. 6.17. Semiautomatic coupler for TGA-GC. (Ref. 113)

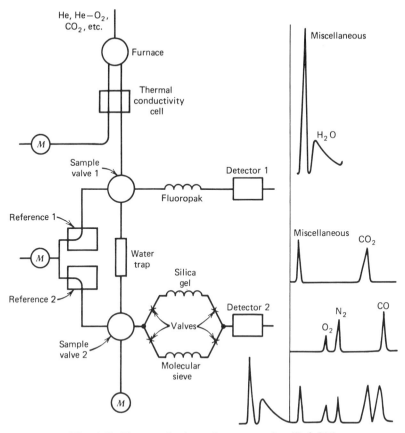

Fig. 6.18. Time-resolved gas chromatography. (Ref. 211)

(693), for destructive differential thermal analysis (360), and for DTA and TG with double-column GC (479). The mass chromatograph mentioned earlier, (347,349,350) contains two gas chromatographs in a parallel arrangement and simplifies identification of unknowns (387). In addition, quantities, retention times, and molecular weights of constituents are obtained. Special devices have been used to detect oil residues on solid surfaces as low as 1 $\mu g/cm^2$ (161), and to discriminate solid-state reactions from gas evolution as part of a lightweight Martian lander (74).

An unusual GC detector consists of a moving aluminum strip coated with Al_2O_3 or silica gel. Part of the GC effluent is sprayed onto the moving strip, and thus is synchronized with chromatographic peaks and can be further analyzed after extraction (49).

With temperature-controlled GC inlet systems or columns, programmed (142) or isothermal EGA can be carried out (86,87,150), and by application of response factors quantitative EGA has been achieved (146).

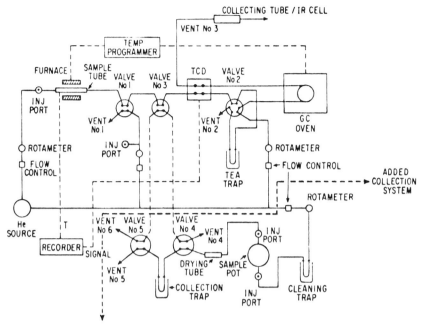

Fig. 6.19. Thermal evolution gas chromatograph. (Ref. 552)

Finally, a gas chromatographic combination method should be mentioned, which often simplifies the EGA analysis of complicated systems. It either makes use of reagents to subtract compounds with various functional groups (62,390,464), or it adds vapor-phase pyrolysis to produce simpler molecules (480), or combustion to obtain characteristic products (142,488).

Beyond GC, two methods are now expanding rapidly with the help of commercial instrumentation. They are gel-Permeation chromatography (GPC) and liquid chromatography (LC). Although primarily designed for analysis of high-molecular-weight materials and polymers, applications for EGA are possible (171). As an example for future expansions, attempts are being made to couple LC to mass spectrometers.

b. Mass Spectrometry

(1) Principle

Although mass spectrometers were invented before 1920 (616,743), they became available commercially only in the 1940s and 1950s, the main reason being the complexity of these instruments and with it their high purchase and maintenance costs. Even though the mass spectrometer can produce very accurate data for compound identification and concentration even within a mixture, many investigators opted for less accurate, more ambiguous, but cheaper analytical tools.

Today, mass spectrometers are available in a very broad range of performance characteristics and operating principles, designed to produce a "spectrum" of ions with high accuracy in mass measurement and high sensitivity (275).

To achieve this accuracy and sensitivity, the test sample must be in the gaseous state and separable according to the masses of its components. Separation is accomplished via ionization of a considerable fraction of the gaseous molecules followed by manipulation of the ions. This process must be carried out at reduced pressures. Finally, the separated ions must be detected and registered according to their mass/charge ratios (m/z):

Consequently, mass spectrometers consist of:

1. A vacuum system
2. A sample inlet system
3. An ion source (for ionization and "draw out")
4. An analyzer section
5. A detector
6. A data-handling system

(a) Vacuum System   In general, mass spectrometers operate at pressures of about 10^{-5} torr or less. For reasons to be explained later, the combination of fore pumps and diffusion pumps (lately also turbomolecular pumps) must be able to maintain the operating pressure accurately during the analysis.

(b) Inlet System   Different physical properties of the samples require different methods of introduction, each without causing significant pressure changes in the instrument. Gases are relatively easy to handle, either by direct injection of very small amounts, as from a syringe through a septum, or for instance, by leaking slowly from a reservoir through some restriction device into the ion source. Restrictions can be in the form of a capillary, a small orifice in a foil (the so-called "molecular leak"), a porous sintered ceramic disk, or even a series of metering valves. Liquids can be vaporized outside the vacuum system or introduced with a "direct probe"—a movable sample holder with differential pumping and vacuum locks.

In many cases, the vacuum of the mass spectrometer is sufficient to vaporize the sample. If it is not, most direct probes can be heated, in addition to the thermal energy supplied by the filament or through source heaters.

(c) Ion Source   From the inlet system, the gaseous sample enters the ionization chamber. The two most important processes are electron impact (EI) and chemical ionization (CI). For EI, electrons emitted from a heated filament are accelerated towards an anode and in their path collide with the sample molecules producing ions according to:

$$e^- + M \longrightarrow M^+ + 2e^-$$

In chemical ionization, a reaction gas is ionized by electron impact, and its ions collide with neutral molecules, producing highly reactive positive ions, which subsequently collide with the sample gas in the ion source. The resulting sample ions are considerably different from those produced by electron impact. The primary chemical reaction is a hydride transfer. In most cases "quasimolecular ions," one unit heavier than the original molecule, are the most abundant ions. Chemical ionization sources operate at considerably higher pressures, most around 1 torr, therefore require tight source configurations and high pumping rates of the vacuum system.

A second function of the ion source is to accelerate and collimate the sample ions into an ion beam. This is accomplished by a series of grids, plates, or slits with various and variable potentials. Several modern instruments can also provide negative sample ion beams.

Field ionization (48) is another method (601,602,617,629,630,631,760) that produces less fragmentation than electron impact. It occurs when a molecule loses an electron to a metal surface through "tunneling" in the presence of a very high electric field. Such fields can be created between a sharp point, thin wire, or a razor blade edge as the anode and a closely spaced cathode with a potential of up to 20 kV. This process creates stable *molecular* ions. The sensitivity, however, is low and sample dependent.

(d) Analyzers   After the ion beam has left the source, it can be separated by different means. The three most important ones are:

1. Magnetic deflection
2. Quadrupole fields
3. Time of flight

Magnetic fields bend the ion beam into circular paths, whose radii are a function of the mass/charge (m/z) ratios. By variation of the accelerating voltage at the ion source or by variation of the magnetic field strength, different ions can be focused onto a detector at the end of the circular path.

In a quadrupole field, ions oscillate along the longitudinal axis of four parallel rods, to which a dc potential and a radiofrequency (rf) field are applied. For a particular dc/rf combination, only ions with a specific m/z ratio are in resonance to traverse the field and strike a detector. To produce a mass spectrum, either or both dc voltage or rf field can be varied.

For time-of-flight instruments, ions are pulsed out of the source into a field-free drift tube, where their flight time is a function of the m/z ratio, resulting at different arrival times at the detector.

(e) Detectors   The oldest and simplest ion collector is the photographic plate. As an integrating device it is highly sensitive, but to take advantage of this sensitivity, relatively long acquisition times are necessary.

Most instruments are now equipped with electron multipliers. The ion beam strikes a target causing ejection of electrons. By combination of a magnetic and electrical field, the electrons are directed towards subsequent targets or along a dynode strip. In each such "jump," more electrons are emitted and detected at the final anode. Thus the major advantage over a photographic plate is the *direct* electrical signal from the electron multiplier, which can be measured, displayed on an oscilloscope screen, or more importantly, fed into a computer for storage and integration.

(f) Data Handling   The output of a *mass spectrometer* is a mass spectrum, which is a series of peaks or lines with various peak heights, arranged in increasing m/z ratios. The scale on the abscissa (m/z) depends on analyzer principles and recording methods. It may be linear or nonlinear (G28). The exact position of a particular ion on this scale may be calculated or measured as, e.g., a voltage or drift time. In general, however, the scale is calibrated with a known sample, and subsequently the unknown ions are determined by interpolation. A major advantage of the calibration method is the elimination of any instrumental deviations from theoretical performance. The ordinate of the mass spectrum represents the ion intensity in arbitrary units. These also can be calibrated, and thereby become a measure of the concentration of a species detected.

Recording of mass spectra can be by photography, either directly or from an oscilloscope, by oscillography, by using recording galvanometers and a strip chart recorder with photosensitive paper, or more and more commonly, by computer techniques.

The most important feature of mass spectrometry is *identification* of a gas or simultaneously of gas mixtures. For this process, the *difference* in mass spectra obtained with *different* ionization techniques can be used to good advantage. For example, in many biological samples, the molecular weight is a characteristic property. Here, chemical ionization (189,213,478,502, 601,617,760) or low-voltage electron-impact ionization (484) are of advantage; however, the sensitivity of these processes may be low.

On the other hand, electron impact at the normal ionizing voltage (70eV) causes fragmentation of molecules. This may complicate interpretations of spectra from mixtures with a multiplicity of fragments originating from the different molecules. However, each molecule has a characteristic "fingerprint-type" fragmentation pattern, which facilitates identification of many organic molecules.

Negative-ion mass spectrometry is helpful for molecules containing electronegative constituents such as the halogens.

Finally, isotope distribution, the most characteristic of all mass spectrometric *phenomena* should be mentioned. Natural abundance of polyisotopic elements creates patterns, which are easily calculated and are unique for any number of these elements within one molecule (454,622).

As an example, chlorine has two isotopes ^{35}Cl and ^{37}Cl with natural abundances of 75.5% and 24.5% respectively. Thus, any molecule containing two chlorine atoms would have an isotope distribution of (75.5 ^{35}Cl +

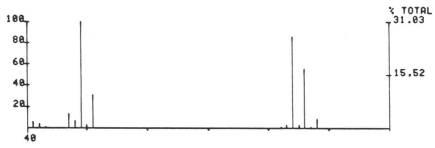

Fig. 6.20. Mass spectrum of methylene chloride.

24.5 ^{37}Cl)^2, corresponding to 100% ^{35}Cl$_2$, 64% ^{35}Cl^{37}Cl, and 10% ^{37}Cl$_2$, as seen in Fig. 6.20 for methylene chloride,

$$m/z\ 84 = {}^{12}C^1H_2\ {}^{35}Cl_2 \qquad\qquad 100\%$$
$$m/z\ 86 = {}^{12}C^1H_2\ {}^{35}Cl^{37}Cl \qquad\qquad 64\%$$
$$m/z\ 88 = {}^{12}C^1H_2\ {}^{37}Cl_2 \qquad\qquad 10\%$$

and its fragment ions

$$m/z\ 48 = {}^{12}C^1H^{35}Cl \qquad\qquad 100\%$$
$$m/z\ 50 = {}^{12}C^1H^{37}Cl \qquad\qquad 32.5\%$$

(2) Pyrolysis Mass Spectrometry

The necessity of vaporizing nongaseous samples for mass spectrometry automatically creates conditions for thermal analysis. Many of the previously described pyrolysis or flash-vaporization techniques (202) have been used inside a mass spectrometer or via a direct attachment (294,614).

In many cases, unfortunately, the thermoanalytical aspects of these evaporation methods have been ignored, even though such data as temperature or applied energy could have been made available.

Thermal effects within the ion source of a mass spectrometer can be utilized in various ways, for example through radiation by the filament itself (585) or by special furnaces and heaters (64,149,356,493), use of electron bombardment (577), direct-filament heating (204,668), including Curie-point pyrolysis (483), or with Tessler coil discharge (197), and spark sources (294). "Linear programmed thermal decomposition" in the ion source of a chemical ionization MS was described in 1968 (198,601,602) and applied to the study of bacteria (760). Focusing properties of the laser were recognized early as a suitable heating device for mass spectrometry (60,295,404), and later developed into the so-called laser probe (89,428,430,431,432), improving the earlier method of flash vaporization with xenon-flash tubes (403).

One particular device deserves special recognition. This is the so-called Knudsen cell, a crucible, mostly heated by electron bombardment, from which a molecular sample beam enters the ion source through a system of aligned orifices (baffles), under equilibrium conditions. High temperatures are attainable. The system inside a mass spectrometer (80,117,582) is mainly used for thermodynamic studies (54) of inorganic compounds (57,59,118, 482,514,589,739) or thermally stable polymers (647,649).

c. Simultaneous Thermal Analysis-Mass Spectrometry

During the early 1960s, developments of truly simultaneous TA-MS methods began. Murphy collected DTA effluents after endotherms or exotherms and subsequently analyzed them in a mass spectrometer (507). He recognized the value of this combination for the study of degradation mechanisms of polymers and complex inorganic materials. Langer and Gohlke developed continuous DTA/MS methods for samples heated either within the ion source of a time-of-flight mass spectrometer, under high vacuum (375), or outside the instrument (376) at pressures up to atmospheric. They pointed out the value of comparing results from both operational modes to detect pressure-dependent reactions (248,376,377,379,383, 384) and later described an automated data recording and display system using analog tape recording (378,380,381). Friedman and co-workers (199, 201,204,205,211) used a stepwise scanning approach for up to 200 mass units with digitized data output for their pyrolysis/MS work. Later they added an advanced, automatic, computer-controlled, data system (206,207,209).

Using a relatively simple mass spectrometer, Wendlandt et. al. reported several EGA-MS investigations (727,728,730,731,G58,G61).

In addition to their earlier work with a Knudsen cell (647,649), Shulman and co-workers later constructed an open-cell pyrolysis system (648,650), similar to that of Langer and Gohlke.

The combination of thermogravimetry and mass spectrometry (208) has been described by Zitomer et. al. (145,270,770), Vaughan (705,706), Wiedemann (737), and Kleineberg et. al. (353) for operations under vacuum as well as under atmospheric pressure.

Linear-programmed thermal degradation-mass spectrometry (LPTD-MS) (601,602,760) is carried out by chemical ionization in the ion source after coating the heating element with a solution or slurry of the sample. Its theory and applicability to biological, polymeric, geological, inorganic, and organic materials has been discussed by Risby et al. (602) in comparison with other pyrolytic and gas chromatographic techniques.

Because of the several ways in which each pyrolysis, thermal analysis, mass spectrometry, and data handling can be carried out, the combination of all four operations leads to an almost infinite number of variations (7,9,11,12,39,53,82,83,84,141,153,155,192,309,335,342,433,603,676,740).

Some publications describe specific interfaces or inlet systems (13,36,37, 45,122,153,166,176,239,358,506,646,662,761,G60), including direct probes for handling solids and liquids (253,377,383,385,491).

Coupling to the hot stage of a microscope (106) and sampling directly from flames (281) has been accomplished. The principle of the so-called partial-pressure mass spectrometer has been explained (14) and applied (153,374). Modulated molecular beams have been employed successfully (180,181), and high-pressure operation has recently been reported (355). High-resolution, double-focusing mass spectrometers have been used for thermal analysis (253,313,314).

Several investigators have directed attention to special precautions and considerations, necessary to avoid problems they encountered in combining thermal analysis with mass spectrometry (78,313,314,490,601,602,683).

Details are beyond the scope of this chapter, however, it is important to understand the general characteristics, advantages, and limitations for the proper choice of a specific combination technique. For this purpose TA-MS may be divided into three operational modes:

(1) Isothermal Pyrolysis

The sample is heated rapidly to or continuously at a specific temperature either inside or very close to the ion source of the mass spectrometer. The principle and operation is the same as in pyrolysis-EGA and pyrolysis-GC, except that here a mass spectrometer is the detector and most analyses are carried out at its operating pressure.

(2) Programmed High-Vacuum Thermal Analysis

Main considerations here are limited or fixed operating conditions rather than flexibility or choice. There is, however, one unique advantage: Heating the sample *inside* the ion source alleviates practically all problems of gas transport from the origin to the detector, as well as those of the pressure reduction necessary for mass spectrometry. Evolved molecules are *immediately* ionized without dilution, without chromatographic separation, and without secondary reactions in the absence of collisions with other molecules or with the walls.

On the other hand, materials with vapor pressures above 10^{-5} to 10^{-7} torr may be lost during evacuation. This may include liquids, adsorbed or chemisorbed gases, and losses by decomposition of metastable compounds as, for instance, a number of hydrates. In the latter category, reaction rates and temperatures often change considerably with pressure. While this may prohibit extrapolations to ambient conditions, to obtain EGA profiles both under high vacuum and at atmospheric pressure is a valuable tool to study such pressure-dependent reactions.

The most important disadvantage for low-pressure thermal analysis is the lack of heat transfer by convection and in part by conduction. This may not only cause problems for the controlled heating of the sample but also for measuring the actual sample temperature. Close contact of the sample with the heat source, also use of radiant heat and highly thermally conductive materials as sample holders (248,377,383,385), are effective counter-measures.

(3) Variable-Pressure Thermal Analysis

This constitutes coupling a mass spectrometer to a thermal analyzer, in which the sample can be subjected to variable pressure. Needless to say, while it is the most versatile and powerful of the combinations, the complexity of simultaneous operations provides a challenge to even the most experienced instrument designer. Both thermal and mass spectrometric analyses must remain unaffected, which makes the interface a critical part of the equipment. It starts with collecting the highest possible concentration of evolved gas at its place of origin and transporting it—almost always with a flowing carrier gas—into the mass spectrometer. This should be accomplished with minimum dilution and delay, which requires short transfer lines with the lowest possible volume. In the process, the pressure reduction must also take place. Both can be achieved at the same time with a capillary, ideally under plug-flow conditions. Yet, even if inert metals are used and the lines are kept meticulously clean, a certain amount of chromatographic separation can be expected. Notoriously bad are very polar molecules as, for instance, water and ammonia, which may plate out on the walls of the interface and sometimes be displaced later by a different reaction product. Highly polished stainless steel or gold-plated metals have been shown to minimize these effects. Condensation of high-boiling substances is avoided by heating the lines.

Alternatively, if for instance the test sample cannot be moved close to the ion source for physical reasons, or if secondary chemical reaction may be expected, sweeping with a fast-moving carrier gas may be a better approach. In this case dilution cannot be avoided, and pressure reduction is coupled with sample enrichment through either jet or membrane separators. The first operates on gas-flow principles, while the second is based on solubility differences. Details will be described below under GC-MS.

In summary, *internal* heating practically guarantees the detection of *all* volatilized species within the operational limits of the mass spectrometer. For known compounds, the method is quantitative, since the amount of ions observed is equivalent to the partial pressure of the species. Simultaneous and overlapping reactions can be recognized. Control and accurate measurement of the sample temperature during the experiment, however, is very difficult. For *external* heating, on the other side, the high accuracy and sensitivity of thermoanalytical instrumentation determines temperature control and measurement, while the detection of evolved species may be subject to delays and unwanted separations and interactions.

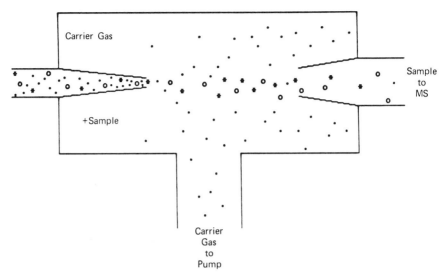

Fig. 6.21. Principle of the jet separator.

d. SIMULTANEOUS THERMAL ANALYSIS-GAS CHROMATOGRAPHY-MASS SPECTROMETRY

Very little more needs to be said about this, probably the most modern of the combination techniques.

The GC-MS principle was first demonstrated in 1957 (293). By 1964 Ryhage (615) described a device, now known as the "jet separator" (47), to interface the GC column with a mass spectrometer. Essentially at the same time, Watson and Biemann (723,724) reported on a fritted glass tube and capillaries combination, to remove preferentially the carrier gas and affect the pressure reduction necessary for mass spectrometry. At present, jet and membrane separators are commercially available. In Fig. 6.21, the mixture of the sample and a light carrier gas (mostly helium) leaves a nozzle at very high velocity. From the nozzle, the lighter molecules expand more rapidly according to Graham's law and are pumped away, while the heavier sample molecules are mostly collected by the so-called skimmer nozzle and subsequently enter the mass spectrometer. For the membrane separator (Fig. 6.22), the gas mixture impacts upon a polymer film (mostly a silicone rubber) which is more permeable to organic gases due to their solubility in the membrane material (411,413). On the other side, the dissolved molecules are released from the membrane and become available for mass spectrometric analysis.

Jet separators are more versatile but are fragile and expensive. Membrane separators are more rugged, do not plug up easily, and are best suited for low-molecular-weight nonpolar organics but not for polar or high-molecular-weight materials.

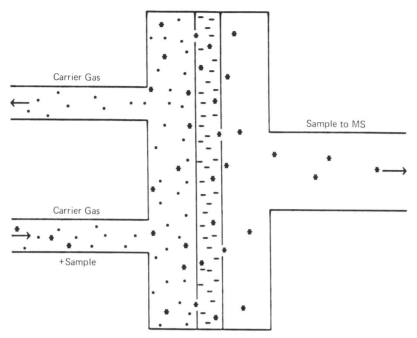

Fig. 6.22. Principle of the membrane separator.

In addition to separators, modern GC-MS systems also come equipped with capillary or microvalve inlets, made possible by very fast and high-volume differential pumping systems. This, in turn, allows for mass spectrometer operations in both electron-impact and chemical-ionization modes (171). Flame-ionization detectors may be attached for a second, quantitative registration of evolved gases.

The triple union TA-GC-MS creates such a large amount of data simultaneously, that computerized processing *and* instrumental control becomes a prerequisite. While completely simultaneous operation may be possible (172), this combination is inherently discontinuous by the operation of the gas chromatograph. For this reason, rapidly successive reactions cannot be studied directly. Instead, trapping of the effluent (102,167,486), multicolumn devices (387), or duplicate analyses may be used.

Finally, at this point, a few specialized techniques should be mentioned. Merritt et. al. developed a qualitative gas chromatographic analysis by vapor-phase pyrolysis (480). The pyrograms are correlated to molecular structure by a diagnostic function based on information theory (481).

Liquid chromatography and gel-permeation chromatography are recent analytical methods, and efforts are being made to combine them with mass spectrometry (751,769) or to use them directly as detectors for thermoanalytical studies (171).

Combustion of evolved gases either from thermal analysis or gas chromatography has yielded qualitative and quantitative information, especially in connection with selective trapping and isotope ratio monitoring mass spectrometry (454,622).

e. SPECIAL INSTRUMENTS AND PRACTICES

Because of the complex problems involved in construction of modern analytical instruments, it should come as no surprise that special equipment for combined techniques is not readily available. Therefore, the literature is full of reports on methods and devices to couple thermal analysis and mass spectrometry, ranging from very simple approaches to the most intricate. As examples, a few such combinations should be mentioned.

The so-called residual gas analyzers are rather inexpensive mass spectrometers with limited scan range, sensitivity, and resolution. Pyrolysis tubes under vacuum can easily be connected to them to identify and measure simple evolved gases of low molecular weight (218).

A "homemade" time-of-flight mass spectrometer has been used to demonstrate similar simple ways for evolved gas analysis at various pressures up to atmospheric (375,376). Recently, a *portable* mass spectrometer for gas analysis has been described (522), while at the other extreme of designs, instruments are available for quantitative analysis of gas mixtures at the part-per-billion level (287).

8. Emanation Thermal Analysis

Generally applied to the study of solid-state phenomena, this is a very elegant gas analytical method. Its basis, first described by Hahn in 1936 (272) and 1949 (273), is the release of a radioactive gas from a matrix. The gas itself may be created by a decay reaction, for example $^{224}Ra \rightarrow {}^{220}Rn$. Therefore, if ^{224}Ra had been evenly distributed throughout a solid sample, the rate of release of radioactive gas would be a function of the properties of that solid. Chemical reactions, solid-state transitions, and other physicochemical changes can be observed (96,463,569,717) by measuring the radioactivity in the gas phase over the sample. Its value as a detector for thermal analysis was recognized soon (20,21,22,25,26,79,97,105,173,174,268,586,587,588, 710) for systems where co-crystallization of an appropriate radioactive element with the test sample was possible. Bussière (96) described an apparatus for TG-DTA-ETA with a scintillation chamber and photomultiplier, which was later improved (196).

In another mode, radioactive gases, such as ^{85}Kr can be diffused into inorganic (700) or organic (277,278,279) solids under pressure and subsequently released during thermal analysis. For the study of organic compounds, ^{14}C has been incorporated and measured as $^{14}CO_2$ (521).

A theory was developed (23) for the relationship between emanation power and physical properties based on a spherical grain.

In several cases the combination of ETA with EGD, TG, dilatometry, and

DTA has shown to give more information than the latter thermoanalytical techniques alone (24,269).

A corollary to ETA was used by Garn (236), who entrapped small amounts of organics into inorganic solids and detected their release during crystal transformations by coupling DTA or TG with a mass spectrometer or other instruments.

9. Thermoparticulate Analysis

In addition to gaseous products, many materials, especially organic polymers, evolve particulate matter under certain conditions of pyrolysis (509,719).

These particles range from relatively large solids (as in smoke), which can be measured directly in photoelectric cells (419) or analyzed after trapping (675), to fogs, composed of liquid droplets, which are more readily measured by light scattering. Particles in the range of $10^{-5}-10^{-7}$ cm, however, cannot be detected directly by either of these optical methods. They do, on the other hand, act as condensation nuclei when exposed to supersaturated water vapor (139,509). The result is a fog, now amenable to counting by light-scattering techniques. When combined with programmed sample heating, this latter technique becomes known as thermoparticulate analysis (162,508,509).

The equipment, which may be quite simple (511), consists of an oven with suitable controls, from which the evolved matter is carried through a heat exchanger to the condensation nuclei counter by a slow stream of filtered carrier gas. In the counter, the sample gas is humidified to approximately 100% and adiabatically expanded into vacuum to cause condensation. As an example, the instrument described by Skala (657) operates at 5 cps for sampling, evacuation, and flushing, and with a gas flow rate of 100 ml/sec. To achieve this fast speed without adverse cooling effects on the sample, auxiliary gas is introduced between heat exchanger and counter. With this system, the growth of a droplet of 0.001 μm diameter to one of 5 μm can be achieved in 26 msec. Despite the dilution with auxiliary gas, in this fashion, 10 condensation nuclei per cubic centimeter can be detected, which translates into a sensitivity of about 4×10^{-17} g of particulates/cm^3.

Thermoparticulate analysis was primarily used for measuring decomposition temperatures of polymers (162,511). But later, Murphy discussed its applicability to a number of materials either directly, or by using conversion techniques (509,510,701). Ammonia, for instance, can be detected after forming condensation nuclei of NH_4Cl in an atmosphere of the HCl/H_2O azeotrope. Irradiation of sulfur dioxide in air leads to oxidation, and the resulting sulfur trioxide forms droplets of sulfuric acid with water vapor (165). Mercury vapor can be oxidized by irradiation in air, and the oxide forms condensation nuclei. Obviously, the table of detectable materials shown below (Table 6.II) (701) can be extended to other gas phase reactions,

TABLE 6.II

Gases Detected by Condensation-Nuclei Techniques[a]

Substance	Conversion process	Detectable concentrations (ppm)
Ammonia	Acid-base	0.005
Benzene	Photochemical	2
Carbon dioxide	Electrochemical	5
Carbon monoxide	Chemical	1
Chlorine	Chemical	1
Ethyl alcohol	Reverse photochemical	5
Freon 12-21	Pyrolysis	2
Hydrogen chloride	Acid-base	0.5
Hydrocarbons	Photochemical	0.1
Methyl mercaptan	Oxidation-photochemical	0.01
Monoethyl amine	Acid-base	0.5
Mercury	Photochemical	0.001
Nitrogen dioxide	Hydrolysis	0.5
Sulfur dioxide	Photochemical	0.001
Toluene	Reverse photochemical	1
Unsym-dimethylhydrazine	Acid-base	0.1

[a]Adapted from Ref. 701.

especially if the products react readily with water. By calibration, the method becomes quantitative (512).

Recently, diazonium compounds (659), metal acetylacetonate chelates (660), and arene sulfonates (331) were investigated, and instrumentation for organic particulate analysis has been described (573) as a way to determine "real" decomposition temperatures.

10. Data Acquisition

The development of a modern analytical instrument constitutes a considerable financial commitment. Consequently, the design must be able to fulfill a variety of needs. For most individual applications, this means that the system supplies more data than are needed or even desirable. Data handling then becomes a very—if not the most—important part of the analysis.

A historical review of the various approaches applied to mass spectrometry shall serve as a framework for an evaluation of the options available to solve problems in evolved gas analysis, regardless of the particular technique being used. This is quite appropriate, because earlier, inherently much simpler, data collection methods are by no means obsolete and still practiced today.

If we remember (see previous section) that analytical data basically carry two properties, quality (or identity) and quantity, it becomes obvious that any data handling technique will have to account for one or the other, or

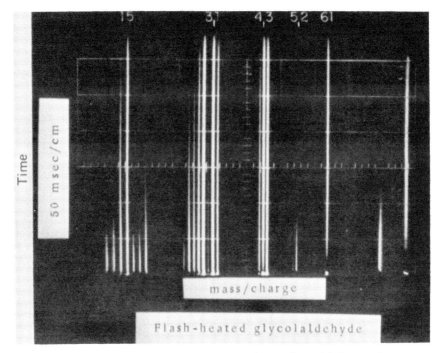

Fig. 6.23. Time resolved mass spectrometry by Z-axis modulation. (Ref. 401)

both. Quantitative data include pressure, volume, weight, or a correlatable signal. Qualitative data mostly appear in spectral form. Two additional variables—time and temperature—will also have to be accounted for in most cases.

The photographic plate was the earliest mass spectrometric recording device. It provided an integrated set of qualitative and quantitative information over the exposure time.

On the other hand, an oscilloscope produces instantaneous display but no permanent record. For many years, this problem occupied the minds of mass spectrometrists. One of the first approaches was simply to take photographic pictures of the oscilloscope screen periodically during an experiment (383,668). This required fast photographic emulsions and manual recording of time. Naturally, it was limited to relatively slow processes, or required high-speed cameras operating in the order of milliseconds per frame (668).

Refinements in oscilloscope techniques allowed Lincoln to display fast reactions with a method called "time resolved mass spectrometry" by "Z-axis modulation" (401,402,405). As seen in Fig. 6.23, each subsequent sweep ($\equiv$ mass spectrum) is vertically displaced and the intensity of the display is representative of the number of ions present.

As an example, the technique was used to study rocket propellants at a heating rate of $1 \times 10^{6}°C/sec$ (668).

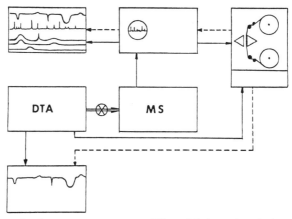

Fig. 6.24. Schematic of a mass spectrometer-differential thermal analysis system, using an Analog tape recorder. (Ref. 380)

Langer et al. (378,380,384) used a combination of a scanning oscilloscope, analog tape, and X, Y recorder to store complete mass spectra at a rate of at least 10/sec (Fig. 6.24).

With this arrangement, any recorded mass spectrum could be reproduced, or any chosen ion intensity plotted as a function of time (equivalent to temperature). Other methods of monitoring selected ion peaks were described (257,593,652). Advances in computer hardware technology allowed Friedman and co-workers (206,207,209) to convert analog signals to digital output and to use computers to identify mass peaks and evaluate their intensities (Fig. 6.25).

The next step was the evolution of computer programs (the so-called software), which not only extended the data-handling capabilities of computers but also added control functions for operating mass spectrometers, gas chromatographs and thermoanalytical instruments by setting up standard conditions maintained by feedback mechanisms (8,88,115,159,160,171, 191,202,216,684). The hardware includes minicomputers, programmable calculators, and curve resolvers (32,287,693). An example for the extent to which computers can control whole combinations of thermoanalytical systems is a commercial GC-MS system (Hewlett-Packard, 5985 GC/MS). It consists of a gas chromatograph, quadrupole mass analyzer, and direct-insertion probe. Temperatures at various parts of the equipment can be set, and both GC oven and probe temperatures are programmable. *All* operations, including the tuning of the mass spectrometer, are controlled by the system's computer. While this increases considerably the reproducibility of results, it limits—in some cases severely—the operator's choice to the "intellectual capacity" of the computer. Since it also allows very few "peeks" into the "black box" during operation, the results depend on almost perfect functioning of the system. Ironically, therefore, the more automated a system becomes, in order to make its use *easier* for the operator, the more

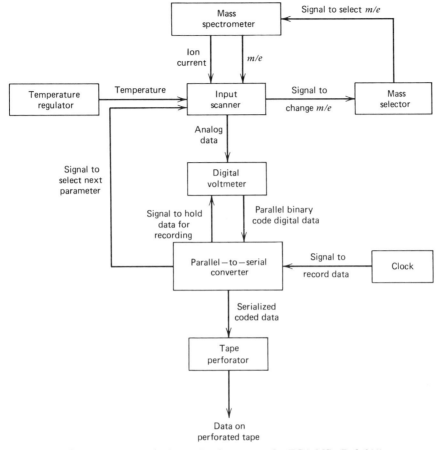

Fig. 6.25. Automatic data-collecting system for EGA-MS. (Ref. 211)

important it is to be aware of the precise logic behind such automation, to be able to evaluate any data output. When the analyst no longer can see a "real" mass spectrum but has to rely on a computer-reconstructed bar graph, he simply must understand how it was obtained. He also must know, if indeed the system was operating properly. This responsibility becomes even greater, when additional computer manipulations are used, e.g., curve smoothing or deconvolution routines, background subtractions, or even literature searches. Improper use can lead not only to unrealistic but also to misleading results. Standardization therefore becomes a critical part of the analytical process.

11. Literature

In addition to the specific references listed throughout the text and the general literature cited, a few reviews should be mentioned that cover particular aspects of evolved gas analysis. Ware wrote a brief review of thermoanalytical techniques including EGA (720). In 1974, Krug reviewed gas analysis methods and applications (367). Evaluations of various detectors were made by Garn et. al. (232,234) and Loveland et. al. (423). Morris (498) and Sĕvčik (641) discussed those for gas and liquid chromatography, and Morgan (494) described a series of specific gas detectors.

The journals covering the subject are *Analytical Chemistry* with its reviews, *Journal of Thermal Analysis, Thermal Analysis Abstracts, Thermochimica Acta;* reviews are also published in trade journals such as *Industrial Research Development* and *American Laboratory.*

IV. APPLICATIONS

A. INTRODUCTION

Up to this point, the discussion centered around equipment, techniques, and methodology developed for analysis of evolved gases. For a look at results obtained, it seems appropriate to classify the materials investigated. The system chosen for this classification is based on thermal properties of matter rather than on a conventional chemical system. Therefore, the release of simple constituent gases or vapors from solids will be considered at the beginning, without any or any significant chemical change, with mostly known products, and often by known reactions. This group includes primarily inorganic materials. The second group, containing organic and polymeric substances, is characterized by significant chemical changes, often by very complicated or even unknown reactions. In the third group, one will find applications in several scientific disciplines where very specific problems are to be solved. Detailed results cannot be comprehensively reported, but the intent is to exemplify the types of questions raised and the nature and quality of answers obtained by the methods discussed in the previous sections.

B. CLASSES OF MATERIALS

1. Inorganic Compounds

a. WATER

It appears, that by far the most common, but at the same time analytically the most difficult to handle, molecule is water. In its simplest association it exists adsorbed on a surface, entrapped in a matrix, or as the so-called water of crystallization, from all of which it may be evolved in a continuous or stepwise fashion. The energies required for this release are generally low and may involve electrostatic interactions, lattice breakdown, or breaking of hydrogen bonds.

As an example for surface-held water, the desorption kinetics from silica gel were recently studied and a mechanism proposed (159). The reaction order was $n = 1$ and the activation energy $E = 36.7 \pm 2.6$ KJ/mol. Similarly, desorption of chemisorbed water from rutile surfaces was investigated by partial pressure mass spectrometry (14).

Probably the most frequently analyzed compound is copper sulfate penta-hydrate, readily available to demonstrate the performance of a newly designed piece of equipment. Therefore, it has been used in discussions involving interpretations, merits of operations, choices of methods, and many others.

Borchardt and Daniels (77) had studied its dehydration together with $CoCl_2 \cdot 6 H_2O$, $MnCl_2 \cdot 4 H_2O$, $SrCl_2 \cdot 6 H_2O$, and $BaBr_2 \cdot 2 H_2O$. They interpreted the sequence as follows:

92.5°C: $CuSO_4 \cdot 5 H_2O(s) \longrightarrow CuSO_4 \cdot 3 H_2O(s) + 2 H_2O(l)$

102–115°C: $2 H_2O(l) \longrightarrow 2 H_2O(g)$

130°C: $CuSO_4 \cdot 3 H_2O \longrightarrow CuSO_4 \cdot H_2O + 2 H_2O$

250°C: $CuSO_4 \cdot H_2O \longrightarrow CuSO_4 + H_2O$

Wendlandt later used the same compound and Mohr's salt to establish that the EGA curve obtained with a thermal conductivity detector, is virtually the mirror image of the DTA trace obtained in helium.

His proposed sequence was:

85°C: $CuSO_4 \cdot 5 H_2O(s) \longrightarrow CuSO_4 \cdot 3 H_2O(s) + 2 H_2O(l)$

115°C: $CuSO_4 \cdot 3 H_2O(s) \longrightarrow CuSO_4 \cdot H_2O(s) + 2 H_2O(g)$

230°C: $CuSO_4 \cdot H_2O \longrightarrow CuSO_4(s) + H_2O(g)$

Langer and Gohlke (375) subsequently established the pressure dependency of this reaction, by carrying out the dehydration at 10^{-6} torr inside the ion source of a time-of-flight mass spectrometer with a specially constructed miniature furnace. They found complete dehydration at room temperature for this pressure.

Paulik, Paulik, and Erdey (554) next demonstrated the effect of particle size or surface area on this reaction with the use of a multiplate (large surface area) sample holder. They observed a three-step dehydration, as compared with a two-step reaction, when a regular crucible was used. A study of these references together with other reports on $CuSO_4 \cdot 5 H_2O$ as a test for specific equipment performance (161,166,211,450,519,593) clearly reveals that EGA results, more than those of the other thermoanalytical techniques, critically depend on the operating conditions and absolute answers are rarely obtainable. Thus, standardization of equipment is of utmost importance.

Other similar hydrates were investigated, showing differences in energies to release the obviously differently held (by geometry, bond, or coordination) constitutional water molecules (13,375,519,545,570,590). In the presence of more than one ligand, as in $Cu(NH_3)_4SO_4 \cdot H_2O$, differences in the type of coordination become even more apparent (379,380,559,567). Complications arise, when red-ox reactions may occur, when hydrolysis takes place, when hydroxide or oxide hydrates are present, or when acid anhydrides may be formed (98,195,222,224,226,228,291,312,416,448,450,528).

It has been shown, that the dehydration of gypsum (475) and $MgSO_4 \cdot 7$ H_2O (414) is a function of the partial pressure of water over the sample, and that pressure dependent thermal dehydrations are indicative of metastable hydrates and diffusion controlled reactions (248,376,383). These latter reports also demonstrate how EGA curves for water can differentiate between a hydroxide, such as $Mg(OH)_2$ and, for instance, triphenyltin hydroxide, which in reality is an oxide hydrate

$$2 \, (C_6H_5)_3SnOH \rightleftharpoons (C_6H_5)_6Sn_2O \cdot H_2O$$

In another example, EGA at different pressures could demonstrate that thermal dehydration of $MgCl_2 \cdot 6 \, H_2O$ cannot produce the anhydrous chloride, due to competing hydrolysis. Only if water is replaced by another ligand—such as dimethyl sulfoxide—that cannot cause hydrolysis, can the anhydrous salt be obtained thermally (383).

As an example of very high-temperature systems, the reaction mechanism of water with lithium oxide was studied by Knudsen cell mass spectrometry (57). Between 1100 K and 1400 K and 0.1 mm H_2O pressure, LiOH was the major species.

b. CALCIUM OXALATE

This compound is being singled out because it provides the best example for a multistep reaction with several evolved gas species. Since each of the products have been known, this reaction mechanism has been studied extensively under varied conditions. Today, it is probably the best understood reaction in thermal analysis and therefore the classical standard for testing and calibrating thermoanalytical instruments (45,106,164,175,176,193,489, 503,593,607,664,733,737). The weight-loss curve in Fig. 6.26 (198) shows the separation of the three reaction steps:

$$CaC_2O_4 \cdot H_2O \longrightarrow CaC_2O_4 + H_2O$$

$$CaC_2O_4 \quad\quad \longrightarrow CaCO_3 + CO$$

$$CaCO_3 \quad\quad\quad \longrightarrow CaO + CO_2$$

After further characterization of the sequence by EGA (607), Wiedemann later conducted a detailed study by TGA, DTA, mass spectrometry, and gas chromatography. At 10^{-4}–10^{-6} torr, he found a decrease of the dehydration temperature to 80°C from 175°C in air. The reversible calcium carbonate decomposition was lowered from 760°C to 650°C in vacuum. On the other hand, the irreversible loss of CO was observed at 470°C at either pressure. He also indicated that the presence of oxygen or water may have an influence on the reactions and noted the oxidation of CO to CO_2 in air.

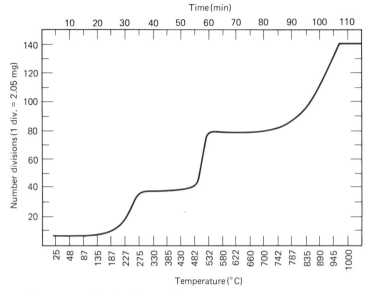

Fig. 6.26. Weight loss from calcium oxalate monohydrate. (Ref. 198)

c. Other Oxalates and Carbonates

Compounds also used for demonstrations and calibrations, as well as kinetic studies, include $NaHCO_3$ (607,688), cobalt oxalate dihydrate and lanthanum oxalate hydrate (232), $CaCO_3$ (216,683) and $MgCO_3$ (164, 338,686,733), double Sr/Ba carbonates and triple Ca/Sr/Ba carbonates (326), coprecipitated carbonates (750), zinc hydrocarbonate (122), alkaline earth carbonates (78,325), $KHCO_3$ (733), $MnCO_3$ (373), $CdCO_3$ (632), as well as Ba, Sr, and Mg oxalates (144). The effect of atmosphere and pressure has been the subject of several reports (16,46,232,373,406,746), and both carbonates and oxalates have been used to demonstrate emanation thermal analysis (21,173,174,269,700). Gallagher and co-workers used a gas partitioner and residual gas analyzer to study the decomposition mechanism for barium and strontium titanyl oxalates, proceeding via mixture of, for example, $BaCO_3$ + TiO_2, and formation of titanate above 600°C (217). They also investigated freeze-dried tantalum and mixed lithium-niobium oxalates and obtained similar results (221). Disproportionation of CO produced carbon during thermolysis of anhydrous lanthanum oxalate (619). By DTA-MS and TG-MS with high selectivity and precision, kinetics of Ag_2CO_3 and $NiC_2O_4 \cdot 2\ H_2O$ decompositions were found to be more complex than previously assumed (39). A comparison between $Gd_2(C_2O_4)_3 \cdot 10\ H_2O$ and $Gd_2(CO_3)_3 \cdot 3\ H_2O$ revealed that the former does not decompose via formation of carbonate (333,334). Germanium dioxide and SiO_2 were found to react with alkali carbonates to release CO_2 (92). On the other hand, in the presence of SnO_2, the carbonates dissociate first, followed by reactions between the oxides.

d. Ammonia and Nitrogen Oxides

Release of ammonia from ammonium salts or from ligands in metal complexes is rather uncomplicated. The mass spectrometer is a sensitive tool for detection of nitrogen-containing species (379,448,534), which sometimes are only present as impurities (228). For quantitative measurements, the thermo-gas-titrimetric method has been successful (559,567). Kinetic investigations were carried out on compounds like $[Ni(NH_3)_6]Cl_2$ (164), NH_4ClO_3 (310), NH_4ClO_4 (357), and NH_4NO_3 (55). Red-ox systems were studied, including ammonium arsenate with reduction of arsenic and oxidation of the ammonium ion (226), urea, biuret, cyanuric acid and compounds of ethylenediamine tetraacetic acid (375,381), and hydrazoic acid (HN_3) with formation of H_2, N_2, NH_3, and undecomposed HN_3 (197). From GaN, only nitrogen was detected (260), whereas lead and silver azides produced nitrogen and ammonia, apparently by interaction with water (644). For ammonium chlorate and perchlorate, a complicated mechanism was observed with formation of oxides of nitrogen and chlorine together with H_2O, HCl, N_2, and NH_3 (17,310,457,707,708), which is affected catalytically by metal oxides (311,395), other additives, as well as particle size and atmosphere (540), or even by collision with walls of the reaction vessel (310).

Similar effects and products were observed for the decomposition of ammonium nitrate (34,55,680) and other nitrates (670), $Mn(NO_3)_2 \cdot 2\ H_2O$, Mn(III) oxynitrate (222), $Mg(NO_3)_2 \cdot 6\ H_2O$ (13), both crystallized and freeze-dried silver and cadmium nitrate (9), and sodium, barium, and strontium nitrate (408). All this work being done by mass spectrometry, in part using a Knudsen cell (117).

e. Halogen Compounds

Surprisingly little can be found in the literature about halide decomposition. The reason is probably difficulties in detection of acidic compounds, such as HCl, by techniques other than trapping or thermo-gas-titrimetry. Of the few studies published, release of HCl was reported from the dihydrate dihydrochloride of ethylenediamine tetraacetic acid (375), $MgCl_2 \cdot 6\ H_2O$ (283), and chlorine-containing extreme pressure oil additives (697). Corrosion of chromium was investigated via volatilization of $CrCl_3$, $CrCl_4$, and CrO_2Cl_2 (599), and kinetics for gas evolution were determined for KIO_4 (164), periodic acid, iodic acid, iodic acid anhydride, and iodine pentoxide (317).

Chlorates and perchlorates are a special case, not as halogen compounds, but as oxidizing agents. In addition to the previously mentioned ammonium salts (310,357,395,457,644,707,708), sublimation and decomposition of nitronium perchlorate was investigated by "thermobarogravimetry" (458), and ammonium perchlorate, hydroxylamine perchlorate, and hydrazinium diperchlorate by "flash mass thermal analysis" (668). Kinetics were studied on 57 wt % $HClO_4$ in water (311), and on $TlClO_2$, $TlClO_3$, and $TlClO_4$ (665). The chlorite exploded even at 45°C, the perchlorate was the most stable, and neither chlorite nor chlorate showed disproportionation. The activation energy of irreversible chromium perchlorate decomposition was obtained by oxygen pressure measurement (591).

f. Oxidizing Agents and Oxidation Reactions

In contrast to oxygen-rich ammonium compounds, a number of inorganic salts decompose simply by release of oxygen. Some reactions are so well known, that not much could be gained by further thermoanalytical work. However, equipment has been tested by oxygen release from HgO and BaO_2 (733), and inorganic thermal emanation analysis was demonstrated with ^{226}Ra-doped βPbO_2 (196). Conversely, oxidation of molybdenum and tungsten was followed between 1500 K and 2600 K inside a mass spectrometer by changes in oxygen pressure (58). Similarly, the sequence

$$Ba(OH)_2 \longrightarrow [BaO] + H_2O \xrightarrow{O_2} BaO_2$$

was observed via oxygen scavenging (383). Oxidation of carbon, catalyzed by inorganic salts, was detected with a thermal conductivity cell (748). The mechanisms for oxygen evolution from $KMnO_4$ in air (285) to form $K_4Mn_7O_{16}$, and the four-stage decomposition $PbO_2 \longrightarrow PbO_x \longrightarrow PbO_y \longrightarrow PbO$ (7) were explained based on mass spectrometric data.

g. Sulfur Compounds

The composition of sulfur vapor itself is strongly dependent on vaporization methods and gas-phase conditions because of equilibria, not only be-

TABLE 6.III
Vapor Composition of Sulfur (S_n) by Mass Spectrometry[a]

F. E.[b]		K. C.[c]		G. I.[d]		MTA[e]	
n	I[f]	n	I	n	I	n	I
8	100	8	100	2	100	2	100
2	70	2	95	1	8	1	51
4	48	4	64	6	7	4	40
5	29	5	41	4	2	8	36
3	15	6	39	8	2	3	25
6	15	3	24	5	2	5	17
1	11	1	10	3	1	6	13
7	<1	7	8	7	<1	7	2
	73[g]		75[g]	Sample:	250g		50–700[g]
				Inlet:	120g		

[a]Adapted from Ref. 383.
[b]F. E. = Free evaporation through baffles of rhombic sulfur.
[c]K. C. = Knudsen cell evaporation of rhombic sulfur.
[d]G. I. = Evaporation through gas inlet system of undefined sulfur.
[e]MTA = Evaporation by "MTA" from sulfur and sulfide containing sample.
[f]I = Intensity (most intense peak ≡ 100).
[g]Temperature °C.

TABLE 6.IV
Mass Spectrum of Fuming Sulfuric Acid (30% SO_3)[a]

m/z	Ion	Intensity
16	S^{2+}	<10 corr.
32	S^+	56 corr.
34	$^{34}S^+$	<5
48	SO^+	100
50	$^{34}SO^+$	<8
64	SO_2^+	100
66	$^{34}SO_2^+$	<8
80	SO_3^+	0
81	HSO_3^+	32
83	$H^{34}SO_3^+$	<5

[a]Adapted from Ref. 249.
Sample temp.: 80–100°C.
Intensities have been normalized and corrected for background errors.

tween S_n molecules but also with the sulfur-producing surfaces and gaseous molecules (59,383).

For quantitative determination of sulfur or sulfides, oxidation to SO_2 or SO_3 is appropriate, as demonstrated for pyrite in bauxite (338), and for heavy-metal sulfonates (477) by thermal gas titration.

A controversy existed for years about thermal decomposition sequences of sulfates, particularly transition metal sulfates. The reason lies in difficulties in interpretation of mass spectra for SO_2 and SO_3 (188,247,249,324, 548,730). This problem surfaced for identification of the last step of the copper sulfate decomposition. It was first postulated to proceed via

$$(1) \ CuSO_4 \longrightarrow CuSO_4 \cdot CuO + SO_3$$

$$(2) \ CuSO_4 \cdot CuO \longrightarrow 2\ CuO + SO_3 \tag{77}$$

When subsequently the SO_2^+ ion was observed in the mass spectrometer,

TABLE 6.V
Mass Spectrum of $CuSO_4 \cdot 5\ H_2O$ at 650°C, 10^{-7} Torr[a]

m/z	Ion	Intensity
32	S^+	29 corr.
48	SO^+	68
50	$^{34}SO^+$	7
64	SO_2^+	100
66	$^{34}SO^+$	5
80	SO_3^+	14

[a]Adapted from Ref. 249.

reaction (1) was interpreted as

$$(3) \quad CuSO_4 \longrightarrow CuO \cdot CuSO_4 + SO_2 + 1/2\ O_2 \qquad (730)$$

Later, doubts were expressed about this conclusion, after the SO_3^+ ion could only be detected in the mass spectrometer by using very special precautions. The reason being the activity of the molecule on surfaces, and especially with water and hydroxylic species, in addition to its fragmentation under electron bombardment (188,249,324).

A recent report again postulates the release of $SO_2 + 1/2\ O_2$, the formation of a basic compound $CuO \cdot SO_3$ (sic), and a final step:

$$2\ CuO \longrightarrow Cu_2O + 1/2\ O_2$$

verified by the melting point of Cu_2O (175). The latter step is in agreement with earlier observations of a red rather than brown-black residue at 1000°C (386).

For $FeSO_4 \cdot 7\ H_2O$ and freeze-dried $FeSO_4 \cdot H_2O$, the following reactions were proposed, based on mass spectrometric observation of the SO^+ ion ($m/z = 48$) from SO_2 and SO_3 (219). Under an inert atmosphere:

$$200\text{–}300°C \quad FeSO_4 \cdot H_2O \longrightarrow FeSO_4 + H_2O$$

$$475\text{–}575°C \quad 2\ FeSO_4 \longrightarrow Fe_2O_2SO_4 + SO_2$$

$$550\text{–}675°C \quad Fe_2O_2SO_4 \longrightarrow Fe_2O_3 + SO_3$$

Under oxidizing conditions, however, two different pathways were postulated between 150°C and 500°C:

$$2\ FeSO_4 \cdot H_2O + 1/2\ O_2 \longrightarrow 2\ FeOHSO_4 + H_2O$$

$$2\ FeOHSO_4 \longrightarrow Fe_2O(SO_4)_2 + H_2O$$

$$\text{or } 2\ FeSO_4 \cdot H_2O \longrightarrow 2\ FeSO_4 + 2H_2O$$

$$2\ FeSO_4 + 1/2\ O_2 \longrightarrow Fe_2O(SO_4)_2$$

followed by

$$Fe_2O(SO_4)_2 \longrightarrow Fe_2O_3 + 2\ SO_3$$

at 600–675°C.

Kinetics and reaction mechanisms were also studied for aluminum sulfate (323,548,689), ammonium aluminum alumn (323,545), potassium aluminum alumn (332), $Hf(SO_4)_2 \cdot 4\ H_2O$ (548), and $BeSO_4 \cdot 4\ H_2O$ (375), all with the same difficulties in identifying SO_3 as a reaction product.

Using a modulated beam, mass-spectrometric sampling from flames allowed studies of thermal decomposition and sublimation of sodium sulfate (281).

h. Hydrides

For the study of hydrides, EGA has been proven to be more valuable than DTA. Due to hydrogen retardation, MgH_2 produced two significant EGA peaks, where DTA only showed a small response (175). In reverse mode, the reduction of MnO_2 (175) was studied in a stream of hydrogen. Uranium hydride (UH_3) decomposed in a 10:1 mixture of He/O_2 (489), decaborane was studied as a component of propellants (17), diborane at low pressure formed BH_3, H_2, and boron (181), while borane carbonyl (BH_3CO) produced large amounts of borane at 250°C (231). The breakdown of triphosphine-5 and diphosphine-4 to phosphine and diphosphine-2 was followed with a magnetic sector mass spectrometer and modulated molecular beam technique (179,180).

i. High-Temperature Inorganic Materials

Several inorganic oxides, or sometimes also halides, are important starting materials for ceramics, catalysts, and glasses. Mass spectrometry has been used in connection with laser vaporization (405), or directly, for identification of oxygen, water, CO_2, and HCl, as products from solid state or gas phase reactions, or as evidence for impurities (11,228,499,535,536,656).

Knudsen cell techniques provided dissociation energies, heats of vaporization, vapor pressures, and equilibrium data on Be/O, W/O, B/O, As/Te, As/Fe, Se/Te, Cd/Zn, Pb/Cd/Cl, Na/Cl, and Al systems (118,482,514, 589,739). The phase diagram of VO_2–$VO_{2.5}$ was determined from O_2 equilibrium pressure (538). Kinetics and decomposition mechanism of ZrO_2, cesium oxide, and MgO (681), as well as hydrous chromium oxide (185), were also determined by EGA.

Emanation thermal analysis has been very effectively applied to the $\alpha \rightarrow$ β transition of silica (700), to check for crystal lattice disorders caused by additives in synthetic rutile (23), to relate density, porosity, surface area, and volatiles of uranyl gels (26), to sintering of UO_2 (710), to decomposition of $Mg(OH)_2$ (588) and sintering of porous MgO (586), to determination of the apparent activation energy of diffusion of ^{220}Rn in MgO (587), and to the study of the reaction of α-Fe_2O_3 with ZnO (24).

j. Metallurgy

Evolved gas analysis in metallurgy is dominated by the work of Bandi and co-workers, after they showed that metal carbides and nitrides burn in oxygen over specific temperature ranges to give CO_2 and N_2 (28), and after they developed a DTA-EGA method for identification and determination of nitrides in steel (29). Later they isolated Fe_3C, amorphous carbon, and graphite from steel, and obtained different reproducible peak temperatures for CO_2 evolution on combustion (362). This led several authors to analyze for carbides in maraging steel (363), and also for nitride phases (30,31,734,735), for the vanadium precipitates VC, VN, and carbonitride

(364), for sulfides (600), and for nitrides and carbides of Cr, Ti, and Nb in steel (31,32).

Degassing of Fe-Ni alloys produced H,H$_2$; N,N$_2$; and O,O$_2$ as a function of Ni content (682). Reduction with graphite and effect of plastic coating was studied by quadrupole mass spectrometry.

To trace the source of evolved oxygen, reduced and re-oxidized metal powders (Co, Cu, Fe) were analyzed by thermogravimetry, and the evolved water was observed by IR and determined by Karl-Fischer titration after absorption in methanol (271).

Dissociation of GaAs and Ga$_7$Al$_3$As during alloying of gold film caused migration of Ga into gold and escape of As through the thin gold film (344).

k. METAL COMPLEXES

Transition metals, using d orbitals, can readily accept electron-donating ligands to form complexes, stable under ambient conditions but subject to thermal decomposition, often via discrete stepwise reactions. Water and ammonia, as such thermally released ligands, have already been discussed earlier. The investigations referred to below were primarily concerned with the effect of the central metal ion on the stability of its complexes.

Of the group-VIII metals, cobalt compounds have been studied by thermal analysis coupled with mass spectrometry (129,131), by manometric measurements on cobalt(III) amine halides (732), by gas chromatography in 10-degree intervals (537), or by derivatography under conventional and quasi-isothermal, quasi-isobaric conditions on cobalt ammonium phosphate hydrates together with those of Mg, Mn, and Cd (559). Nitrates, sulfates (126), halides, (17,128,130), and thiocyanates (130) of ethylenediamine Co(III) complexes showed different and complicated patterns in vacuum and under nitrogen, involving structural rearrangements and red-ox reactions between the metal and organic ligands. Similar investigations were carried out on mercury compounds (127), copper complexes (247,249,379,380, 563,661), hexammine chromium(III) compounds (728), and cyclopentadienyl-triphenylphosphine complexes of nickel(II) (378,379,380). The decomposition process was discussed and kinetics calculated for Co(II), Ni(II), and Cu(II) salicylaldoximates, 2-indole carboxylates, and 2-thiophene carboxylates (434,693).

Catalytic effects of tin in its complexes and its participation in reduction and oxidation of its ligands accounts for complicated decomposition pathways of barium bisoxalatostannate(II) and tetraoxalatostannate(IV) (223), tetrakis(diethyldithiocarbamato)tin(IV) (83), dihalotin(IV) *bis*(diethyldithiocarbamates) (82), and for the effect of oxygen on the two-stage decomposition of tin(IV)-dithiocarbamate complexes (84). Other widely studied ligands were oxalate (513), acetylacetone (660), salicylaldimine (693), ethylenediaminetetraacetate (375,383,523,524), nitrilotriacetate (523), and dimethylsulfoxide (383). Amphoteric elements were studied in the form of salts like 5(NH$_4$)$_2$O · 12 WO$_3$ · 5 H$_2$O, 5(NH$_4$)$_2$O · 12 WO$_3$ · 11 H$_2$O, spray-dried ammonium paratungstate (43), and also as $UO_4 \cdot 2\ NH_3 \cdot 2\ HF$ (753), an example of uranium complexes that may be considered for separation techniques.

Investigations of a number of less conventional systems were reported by Gallagher et. al., such as the dehydration of $BaPt(OH)_6 \cdot 0.5\ H_2O$ in oxygen to produce $BaPtO_3$ (220,227), decomposition of $K_2Pt(CN)_4Br_2$ and $K_2Pt(CN)_4 \cdot 8\ H_2O$ by EGA-MS with formation of cyanogen, platinum, and KBr, or respectively cyanogen, platinum, and KCN (225). The work also covers thermal analysis of boracites (230) of Cu, Ni, and Fe in air and vacuum, and europium hexacyanoferrate(III) and ammonium europium hexacyanoferrate(II) with detection of intermediate phases (218).

In connection with metal complexes, investigations on intercalates and clathrates should be mentioned. Urea clathrates with n-paraffins were used to study kinetic parameters (525). Clathrates of copper and nickel cyanides and chlorides with ethylenediamine and benzene were monitored for release of benzene, ethylenediamine, ammonia, and water (666). The graphite intercalates of IF_7 and IF_5 + HF were shown by mass spectrometry to produce only IF_5 and fluorocarbons. On the other hand, those of ClF_5, ClF_3, BrF_5, and BrF_3, fluorinate the lattice carbon (638). Krypton difluoride reacts directly with graphite to form intercalates $C_mKrF_n (n > 2)$ (636). EGA by MS detected Kr and fluorocarbons in contrast to xenon fluoride, which shows xenon fluoride fragments, indicating that the graphite intercalation compound may be the most stable form of KrF_2 (636). Other graphite intercalates studied were XeF_6, XeF_4, $XeOF_4$, and AsF_5 (637). That of UF_6 produced CF_4 and some C_2F_6, as the major species found by mass spectrometry (65).

l. Catalysts

Studies of catalysts or catalytic reactions may be the youngest field for EGA, but also possibly the most rapidly expanding. The range of investigations covers desorption from surfaces of catalysts or catalyst supports, their analysis by vaporization of components, and identification and measurements of reactants and products.

Spark-source mass spectrometry has been applied to determine 22 elements in compressed ground catalysts at concentrations of 0.19–160 ppm (116). Iron, cobalt, nickel, and vanadium have been vaporized for mass spectrometric analysis by a 5–6 KeV electron beam, and sodium was determined in a Y-zeolite with 6% accuracy (577).

Systems for catalytic hydrogenation have been analyzed through different approaches. Kinetics of adsorption and release of hydrogen on $LaNi_5$ have been followed with a differential manometer (52), desorption of hydrogen from Ni-Al, Ni-Cu-Al, Ni-Mg, Ni-Zn, and Raney nickel with a platinum-resistance thermometer (568), that of hydrogen, water, and noncondensible gases alone with an ionization gauge, Pirani gauge, or MKS Baraton gauge (72), and of HCl by absorption and conductivity measurements—all from alumina-supported Pt, Re, Pt-Re, Pt-Ir, or Pt-Pb bimetallic catalysts (71,72). Thermal desorption analysis of catalysts by mass spectrometry followed the

HCN yield from reduction of NO over Pt, Pd, Ru, Monel, and Perovskite catalysts with the possibility of intermediate formation of isocyanates on the surfaces (714).

For oxide systems, kinetics of the water–gas-shift reaction over MnO was determined by gas chromatography for CO, H_2O, and CO_2, using a thermal conductivity bridge (369). Ammoxidation of propene over antimony oxide/ferric oxide was investigated (628). Yields of acrylonitrile, CO, and CO_2 were related to the closeness of hydroxyl groups on the solid surfaces. Formation of a nickel-molybdenum oxide catalyst from $(NH_4)_4[NiMo_6O_{18}H_6]$ $\cdot$ 5 H_2O was observed with a Derivatograph and attached EGD equipment (267). The importance of EGA was exemplified for catalytic decomposition of formic acid on manganese(II) formate, which may lead to either

$$HCOOH \longrightarrow H_2 + CO_2$$

$$\text{or } HCOOH \longrightarrow H_2O + CO$$

depending on the presence of metallic phases or unstable manganese(II) formate (154). The thermodynamics of the acetonitrile-nickel(II) chloride system were studied by equilibrium vapor pressure determination (33).

Zeolites have been characterized successfully by thermal desorption (759), dehydroxylation, and deammonation (76,596). The latter was also used for determination of surface acidity (50). Complex curves for the release of H_2O from zeolites NaA, MgA, and LiX have been resolved by computer (160). Sorption and surface reactions of cumene, ammonia, and pyridine on sodium and cerium Y-zeolites have been analyzed by intermittent sampling for GC (447). With a capillary over a sample of Na X-zeolite in the Derivatograph, dehydration of alcohol was followed by GC (371). In the product stream from fluid catalytic cracking, the peak intensity of 15 selected hydrocarbon components were recorded by a low-resolution mass spectrometer, interfaced with an on-line computer (191).

m. MINERALS

Had it not been for minerals, thermal analysis may not have enjoyed the rapid rise we have observed. Yet, it appears, that EGA has not found its way into mineralogy to the extent it could have. A few systems that have been investigated for practical purposes are dehydrations of the porcelain ore kaolin (437,579,679) up to 1350°C. With a quadrupole mass spectrometer, H_2, H_2O, CO, CO_2, and "crack-products" were identified, and in the presence of NaCl, HCl was released above 400°C. The kinetics of kaolinite dehydroxylation were determined by gas chromatography of water in self-generated atmospheres (10).

Impurities of pyrite, sulfates, carbonates, and organic substances have been analyzed in bauxites by thermo-gas-titrimetry of CO_2, SO_3, and NH_3, and also SO_2 after oxidation with peroxide (562). For clays, specifically Opalinustron from Switzerland, H_2O and CO_2 from organic material was formed at 335°C, SO_2 from pyrite at 412°C and 469°C, and CO_2 from Siderite at 529°C, as identified by mass spectrometry (501). Earlier work had suggested the formation of FeS and iron sulfates from pyrite in oxidizing atmospheres. This was documented by weight gain and later evolution of SO_3. The absence of either at slow heating rates in this investigation and the observation of a single SO_2 peak points to diffusion processes as the cause for multistage decomposition. Using nondispersive IR analyzers for H_2O and CO_2 and an electrochemical cell for SO_2, quantitative compositions and sequences of thermal reactions were determined by DTA-EGA for clays and shales containing organic matter and sulfides and also for hydroxy carbonates such as malachite and stichtite (494). The carrier gas was $2:1\ N_2:O_2$. By this method, the authors observed the dissociation of magnesite at 576°C, that of dolomite in two stages at 762°C and 884°C, $-OH$ loss from chlorite between 599°C and 828°C and that from talc at 934°. Combustion of graphite-calcite schist took place just below 500°. At 800–900°C, CO_2 was released by calcite dissociation. Pyrite in shale decomposed at 350°C and 600°C by reactions postulated as follows:

$$FeS_2 \longrightarrow Fe_2O_3 + SO_2$$

$$FeS_2 \longrightarrow FeS + SO_2, \text{ and after oxidation,}$$

$$700°C \quad FeSO_4 \longrightarrow Fe_2O_3 + SO_3 \longrightarrow SO_2 + O_2$$

$$919°C \quad Fe_2(SO_4)_3 \longrightarrow Fe_2O_3 + SO_3 \longrightarrow SO_2 + O_2$$

Siderite was studied by both mass spectrometry and gas chromatography (372) in air, helium, carbon dioxide, and under vacuum. The reactions were found to be overlapping:

$$FeCO_3 \longrightarrow FeO + CO_2 \longrightarrow Fe_3O_4 + CO$$

$$\longrightarrow \gamma\ Fe_2O_3 \longrightarrow \alpha\ Fe_2O_3.$$

By monitoring the H_2O peaks for dehydration of monetite ($CaHPO_4$) to the pyrophosphate, an activation energy of 49 ± 1 kcal/mole was calculated (741).

Sodium aluminum hydrosilicates and also carbonates and sulfates in synthetic carbonate-sodalite and "red mud" were analyzed by thermo-gas-titrimetry of CO_2 and SO_3 (339). Iceland spar, limestone, and cement powder have also been investigated by IR at fast-heating rates (711), and calcium hydrosilicates gas volumetrically after fixation of all gases with calcium hydride (98). Repetitive EGA of ocean sediments has provided essential information in a fast and inexpensive manner (235).

2. Organometallics

This class of compounds is one of the hardest to study, not only because of interactions between the particular metals and the organic constituents, but also because of the interference of metals in several analytical processes. This is especially true in mass spectrometry, where plating out of metals in the ion source is the operator's nightmare. Similar effects may be observed for chromatographic techniques.

Carbonyls, probably the simplest form of organometallics, however, have been studied successfully. Those of Cr, Mo, and W were heated up to 400°C, close to the molecular leak of a mass spectrometer, and Mn and Re pentacarbonyl halides could be introduced by direct probe (676). Organometallic compounds of Fe, Ni, Cr, Mo, W, and Co with CO, C_6H_6, C_5H_5, NO, and C_2H_5 ligands have been isothermally decomposed inside the ion source at 250°C (574). Dimethyltitanocene was pyrolyzed in an interfaced thermal analysis system, allowing for combustion, separation of products, and computerized IR and GC analysis (693). The difference between trivinyltin hydroxide—as a true hydroxide—and triphenyltin hydroxide—as an oxide hydrate—was established by the pressure-dependent dehydration observed by DTA-MS on fast heating (380,383).

For slow heating rates, on the other hand, a competing reaction took place for $(CH_2{=}CH)_3SnOH$:

$$(CH_2{=}CH)_3SnOH \longrightarrow (CH_2{=}CH)_2SnO + CH_2{=}CH_2,$$

confirmed by the appearance of $CH_2{=}CH_2$ ($m/z = 28$) in the mass spectrometer (380).

3. Organic Compounds

With essentially unlimited combinations of elements in organic chemistry, EGA is applicable in many different ways to a large variety of problems, covering compounds with vastly different properties. In contrast to inorganic materials, the majority of organic compounds are inherently volatile or can readily be converted into vaporizable products. This makes them available for identification by most gas analytical techniques, either directly, by vaporization of fragments after pyrolysis (284,614,767), or as products of combustion (453). The latter two processes, however, are primarily used to obtain additional data as, for instance, thermal stability, or nature, number, and amounts of reaction products. Such information is especially important for polymeric materials in terms of mechanical strength, toxicity, and flammability. A fourth area, which in the opinion of this author is still very much underutilized, is the study of reactions themselves. The exception is kinetics of *known* reaction; but evolved gas analysis and thermal chemistry is still badly neglected as a tool for the synthetic organic chemist.

The organization of this section is primarily by functional groups, as they behave similarly in analytical processes. Consequently, polymers are grouped by chemical criteria and not by properties. A few noteworthy specialties are collected under general headings.

TABLE 6.VI
Postulated Decomposition Sequences of Alkaline Earth
Salts of Formic Acid[a]

Calcium Formate

$$CaC_2H_2O_4 \xrightarrow[220°C]{in\ N_2} CaC_2O_4 + H_2 \qquad\qquad\qquad I$$

$$CaC_2O_4 \xrightarrow[465°C]{in\ N_2} CaCO_3 + CO \qquad\qquad\qquad II$$

$$CaCO_3 \xrightarrow[600-800°C]{} CaO + CO_2 \qquad\qquad\qquad III$$

Strontium Formate Dihydrate

$$SrC_2H_2O_4 \cdot 2\ H_2O \xrightarrow[100°C]{} \alpha\text{-}SrC_2H_2O_4 \qquad\qquad\qquad I$$

$$\alpha\text{-}SrC_2H_2O_4 \xrightarrow[225°C]{} \beta\text{-}SrC_2H_2O_4 \qquad\qquad\qquad II$$

$$\beta\text{-}SrC_2H_2O_4 \xrightarrow[450°C]{} CO + H_2 + SrCO_3 \qquad\qquad\qquad III$$

$$CO + H_2 \xrightarrow[450°C]{} CH_2O \qquad\qquad\qquad III\text{-}A$$

$$2CO \underset{450°C}{\rightleftharpoons} CO_2 + C \qquad\qquad\qquad III\text{-}B$$

$$SrCO_3 \xrightarrow[600-1000°C]{} SrO + CO_2 \qquad\qquad\qquad IV$$

Barium Formate

$$BaC_2H_2O_4 \xrightarrow{340°C} \alpha\text{-}BaC_2H_2O_4 \qquad\qquad\qquad I$$

$$\alpha\text{-}BaC_2H_2O_4 \xrightarrow{380°C} \beta\text{-}BaC_2H_2O_4 \qquad\qquad\qquad II$$

$$\beta\text{-}BaC_2H_2O_4 \xrightarrow{420°C} CO + H_2 + BaCO_3 \qquad\qquad\qquad III$$

---

[a]Adapted from Ref. 155.

a. ORGANIC SALTS AND ACIDS

Because of its special standing in thermal analysis, calcium oxalate had been singled out under Inorganics, where carbonates also were discussed.

Alkaline earth salts of formic acid, the first homolog of aliphatic carboxylic acids, probably decompose via intermediate formation of carbonates and/or oxalates. The postulated sequences (155) in Table 6.VI are based on TA-MS data.

Strontium formate dihydrate was also studied isothermally at 370°K under nitrogen and at reduced pressure (529,530). The lithium, sodium, potassium, rubidium, and cesium salts of formic, acetic, propionic, and n-butyric acid were analyzed by a MOM Derivatograph in air and in inert gas (476). The products were determined by GC, IR, and chemical methods. Only lithium salts and formates were found to change their decomposition temperatures with composition, while all other peak temperatures remained essentially the same. The reactions were

$$2\, RCOOM \xrightarrow{\Delta} RCOR + M_2CO_3$$

$$M_2CO_3 \xrightarrow{\Delta} M_2O + CO_2$$

and simultaneously

$$2\, RCOOM \xrightarrow{\Delta} M_2CO_3 + \text{alkanes, alkenes, } H_2, CO_2, CO, CH_4, \text{etc.}$$

On the other hand, for formates:

$$2\, HCOOM \xrightarrow{\Delta} H_2 + (COOM)_2$$

$$(COOM)_2 \xrightarrow{\Delta} M_2CO_3 + CO$$

By quadrupole mass spectrometry, potassium acetate was shown to produce CH_3COCH_3, H_2, $HC{\equiv}CH$, and CO_2 (578). Acetic acid was detected, probably after hydrolysis with residual water, and ketene either as a fragment ($m/z = 42$) or an intermediate. The thermal reaction ends at about 900°C with decomposition of K_2CO_3. For other acetates, the products were different for different cations (327). Ketonization of acetic acid from $3\,d$ metal acetates was observed on the surfaces of residual V_2O_3, Cr_2O_3, MnO, Fe_3O_4(Y-Fe_2O_3), Co, and Ni (61). In contrast, chromium acetate yields CH_4, CO, HCHO, and H_2O in addition to CH_3COCH_3 and CO_2. This finding makes the previously determined activation energy (677) probably erroneous. At elevated pressure in a stainless steel bomb, C_6H_5OK and p-$HOC_6H_4CO_2K$ were studied at a heating rate of 2 ± 0.5°C/min, by detecting pressure changes of as little as 0.1 kg/cm^2 (541).

Preliminary analysis by chromatography, etc. [sic], revealed very complex reaction products from calcium stearate at 365°C (93), which had previously been thought to be simple (420). By simultaneous DTA-MS and TG-MS, three oxidative exothermic processes were observed by a decrease in the level of oxygen and an increase in the pressure of CO_2 accompanied by small fragments of $m/z < 90$ (37).

Thermal dissociation is increased for ammonium salts of fatty acids with increasing chain length, while amide formation is decreased (610). Gaseous decomposition products from thallium(I) salts of fatty acids were chemically absorbed after DTA and TG, converted into hydrazones, and subsequently identified by thin-layer chromatography (642). The acetate released acetone, the propionate formed symmetric and the n-butyrate formed asymmetric ketones.

Diffusion properties of organic solids can be characterized by emanation thermal analysis, as described for barium salts of monocarboxylic acids (benzoic, salicyclic, aminobenzoic, 2-chlorobenzoic, and 2-iodobenzoic) and of dicarboxylic acids (phthalic, isophthalic, and terephthalic) between 298 K and 373 K (25). Dehydration and polymorphic transformations of barium salts of o-, m-, and p-nitrobenzoic acids were also determined by ETA (105).

Potassium ethylsulfate and potassium methylsulfate undergo several endothermic transitions during DTA (727,731). In each case, one of the transitions also gives a response for EGA. The product was identified by mass spectrometry as dimethyl ether for the methyl compound.

By chemical analysis of gaseous products from alkali metal picrates, only cyanide and carbonate were detected in larger amounts (667). Separate mass spectrometry showed CO_2, HCN, CN^-, and H_2O.

Copper dimethyl- and diethyl dithiocarbamates revealed different thermal decomposition products by GC-MS. They were CS_2 and $(CH_3)_2NC(S)$-$N(CH_3)_2$ for the methyl, but CS_2 and $(C_2H_4)_2N$ for the ethyl complex (627).

Evolved gas analysis of several compounds of hydrazido carbonic acid (672) in air, carbon dioxide, and argon was reported to be in progress (436).

b. AMINES, AMINE SALTS, AND AMIDES

Apart from dissociations of their salts (187,610), most amines are quite stable and distill generally with little decomposition. In the presence of strong oxidizing agents, which may be the anions of their own salts, rather complicated reactions have been observed. For example, the thermal decomposition of hydroxylamine and methoxyamine was investigated *in vacuo* by solid probe, double-focusing mass spectrometry at 0.5°C/min (253). The hydroxylamine salt formed water, oxides of nitrogen, oxygen, and hydrogen, while methoxyamine perchlorate produced methyl chloride, methyl alcohol, hydrogen chloride, hydrogen cyanide, methyl perchlorate, water, and oxides of carbon and nitrogen. Activation energies of the proposed steps are as follows:

TABLE 6.VII
Activation Energies
Hydroxylamine Perchlorate[a]

	Ea, kcal/mol
$NH_3OHClO_4 \longrightarrow NH_2OH + HOClO_3$	20.7
$HO\text{-}ClO_3 \longrightarrow HO + ClO_3$	46.4
$NH_2OH + OH \longrightarrow NH_2O + H_2O$	28.6
$NH_2O + ClO_3 (HOClO_3) \longrightarrow$ Oxidation Products	28–32

[a]Adapted from Ref. 253.

TABLE 6.VIII
Activation Energies
Methoxyamine Perchlorate[a]

$CH_3ONH_3ClO_4 \longrightarrow CH_3ONH_2 + HOClO_3$	Ea, 30 kcal/mol
$HO\text{-}ClO_3 \longrightarrow HO + ClO_3$	Ea, 44.2 kcal/mol
$CH_3ONH_2 \longrightarrow CH_3O^+ + NH_2$	A.P. = 19 0.5 ev
$CH_3ONH_2 \longrightarrow CH_3 + {}^+ONH_2$	A.P. = 17 0.5 ev
CH_3ONH_2 dissociation products + $ClO_3(HOClO_3)$	
$\longrightarrow$ reaction and oxidation products	Ea, 28–32 kcal/mol

[a]Adapted from Ref. 253.

The first step is dissociation of the salt via proton transfer, which is followed by decomposition of perchloric acid to the reactive ClO_3 radical. In reverse fashion, methylamine was produced from azomethane between 632.8 K and 754.8 K in a heated flow reactor with leak into a time-of-flight mass spectrometer (716). Aromatic diamines form polymers with bis-diazonium salts by azocoupling (12). Identification (MS) of nitrogen, little ammonia, m-phenylene diamine, aniline, and benzene at 330°C pointed to bond rupture at the azo groups.

Condensation of urea to biuret and cyanuric acid

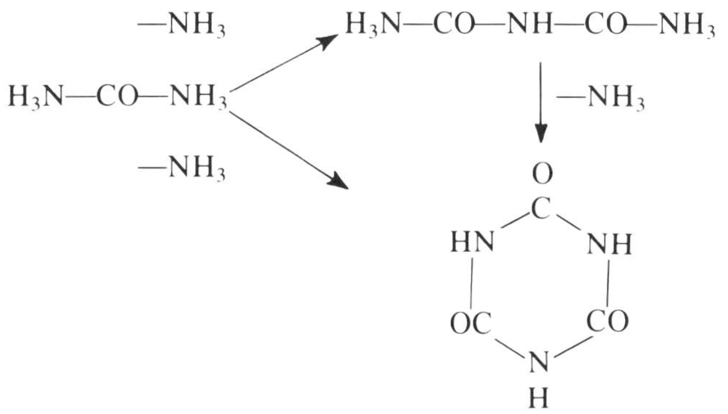

has been followed at 10^{-5} torr by recording the ions $HNCO^+$ ($m/z = 43$), CO^+ ($m/z = 28$), and NH_3^+ ($m/z = 17$). The condensation reactions were found to overlap with themselves and with depolymerization and decomposition (381).

c. AMINO ACIDS

Amino acids, peptides, and proteins have only recently been subject of thermoanalytical investigations. Most are very sensitive to temperature elevations, which makes it both important and difficult to understand their decomposition behavior. The major reactions are deamination and decarboxylation and also dehydration with polymerization and intramolecular or bimolecular ring closure. The latter has been observed for DL-β-phenyl-α-alanine and DL-β-aminophenyl-α-alanine to give dioxopiperazines, as identified by mass spectrometry, DTA, and IR (266).

However, gaseous products often only appear as the last in line of the thermal breakdown and therefore are of little value for any direct characterization. Modern techniques promise to have an impact on this field, so fundamental to all the life sciences. One of the methods is ^{14}C labeling. For several amino acids, release of $^{14}CO_2$ was monitored by a flow proportional counter from 25°C to 950°C (521). Several maxima observed for the evolved activity are helpful to explain decomposition pathways. Use of thermal effects, including rapid heating inside mass spectrometers, especially in tandem and chemical ionization modes, also hold promise. Examples are studies of cyclic depsipeptides from 200–240°C (585), which thermally form morpholines, and also of underivatized arginine-containing peptides (64). In this study, fracture of the peptide links produced small, stable, neutral molecules, indicative of the structure. It was possible to deduce the unique amino acid sequence for bradykinin: Arg-Pro-Pro-Gly-Phe-Ser-Pro-Phe-Arg. Other biological samples will be discussed later.

d. HALOGEN COMPOUNDS

The majority of the work with halogen-containing compounds deals with fire-retardant properties. These reagents will be found under the heading *polymers*.

Pyrolysis of brominated methane derivatives used in halogen lamps was investigated in heterogeneous and homogeneous reactions (335). The first utilized a reactor within a mass spectrometer ion source at 10^{-2} torr. This makes collisions in the gas phase negligible, and the observed decomposition is due to collision with the tungsten filament. The latter was carried out in a Chevenard quartz oven, and the products, except for Br_2 and H_2, were analyzed by IR. The activation energies for both methyl bromide and methylene bromide decomposition were lower by 10–20 kcal/mol on the tungsten wire, due to a catalytic effect.

In the area of kinetic studies, the volatilization—simple sublimation—of *p*-dichlorobenzene is an example (305), and for high sensitivity the trace analysis of 2,3,7,8-tetrachloro *p*-dibenzodioxin by high-resolution GC and selected ion monitoring MS should be mentioned (95,135).

e. MISCELLANEOUS ORGANICS

To complete the treatment of organic compounds, the following examples were chosen either to point out successful applications of specific methods, or observations of unusual chemical phenomena.

A continuous thermo-gas-analytical method was used for direct determination of water, evolved during thermal decomposition of sugars. It involved an attachment for quantitative absorption, followed by Karl-Fischer titration

(450). Trapped volatiles from mono-o-methyl-d-glucose derivatives were analyzed by gas chromatography (361). Release of water and aldehydes was a function of the position of substituents. From acetaldehyde, water and C_2H_2, but no CH_4, were observed by time-of-flight mass spectrometry (TOF-MS) (124). The intermediate CH_3CHOH was postulated for a first-order reaction with an activation energy of 57 kcal/mol. Also by TOF-MS, C_3H_3 and C_6H_6 radicals were detected from quartz tube pyrolysis of biphenyl above 450°C (123). Already at 200°C, ethyl nitrite produced CH_3 radicals, EtOH and AcH, followed by release of CH_4 and CO at 350°C, presumably by dehydration of EtOH and AcH. Thermal decomposition of another ester system, isobutyl-benzyl acetate, was studied by time-resolved chemical ionization mass spectrometry (478). Benzyl ions polymerized to $[C_7H_7 \cdot C_7H_6]^+$, $[C_7H_7 \cdot 2C_7H_6]^+$, and $[C_7H_7 \cdot 3C_7H_6]^+$, which formed protonated anthracene ions by loss of H_2. Two unimolecular reactions produced isobutene and isopropanol or propene and t-butanol from t-butylisopropyl ether between 409°C and 475°C, as seen by GC-MS (138). Hydrocarbons were analyzed as components of a product stream from fluid catalytic cracking (191), and by TG-DTA-HRMS (high-resolution mass spectrometry), as an example to distinguish benzene from pyridine ($m/z = 79$; $\Delta m/z = 0.0081$) (314). Fragments from poly(diene sulfones) were observed by field-ion mass spectrometry (629), and at 80–85°C the thermal decomposition of the dimsyl ion—as sodium dimsyl in DMSO—produced methylated butadienes, observed by GC after trapping at -78°C (583). Solid white precipitates consisted of sodium methane sulfonate, sodium methane sulfinate, and sodium methyl mercaptide.

Facile, or highly reactive, systems investigated by special techniques include flash photolysis of ketene (352) (which formed ethylene and CO, probably from an intermediate C_3H_4O), thermobarogravimetry of the explosives pentaerythritol tetranitrate (PETN), 1,3,5-trinitro-1,3,5-triazacyclohexane (RDX), and trinitrotoluene (TNT) (459), and of a number of propellants and formulations with nitrocellulose, nitroglycerine, isopropyltetrazole, and others (17,668). Subtractive gas chromatography has been demonstrated with alcohols, aldehydes, and ketones (390,464). Phosphines and phosphine complexes were studied by high-vacuum pyrolysis in the mass spectrometer (262,382).

4. Polymers

a. GENERAL

Thermal stability is probably the most critical property of any given polymeric material. On one hand, it allows for taking advantage of convenient fabrication processes, especially molding, extrusion, and foaming. On the other hand, if lacking, it affects adversely all other, mostly mechanical, assets of articles manufactured from polymeric chemicals. Temperature-dependent changes may be abrupt, thus representing a limit for heat exposure, or gradual, but nevertheless thermal, in which case durability is the performance characteristic of concern. Almost all thermal degradation reactions of polymers occur at temperatures, at which the resultant products are in the gaseous state. Hence, evolved gas analysis has become an indispensible tool in polymer analysis (85,107,112,134,137,147,200,210,251,254,300, 340,345,396,492,506,515,527,597,653,703,762,765,767,771).

The aspects to be considered determine the choice of the method. Rapid reactions are easily recognized by dynamic measurements, DTA, DSC, or TGA. Products are abundant and therefore do not generally require very sensitive gas analytical methods. For slow processes, thermodynamics and kinetics become important, and identification of small amounts of reaction products at the onset of the reaction is valuable. Here the mass spectrometer is often the instrument of choice, and isothermal modes are employed. Two specific properties are of concern for degradation products of polymers, or better of manufactured plastics. They are flammability and toxicology. In these cases, the need for identification and quantitation of components at very low concentrations demands high sensitivity, accuracy, and reproducibility of the analytical technique.

Under the flammability heading, two processes may be separated: pyrolysis and combustion, with the latter obviously being of more direct consequence. For toxicology, the opposite is true. While it would be desirable to investigate each system by both processes, (171,213,280), instrumental restrictions seldom make it possible, and extrapolations to different conditions become necessary (100).

Investigations have been carried out with representative samples of the major polymer systems and by a variety of methods (100,171,391,488, 526,547), for identification of low-molecular-weight products, including hydrogen, hydrocarbons, CO, CO_2, H_2O, HCl, SO_2, SO, CS_2, and COS, several specifically to detect potentially toxic gases (328,391,547). An interesting problem, bridging several disciplines of analytical chemistry and even EGA, is the generation of smoke. It is a recognized health hazard. Directly, it reduces visibility and irritates the respiratory system; indirectly, however, it is far more hideous and deadly. By itself, it can cause chronic, edemic effects but in addition it may act as an adsorbent to toxic pyrolysis products. Direct photometric measurements (419) and previously discussed desorption EGA can contribute towards a solution of this problem.

Special attention has been focused on the mechanism of brominated fire retardants. The main effect is generally understood to be an interruption of the flame propagation reaction by competition of the bromine radical for the OH radical, e. g.,

$$\cdot Br + CH_4 \longrightarrow HBr + \cdot CH_3 \text{ and}$$

$$\cdot Br + \cdot CH_3 \longrightarrow CH_3Br \text{ for } \cdot OH + CH_4 \longrightarrow H_2O + CH_3 \cdot$$

(288,290,381).

Other effects are thought to include dilution by "nonburning" halogen-containing molecules (389) and energy absorption for C–Br bond scission (571). Knowledge of the composition of the gas phase with respect to combustible molecules, bromine-containing materials, and oxygen has provided support for these assumptions (381). Other reports mention the effect of flame-retardant (FR) additives on pyrolysis of polymers, as observed by EGA (693,698,749). Of special interest have been the so-called **synergists** to brominated FR agents, primarily antimony oxide. As an example, in poly-vinyl chloride, Sb_2O_3 was shown to form $SbCl_3$ by reaction with HCl (429), while MoO_3 catalyzed the dehydrochlorination at lower temperatures and at an increased rate (431).

Additives, in the form of plasticizers, lubricants, antioxidants, or residual monomers are amenable to EGA by sensitive methods (330,764,765), and a number of polymeric materials have been characterized for forensic applications (527,618).

As an extension of or as an indirect thermoanalytical method, the reader should be aware of photolysis experiments on polymers (330).

b. FUNCTIONAL

Below, emphasis is placed on typical reactions of polymers, containing functional chemical units, and studies involving specifically this functionality.

(1) Polyolefins

The most important polyolefins are polyethylene and polypropylene, both homopolymers. Their decomposition is straightforward depolymerization. Thermal analysis, therefore, is concentrated on structural problems or reaction mechanisms. Branch length and the number of side chains in polyethylene were determined by flash pyrolysis combined with in-line hydrogenation and gas chromatography (633,634). Sequence, length distribution, and homogeneity of polyolefins were investigated by pyrolysis gas chromatography (635), and pyrolysis–molecular-weight chromatography was applied to thermal decompositions of polyethylene, polypropylene, and polyiso-butylene (348). Gas chromatography, with or without mass spectrometry, was used to measure CO and CO_2 at various temperatures (496). From that, activation energies of 13.5 and 11.4 kcal/mol were obtained for unirradiated and irradiated polyethylene. Thermal dekryptonation analysis of polyethylene and polypropylene with [85]Kr revealed several steps with retention of krypton, even after large portions of the polymers had already melted (277,278,279). Early work with Bunsen burner combustion of ethylene-propylene copolymers, starting at 520°C and dropping to 470°C, allowed an estimate of the propylene content by mass spectrometry, using an empirical ratio of mass intensities (91). During laser pyrolysis (694), the presence of $NaBH_4$ increased the yield of monomer from polyethylene and polypropylene and produced methane by side chain cleavage. Pyrolysis GC-MS was used for structure studies of α-irradiated 1,2-polybutadiene (713). Both polybutadiene and poly(diene sulfones) formed molecular-weight-related fragments by field-ion mass spectrometry with little secondary fragmentation (629). This not only suggests the decomposition mechanism, but also yields structural information on copolymers. Electron impact fragmentation could also be suppressed by lowering the ionization energy during direct pyrolysis of poly(oxy-1,4-phenylene), poly(thio-1,4-phenylene), and poly(dithio-1,4-phenylene) in the mass spectrometer ion source (493).

Determination of volatiles in rubber samples was accomplished by outgassing in the ion source of a high-resolution mass spectrometer at various temperatures (654). The sensitivity was monitored by a constant influx of argon. A physico-chemical analysis of gas evolved from rubber has been reported (163), and multicomponent gases that evolved during continuous vulcanization of rubber articles without pressure have been analyzed (316).

(2) Polystyrene

Early pyrolysis studies were carried out with thin polystyrene film in sealed tubes, connected by an inlet leak to a magnetic mass spectrometer (768). Stepwise temperature elevation first released solvent, monomer, and impurities; decomposition followed above about 230°C. Main products over the 10-day experiment were styrene, benzene, and toluene, as a function of temperature. Straight-line plots of ion intensities versus $1/T$ gave activation energies of 57 kcal/mol for benzene and toluene and 65 kcal/mol for styrene.

In dynamic systems, however, polystyrene, like the polyolefins, depolymerizes to monomers by "unzipping." For the homopolymer, decomposition takes place at lower temperatures than for polyolefins (381). It should not be surprising, therefore, that many reports on new instrumentation, techniques or methods rely on polystyrene for demonstration purposes. Examples include construction of a "quasimolecular beam former" with modified DTA cell and MS ion source (239), a "mass spectrometric kinetic flow system" (358), a "dynamic molecular still" (474) for determination of activation energies, a fast-heating system to obtain "essentially isothermal pyrolysis" up to 1125°C (392), "pyrolysis-molecular weight chromatography" (347,349), and a rapid successive DTA-GC method (756).

Computer simulation allowed an evaluation of kinetic parameters for polystyrene decomposition (604). The extent of branching was determined by pyrolysis-GC (5). Release of benzene by side-chain cleavage was increased in the presence of $NaBH_4$, and ammonium perchlorate reacted with polystyrene in two stages, one with evolution of styrene oligomers, CO_2, and HCl, the other generating CO_2, CO, and HCl (487). "Field ion mass pyrograms" (630) consist of characteristic ions for polystyrene. In styrene-vinyl chloride copolymers, they are the same as in the homopolymer (617). Additional fragments are indicative of block character of the copolymer, and slow pyrolysis between 350°C and 400°C was attributed to pyrolysis starting within vinyl chloride sequences. Quantitative determination of the styrene content of styrene-butadiene and acrylonitrile-butadiene-styrene rubbers was achieved by GC with a special pyrolysis accessory operating at 600°C and 800°C (119), while the distribution of polystyrene chains in butadiene-styrene block copolymers was deduced from pyrolysis-GC analysis of poly-styrene-polybutadiene mixtures (692).

Cross-linked polystyrene, mainly by divinylbenzene, plays a special role as ion exchange resins. Filament or Curie-point pyrolyzers in combination with gas chromatography have been used to characterize them by the amount of xylene (549) or ethyl benzene and styrene (550) released. Both pyrolysis-GC and pyrolysis-MS were applied to estimate the degree of cross-linking and to determine type of functional groups and networks and also the positions of substitutions (68,69).

(3) Vinyl Polymers

This group contains polymers of vinyl chloride, acetate, and alcohol, their blends and copolymers. A good summary of their characterization by evolved gas analysis is found in a recent publication by Chiu et al. (114).

Dehydrochlorination is the most important aspect of polyvinyl chloride decomposition. Mass spectrometry is convenient to monitor its products (103,546). Kinetics have been determined by HCl titration (715), and it was found that retardation is increasing with sample weight. In nitrogen at 190°C, conductometric measurements indicated that the reaction was initiated at allylic chlorines (1). With d-labeled polyvinyl chloride (PVC), it was shown that benzene formation is due to intramolecular cyclization, rather than to intermolecular Diels-Alder condensation (533). By HCl titration at constant temperature, the effect of oxygen concentration during polymerization on the stability of the product was studied (238). A series of papers describes laser-probe analysis coupled with modulated molecular-beam mass spec-trometry (428,430). Benzene, toluene, HCl, and heavier hydrocarbons were released concurrently but no monomer was observed. Plasticizers and fire-retardant additives, especially Sb_2O_3, moderated the HCl release. Polymers, formed by Friedel-Crafts addition of polyvinyl chloride and aromatic com-pounds to polystyrene, were analyzed by pyrolysis-GC (101). Morphology and crystal structure of PVC were studied by thermal dekryptonation (277,278,279). Leather substitutes, containing polyvinyl alcohol, polyester, and polyvinyl chloride were investigated by multiple technique DTA-GC, using a commercial evolved gas analyzer (754).

Water and acetaldehyde from polyvinyl alcohol were detected by DTA-MS, and acetic acid was the main product from polyvinyl acetate (44). The kinetics of the last reaction depend on internal structure and polydispersity. Ethylene-vinylacetate copolymers produced acetic acid during the first stage and ethylene, propylene, butene-1, etc., in the second (470).

(4) Polyesters

Kinetics and mechanisms of thermal degradation of aliphatic polyesters were investigated by time-of-flight mass spectrometry (205). Many products were evolved concurrently in a single-stage reaction, which made it difficult to differentiate between all without the help of gas chromatography. Laser microprobe-dynamic mass spectrometry and temperature-programmed pyrolysis-quadrupole mass spectrometry was applied to polyester decomposition (430,432,433). Polybutylene terephthalate decomposed by a complex multistage process. First, tetrahydrofuran was formed by an anionic mechanism. Then, concerted ester pyrolysis, involving an intermediate cyclic transition state, produced 1,3-butadiene with simultaneous decarboxylation. Finally, CO evolved together with organics such as toluene, benzoic acid, and terephthalic acid. Activation energies were 27.9 kcal for THF and 49.7 kcal/mol for butadiene. Polycarbodiimide, as an additive, increased the thermal stability of the polyester by approximately 20°C through an increase in resistance to acid-catalyzed hydrolysis.

(5) Polyacrylates

This group can be further divided into acrylic acid polymers, with their derivatives methacrylates and methyl methacrylates, and acrylonitrile polymers. Characteristic decomposition products for the first category are CH_4, H_2, CO, CO_2, and H_2O, as identified by Derivatograph-GC analysis of sodium and potassium methacrylates and polymethacrylates (370). A change in prepyrolytic decomposition mechanism of poly(methacrylic acid) at about 170°C was detected by isothermal monitoring of the $CO_2 : H_2O$ ratios by mass spectrometry (412). By isothermal pyrolysis with fast-heating rates of polymethyl methacrylate, the monomer was identified below 450°C by GC (392). At higher temperatures, gaseous and liquid products were formed. With an all-metal coupler for TG-GC, methyl methacrylate from its polymers could be measured with 2% peak height and 4% peak area reproducibility (113), allowing the identification of the polymer in an aerosol spray. Kinetic parameters were determined by mass spectrometry (620) and computer simulation (604). Three "thermal evolution" peaks were found by differential trapping, followed by mass spectrometric analysis (115). The first was due to residual monomer and moisture, the second resulted from degradation of unstable fractions at the chain ends, while the third was caused by random chain scission of stable fractions, in each case producing monomer. The method was also used to identify methyl methacrylate and methacrylate as copolymers in commercial acrylic resins. Stepwise thermal degradation in nitrogen between room temperature and 1000°C was applied to polymethyl methacrylate and methyl methacrylate-styrene copolymers, using a combination of an infrared image furnace, a thermal balance, and gas chromatograph (691). By recording the monomer concentration, ionically and radically polymerized poly(methyl methacrylate) could be differentiated. The thermal stability of the copolymers was higher for longer styrene chain lengths but lower for longer methyl methacrylate chains. Similarly, appearance of dimer and trimer peaks from methyl acrylate-styrene copolymers was related to sequence distribution (690). The effect of morphology in powder, and thin and thick films of poly(methacrylates) was investigated by "thermal volatilization analysis" (470). The microstructure of methyl methacrylate-vinyl chloride and acrylonitrile-vinyl chloride (685) and also acrylonitrile-methyl methacrylate (485) copolymers was studied by pyrolysis-GC. Pyrolysis of acrylonitrile block copolymers was studied by GC (643), that of ethylene-acrylic acid copolymers by pressure gauge-mass spectrometry combination (462), acrylics with detector tubes and mass spectrometry (761), and polyacrylonitrile resins by DTA-GC-MS (504,505). The products were H_2O, CO_2, and hydrocarbons above 200°C (462), and also HCN and CH_2CHCN (359). In oxygen-containing atmospheres, the amounts of gaseous products from polyacrylonitrile were higher than in inert gases, and the evolution started at lower temperatures (504,505). They consisted of HCN, methylcyanide, allylcyanide, pentane, cyclohexene, cyclohexane, toluene, methylbutene, acetonitrile, acrylonitrile, propane nitrile, and 2-methylpropane nitrile.

(6) Phenolics

Condensation of phenol with formaldehyde leads to the well known hard, durable, and thermally rather stable thermoset resins. Pyrolysis involves uncomplicated chemistry, therefore investigations centered around thermal and mechanistic effects and ablating properties (203). Examples of the techniques used are isothermal pyrolysis at 450°C, 550°C, and 650°C in argon and air, followed by GC (286), or trapping of reaction products from furnace pyrolysis in vacuum or inert atmosphere up to 1200°C, subsequent trap-to-trap distillation and mass spectrometric analysis (444). From rapidly scanned time-of-flight mass spectra, peak heights of several ions evolved from a Knudsen cell were plotted as a function of temperature (647). At linear heating of 12.5°/min, 28.5°/min, and 29°/min, the mass range m/z 1–200 was scanned each minute to obtain kinetic and mechanistic information (649). Flash pyrolysis with xenon flash lamps and time-of-flight mass spectrometry gave reproducible spectra within 10 sec after the flash (202). Electrical heating in a similar arrangement was successful at a heating rate of 930°C/sec and a scan rate of 0.2 seconds from m/z 1–200 (204). Pyrolysis-GC of cross-linked phenol-formaldehyde polymers indicated cleavage of bridges between rings, to give phenol, o- and m-cresol, and 2,4,6-trimethylphenol (584).

(7) Polyamides and Polyimides

This group includes the more flame-resistant synthetic fibers. On the other hand, pyrolysis may produce toxic gases, such as HCN, which was detected together with NH_3 by DTA-GC-MS of Nylon 6 in inert gas and oxygen-containing atmospheres (504,505). Pyrolysis-MS (201), TG-MS (270,353), and TG-GC (113) were applied to analyze evolved gases from polyamides and polybenzimidazoles. Thermal degradation products from truxillic and truxinic polyamides were determined by mass spectrometry (99). Ion-source pyrolysis detected thermal and electron impact-induced fragments from different types of cyclobutane ring cleavage. Oxidative thermal degradation of poly(dodecaneamide) was studied with GC by simultaneous measurement of oxygen absorption and volatile products release (516). Antioxidants increase the thermal stability of phenylone[poly(m-phenyleneisophthalamide)], which starts at about 360°C with an activation energy of 59 kcal/mol (696). The products were identified by MS as H_2O, CO, CO_2, NH_3, C_6H_6, HCN, and C_6H_5CN. Gas evolution during thermal degradation of polyterephthalamides could not be used to evaluate the reaction because of cross-linking via ether bridging by preliminary oxidation of methylene groups (178). Pyrrone and polyimide prepolymers were heated in helium at 2°/min from 25–400°C (678). Conversions were less than 80% and 100%, respectively, with significant amounts of CO_2 losses detected by GC-MS during pyrrone polymer formation.

Kinetics of gas evolution were determined from N-phenylphthalimide, 4,4-diaminodiphenylether, N-phenylbenzamide, benzoic, and pyromellitic acid, as models for polyimides (625). Pyrolysis was carried out in sealed ampuls and gas analysis by MS and GC.

By high-vacuum DTA and high-resolution double-focusing mass spectrometry, CO, CO_2, H_2O, HCN, H_2, some benzonitrile, benzene, methane, and ammonia were detected from polyimide resin (239). On the basis of the gas evolution, a mechanism was proposed that explains the presence of CO_2 by an intramolecular oxygen rearrangement in the imide ring under high vacuum. By a combination of Derivatograph-GC, polypyromellitimides were found to release CO, CO_2, H_2O, H_2, and traces of HCN (772). Methane and ethane evolved from methoxy side groups at the beginning of degradation, later from methyl-substituted ends, and at all times from methylene bridges.

Other reports include a MS study of pyrolysis products from polyimide and poly-2,2'-(m-phenylene)-5,5'-bibenzimidazol (250), another of the latter compound by EGA-MS (651) and its formation from diphenyl isophthalate and 2,2',3,3'-tetraaminobisphenyl, by recording the evolution of phenol and water as a function of temperature (256). Polybenzimidazoles were also studied by Knudsen cell with time-of-flight mass spectrometry (647) and by MS at three to four temperatures between 200–650°C with release of organics, CO_2, CO, HCN, H_2O, NH_3, and H_2 (170). During toxicological studies by TG-TOF-MS of materials used in aircraft (353), polybenzimidazole fabric released H_2O, CO_2, and some NO, while H_2O, then CO_2, and possibly N_2O was observed from the aromatic polyamide fiber Nomex®.

(8) Polyurethanes

As for polyimides, one of the concerns for polyurethane decomposition is the release of HCN. It has been detected by selected ion-monitoring MS (761), and a tentative mechanism by thermal oxidation of two aromatic and one segmented polyurethanes has been proposed for its formation, which is inhibited by copper or its oxides (319). Kinetics and reaction mechanism for toluene diisocyanate and methylene *bis*-4-phenyl isocyanate-based poly-urethanes were determined by a directly-coupled thermal balance with a quadrupole mass spectrometer, avoiding secondary reactions (490). Component monomers, blowing agents, and other additives released by heating polyurethane foams were identified by TOF-MS (594), and mass spectrometry was also used to monitor polyurethane foam pyrolysis for toxicological evaluation (289).

(9) Oxide Polymers and Epoxides

Starting with epichlorohydrin, epoxides are formed by elimination of HCl. Low-molecular-weight epoxides can be "advanced" to form higher-molecular-weight epoxy resins or polymerized to polyglycols. Subsequently, epoxy resins are "cured" by reactions with amines. Evolved gas analysis is applicable to each step of this sequence.

Thermal decomposition of polyepichlorohydrin and epichlorohydrin-ethylene oxide copolymers was carried out isothermally between 175°C and 225°C or at a heating rate of 2°C/min (495). Rapid hydrogen chloride release from side chains at 200°C was measured with a glass electrode and recorded. Proper formulation with acid acceptors can keep the HCl evolution low, up to 175°C. By temperature-programmed pyrolysis and subsequent mass spectrometry, the decomposition products of epoxy resin were identified and kinetics for formation were determined (342). Chemical and structural differences were elucidated by gas chromatography of volatile products from pyrolysis of bisphenol-A diglycidyl ether and bisphenol-A epichlorohydrin resin samples for 7.5 sec at 800°C (725). Fourteen peaks were recorded. Thermal analysis with mass spectrometry detected the presence of HCl and other impurities in amine-cured "EPON 827" resin and allowed for a discussion of kinetics and reaction mechanism (621).

Relative amounts of ethylene oxide and propylene oxide in polyethylene-polypropylene glycols were determined by pyrolysis-GC calibrated with polyethylene glycol and polypropylene glycol standards (766).

(10) Fluorine Polymers

After several ions had been recorded from polytetrafluoroethylene by Knudsen cell time-of-flight mass spectrometry (647), its thermal degradation in vacuo was found to be a two-stage process at 630–760 K to form tetrafluoroethylene with activation energies of $E_1 = 42$, $E_2 = 85$ kcal/mol (456). Pyrolysis-GC of polytetrafluoroethylene and tetrafluoroethylene-hexafluoropropylene copolymers produced 23 GC peaks in helium at about 700°C (497). No differences were noted between the samples, all yielded perfluoroparaffins, olefins, and fluorinated cyclic compounds together with SiF_4 by reaction with the pyrolysis tube, and CO, and CO_2 from oxygen impurities. In air, COF_2, CF_4, and CO_2 were identified. By TG-MS at 10^{-4} torr in air and 10^{-5} torr in helium, mostly monomer and some perfluoropropane was observed from the homopolymer with activation energies of 86.4 kcal/mol for

C_2F_4 and 89.7 kcal/mol for C_3F_6. The copolymer decomposed in two stages, the first to form C_3F_6 (70.5 kcal/mol) and C_4F_8 by first order and the second to give C_3F_6 (85.6 kcal/mol) and CF_4 (84.2 kcal/mol) by 0.5 order reaction. In air, CO_2 and COF_2 dominated for both systems with activation energies below 50 kcal/mol by first order. With thermal evolution analysis by differential trapping, quantitative determination of Teflon 6® by peak-area measurement has been demonstrated (115). With a fluoride ion specific electrode, HF was found to be the major breakdown product from polyvinyl fluoride and polyvinylidene fluoride but not from poly trifluoroethylene, which formed only small amounts of HF (186). The release of HF from the main chain of hydrofluoro elastomers was recorded via a glass electrode and was strongly dependent on the composition or formulation (354,495,752). From perfluoroalkylene-linked polyimides, CO and CO_2 and also SiF_4 by secondary reaction of HF was observed (133), and carboxynitroso rubber

$$CF_3$$
$$|$$

with building blocks of $[CF_2{-}N{-}O{-}CF_2]$ formed carbonyl fluoride and perfluoro-*N*-methyl methylene imine and subsequently CO_2, HF, and an isocyanate derivative by reaction of water vapor, and finally SiF_4 from reaction of HF with glass (353).

5. Biological Matter

Possibly the first application of pyrolysis-mass spectrometry for biological materials was reported in 1952 (767). By coating the materials onto helical heater wire and rapidly heating (up to 1700°C), "fingerprint" spectra of albumin and pepsin were obtained. Also by TA-MS, different forms of water in biological tissues could be differentiated (572). Examples include mouse liver hepatoma, burn edema from Wistar rats, and water in grain. Based on pyrolysis-GC analysis of bacteria, it was suggested, that characteristic patterns may be expected from proteins of plant and animal sources (206,542,598,767). This work was extended by recording release patterns of CO_2 from bacteria, with promising results towards differentiation (207). Uptake and release of water from bacterial species was measured, followed by analysis of organic carbon released during thermogravimetry (258). More recent work on bacterial pyrolysis products involved high-resolution field-ionization mass spectrometry (631), Curie point pyrolysis-quadrupole MS (483) with reproducible "fingerprinting" and linear-programmed thermal degradation mass spectrometry (601). The latter was carried out on 10 bacteria with 77 ions profiled, using chemical ionization in a quadrupole mass spectrometer and a solid direct insertion probe programmed at 20°C/min. A high degree of reproducibility was found within an organism and a high degree of separation between closely related organisms. A relationship was found with the taxonomy of the bacteria studied. A review of automatic identification of microbes by pyrolysis-spectrometry was recently published (455). State-of-the-art reports on mass spectrometry of biological samples deal with rapid heating of thermally fragile molecules, e.g., Na_2ATP from an oxalic acid matrix, in the order of msecs (214), and protein sequencing by GC-MS (298). Drugs and metabolites are being analyzed by similar techniques in blood, serum, and other body fluids or tissues. Biopolymers, microorganisms, cells and tissues, sludges, humic substances, and geopolymers were investigated by on-line Curie point pyrolysis, low-voltage electron impact MS, and computerized data analysis (484).

6. Wood, Cellulose, and Agricultural Products

Pyrolysis of wood is characterized by that of its components—low-order cellulose (hemicellulose), highly crystalline cellulose, and lignin-containing char. Cellulose can follow two different reaction paths (343). At low temperatures (200–280°C), dehydration occurs, followed by charring and release of H_2O, CO, CO_2, etc. Between 280°C and 340°C, on the other hand, the product is a tar, consisting primarily of levoglucosan. A large number of cellulose pyrolysis products were found by GC, many of which have been identified (399). Mass spectra were recorded after laser vaporization (404), and kinetics were determined by isothermal pyrolysis together with GC-MS analysis of the degradation products (410). The decomposition mechanism of untreated and fire-retarded cellulose was found to be similar between 315–360°C and 276–298°C. Little difference was seen by GC-MS for three stages of decomposition (409). Addition of *tris*(2,3-dibromopyropyl)phosphate to cellulose reduced the yield of compounds normally observed in cellulose pyrolysis (237).

The amount of acetic acid, determined by GC after pyrolysis of cellulose triacetate in helium at 450°C, agreed with the chemically determined acetyl content of the polymer (315).

For fire-retardant research on wood, the amount of char formed, and that of water released is most important (655). By gas titrimetry and GC, it was found that fire retardant additives lower the carbonization temperature of cellulose (655).

The thermal stability of β-ether bonds in lignin and model compounds was studied by TA-GC and TA-MS (157). In decidious and coniferous wood, lignin was found to activate cellulose decomposition and levoglucosan formation (158). On the other hand, hemicellulose impedes the destruction of lignin and levoglucosan. Products included acetol, furfural, methylfurfural, hydroxymethylfurfural, furfuryl alcohol, methylcyclopentenolone, levoglucosenone, and others. In the distillate from rapid thermolysis at 475°C of alkali, hydrochloric acid, and sulfuric acid lignins from aspen and spruce, phenolic compounds were identified by gas chromatography (156). Sulfuric acid lignins—regardless of wood source—were thermally the most stable.

To demonstrate the potential application for EGA in other agricultural areas, three examples have been chosen. The vapor pressure of agricultural chemicals has been measured with a flame-ionization detector (66). Gases from the thermal decomposition of tobacco at 2°C/sec in an inert atmosphere were hydrogen, methane, and propene in one temperature region, while CO and CO_2 were observed between 100 and 450°C and again from 550–900°C (19). The last example is GC-MS analysis of volatile flavor compounds of roast beef. After work-up and trapping, 125 compounds were identified (302).

7. Geological Specimens

Aside from minerals, which were grouped with inorganics, practically only the energy-producing geological materials oil, oil shale, asphaltenes, peat, and coal are amenable to evolved gas analysis. With the acute energy crisis, petroleum plays an ever-increasing role, and substitutes are eagerly sought for the eventual production of energy or hydrocarbons as chemical feedstocks.

Review articles for petroleum deal with physical properties, flammability, and GC and GC-MS (388) of nonmetallic elements and compounds (274), especially gaseous sulfur compounds being analyzed by gas chromatography, mass spectrometry, and other chemical and instrumental methods. Also described were analytical methods for hydrocarbons (81), as well as analytical (4) and process instrumentation (423), with an extensive discussion on sampling techniques, applicability of instruments, and special analyses.

Gases dissolved in oil were identified by GC-MS as nitrogen, oxygen, carbon monoxide, CH_4, C_2H_2, C_2H_4, C_2H_6, and C_3H_8 (393). "Syncrudes" as well as oil, oil shale, and coal have been characterized by compound types using gel permeation chromatography and mass spectral correlations (751).

The so-called Green River shale has received considerable attention as a potential energy source. With simultaneous DTA, TG, and low-ionization quadrupole mass spectrometry, evolution of C_2H_6, C_3H_8, C_4H_{10}, C_5H_{12}, and corresponding olefins, H_2, CH_4, NH_3, H_2S, CO, CO_2, and H_2O was studied (662,663). By similar combination methods, mineral components, carbonates, clays, and zeolites can be investigated (322,664). A major constituent, Shortite ($Na_2CO_3 \cdot 2\ CaCO_3$), dissociates endothermically at 470°C, representing a significant absorption of heat for a retorting process (321). Conventional pyrolysis (394), thermal analysis (595), and laser pyrolysis-gas chromatography (276) have been applied to oil shales. By flame ionization and thermal conductivity detectors, evolution was registered between 250°C and 550°C (595), and subsequent gas chromatography of differentially trapped products indicated an increase in low-molecular-weight hydrocarbons with increasing temperatures. Colorado oil shale, containing Dawsonite [$NaAl(OH)_2CO_3$], released water and CO_2 at 180°C and 395°C, "crack products" of masses 2, 14, 15, 26, 27, 29, 39, 40, 41, 42, and 43, and some water and CO_2 at 490°C and 670°C (500). The amounts of hydrocarbons have also been determined from other oil and gas parent rocks (110).

Asphaltenes from petroleum lake muds and from peat bituminoids were pyrolyzed between 200°C and 900°C (109). Those from petroleum sources were more condensed and aromatized. Products from asphaltenes were C_1 to C_7 light hydrocarbons, naphthenic-paraffinic compounds of about 350 molecular weight, and gases with predominantly methane, followed by ethane and propane (449). Techniques include pyrolysis-GC (452), DTA, TG and gas-volumetry (580), slow programmed heating up to 873 K (736), and "total-recovery thermal analysis" of carbonaceous materials for yields of products with boiling points up to 1600°F (365). Essentially complete recovery of organics from bitumens was achieved by pyrolysis-flame detection, even before commercial thermal evolution analyzers had been developed (169).

For investigations of carbonaceous materials, partial-pressure mass spectrometry of graphite (374), laser vaporization of graphite coupled with time-resolved time-of-flight mass spectrometry (762), and laser vaporization of coal (704) have been reported. Theoretical predictions for multiple reactions during coal pyrolysis were tested by TG and a volumetric technique, yielding an average activation energy of 50 kcal/mol for the primary degasification step (108). Mass spectrometry was used for determination of the total evolved gas from carbon gasification (15,90) and with a Knudsen cell for carbon oxidation (153). Oxidation of petroleum coke by 15% oxygen in nitrogen was inhibited by 1% and 3% SO_2 (757). Sulfur had deposited on cooler walls, and COS and CO_2 by reaction of SO_2 with stable surface oxides were detected by the mass spectrometer. Carbon monoxide, carbon dioxide, and oxygen were monitored from carbons and graphite by desorption or decomposition (152). Oxidation proceeds via chemisorption of oxygen, formation of surface oxide, and release of CO and CO_2. Steam-hydrogen mixtures were applied to "hard coal" and "brown coal" (182), and hydrogenolysis of coal was studied by high-pressure DTA-differential pressure analysis in a flow-type apparatus (446). The area under the DTA exotherm could be correlated with the amount of water produced.

A complicated sequence was observed for Siberian peat (51). Starting with removal of water, the second and third reactions involved decomposition of organics. Carbon monoxide was observed throughout the whole temperature range up to 900°C, carbon dioxide from beginning decomposition to 600°C, hydrogen above 300°C, and methane from 300°C to 700°C.

From soil gases, isothermal chromatographic separation was achieved for H_2, O_2, N_2, NO, CH_4, CO_2, N_2O, and CO (111). By appropriate sample preparation and instrumentation, carbon isotopes in natural gas were determined by mass spectrometry (520).

Evolved gas analyses by mass spectrometry and infrared analysis from solid and gaseous fuels have been reviewed (282).

8. Environmental Samples

The search for sources of energy and its efficient utilization has in the past often excluded consideration for its impact on our environment. While thermal analysis by definition plays a role in energy-related problems (67), the steadily increasing concern over protecting the environment opens more and more opportunities for evolved gas analysis. An article on analytical and process instrumentation, particularly for petroleum chemistry, reviews techniques for sampling and detection of air and water pollutants (423).

Nondispersive infrared analysis has been used for simultaneous determination of CO and hydrocarbons (320) and for nitric oxide (183). A passive infrared sensor was applied to SO_2 emission (41). Oxides of nitrogen were analyzed by automatic IR measurements (301), spectrophotometric and electrochemical procedures (336,758), chemiluminescence, UV, and dispersive and nondispersive IR spectrometry (183,645).

Release of potential carcinogens—especially polycyclic aromatic hydrocarbons—from soot has been investigated by various combinations of TG, MS and GC-MS, IR, and UV (148,149,150). By GC alone, only lower-molecular-weight species were detected, while direct sampling into a quadrupole mass spectrometer with a solids probe up to 500°C provided a material balance for the observed TG curve. In-line GC identified release of hydrogen between 600°C and 1000°C.

Paint samples were analyzed for forensic applications by laser probe mass spectrometry (89), the physics of combustion and explosion were studied with a special device in the ion source of a time-of-flight mass spectrometer (356), and evolved gases from the exothermic decomposition of discarded tires were examined for their potential fuel value (132).

9. Space-Related Materials

Outside the earth's environment, space exploration depends heavily on evolved gas analysis. Gas evolution profiles from DTA of propellants of atmospheric pressure were obtained with a thermal conductivity cell (17). Mainly low-molecular-weight hydrocarbons were identified by mass spectrometry from polybutadiene propellant binders at 500°C in nitrogen (687).

Ablating plastics for re-entry shields, such as glass- and nylon-reinforced resins, were pyrolyzed in high vacuum with an arc image furnace to obtain the composition of gaseous degradation products and information on the mechanism (203). Products from polyimides as ablative polymers have been mentioned before (239).

Inside spacecrafts, oxidative thermal stability of polymeric materials and potential toxicity of their pyrolysis products are of vital importance. Twenty-five polymers from five different groups were evaluated by GC, GC-MS, and MS, directly or after trapping (613). With the exception of Delrin, which released formaldehyde, mainly CO_2 and water were observed; carbon monoxide could not be determined with this method.

For operations in space, a DTA-EGA combination was developed with pressure gauge and individual chemical detectors (74). Gas evolution, total organics, water, CO, CO_2, total gas, and C, H, N from organics can be determined from ambient temperature to 1600°C. With a photoionization detector, this combination is applicable to studies of minerals, meteorites, carbonaceous chondrites, and organic materials indicative for the presence of extraterrestrial life (75).

The release sequence of volatiles from the *Orgueil* carbonaceous chondrite was determined by TG-quadrupole-MS (241). Adsorbed water, a minor amount of adsorbed CO_2, and traces of nitrogen were detected. In addition, water was released from Limonite (α-$Fe_2O_3 \cdot H_2O$) and from interlayers of clay minerals, SO_2, and H_2O from Epsomite ($MgSO_4 \cdot 7 H_2O$) and Gypsum ($CaSO_4 \cdot 2 H_2O$), CO_2 from Siderite ($FeCO_3$), Breunnerite [(Fe, Mg)CO_3], and dolomite [$CaMg(CO_3)_2$], and CO from carbonaceous matter.

Isothermal pyrolysis was carried out on moon soil (240), and the release of low-molecular-weight compounds of organogenic elements of lunar samples has been reviewed (244). At a heating rate of 6°/min, up to 1400°C under vacuum, H, He, H_2O, CO, N, O, H_2S, CO_2, and SO_2 were observed from Apollo 14 and 15 lunar soils (243). The origin was thought to be atmospheric contamination, derived from solar wind, produced by chemical reaction, released from vesicles and inclusions, or exsolved from the melt. Subsurface Apollo 16 soil was richer in volatiles than any other soil, possibly due to cometary impact (246). Between 175°C and 350°C, 0.03% weight loss as H_2O, CO_2, CH_2, HCN, H, and minor amounts of hydrocarbons was detected by TA-EGA. With volatilization by stepwise heating, the total abundances of S and C in Apollo 14, 15, and 16 soil samples were determined (245). From lunar basalts, breccias, and soil, rubidium was lost below 950°C, and potassium and sodium at higher temperatures; lithium, barium, strontium, and the rare earths were not released below 1400°C (242).

The quest for life in space has always been part of exploratory missions. As early as 1963, pyrolysis-GC with a hydrogen flame detector has been suggested for this purpose, based on characteristic patterns to be observed from proteins of plants and animals (542).

V. CONCLUSIONS

On the surface, EGA may appear to be a confusing conglomeration of techniques and practices, rather than an increasingly important branch of thermal analysis. This may be true, since it cannot be precisely defined in terms of methodology. Rather, it is left to the analyst, to determine how best to combine individual methods to obtain the most and most meaningful information. And herein perhaps lies the strength of EGA—it requires stepping out of the routine of operations and fitting the method to the problem at hand.

GENERAL REFERENCES

G1. Anon, *Calorimetry, Thermometry, and Thermal Analysis—1970 Edition,* Society of Calorimetry and Thermal Analysis, Ed., Kagaku Gijutsu Sha, Tokyo, 1970.

G2. Anon, *Proc. Apollo 11 Lunar Sci. Conf.,* Pergamon, New York, 1970.

G3. Anon, *Proc. 4th Lunar Sci. Conf.,* Pergamon, New York, 1973.

G4. Anon, Proc., *25th Annual Conf. on Mass Spectrometry and Allied Topics,* 1977.

G5. Berg, L. G., Ed., *Trudy Vtorogo Soveshchaniya Po Termografii* (Transactions of the 2nd Conf. on Thermal Analysis), Akad. Nauk SSR, 1961.

G6. Berg, L. G., Ed., *Trudy Pervogo Soveshchaniya Po Termografii, Kazan, 1953* (Transactions of the 1st Conf. on Thermal Analysis, Kazan, 1953), Izd. Akad. Nauk SSR, Moscow-Leningrad.

G7. Berg, L. G., *Vvedenia v Termografiyu* (Introduction to Thermal Analysis), Izd. Nauka, Moscow, 1969.

G8. Blazek, A., *Thermal Analysis,* Van Nostrand-Reinhold, London, 1974.

G9. Burlingame, A. L., C. H. L. Shackleton, I. Howe, and O. S. Chizhov, *Anal. Chem.,* **50,** 346R (1978).

G10. Buzás, I., Ed., *Thermal Analysis,* Proc. 4th ICTA, Budapest, 1974, Akadémiai Kiadó, Budapest, 1975.

G11. Chihara, H., Ed., *Thermal Analysis,* Proc. 5th ICTA, Kyoto, 1977, Kagaku Gijutsu-Sha, 1977, Heyden, London, 1977.

G12. Chiu, J., Ed., *Polymer Characterization by Thermal Methods of Analysis,* Marcel-Dekker, New York, 1974.

G13. Cobler, J. G., and C. D. Chow, *Anal. Chem.,* **49,** 159R (1977).

G14. Collins, L. W., and L. D. Haws, *Thermochim. Acta,* **21,** 1 (1977).

G15. Cram, S. P., and T. H. Risby, *Anal. Chem.,* **50,** 213R (1978).

G16. Daniels, T., *Thermal Analysis,* Halsted, New York, 1973.

G17. Daniels, T., *Thermal Analysis,* John Wiley & Sons, New York, 1973, p. 186.

G18. Dollimore, D., Ed., *Proceedings of the First European Symposium on Thermal Analysis,* Heyden, London, 1976.

G19. Einhorn, I., Ed., *Thermal Analysis,* Polymer Conf. Series, University of Utah, 1970.

G20. Ettre, L. S., and A. Zlatkis, Eds., *The Practice of Gas Chromatography,* Interscience, New York, 1967.

G21. Fraser, J. M., *Anal. Chem.,* **49** (5), 231R (1977).

G22. Freel, J., *Anal. Chem.,* **49** (5), 243R (1977).

G23. Friedman, H. L., *Treatise Analytical Chemistry,* 1976, Pt. 3, Vol. 3, p. 393.

G24. Gal, S., J. Simon, and L. Erdey, *Proceedings of the Third Analytical Chemistry Conference,* Budapest, 1970, p. 243.

G25. Garn, P. D., *Thermoanalytical Methods of Investigation,* Academic, New York, 1965.

G26. Gose, W. A., Ed., *Proc. 5th Lunar Sci. Conf.,* Pergamon, Elmsford, 1974.

G27. Grob, R. L., Ed., *Modern Practice of Gas Chromatography,* Wiley-Interscience, New York, 1977.

G28. Hamming, M. C., and N. G. Foster, *Interpretation of Mass Spectra of Organic Compounds,* Academic, New York, 1972.

G29. Henning, F., in H. Moser, Ed., *Temperatur Messung,* 3rd Ed., Springer-Verlag, Berlin, 1977.

G30. Heymann, D., Ed., *Proc. 3rd Lunar Sci. Conf.,* MIT, Cambridge, 1972.

G31. Kambe, H., and P. D. Garn, *Thermal Analysis: Comparative Studies on Materials,* John Wiley & Sons, New York, 1974.

G32. Keattch, C. J., and D. Dollimore, *An Introduction to Thermogravimetry,* 2nd ed., Heyden, London, 1975.

G33. Kenyon, A. S., in P. E. Slade, Jr. and L. T. Jenkins, Eds., *Techniques and Methods of Polymer Evaluation,* Marcel-Dekker, New York, 1966.

G34. Krug, D., *Chem.-Ing.-Tech.,* **46,** 839 (1974).

G35. Levinson, A. A., Ed., *Proc. 2nd Lunar Sci. Conf.,* MIT, Cambridge, 1970.

G36. Liptay, G., *Atlas of Thermoanalytical Curves,* Vol. 5, Akadémiai Kiadó, Budapest, 1976.

G37. Lodding, W., Ed., *Gas Effluent Analysis,* Marcel Dekker, New York, 1967.

G38. Lombardi, G., *For Better Thermal Analysis,* Instituto di Mineralogiae Petrographia, University of Rome, Rome, Italy, 1977.

G39. Loveland, J. W. and C. N. White, *Anal. Chem.,* **49** (5), 262R (1977).

G40. McAdie, H. G., Ed., *Proceedings of the 1st Toronto Symposium on Thermal Analysis,* Chemical Institute of Canada, Toronto, 1965.

G41. McAdie, H. G., Ed., *Proceedings of the 2nd Toronto Symposium on Thermal Analysis,* Chemical Institute of Canada, Toronto, 1967.

G42. McAdie, H. G., Ed., *Proceedings of the 3rd Toronto Symposium on Thermal Analysis,* Chemical Institute of Canada, Toronto, 1969.

G43. Menis, O., Ed., *Status of Thermal Analysis,* National Bureau of Standards Special Publication 338, U.S. Govt. Printing Office, Washington, D.C., 1970.

G44. Murphy, C. B., *Anal. Chem.,* **46,** 451R (1974).

G45. Murphy, C. B., *Anal. Chem.,* **50,** 143R (1978).

G46. Ozawa, T., R. Sakamoto, and Y. Takahashi, *Shinku,* **16,** 240 (1973); *Chem. Abstr.,* **79,** 137617 (1973).

G47. Paulik, J., and F. Paulik, *Proceedings of the Third Analytical Conference,* Budapest, 1970, p. 225.

G48. Porter, R. S., and J. F. Johnson, Eds., *Analytical Calorimetry,* Plenum, New York, 1968.

G49. Redfern, J. P., Ed., *Thermal Analysis '65,* Proc. 1st ICTA, Aberdeen, 1965, MacMillan & Co., Ltd., London, 1965.

G50. Redfern, J. P., in H. G. Wiedemann, Ed., *Thermal Analysis,* Vol. 1, Birkhäuser Verlag, Basel, 1972, p. 615.

G51. Schultze, D., *Differential Thermoanalyse,* Verlag Chemie, Weinheim, 1969.

G52. Schwenker, R. F., Jr., and P. D. Garn, Eds., *Thermal Analysis,* Proc. 2nd ICTA, Worcester, Mass., 1968, Academic, New York, 1969.

G53. Sevcik, J., *Detectors in Gas Chromatography,* Elsevier, Amsterdam, 1975.

G54. Slade, P. E., and L. T. Jenkins, Eds., *Techniques and Methods of Polymer Evaluation,* Vol. 1, *Thermal Analysis,* Arnold, London, 1966.

G55. Takeuchi, T., *Thermal Analysis,* Kyoritsu Shuppan, Tokyo, 1968.

G56. Todor, D. N., *Thermal Analysis of Minerals,* Abacus Press, Turnbridge Well, Kent, England, 1976.

G57. Ware, R. K., "Thermal Analysis," in L. L. Hench, Ed., *Character. Ceram., 1971,* Marcel Dekker, New York, N.Y., 1971, p. 273; *Chem. Abstr.,* **77,** 39819 (1972).

G58. Wendlandt, W. W., *Thermal Methods of Analysis,* Interscience, New York, London, Sidney, 1964.

G59. Wendlandt, W. W., and J. P. Smith, *The Thermal Properties of Transition Metal Amine Complexes,* Elsevier, New York, 1967.

G60. Wendlandt, W. W., *Handbook of Commercial Scientific Instruments*, Vol. 2: *Thermoanalytical Techniques*, Marcel Dekker, New York, 1974.

G61. Wendlandt, W. W., *Thermal Methods of Analysis*, 2nd ed., John Wiley & Sons, New York, 1974.

G62. Wendlandt, W. W., and L. W. Collins, *Benchmark Papers in Analytical Chemistry*, Vol. 2, *Thermal Analysis*, Dowden, Ross and Hutchinson, Stroudsburg, Penna., 1976.

G63. Wendlandt, W. W., and L. W. Collins, *Thermal Analysis*, Wiley, Chichester, England, 1977.

G64. Wiedemann, H. G., Ed., *Thermal Analysis*, Proc. 3rd ICTA, Davos, 1971; Birkhäuser Verlag, Basel, 1972.

REFERENCES

1. Abbas, K. B., and E. M. Sorvik, *J. Appl. Polym. Sci.*, **20**, 2395 (1976).

2. Adonyi, Z., and G. Korosi, in D. Dollimore, Ed., *Proc. Eur. Symp. Therm. Anal., 1st*, Heyden, London, 1976, p. 200.

3. Adonyi, Z., and G. Korosi, in I. Buzás, Ed., *Thermal Analysis*, Vol. 2, Akadémiai Kiadó, Budapest, 1975, p. 453.

4. Afanas'ev, M. I., S. P. Kozlov, V. M. Lozovskii, and A. A. Datskevich, *Tr. Vses. Nauchno-Issled. Proektno-Konstr. Inst. Kompleksn. Avtom. Neft. Gazov. Prom-Sti., 1973* (5), 289; *Chem. Abstr.* **81**, 155429 (1974).

5. Ahlstrom, D. H., S. A. Leibman, and K. B. Abbas, *J. Polym. Sci., Polym. Chem. Ed.*, **14**, 2479 (1976).

6. Allen, J. D., J. Billingsley, and J. T. Shaw, *J. Inst. Fuel*, **47** (393), 275 (1974).

7. Alexandrov, V. V., V. V. Boldyrev, and V. G. Morozov, in D. Dollimore, Ed., *Proc. Eur. Symp. Therm. Anal., 1st*, Heyden, London, 1976, p. 301.

8. Amstutz, D., in H. G. Wiedemann, Ed., *Thermal Analysis*, Vol. 1, Birkhäuser Verlag, Basel, 1972, p. 415.

9. Anderton, D. J., and F. R. Sale, in D. Dollimore, Ed., *Proc. Eur. Symp. Therm. Anal., 1st*, Heyden, London, 1976, p. 278.

10. Anthony, G. D., and P. D. Garn, *J. Am. Ceram. Soc.*, **57**, 132 (1974).

11. Armitage, G. M., and S. J. Lyle, *Talanta*, **20**, 315 (1973).

12. Asseva, R. M., A. A. Berlin, Z. S. Kazakova, and S. M. Mezhikovsky, in I. Buzás, Ed., *Thermal Analysis*, Vol. 2, Akadémiai Kiadó, Budapest 1975, p. 183.

13. Aspinal, M. L., H. J. Madoc-Jones, E. L. Charsley, and J. P. Redfern, in H. G. Wiedemann, Ed., *Thermal Analysis*, Vol. 1, Birkhäuser Verlag, Basel, 1972, p. 303.

14. Austin, F. E., J. Dollimore, and B. H. Harrison, in R. F. Schwenker, Jr., and P. D. Garn, Eds., *Thermal Analysis*, Vol. 1, Academic, New York, 1969, p. 311.

15. Austin, F. E., J. G. Brown, J. Dollimore, C. M. Freedman, and B. H. Harrison, *Analyst*, **96**, 110 (1971).

16. Ayres, W. M., and E. M. Bens, *Compt. Rend.*, **251**, 2961 (1960).

17. Ayres, W. M., and E. M. Bens, *Anal. Chem.*, **33**, 568 (1961).

18. Babu, S. V., and Y. V. Chalapati Rao, *Chem. Phys. Lett.*, **37** (2), 249 (1976).

19. Baker, R. R., in D. Dollimore, Ed., *Proc. Eur. Symp. Therm. Anal., 1st*, Heyden, London, 1976, p. 219.

20. Balek, V., *J. Mater. Sci.*, **4**, 919 (1969).

21. Balek, V., and K. Habersberger, in H. G. Wiedemann, Ed., *Thermal Analysis*, Vol. 2, Birkhäuser Verlag, Basel, 1972, p. 501.

22. Balek, V., *Anal. Chem.*, **42**, 16A (1970).

23. Balek, V., in I. Buzas, Ed., *Thermal Analysis*, Vol. 2, Akademiai Kiado, Budapest, 1975, p. 551.

24. Balek, V., *Termanal. '76, Celostatna Konf. Term. Anal., [Pr], 7th*, 1976, Al; *Chem. Abstr.*, **87**, 94757 (1977).

25. Balek, V., J. Krouda, and M. Prachar, *Radiochem. Radioanal. Lett.*, **28**, 279 (1977), *Chem. Abstr.*, **86**, 164677 (1977).

26. Balek, V., H. Landsperksy, and M. Voboril, *Radiochem. Radioanal. Lett.*, **28**, 289 (1977); *Chem. Abstr.*, **86**, 147415 (1977).

27. Bancroft, G. M., and H. D. Gesser, *J. Inorg. Nucl. Chem.*, **27**, 1537 (1965).

28. Bandi, W. R., H. S. Karp, W. A. Straub, and L. M. Melnick, *Talanta*, **11**, 1327 (1964).

29. Bandi, W. R., W. A. Straub, E. G. Buyok, and L. M. Melnick, *Anal. Chem.*, **38**, 1336 (1966).

30. Bandi, W. R., E. G. Buyok, G. Krapf, and L. M. Melnick, in R. F. Schwenker, Jr., and P. D. Garn, Eds., *Thermal Analysis*, Vol. 2, Academic, New York, 1969, p. 1363.

31. Bandi, W. R. and G. Krapf, *Anal. Chem.*, **49**, 649 (1977).

32. Bandi, W. R., *Science*, **196**, 136 (1977).

33. Banewicz, J. J., J. A. Maguire, M. E. Munnell, and M. L. Moore, *Thermochim. Acta*, **12**, 377 (1975).

34. Barclay, K. S., and J. M. Crewe, *J. Appl. Chem.*, **17**, 21 (1967).

35. Barlow, A., R. S. Lehrle, and J. C. Robb, *Polymer*, **2**, 27 (1961).

36. Barnes, P. A., in D. Dollimore, Ed., *Proc. Eur. Symp. Therm. Anal., 1st*, Heyden, London, 1976, p. 31.

37. Barnes, P. A., B. V. Burnley, and J. T. Pearson, in D. Dollimore, Ed., *Proc. Eur. Symp. Therm. Anal., 1st*, Heyden, London, 1976, p. 244.

38. Barnes, P. A., and R. M. Tomlinson, *J. Therm. Anal.*, **7**, 469 (1975).

39. Barnes, P. A., in J. Wood, O. Lindqvist, and C. Helgesson, Eds., *React. Solids, [Proc. Int. Symp.] 8th 1976*, Plenum, New York, 1977, p. 663.

40. Barrall, E. M., II, and J. A. Logan, *Thermochim. Acta*, **9**, 205 (1974).

41. Bartle, E. R., and E. A. Meckstroth, U.S. NTIS, PB Rep., 1975, PB-243478, 55 pp. Avail NTIS.

42. Basden, K. S., *Fuel*, **39**, 3 (1960).

43. Basu, A. K., and F. R. Sale, *J. Mater. Sci.*, **12**, 1115 (1977); *Chem. Abstr.*, **87**, 56319 (1977).

44. Bataille, P., and B. T. Van, *J. Therm. Anal.*, **8**, 141 (1975).

45. Baumgartner, E., and E. Nachbaur, in D. Dollimore, Ed., *Proc. Eur. Symp. Therm. Anal.*, 1st, Heyden, London, 1976, p. 35.

46. Bayer, G., and H. G. Wiedemann, in D. Dollimore, Ed., *Proc. Eur. Symp. Therm. Anal.*, 1st, Heyden, London, 1976, p. 256.

47. Becker, E. W., *Separation of Isotopes*, George Newnes, Ltd., London, 1961, p. 360.

48. Beckey, H. D., N. Knöppel, G. Metzinger, and P. Schulze, "Advances in Experimental Techniques, Applications, and Theory of Field Ion Mass Spectrometry," in W. L. Mead, Ed., *Advances in Mass Spectrometry*, Vol. 3, The Institute of Petroleum, London, 1966, p. 35.

49. Bednarski, V. N., German (W) Pat. 2,246,836 (Mar. 28, 1974); British Pat. 1,378,807 (Dec. 27, 1974).

50. Beglarlyan, A. A., and B. V. Romanovskii, *Int. Chem. Eng.*, **15**, 613 (1975).

51. Belichmaer, Ya. A., V. M. Ikrin, and S. I. Smolyaninov, in I. Buzás, Ed., *Thermal Analysis*, Vol. 3, Akadémiai Kiadó, Budapest, 1975, p. 265.

52. Belkbir, L., E. Joly, and N. Gerard, in D. Dollimore, Ed., *Proc. Eur. Symp. Therm. Anal., 1st,* Heyden, London, 1976, p. 87.

53. Bell, A. E., J. Pritchard, and K. W. Sykes, *2nd Conf. Industrial Carbon and Graphite Soc., Chem. 2nd,* London, 1966, p. 214.

54. Belton, G. R., and R. J. Fruehan, *J. Phys. Chem.*, **71**, 1403 (1967).

55. Bennett, D., *J. Appl. Chem. Biotechnol.*, **22**, 973 (1972); *Chem. Abstr.*, **77**, 156928 (1972).

56. Berg, L. G., *Izv. Sektora. Fiz. Khim. Analiza. Inst. Obshch. Neorgan. Khim. Akad. Nauk. SSSR*, **19**, 249 (1949); *Chem. Abstr.*, **45**, 1830 (1951).

57. Berkowitz, J., D. J. Meschi, and W. A. Chupka, *J. Chem. Phys.*, **33**, 533 (1960).

58. Berkowitz-Mattuck, J. B., A. Büchler, J. L. Engelke, and S. N. Goldstein, *J. Chem. Phys.*, **39**, 2722 (1963).

59. Berkowitz, J., and W. A. Chupka, *J. Chem. Phys.*, **40**, 287 (1964).

60. Berkowitz, J., and W. A. Chupka, *J. Chem. Phys.*, **40**, 2735 (1964).

61. Bernal, S., J. Cornejo, J. M. Criado, and J. M. Trillo, in D. Dollimore, Ed., *Proc. Eur. Symp. Therm. Anal., 1st,* Heyden, London, 1976, p. 121.

62. Beroza, M., and M. N. Inscoe, in L. S. Ettre, and H. M. McFadden, Ed., *Ancilliary Techniques of Gas Chromatography*, Wiley-Interscience, New York, 1969, p. 89.

63. Berthold, R., Ger. Offen., 2,346,792 (CI. GO1N); *Chem. Abstr.* **83**:107911 (1975).

64. Beuhler, R. J., E. Flanigan, L. J. Greene, and L. Friedman, *J. Am. Chem. Soc.*, **96**, 3990 (1974).

65. Binenboym, J., H. Selig, and S. Sarig, *J. Inorg. Nucl. Chem.*, **38**, 2313 (1976).

66. Blaine, R. L., and P. F. Levy, *Anal. Calorimetry*, **3**, 185 (1974).

67. Blaine, R. L., *Ind. Res.*, **17**(4), 56 (1975).

68. Blasius, E., and H. Haeusler, *Fresenius' Z., Anal. Chem.*, **277**(1), 9 (1975); *Chem. Abstr.* **83**,206850 (1975).

69. Blasius, E., H. Haeusler, and H. Lander, *Talanta*, **23**(4), 301 (1976).

70. Block, J., and A. P. Gray, *Inorg. Chem.*, **4**, 304 (1965).

71. Bolivar, C., H. Charcosset, R. Frety, G. Leclercq, B. Neff, and J. Varloud, in D. Dollimore, Ed., *Proc. Eur. Symp. Therm. Anal., 1st,* Heyden, London, 1976, p. 55.

72. Bolivar, C., H. Charcosset, R. Frety, G. Leclercq, and L. Tournayan, in D. Dollimore, Ed., *Proc. Eur. Symp. Therm. Anal., 1st,* Heyden, London, 1976, p. 117.

73. Boll, R. H., Ger. Offen., 2,416,672 (Cl. GO ln), 24 Oct., 1974; *Chem. Abstr.* **82**, 678719 (1975).

74. Bollin, E. M., in Vol. 1, R. F. Schwenker, Jr., and P. D. Garn, Eds., *Thermal Analysis*, Academic Press, New York, 1969, p. 255.

75. Bollin, E. M., in R. F. Schwenker, Jr., and P. D. Garn, Eds., *Thermal Analysis*, Vol. 2, Academic Press, New York, 1969, p. 1387.

76. Bolton, A. P., *J. Catalysis*, **18**, 154 (1970).

77. Borchardt, H. J., and Daniels, F., *J. Phys. Chem.*, **61**, 917 (1957).

78. Bouwknegt, A., J. deKok, and J. A. W. deKock, *Thermochim. Acta*, **9**, 399 (1974).

79. Bowen, D. O., *Mod. Plast.*, **44** (12), 127, 163 (1967).

80. Bowles, R., in C. D. Price and J. E. Williams, Eds., *Time-of-Flight Mass Spectrometry*, Pergamon, Oxford, 1969, p. 211.

81. Bradley, M. P. T., *Anal. Chem.*, **49**(5), 249R (1977).

82. Bratspies, G. K., J. F. Smith, J. O. Hill, and R. J. Magee, *Thermochim. Acta*, **19**, 335 (1977).

83. Bratspies, G. K., J. F. Smith, J. O. Hill, and R. J. Magee, *Thermochim. Acta*, **19**, 361 (1977).

84. Bratspies, G. K., J. F. Smith, and J. O. Hill, *Thermochim. Acta*, **19**, 373 (1977).

85. Brauer, G. M., *J. Polym. Sci.*, **8**, 3 (1965).

86. Braun, J. M., and J. E. Guillet, *J. Polym. Sci., Polym. Chem. Ed.*, **13**, 1119 (1975).

87. Braun, J. M., A. Lavoie, and J. E. Guillet, *Macromolecules*, **8**, 311 (1975).

88. Briggs, P., D. Dix, D. Glover, and R. Kleinman, *Amer. Lab.*, **4**(9), 57 (1972).

89. Brochard, G., and J. F. Eloy, presented at the *25th Annual Conference on Mass Spectrometry and Allied Topics*, Washington, D.C., 1977.

90. Brown, J. G., J. Dollimore, C. M. Freedman, and B. H. Harrison, *Thermochim. Acta*, **1**, 499 (1970).

91. Bua, E., and P. Manaresi, *Anal. Chem.*, **31**, 2022 (1959).

92. Burmistrova, N. P., A. V. Bardimova, and R. G. Fitseva, in D. Dollimore, Ed., *Proc. Eur. Symp. Therm. Anal., 1st,* Heyden, London, 1976, p. 101.

93. Burnley, B. V., and J. T. Pearson, in D. Dollimore, Ed., *Proc. Eur. Symp. Therm. Anal., 1st,* Heyden, London, 1976, p. 90.

94. Burns, E. A., *Anal. Chem.*, **35**, 1106 (1963).

95. Buser, H.-R., *Anal. Chem.*, **49**, 918 (1977).

96. Bussière, P., B. Claudel, J. P. Renouf, Y. Tramibouze, and M. Prettre, *J. Chim. Phys.*, **58**, 668 (1961).

97. Bussière, P., *Fine Part., Int. Conf. Pap., 2nd, 1973* (Pub. 1974), p. 69.

98. Butt, Yu. M., V. V. Timashev, V. S. Bakshutov, M. K. Grineva, and V. V. Ilyukhin, in H. G. Wiedemann, Ed., *Thermal Analysis,* Vol. 3, Birkhäuser Verlag, Basel, 1972, p. 513.

99. Caccamese, S., P. Maravigna, G. Montaudo, and M. Przybylski, *J. Polym. Sci., Polym. Chem. Ed.*, **13**, 2061 (1975).

100. Carroll-Porczynski, C. S., in H. G. Wiedemann, Ed., *Thermal Analysis,* Vol. 3, Birkhäuser Verlag, Basel, 1972, p. 273.

101. Cascaval, C. N., I. A. Schneider, and I. C. Poinescu, *J. Polym. Sci., Polym. Chem. Ed.*, **13**, 2259 (1975).

102. Chang, T., and T. E. Mead, *Anal. Chem.*, **43**, 534 (1971).

103. Chang, E. P., and R. Salovey, *J. Polym. Sci., Polym. Chem. Ed.*, **12**, 2957 (1974).

104. Chantret, F. in H. G. Wiedemann, Ed., *Thermal Analysis,* Vol. 1, Birkhäuser Verlag, Basel, 1972, p. 313.

105. Charbonnier, F., V. Balek, and P. Bussiere, *J. Therm. Anal.*, **7**, 373 (1975).

106. Charsley, E. L., and A. C. F. Kamp, in H. G. Wiedemann, Ed., *Thermal Analysis,* Vol. 1, Birkhäuser Verlag, Basel, (1972), p. 499.

107. Chen, Tung-Ling, *Appl. Chem., Hua Hsueh Tung Pao* (2), **90–9**, 76 (1975).

108. Chermin, H. A. G., and D. W. van Krevelen, *Fuel*, **36**, 85 (1957).

109. Chernova, T. G., E. P. Shishenina, and I. L. Maryasin, *Khim. Tekhnol. Topl. Masel*, **18**(11), 59 (1973).

110. Chetverikova, O. P., and M. K. Kalinko, *Tr. Vses. N.-i. Geologorazved. Neft. In-t*, **196**, 98 (1976); *Chem. Abstr.*, **87**, 70481 (1977).

111. Chiang, Ching-Tsun, *Chung Kuo Nung Yeh Hua Hsueh Hui Chih*, **7**, 69 (1969); *Chem. Abstr.*, **73**, 55042 (1970).

112. Chiu, J., *Anal. Chem.*, **40**, 1516 (1968).

113. Chiu, J., *Thermochim. Acta*, **1**, 231 (1970).

114. Chiu, J., and E. F. Palermo, *Anal. Chim. Acta*, **81**, 1 (1976).

115. Chiu, J., and A. J. Beattie, *Thermochim. Acta*, **21**, 263 (1977).

116. Chupakhin, M. S., L. O. Kogan, and A. A. Polyakova, *Zh. Anal. Khim.*, **29**, 1028 (1974).

117. Chupka, W. A., and M. G. Inghram, *J. Phys. Chem.*, **59**, 100 (1955).

118. Chupka, W. A., J. Berkowitz, and C. F. Giese, *J. Chem. Phys.*, **30**, 827 (1959).

119. Cianetti, E., and G. F. Pecci, in H. G. Wiedemann, Ed., *Thermal Analysis*, Vol. 3, Birkhäuser Verlag, Basel, (1972), p. 255.

120. Clancey, V. J., *Nature*, **166**, 275 (1950).

121. Clark, J. E., *Polym. Eng. Sci.*, **7**, 137 (1967).

122. Clinckemaillie, A., and C. Hofmann, in H. G. Wiedemann, Ed., *Thermal Analysis*, Vol. 1, Birkhäuser Verlag, Basel (1972), p. 337.

123. Collin, J. E., Rept. No. EUR-2114.e., AEC No. 17941, 1964; *Chem. Abstr.*, **64**, 12020 (1966).

124. Collin, J. E., and A. Delplace, *Bull. Soc. Chim. Belges*, **75**, 304 (1966); *Chem. Abstr.*, **65**, 13508 (1966).

125. Collins, L. W., and W. W. Wendlandt, *Thermochim. Acta*, **7**, 201 (1973).

126. Collins, L. W., W. W. Wendlandt, E. K. Gibson, and G. W. Moore, *Thermochim. Acta*, **7**, 209 (1973).

127. Collins, L. W., E. K. Gibson, and W. W. Wendlandt, *Thermochim. Acta*, **11**, 177 (1975).

128. Collins, L. W., W. W. Wendlandt, and E. K. Gibson, *Thermochim. Acta*, **8**, 205 (1974).

129. Collins, L. W., W. W. Wendlandt, and E. K. Gibson, *Thermochim. Acta*, **8**, 303 (1974).

130. Collins, L. W., and W. W. Wendlandt, and E. K. Gibson, *Thermochim. Acta*, **8**, 307 (1974).

131. Collins, L. W., and W. W. Wendlandt, *Thermochim. Acta*, **8**, 315 (1974).

132. Collins, L. W., W. R. Downs, E. K. Gibson, and G. W. Moore, *Thermochim. Acta*, **10**, 153 (1974).

133. Cotter, J. L., G. J. Knight, and W. W. Wright, in I. Buzás, Ed., *Thermal Analysis*, Vol. 2, Akadémiai Kiadó, Budapest, 1975, p. 163.

134. Courval, G., and D. G. Gray, *Macromolecules*, **8**, 326 (1975).

135. Crummett, W. B. and H. R. Stehl, *Environ. Health Perspect.*, **5**, 15 (1973).

136. Cukor, P., and E. W. Lanning, *J. Chromatogr. Sci.*, **9**, 487 (1971).

137. Cukor, P., and C. Persiani, in J. Chiu, Ed., *Polymer Characterization by Thermal Methods of Analysis*, Marcel Dekker, New York, 1974, p. 105.

138. Daly, N. J., and F. J. Ziolkowski, *Aust. J. Chem.*, **23**, 541 (1970); *Chem. Abstr.*, **72**, 99848 (1970).

139. DasGupta, N. N., and S. K. Ghosh, *Rev. Mod. Phys.*, **18**, 225 (1946).

140. David, D. J., *Thermochim. Acta*, **3**, 277 (1972).

141. Davies, W. D., and T. A. Vanderslice, *Proc. 7th Nat. Symp. Am. Vacuum Soc.*, Pergamon Press, 417 (1960).

142. Davis, C. E., and A. E. Krc, et. al., *API Publ.*, **4245**, 93 (1975).

143. Davison, W. H. T., S. Slaney, and A. L. Wragg, *Chem. Ind.* (London), 1356 (1954).

144. Derouane, E. G., Z. Gabelica, R. Hubin, and J. Hubin-Franskin, *Thermochim. Acta*, **11**, 287 (1975).

145. DiEdwardo, A. H., and F. Zitomer, presented at the 17th Annual Conference on Mass Spectrometry and Allied Topics, Washington, D.C., 1969.

146. Dietz, W. A., *J. Gas Chromatog.*, **5**, 68 (1967).

147. DiGiovine, S. J., *Proc. R. Aust. Chem. Inst.*, **42**(1), 20 (1975).

148. DiLorenzo, A., S. Masi, and A. Pennacchi, in D. Dollimore, Ed., *Proc. Eur. Symp. Therm. Anal., 1st*, Heyden, London, 1976, p. 37.

149. DiLorenzo, A., S. Masi, and R. Guerrini, presented at Chemical Panel of the Int. Flame Res. Found., Karlsruhe, Germ., Sept., 1975.

150. DiLorenzo, A., S. Masi, and D. Paparone, in I. Buzás, Ed., *Thermal Analysis*, Vol. 3, Akadémiai Kiadó, Budapest (1975), p. 273.

151. Dollimore, D., and G. R. Heal, *Carbon*, **5**, 65 (1967).

152. Dollimore, J., C. M. Freedman, and B. H. Harrison, presented at the 17th Annual Conference on Mass Spectrometry and Allied Topics, Washington, D.C., 1969, p. 148.

153. Dollimore, J., and B. H. Harrison, in H. G. Wiedemann, Ed., *Thermal Analysis*, Vol. 1, Birkhäuser Verlag, Basel, 1972, p. 343.

154. Dollimore, D., B. W. Krupay, and R. A. Ross, in D. Dollimore, Ed., *Proc. Eur. Symp. Therm. Anal., 1st*, Heyden, London, 1976, p. 125.

155. Dollimore, D., J. P. Gupta, and D. V. Nowell, in *Proc. Eur. Symp. Therm. Anal., 1st*, D. Dollimore, Ed., Heyden, London, (1976), p. 233.

156. Domburg, G. E., V. N. Sergeeva, and A. I. Kalninsh, in H. G. Wiedemann, Ed., *Thermal Analysis*, Vol. 3, Birkhäuser Verlag, Basel, 1972, p. 327.

157. Domburg, G., G. Rossinskaya, and V. Sergeeva, in I. Buzás, Ed., *Thermal Analysis*, Vol. 2, Akadémiai Kiadó, Budapest, 1975, p. 211.

158. Domburg, G., V. Sergeeva, I. Kirschbaum, T. Sharadova, and T. Skripchenko, in H. Chihara, Ed., *Thermal Analysis*, Heyden, London, 1977, p. 304.

159. Dondur, V. T., and D. B. Vucelic, *Glas. Hem. Drus., Beograd*, **41**, 11 (1976); *Chem. Abstr.*, **85**, 198712 (1976).

160. Dondur, V. T., D. R. Vucelic, and N. O. Juranic, *Glas. Hem. Drus., Beograd*, **41**, 91 (1976); *Chem. Abstr.*, **86**, 60845 (1977).

161. Dorsey, G. A. *Anal. Chem.*, **41**, 350 (1969).

162. Doyle, C. D., *Evaluation of Experimental Polymers*, U.S. Air Force Rept. WADD-TR-60-283, Wright Patterson Air Force Base, 1960.

163. Drugov, Yu. S., and G. V. Murav'eva, *Proizvod. Shin, Rezinotekh. Asbestotekh. Izdelii, Nauchno-Tekh. Sb.*, **10**, 20 (1973); *Chem. Abstr.*, **81**, 154094 (1974).

164. Dubik, M., and V. Jesenak, *Thermanal '76, Celostatna Conf. Term. Anal.*, ±*PR.1, 7th*, 45 (1976); *Chem. Abstr.*, **86**, 128073 (1977).

165. Dunham, S. B., *Nature*, **188**, 51 (1960).

166. Dünner, W., and H. Eppler, in I. Buzás, Ed., *Thermal Analysis*, Vol. 3, Akadémiai Kiadó, Budapest, 1975, p. 1049.

167. Ebert, A. A., *Anal. Chem.*, **33**, 1865 (1961).

168. Eggertsen, F. T., and F. H. Stross, *J. Appl. Polym. Sci.*, **10**, 1171 (1966).

169. Eggertsen, F. T., H. M. Joki, and F. M. Stross, in R. F. Schwenker, Jr., and P. D. Garn, Eds., *Thermal Analysis*, Vol. 1, Academic, New York, 1969, p. 341.

170. Ehlers, G. F. L., and K. R. Fisch, in H. G. Wiedemann, Ed., *Thermal Analysis*, Vol. 3, Birkhäuser Verlag, Basel, (1972), p. 187.

171. Einhorn, I. N., D. A. Chatfield, K. J. Voorhees, F. D. Hileman, R. W. Mickelson, S. C. Israel, J. H. Futrell, and P. W. Ryan, *Fire Research*, **1**, 41 (1977).

172. Elliott, R. M., and W. J. Richardson, presented at 15th Annual Conf. Mass Spec. and Allied Topics, Denver, 1967.

173. Emmerich, W.-D., and V. Balek, in H. G. Wiedemann, Ed., *Thermal Analysis*, Vol. 1, Birkhäuser Verlag, Basel, 1972, p. 475.

174. Emmerich, W.-D., *Proc. Conf. on Thermal Analysis,* Helsinki, 1972, p. 175.

175. Emmerich, W.-D., and K. Bayreuther, in I. Buzás, Ed., *Thermal Analysis,* Vol. 3, Akadémiai Kiadó, Budapest, 1975, p. 1017.

176. Emmerich, W.-D., and E. Kaisersberger, in H. Chihara, Ed., *Thermal Analysis,* Heyden, London, 1977, p. 67.

177. Fanter, D. L., R. L. Levy, and C. J. Wolf, *Anal. Chem.,* **44,** 43 (1972).

178. Fedotova, O. Ya., V. I. Gorokhov, and V. V. Korshak, *Vysokomol. Soedin,* Ser. A, **16,** 1228 (1974).

179. Fehlner, T. P., *J. Am. Chem. Soc.,* **89,** 6477 (1967).

180. Fehlner, T. P., *J. Am. Chem. Soc.,* **90,** 4817 (1968).

181. Fehlner, T. P., and S. A. Fridmann, *Inorg. Chem.,* **9,** 2288 (1970).

182. Feistel, P. P., K. H. van Heek, and H. Jüntgen, in D. Dollimore, Ed., *Proc. Eur. Symp. Therm. Anal., 1st,* Heyden, London, 1976, p. 361.

183. Feldman, J., *Air Qual. Instrum.,* **2,** 147 (1974).

184. Felton, H. R., and A. A. Buehler, *Anal. Chem.,* **30,** 1163 (1958).

185. Fenerty, J., and K. S. W. Sing, in D. Dollimore, Ed., *Proc. Eur. Symp. Therm. Anal., 1st,* Heyden, London, 1976, p. 304.

186. Fennell, T. R. F. W., G. J. Knight, and W. W. Wright, in H. G. Wiedemann, Ed., *Thermal Analysis,* Vol. 3, Birkhäuser Verlag, Basel, 1972, p. 245.

187. Ferrari, H., in R. F. Schwenker, Jr., and P. D. Garn, Eds., *Thermal Analysis,* Vol. 1, Academic, New York, 1969, p. 41.

188. Ficalora, P. J., O. M. Uy, D. W. Muenow, and J. L. Margrave, *J. Am. Ceram. Soc.,* **51,** 574 (1968).

189. Field, F. H., *Acc. Chem. Res.,* **1,** 43 (1968).

190. Findeis, A. F., K. D. W. Rosinski, P. P. Petro, and R. E. W. Earp, *Thermochim. Acta,* **1,** 383 (1970).

191. Fisher, I. P., and A. Johnson, *Anal. Chem.,* **47,** 59 (1975).

192. Flowers, W. T., R. N. Haszeldine, E. Henderson, A. K. Lee, and R. D. Sedgwick, *J. Polym. Sci.,* **10,** 3489 (1972).

193. Flügge, S., and K. E. Zimens, *Z. Physik. Chem.,* **B32,** 179 (1939).

194. Flynn, J. H., and L. E. Wall, *J. Res. Natl. Bur. Stand.,* **70A,** 487 (1966).

195. Forrester, J. A., *Chem. and Ind.,* 1244 (1969).

196. Fouque, D., P. Fouilloux, P. Bussière, D. Weigel, and M. Prettre, *J. Chim. Phys.,* **62,** 1088 (1965).

197. Franklin, J. L., J. T. Herron, P. Bradt, and V. H. Dibeler, *J. Am. Chem. Soc.,* **80,** 6188 (1958).

198. Freeman, E. S., and B. Carroll, *J. Phys. Chem.,* **62,** 394 (1958).

199. Friedman, H. L., and G. A. Griffith, in J. P. Redfern, Ed., *Thermal Analysis '65,* MacMillan, London 1965, p. 22.

200. Friedman, H. L., *J. Macromol. Sci.,* **A1**(1), 57 (1967).

201. Friedman, H. L., *The Relationship Between Structure and Thermal Stability of New High Temperature Polymers,* ML-TDR-64-274, Air Force Materials Lab., Wright-Patterson Air Force Base, Ohio, August 1964.

202. Friedman, H. L., *J. Appl. Polymer Sci.,* **9,** 651 (1965).

203. Friedman, H. L., *J. Appl. Polymer Sci.,* **9,** 1005 (1965).

204. Friedman, H. L., H. W. Goldstein, and G. A. Griffith, presented at 15th Annual Conference on Mass Spectrometry and Allied Topics, Denver, 1967.

205. Friedman, H. L., H. W. Goldstein, and G. A. Griffith, *Kinetics and Mechanisms of Thermal Degradation of Polymers Using Time-of-Flight Mass Spectrometry for Continuous Gas Analysis*, AFML-TR-68-111, Air Force Materials Laboratory, Wright-Patterson Air Force Base, Ohio, May, 1968.

206. Friedman, H. L., presented at 17th Annual Conference on Mass Spectrometry and Allied Topics, Dallas, 1969.

207. Friedman, H. L., G. A. Griffith, and H. W. Goldstein, in R. F. Schwenker, Jr., and P. D. Garn, Eds., *Thermal Analysis,* Vol. 1, Academic, New York, 1969, p. 405.

208. Friedman, H. L., *Thermochim. Acta,* **1,** 199 (1970).

209. Friedman, H. L., G. A. Griffith, J. R. Mallin, and N. M. Jaffe, *Thermochim. Acta,* **8,** 119 (1974).

210. Friedman, H. L., "Thermal Aging and Oxidation with Emphasis of Polymers," in I. M. Kolthoff, P. J. Elving, and F. H. Stross, Eds., *Treatise on Analytical Chemistry,* Part III, Vol. 3, Section D-1, Wiley-Interscience, New York, 1976, p. 398.

211. Friedman, H. L., ibid., p. 503.

212. Friedman, H. L., ibid., p. 526.

213. Futrell, J. H., D. A. Chatfield, F. D. Hileman, K. H. Voorhees, and I. N. Einhorn, *Am. Chem. Soc. Polym. Prepr.,* **17**(2), 767 (1976).

214. Gaffney, J. S., and L. Friedman, presented at 25th Annual Conference on Mass Spectrometry and Allied Topics, Washington, D.C., 1977.

215. Galbács, M. Z., and L. J. Csányi, *Anal. Chem.,* **45,** 1784 (1973).

216. Gallagher, P. K., *Thermochim. Acta.* (in press).

217. Gallagher, P. K., and J. Thomson, Jr., *J. Am. Ceram. Soc.,* **48,** 644 (1965).

218. Gallagher, P. K., and B. Prescott, *Inorg. Chem.* **9,** 2510 (1970).

219. Gallagher, P. K., D. W. Johnson, and F. Schrey, *J. Am. Ceram. Soc.,* **53,** 666 (1970).

220. Gallagher, P. K., D. W. Johnson, Jr., E. M. Vogel, G. K. Wertheim and F. J. Schnettler in H. Chihara, Ed., *Thermal Analysis,* Heyden, London, 1977, p. 194.

221. Gallagher, P. K., and F. Schrey, *Thermochim. Acta,* **1,** 465 (1970).

222. Gallagher, P. K., F. Schrey, and B. Prescott, *Thermochim. Acta,* **2,** 405 (1971).

223. Gallagher, P. K., and F. Schrey in H. G. Wiedemann, Ed., *Thermal Analysis,* Vol. 2, Birkhäuser Verlag, Basel, 1972, p. 623.

224. Gallagher, P. K., W. R. Sinclair, R. A. Fastnacht, and J. P. Luongo, *Thermochim. Acta,* **8,** 141 (1974).

225. Gallagher, P. K., and J. P. Luongo, *Thermochim. Acta,* **12,** 159 (1975).

226. Gallagher, P. K., *Thermochim. Acta,* **14,** 131 (1976).

227. Gallagher, P. K., D. W. Johnson, E. M. Vogel, G. K. Wertheim, and F. J. Schnettler, *J. Solid State Chem.,* **21,** 277 (1977).

228. Gallagher, P. K., Private Comm., 1978.

229. Gallagher, P. K., *Thermochim. Acta* (in press).

230. Gallagher, P. K., *Thermochim. Acta,* **29,** 165 (1979).

231. Ganguli, P. S., and H. A. McGee, Jr., *J. Chem. Phys.,* **50,** 4658 (1969).

232. Garn, P. D., and J. E. Kessler, *Anal. Chem.* **33,** 952 (1961).

233. Garn, P. D., *Talanta,* **11,** 1417 (1964).

234. Garn, P. D., and G. D. Anthony, *Anal. Chem.,* **39,** 1445 (1967).

235. Garn, P. D., and G. D. Anthony, *Inst. Environ. Sci., Tech. Meet., Proc. 16th,* p. 308, 1970; *Chem. Abstr.,* **74,** 56268 (1971).

236. Garn, P. D., and R. L. Tucker, *J. Therm. Anal.,* **5,** 483 (1973).

237. Garn, P. D., and C. L. Denson, *Text. Res. J.*, **47**, 485 (1977); *Chem. Abstr.*, **87**, 69649 (1977).

238. Garton, A., and M. H. George, *J. Polym. Sci., Polym. Chem. Ed.*, **12**, 2779 (1974).

239. Gaulin, C. A., F. M. Wachi, and T. H. Johnston, in R. F. Schwenker, Jr., and P. D. Garn, Eds., *Thermal Analysis*, Vol. 2, Academic, New York, 1969, p. 1453.

240. Gibson, E. K., Jr., and S. M. Johnson, *Proc. Lunar Sci. Conf.*, **2**, 1351 (1971).

241. Gibson, E. K., Jr., and S. M. Johnson, *Thermochim. Acta*, **4**, 49 (1972).

242. Gibson, E. K., Jr., and N. J. Hubbard, in D. Heymann, Ed., *Proc. Lunar Sci. Conf., 3rd*, **2**, 2003 (1972), MIT:Cambridge, Mass; *Chem. Abstr.*, **78**, 87320 (1973).

243. Gibson, E. K., Jr., and G. W. Moore, *Proc. Lunar Sci. Conf.*, **3**, 2029 (1972); *Chem. Abstr.*, **78**, 87322 (1973).

244. Gibson, E. K., Jr., and C. B. Moore, *Space Life Sci.*, **3**, 404 (1972); *Chem. Abstr.*, **78**, 100314 (1973).

245. Gibson, E. K., Jr., and G. W. Moore, *Proc. Lunar Sci. Conf., 4th*, **2**, 1577 (1973), Pergamon, New York, 1973; *Chem. Abstr.*, **81**, 66767 (1974).

246. Gibson, E. K., Jr., and G. W. Moore, *Science*, **179**, 69 (1973).

247. Gohlke, R. S., and H. G. Langer, *Anal. Chim. Acta*, **35**, 333 (1966).

248. Gohlke, R. S., and H. G. Langer, *Anal. Chem.*, **37**, 25A (1965).

249. Gohlke, R. S., and H. G. Langer, *Anal. Chim. Acta*, **36**, 530 (1966).

250. Goldstein, H. W., *Mass Spectrometric Analysis of the Pyrolysis Products of Polymeric Materials*, Rept. No. UCRL-13332, Lawrence Radiation Lab., Livermore, 1967; *Mass. Spectrometric Thermal Analysis of Polymers*, Rept. No. UCRL-13398, 1968.

251. Goldstein, H. W., Rept. No. UCRL-13337, Lawrence Radiation Lab., Livermore, Calif., 1967.

252. Gorbachev, V. M. in D. Dollimore, Ed., *Proc. Eur. Symp. Therm. Anal., 1st*, Heyden, London, 1976, p. 92.

253. Goshgarian, B. B., presented at the 17th Annual Conference on Mass Spectrometry and Allied Topics, Dallas, 1969.

254. Grassie, N., *The Chemistry of High Polymer Degradation Processes*, Butterworth's, London, 1956.

255. Grassie, N., *Degradation Stab. Polym. Proc. Plenary Main Lect. Int. Symp.*, 1974 (Pub. 1975), 1–22.

256. Gray, D. N., G. P. Shulman, and R. T. Conley, *J. Macromol. Sci.* (Chem.), **A1**, 395 (1967).

257. Grayson, M. A., and R. J. Conrads, *Anal. Chem.*, **42**, 456 (1970).

258. Grecz, N., S. Gal, and J. Sztatisz in D. Dollimore, Ed., *Proc. Eur. Symp. Therm. Anal., 1st*, Heyden, London, (1976), p. 57.

259. Griffiths, P. R., *Amer. Lab.*, **7**(3), 37–8, 40, 42, 44–5 (1975).

260. Groh, R., Gy. Gerey, L. Bartha, and J. I. Pankove, in I. Buzás, Ed., *Thermal Analysis*, Vol. 1, Akadémiai Kiadó, Budapest, 1975, p. 909.

261. Groten, B., "Pyrolysis-Gas Chromatography Effluent Analysis," in W. Lodding, Ed., *Gas Effluent Analysis*, Marcel Dekker, New York, 1967, p. 101.

262. Gruetzmacher, H. F., Silhan, and U. Schmidt, *Chem. Ber.*, **102**, 3230 (1969); *Chem. Abstr.*, **71**, 91584 (1969).

263. Gudzinowicz, B. J., "Detectors," in L. S. Ettre and A. Zlatkis, Eds., *The Practice of Gas Chromatography*, Interscience, New York, 1967, Chapter 5.

264. Guyot, A., M. Bert, and A. Michel, *Eur. Polym. J.*, **7**, 471 (1971).

265. Guyot, A., and M. Bert, *J. Appl. Polym. Sci.*, **17**, 753 (1973).

266. Györe, J., and M. Ecet, in I. Buzás, Ed., *Thermal Analysis*, Vol. 2, Akadémiai Kiadó, Budapest, 1975, p. 387.

267. Habersberger, K., and E. Alsdorf, in I. Buzás, Ed., *Thermal Analysis*, Vol. 3, Akadémiai Kiadó, Budapest (1975), p. 203.

268. Habersberger, K., and V. Balek, *Thermochim. Acta*, **4**, 457 (1972).

269. Habersberger, K., V. Balek, and J. Sramek, *Radiochem. Radioanal. Lett.*, **28**, 301 (1977); *Chem. Abstr.*, **86**, 164678 (1977).

270. Haddon, W. F., A. H. DiEdwardo, and F. Zitomer, presented at 17th Annual Conference on Mass Spectrometry and Allied Topics, Dallas, 1969.

271. Haglund, B. O. in D. Dollimore, Ed., *Proc. Eur. Symp. Therm. Anal., 1st*, Heyden, London, 1976, p. 415.

272. Hahn, O., *Applied Radiochemistry*, Cornell University Press, New York, 1936, p. 191.

273. Hahn, O., *J. Chem. Soc.*, 259 (1949).

274. Haines, W. E., and D. R. Latham, *Anal. Chem.*, **49**(5), 256R (1977).

275. Hamming, M. C., and N. G. Foster, *Interpretation of Mass Spectra of Organic Compounds*, Academic, New York, 1972.

276. Hanson, R. L., N. E. Vanderborgh, and D. G. Brookins, *Anal. Chem.*, **47**, 335 (1975).

277. Harangozo, M., J. Tolgyessy, T. Dillingerova, P. Naoum, and M. Magdi, *Radiochem. Radioanal. Lett.*, **23**(4), 333 (1975).

278. Harangozo, M., J. Tolgyessy, T. Dillingerova, P. Dillinger, and M. Kosik, *Radiochem. Radioanal. Lett.*, **28**, 309, (1977); *Chem. Abstr.*, **86**, 156117 (1977).

279. Harangozo, M., J. Tolgyessy, T. Dillingerova, P. Dillinger, and M. Kosik, *Radiochem. Radioanal. Lett.*, **28**, 315 (1977); *Chem. Abstr.*, **86**, 156118 (1977).

280. Hassel, R. L., *Amer. Lab.*, **9**(1), 35 (1977).

281. Hastie, J. W., D. W. Bonnell, and D. M. Sanders, presented at the 25th Annual Conference on Mass Spectrometry and Allied Topics, Washington, D.C., 1977, p. 217.

282. Hattman, E. A., H. Schultz, and W. E. McKinstry, *Anal. Chem.*, **49**(5), 176R (1977).

283. Heijne, L., and A. T. Vink, *Philips Tech. Rev.*, **30**(617), 166 (1969); *Chem. Abstr.*, **72**, 57049 (1970).

284. Herain, J., and M. Mokra, *Ropa Uhlie*, **15**, 547 (1973); *APIA*, **21**, 21-6238 (1974).

285. Herbstein, F. H., G. Ron, and A. Weissman, in H. G. Wiedemann, Ed., *Thermal Analysis*, Vol. 2, Birkhäuser Verlag, Basel, 1972, p. 281.

286. Heron, G. F., *Soc. Chem. Ind. (London) Monogr.*, **13**, 475 (1961).

287. Herzog, L. F., T. J. Eskew, M. Soble, J. Humenic, and J. P. Mannaerts, presented in the 25th Annual Conference on Mass Spectrometry and Allied Topics, Washington, D.C., 1977, p. 101.

288. Hilado, C. J., *Flammability Handbook for Plastics*, Technomic Publ. Co., Stamford, Conn., 1969.

289. Hileman, F. D., K. J. Voorhees, L. H. Wojcik, M. M. Birky, P. W. Ryan, and I. N. Einhorn, *J. Polym. Sci., Polym. Chem. Ed.*, **13**, 571 (1975).

290. Hindersinn, R. R. and G. Witschard, "The Importance of Intumescence and Char in Polymer Fire Retardance," in W. C. Kuryla and A. J. Papa, Eds., *Flame Retardancy of Polymeric Materials*, Vol. 4, Marcel Dekker, New York and Basel, 1978, p. 13.

291. Hitchcock, J. L., and P. F. Pelter, in I. Buzás, Ed., *Thermal Analysis*, Vol. 1, Akadémiai Kiadó, Budapest, 1975, p. 979.

292. Hobbs, A. P., "Pressure Measurement and Control," in F. D. Snell and C. L. Hilton, Eds., *Encyclopedia of Industrial Chemical Analysis*, Interscience, New York, 1966, p. 222.

293. Holmes, J. C., and F. A. Morrell, *Appl. Spectry.*, **11**, 86 (1957).

294. Honig, R. E., "Analysis of Solids by Mass Spectrometry," in W. L. Mead, Ed., *Advances in Mass Spectrometry*, Vol. 3, The Institute of Petroleum, London, 1966, p. 101.

295. Honig, R. E., and J. R. Woolstron, *Appl. Phys. Letters*, **2**, 138 (1963).

296. Horowitz, H. H., and G. Metzger, *Anal. Chem.*, **35**, 1464 (1967).

297. Hudgens, C. R., and G. Pish, *Anal. Chem.*, **37**, 414 (1965).

298. Hudson, G., and K. Biemann, presented at the 25th Annual Conference on Mass Spectrometry and Allied Topics, Washington, D.C., 1977, p. 549.

299. Hummel, H., German (W) Pat. 2,253,837 (May 9, 1974).

300. Hummel, D. O., H. D. Schueddemage, and K. Ruebenacker, *Monogr. Mod. Chem.*, **6**, 355 (1974).

301. Ichihara, A., *Kagaku Kojo*, **19**(12), 31 (1975); *Chem. Abstr.*, **84**, 139987 (1976).

302. Ina, K., D. B. Min, R. Peterson, and S. S. Chang, presented at the 25th Annual Conference on Mass Spectrometry and Allied Topics, Washington, D.C. 1977, p. 530.

303. Ingraham, T. R., in H. G. McAdie, Ed., *Proc. 1st Toronto Symp., Therm. Anal.*, Chem. Inst. Can., 1965, p. 81.

304. Ingraham, T. R., in H. G. McAdie, Ed., *Proc. 2nd Toronto Symp., Therm. Anal.*, Chem. Inst. Can., 1967, p. 21.

305. Ingraham, T. R., and D. Fraser, in H. G. McAdie, Ed., *Proc. 3rd Toronto Symp., Therm. Anal.*, Chem. Inst. Can., 1969, p. 101.

306. Ingraham, T. R., and P. Marier, *Can. J. Chem. Eng.*, **41**, 170 (1963).

307. Ingraham, T. R., "Thermal Conductivity Detectors and Their Application to Some Thermodynamic and Kinetic Measurements," in W. Lodding, Ed., *Gas Effluent Analysis*, Marcel Dekker, New York, 1967, p. 25.

308. Innes, W. B., *Adv. Chem. Ser.*, **143**, 14 (1975).

309. Isaev, R. N., Yu. A. Zakharov, and V. V. Bordachev, *Zh. Fiz. Khim.*, **41**, 2398 (1967); *Chem. Abstr.*, **67**, 120551 (1967).

310. Isaev, R. N., Yu. A. Zakharov, N. M. Kraskovich, and V. N. Chizhov, *Isv. Tomsk. Poltekh. Inst.*, **251**, 155 (1970); *Chem. Abstr.*, **75**, 122717 (1971).

311. Isaev, R. N., Yu. A. Zakharov, *Isv. Tomsk. Poltekh. Inst.*, **251**, 163 (1970); *Chem. Abstr.*, **75**, 80701 (1971).

312. Ishikawa, T., and K. Inouye, *J. Therm. Anal.*, **10**, 399 (1976).

313. Ishimura, H., C. Fukuhara, and K. Nakajima, *Mass Spectroscopy Japan*, **22**(2), 163 (1974).

314. Ishimura, H., and K. Isa, in H. Chihara, Ed., *Thermal Analysis*, Heyden, London, 1977, p. 488.

315. Isobe, E., and T. Nakajima, *Sen'i Gakkaishi*, **31**(3), T101-T103 (1975); *Chem. Abstr.*, **82**, 171610 (1974).

316. Izmailova, T. I., L. A. Vodol'skaya, M. P. Ryazanov, A. V. Popov, and A. A. Stepanenko, *Proizvod Shin, Rezinotekh, Asbesto-Tekh. Izdelii, Nauchno-Tekh. Sb.*, **8**, 23 (1973); *Chem. Abstr.*, **81**, 154171 (1974).

317. Jaky, K., and F. Solymosi, in I. Buzás, Ed., *Thermal Analysis*, Vol. 1, Akadémiai Kiadó, Budapest, 1975, p. 433.

318. Jellinek, H. G., and J. E. Clark, *Can. J. Chem.*, **41**, 355 (1963).

319. Jellinek, H. H. G., and K. Takada, *J. Polym. Sci., Polym. Chem. Ed.*, **15**, 2269 (1977).

320. Jeunehomme, M. L., and M. C. Johnson, U.S. Pat. 3,851,176 (Cl.250-343; GO1m) (Nov. 26, 1974).

321. Johnson, D. R., J. W. Smith, and A. W. Robb, "Thermal Characteristics of Shortite," *Bur. Mines Rept. Invest., R.I. 7862*, 9pp, 1974.

322. Johnson, D. R., and J. W. Smith, Rept. NTIS 196676, 1970; *Chem. Abstr.*, **75**, 51119 (1971).

323. Johnson, D. W., Jr., and P. K. Gallagher, *J. Am. Ceram. Soc.*, **54**, 461 (1971).

324. Johnson, D. W., Jr., and P. K. Gallagher, *Thermochim. Acta*, **6**, 333 (1973).

325. Judd, M. D., and M. I. Pope, in R. F. Schwenker, Jr., and P. D. Garn, Eds., *Thermal Analysis*, Vol. 2, Academic, New York, 1969, p. 1423.

326. Judd, M. D., and M. I. Pope, in H. G. Wiedemann, Ed., *Thermal Analysis*, Vol. 2, Birkhäuser Verlag, Basel, (1972), p. 777.

327. Judd, M. D., B. A. Plunkett, and M. I. Pope, *J. Therm. Analysis*, **5**, 555 (1974).

328. Junod, L. L., *Rept. NASA TND-8338, Oct. 1976*, p. 67, Lewis Res. Center, Cleveland, Ohio.

329. Jüntgen, H., and K. H. van Heek, *Fortschr. Chem. Forsch.*, **13**, 601 (1970).

330. Juvet, R. S., Jr., J. L. S. Smith, and K.-P. Li, *Anal. Chem.*, **44**, 49 (1972).

331. Kaczmarek, T. D., D. C. Phillips, and J. D. B. Smith, *Microchem. J.*, **22**, 15 (1977).

332. Kahul-Chlebowska, E., and J. Pysiak, in I. Buzás, Ed., *Thermal Analysis*, Vol. 1, Akadémiai Kiadó, Budapest, 1975, p. 671.

333. Kaneko, H., Y. Saito, M. Umeda, and K. Nagai, *Nippon Kagaku Kaishi*, **6**, 792 (1977).

334. Kaneko, H., Y. Saito, M. Umeda, and K. Nagai, *Nippon Kagaku Kaishi*, **6**, 798 (1977).

335. Kaposi, O., A. B. Kiss, M. M. Riedel, and T. Deutsch, in I. Buzàs, Ed., *Thermal Analysis*, Vol. 2, Akadémiai Kiadó, Budapest, 1975, p. 297.

336. Katagiri, Y., T. Shimada, S. Fukui, and S. Kanno, *Eisei Kagaku*, **20**(6), 322 (1974); *Chem. Abstr.*, **84**, 34800 (1976).

337. Keattch, C. J., *Analysis of Calcareous Materials*, Soc. of Chem. Ind. Monograph, No. 18, London, 1964, p. 279.

338. Kenyeres, S., P. Gado, M. Sajó, and K. Solymár, in I. Buzás, Ed., *Thermal Analysis*, Vol. 2, Akadémiai Kiadó, Budapest, 1975, p. 531.

339. Kenyeres, S., K. Solymár, M. Orbán-Kelemen, and K. Jónás, in I. Buzás, Ed., *Thermal Analysis*, Vol. 2, Akadémiai Kiadó, Budapest, 1975, p. 541.

340. Kenyon, A. S., in P. E. Slade, Jr., L. T. Jenkins, Eds., *Techniques and Methods of Polymer Evaluation*, Vol. 1, Marcel Dekker, New York, 1966, p. 228.

341. Kerr, G. T., and A. W. Chester, *Thermochim. Acta*, **3**, 113 (1972).

342. Khmel'nitskii, R. A., I. M. Lukashenko, G. A. Kalinkevich, V. A. Konchits, and E. S. Brodskii, *Izv. Timiryazevsk. Skh. Akad.* (6), 170 (1975); *Chem. Abstr.* **84**, 45053 (1976).

343. Kilzer, F. J. and A. Broido, "Speculations on the Nature of Cellulose Pyrolysis," WSS/CI Paper 64-4, U.S. Department of Agriculture, Washington, D.C., 1964.

344. Kinsbron, E., P. K. Gallagher, and A. T. English, Private Comm. (1978).

345. Kiran, E., and J. K. Gillham, in J. Chiu, Ed., *Polymer Characterization by Thermal Methods of Analysis*, Marcel Dekker, New York, 1974, p. 211.

346. Kiran, E., and J. K. Gillham, *Anal. Chem.*, **47**, 983 (1975).

347. Kiran, E., and J. K. Gillham, *J. Appl. Polym. Sci.*, **20**, 931 (1976); *Chem. Abstr.* **84**, 151152 (1976).

348. Kiran, E., and J. K. Gillham, *J. Appl. Polym. Sci.*, **20**, 2045 (1976); *Chem. Abstr.* **85**, 124573 (1976).

349. Kiran, E., J. K. Gillham, and E. Gibstein, *J. Appl. Polym. Sci.*, **21**, 1159 (1977).

350. Kiran, E., and J. K. Gillham, *J. Macromol. Sci., Chem.*, **8**, 211 (1974).

351. Kiss, A. B., *Acta Chim. Acad. Sci. Hung.*, **61**, 207 (1969).

352. Kistiakowsky, G. B., and P. H. Kydd, *J. Am. Chem. Soc.*, **79**, 4825 (1957).

353. Kleineberg, G. A., and D. L. Geiger, in H. G. Wiedemann, Ed., *Thermal Analysis*, Vol. 1, Birkhäuser Verlag, Basel, 1972, p. 325.

354. Knight, G. J., and W. W. Wright, *Br. Polym. J.*, **5**, 396 (1973).

355. Kohl, F. J., R. A. Miller, C. A. Stearns, and G. C. Fryburg, presented at the 25th Annual Conference on Mass Spectrometry and Allied Topics, Washington, D.C., 1977, p. 223.

356. Korobeinichev, O. P., V. V. Boldyrev, and Yu. Ya. Karpenko, *The Physics of Combustion and Explosion, USSR*, **4**, 33 (1968).

357. Korobeinichev, O. P., and Yu. Ya. Karpenko, *Isv. Akad. Nauk. SSSR, Ser. Khim.*, 1557 (1971); *Chem. Abstr.*, **75**, 122718 (1971).

358. Korobeinichev, O. P., A. S. Shmelev, V. G. Voronov, and G. I. Anisiforov, in I. Buzás, Ed., *Thermal Analysis*, Vol. 1, Akadémiai Kiadó, Budapest, 1975, p. 77.

359. Koroskys, M. J., *Am. Dye Rep.*, March 24, 1969, p. 15.

360. Košik, M., M. Vanko, L. Sismis, and J. Spanik, *Czech.* **135**, 842, 15 Mar 1971; *Chem. Abstr.*, **74**, 134684 (1971).

361. Košik, M., V. Reiser, and P. Kováč, in I. Buzás, Ed., *Thermal Analysis*, Vol. 2, Akadémiai Kiadó, Budapest, 1975, p. 229.

362. Krapf, G., J. L. Lutz, L. M. Melnick, and W. R. Bandi, *Thermochim. Acta*, **4**, 257 (1972).

363. Krapf, G., W. R. Bandi, and L. M. Melnick, *J. Iron Steel Inst.*, **211**, 890 (1973).

364. Krapf, G., E. G. Buyok, and W. R. Bandi, *Thermochim. Acta*, **13**, 47 (1975).

365. Krc, A. E., U.S. Pat. 3,861,874 (Jan. 21, 1975).

366. Krishen, A., and R. G. Tucker, *Anal. Chem.*, **46**, 29 (1974).

367. Krug, D., *Chem.-Ing.-Tech.*, **46**, 839 (1974); *Chem. Abstr.* **82**, 22510 (1975).

368. Krug, D., and W. Haedrich, *Thermochim. Acta*, **15**, 179 (1976).

369. Krupay, B. W., and R. A. Ross, *Can. J. Chem.*, **51**, 3520 (1973).

370. Krzyszowska, R., in I. Buzàs, Ed., *Thermal Analysis*, Vol. 2, Akadémiai Kiadó, Budapest, 1975, p. 159.

371. Krzyzanowski, S., M. Malinowski, and P. Wierzchowski, in I. Buzás, Ed., *Thermal Analysis*, Vol. 3, Akadémiai Kiadó, Budapest, 1975, p. 233.

372. Kubas, Z., and M. Szalkowicz, in H. G. Wiedemann, Ed., *Thermal Analysis*, Vol. 2, Birkhäuser Verlag, Basel, 1972, p. 447.

373. Kubas, Z., and J. Orewczyk, in I. Buzás, Ed., *Thermal Analysis*, Vol. 1, Akadémiai Kiadó, Budapest, 1975, p. 517.

374. Laine, N. R., F. J. Vastola, and P. L. Walker, *Proc. Conf. Carbon, 5th*, **2**, 211 (1963).

375. Langer, H. G., and R. S. Gohlke, *Anal. Chem.*, **35**, 1301 (1963).

376. Langer, H. G., R. S. Gohlke, and D. H. Smith, *Anal. Chem.*, **37**, 433 (1965).

377. Langer, H. G., and R. S. Gohlke, "Mass Spectrometric Identification of Gaseous Products from Thermal Analysis," in W. Lodding, Ed., *Gas Effluent Analysis*, Marcel Dekker, New York, 1967, p. 71.

378. Langer, H. G., and F. Karle, presented at the 15th Annual Conference on Mass Spectrometry and Allied Topics, Denver, 1967.

379. Langer, H. G., presented at the 17th Annual Conference on Mass Spectrometry and Allied Topics, Dallas, 1969, p. 219.

380. Langer, H. G., and T. P. Brady, in R. F. Schwenker, Jr., and P. D. Garn, Eds., *Thermal Analysis*, Vol. 1, Academic, New York, 1969, p. 295.

381. Langer, H. G. and T. P. Brady, *Thermochim. Acta*, **5**, 391 (1973).

382. Langer, H. G., T. P. Brady, M. D. Rausch and H. B. Gordon, *Proc., 16th Annual Conf. on Mass Spectrometry and Allied Topics*, Pittsburgh, Penna., 1968, p. 138.

383. Langer, H. G., and R. S. Gohlke, *Fortschr. Chem. Forsch.*, **6**, 515 (1966).

384. Langer, H. G. in H. G. McAdie, Ed., *Proc. 2nd Toronto Symp., Therm. Anal.*, Chem. Inst. Can., 1967, p. 137.

385. Langer, H. G., et. al. U.S. Pat. 3,560,627; 3, 623,355; 3,629,888; 3,629,889; 3,634,591; 3,667,278; 3,667,279; 3,685,344; 3,773,963; 3,773,964; 3,888,107.

386. Langer, H. G. Unpublished data.

387. Lanser, A. C., J. O. Ernst., W. F. Kwolek, and H. J. Dutton, *Anal. Chem.*, **45**, 2344 (1973).

388. Lambert, N. W., *Anal. Chem.*, **49**(5), 246R (1977).

389. Larsen, E. R., *J. Fire and Flammability/Fire Retardant Chemistry*, **1**, 4 (1974).

390. Leathard, D. A., and B. C. Shurluck, *Identification Techniques in Gas Chromatography*, Wiley-Interscience, New York, 1970, p. 66.

391. Lehmann, E. I., and D. M. Cavagnaro, Rept. NTIS/P5-76/0915, Nov. 1976.

392. Lehmann, F. A., and G. M. Brauer, *Anal. Chem.*, **33**, 673 (1961).

393. Leigh, D., and N. Lynaugh, *Adv. Mass Spectrom*, **6**, 463 (1974).

394. LePlat, G. P. A., German (W) Pat. 2,449,143 (Oct. 16, 1973).

395. Levy, A., J. M. Bregeault, and G. Pannetier, *Analysis*, **1**, 278 (1972); *Chem. Abstr.*, **78**, 32220 (1973).

396. Levy, R. L., in M. Lederer, Ed., *Chromatographic Reviews*, Vol. 8, Elsevier, Amsterdam, 1966, p. 48.

397. Levy, R. L., and D. L. Fanter, *Anal. Chem.*, **41**, 1465 (1969).

398. Levy, R. L., D. L. Fanter, and C. J. Wolf, *Anal. Chem.*, **44**, 38 (1972).

399. Lewin, M. and A. Basch, "Structure, Pyrolysis and Flammability of Cellulose," in M. Lewin, S. M. Atlas, and E. M. Pearce, Eds., *Flame-Retardant Polymeric Materials*, Vol. 2, Plenum, New York and London, 1978.

400. Liberti, A., L. Conti, and V. Crescenzi, *Nature*, **178**, 1067 (1956).

401. Lincoln, K. A., *Rev. Sci. Instrum.*, **35**, 1688 (1964).

402. Lincoln, K. A., USNRDL Rept. No. TR-731, 1964.

403. Lincoln, K. A., *Anal. Chem.*, **37**, 541 (1965).

404. Lincoln, K. A., and D. Werner, presented at 15th Annual Conf. on Mass Spec. and Allied Topics, Denver, 1967.

405. Lincoln, K. A., *Int. J. Mass Spect. Ion Phys.*, **2**, 75 (1969).

406. Lindsey, J. W., H. N. Robinson, H. L. Bramlet, and A. J. Johnson, *J. Inorg. Nucl. Chem.*, **32**, 1559 (1970).

407. Link, W. T., and E. A. McClatchie, U.S. Pat. 3,864,613 (C1.250-343; GOln) (Mar. 4, 1975).

408. Lippiatt, J. H., D. Price, R. W. Brown, and D. C. A. Izod, in D. Dollimore, Ed., *Proc. Eur. Symp. Therm. Anal., 1st*, Heyden, London, 1976, p. 280.

409. Lipska, A. E., and F. A. Wodley, *U.S. Clearinghouse Fed. Sci. Tech. Inform., AD 1968*, AD 676351; *Chem. Abstr.*, **70**, 107630 (1969).

410. Lipska, A. E., and F. A. Wodley, *J. Appl. Polymer Sci.*, **13**, 851 (1969).

411. Lipsky, S. R., C. G. Horvath, and W. J. McMurray, *Anal. Chem.*, **38**, 1585 (1966).

412. Lleras, J., M. Bernard, and S. Combet, *C. R. Hebd. Seances Acad. Sci., Ser. C*, **283**, 405 (1976); *Chem. Abstr.*, **86**, 30267 (1977).

413. Llewellyn, P. M., and D. P. Littlejohn, presented at Pittsburgh Conference on Analytical Chemistry and Applied Spectroscopy, February 1966.

414. Locke, C. E., and R. L. Stone, in R. F. Schwenker, Jr., and P. D. Garn, Eds., *Thermal Analysis*, Vol. 2, Academic, New York, London, 1969, p. 963.

415. Lodding, W., and L. Hammell, *Rev. Sci. Instrum.*, **30**, 885 (1959).

416. Lodding, W., and L. Hammell, *Anal. Chem.*, **32**, 657 (1960).

417. Lodding, G. W., "Principles and General Instrumentation," in W. Lodding, Ed., *Gas Effluent Analysis*, Marcel Dekker, New York, 1967, p. 1.

418. Lodding, W., "Selective Sorption and Condensation of Effluent Gases," in W. Lodding, Ed., *Gas Effluent Analysis*, Marcel Dekker, New York, 1967, p. 143.

419. Loehr, A. A., and P. F. Levy, *Amer. Lab.*, **4**, (1), 11 (1972).

420. Lorant, B., *Seifen-Öle-Fette-Wachse*, **93**, 547 (1967).

421. Lord, H. C., *Anal. Methods Appl. Air Pollut. Meas.*, 233 (1974).

422. Lord, F. M., and J. S. Kittelberger, *Surf. Sci.*, **43**, 173 (1974); *Chem. Abstr.*, **81** 6435 (1974).

423. Loveland, J. W., and C. N. White, *Anal. Chem.*, **49**(5), 262R (1977).

424. Lovelock, J. E., *J. Chromatogr.*, **1**, 35 (1958).

425. Lovelock, J. E., *Nature*, **182**, 1663 (1958).

426. Lovelock, J. E., and S. R. Lipsky, *J. Am. Chem. Soc.*, **82**, 431 (1960).

427. Low, M. J. D., "Analysis of Gas Effluent Streams by Infrared Absorption," in W. Lodding, Ed., *Gas Effluent Analysis*, Marcel Dekker, New York, 1967, p. 155.

428. Lum, R. M., *J. Appl. Polym. Sci.*, **20**, 1635 (1976).

429. Lum, R. M., *J. Appl. Polym. Sci., Polym. Chem. Ed.*, **15**, 489 (1977).

430. Lum, R. M., *Thermochim. Acta*, **18**, 73 (1977).

431. Lum, R. M., *J. Appl. Polym. Sci.*, **23**, 1247 (1979).

432. Lum, R. M., *J. Appl. Polym. Sci., Polym. Chem. Ed.*, **17**, 203 (1979).

433. Lum, R. M., *J. Appl. Polym. Sci., Polym. Chem. Ed.*, **17**, 3017 (1979).

434. Lumme, P., and M. L. Korvola, *Thermochim. Acta*, **13**, 419 (1975).

435. MacCallum, J. R., and C. K. Schoff, *Polym. Lett.*, **9**, 395 (1971).

436. Macek, J., A. Rahten, and J. Slivnik, in D. Dollimore, Ed., *Proc. Eur. Symp. Therm. Anal., 1st*, Heyden, London, (1976), p. 161.

437. MacKenzie, K. J. D., *J. Inorg. Nucl. Chem.*, **32**, 3731 (1970).

438. MacKenzie, R. C., *Talanta*, **16**, 1227 (1969).

439. MacKenzie, R. C., *Differential Thermal Analysis*, Vol. 1, Academic Press, London, 1970.

440. MacKenzie, R. C., *J. Polym. Sci., Polym. Lett. Ed.*, **12**, 523 (1974).

441. MacKenzie, R. C., C. J. Keatch, T. Daniels, D. Dollimore, J. A. Forrester, J. P. Redfern, and J. H. Sharp, *J. Therm. Anal.*, **8**, 197 (1975).

442. MacKenzie, R. C., in H. Chihara, Ed., *Thermal Analysis*, Heyden, London, 1977, p. 559.

443. Madorsky, S. L., V. E. Hart, S. Straus, and V. A. Sedlak, *J. Res. Nat. Bur. Stand.*, **51**, 327 (1953).

444. Madorsky, S. L., and S. Straus, *Soc. Chem. Ind. (London) Monogr.*, **13**, 60 (1961).

445. Madorsky, S. L., *Thermal Degradation of Organic Polymers*, Interscience, New York, 1964.

446. Makino, K., T. Takekawa, and K. Ouchi, in H. Chihara, Ed., *Thermal Analysis*, Heyden, London, 1977, p. 52.

447. Malinowski, M., and S. Krzyzanowski, in D. Dollimore, Ed., *Proc. Eur. Symp. Therm. Anal., 1st*, Heyden, London, 1976, p. 128.

448. Manning, N. J., and D. V. Nowell, in D. Dollimore, Ed., *Proc. Eur. Symp. Therm. Anal., 1st,* Heyden, London, 1976, p. 317.

449. Maragil, R. Z., and L. E. Ovintitskikh, *Tr. Tyumen. Ind. Inst., 1972,* p. 305.

450. Marik, P., E. Buzagh, J. Inczedy, J. Paulik, and L. Erdey, *Proc. Anal. Chem. Conf., 3rd,* **2,** 235 (1970); *Chem. Abstr.,* **74,** 71371 (1971).

451. Martin, A. J. P., and A. T. James, *Biochem. J.,* **63,** 138 (1956).

452. Martynov, A. A., G. M. Usacheva, V. G. Khasanov, G. P. Kurbskii, and M. S. Vigder-gauz, *Usp. Gazov. Khromatogr.,* 1973, p. 133.

453. Mathew, R. D., O. I. Smith, N. J. Brown, and R. F. Sawyer, West. States Sect., Combust. Inst., (Pap.), WSCI 75-16, 21 pp. (1975).

454. Matthews, D. E., and J. M. Hayes, presented at the 25th Annual Conference on Mass Spectrometry and Allied Topics, Washington, D.C., 1977, p. 98.

455. Maugh, T. H., II, *Science,* **194,** 1403 (1976).

456. Mavlyanov, A. M., T. M. Muinov, and Z. A. Kobilov, *Dokl. Akad. Nauk Tadzh. SSR,* **14,** 24 (1971); *Chem. Abstr.,* **76,** 141470 (1972).

457. Maycock, J. N., V. R. P. Verneker, and P. W. M. Jacobs, *J. Chem. Phys.,* **46,** 2857 (1967).

458. Maycock, J. N., and V. R. P. Vernecker, *Anal. Chem.,* **40,** 1935 (1968).

459. Maycock, J. N., and V. R. P. Vernecker, *Thermochim. Acta,* **1,** 191 (1970).

460. McAdie, H. G., *Anal. Chem.,* **44,** 640 (1972).

461. McGaugh, M. C., and S. Kottle, *Polym Lett.,* **5,** 719 (1967).

462. McGaugh, M. C., and S. Kottle, *J. Appl. Polym. Sci.,* **12,** 1981 (1968).

463. McGuchan, R., and I. C. McNeill, *Eur. Polym. J.,* **3,** 511 (1967).

464. McKeag, R. G., and F. W. Hougen, *Anal. Chem.,* **49,** 1078 (1977).

465. McKewan, W. M., "Kinetics of Reduction of Iron Ores," in *The Chipman Conference on Steel Making,* M.I.T. Press, Cambridge, Mass., 1962.

466. McNeill, I. C., *J. Polym. Sci.,* Part A-4, 2479 (1966).

467. McNeill, I. C., *Eur. Polym. J.,* **3,** 409 (1967).

468. McNeill, I. C., in R. F. Schwenker, Jr., and P. D. Garn, Eds., *Thermal Analysis,* Vol. 1, Academic, New York, 1969, p. 417.

469. McNeill, I. C., *Eur. Polym. J.,* **6,** 373 (1970).

470. McNeill, I. C., in H. G. Wiedemann, Ed., *Thermal Analysis,* Vol. 3, Birkhäuser Verlag, Basel, 1972, p. 229.

471. McNeill, I. C., L. Ackerman, S. N. Gupta, M. Zulfiqar, and S. Zulfiqar, *J. Polym. Sci., Polym. Chem. Ed.,* **15,** 2381 (1977).

472. McNeill, I. C., and D. Neil, in R. F. Schwenker, Jr., and P. D. Garn, Eds., *Thermal Analysis,* Vol. 1, Academic, New York, 1969, p. 353.

473. McWilliam, I. G., and R. A. Dewar, *Nature,* **181,** 760 (1958).

474. Mehmet, Y., R. S. Roche, *J. Appl. Polym. Sci.,* **20,** (7), 1955 (1976).

475. Mehta, S. K., *Trans. Indian Ceram. Soc.,* **34,** 65 (1976); *Chem. Abstr.,* **85,** 98476 (1976).

476. Meisel, T. and Z. Halmos, in H. G. Wiedemann, Ed., *Thermal Analysis,* Vol. 3, Birk-häuser Verlag, Basel, 1972, p. 43.

477. Meisel, T., Cs. Mélykuti, and Z. Halmos, in I. Buzás, Ed., *Thermal Analysis,* Vol. 2, Akadémiai Kiadó, Budapest, 1975, p. 371.

478. Meot-ner, M., E. P. Hunter and F. H. Field, Proc. 25th Annual Conference on Mass Spectrometry and Allied Topics, Washington, D.C. 1977, p. 616.

479. Mercier, J. C. in I. Buzás, Ed., *Thermal Analysis,* Vol. 3, Akadémiai Kiadó, Budapest, 1975, p. 1041.

480. Merritt, C., Jr., and C. DiPietro, *Anal. Chem.*, **44**, 57 (1972).

481. Merritt, C., Jr., and D. H. Robertson, *Anal. Chem.*, **44**, 60 (1972).

482. Meschi, D. J., W. A. Chupka, and J. Berkowitz, *J. Chem. Phys.*, **33**, 530 (1960).

483. Meuzelaar, H. L. C., and P. G. Kistemaker, *Anal. Chem.*, **45**, 587 (1973).

484. Meuzelaar, H. L. C., P. G. Kistemaker, W. Eshuis, and M. A. Posthumus, presented at the 25th Annual Conference on Mass Spectrometry and Allied Topics, Washington, D.C., 1977.

485. Milina, R., and M. Pankova, *Faserforsch. Textiltech.*, **28**(5), 217 (1975).

486. Miller, D. O., *Anal. Chem.*, **35**, 2033 (1963).

487. Mills, K. C., *J. Chem. Soc., Faraday Trans. 1*, **70**, 2224 (1974).

488. Miromoto, T., K. Takeyama, and F. Konishi, *J. Appl. Polym. Sci.*, **20**, 1967 (1976).

489. Mizutani, N., and M. Kato, *Anal. Chem.*, **47**, 1389 (1975).

490. Mol, G. J., *Thermochim. Acta*, **10**, 259 (1974).

491. Mol, G. J., R. J. Gritter, and G. E. Adams, *Am. Chem. Soc. Polym. Prepr.*, **17**(2), 758 (1976).

492. Mol, G. J., 1977 ACS/SAS Pacific Conf., "Evolved Gas Analysis of Greases and Polymers by Mass Spectrometry."

493. Montaudo, G., M. Przybski, and H. Ringsdorf, *Makromol. Chem.*, **176**(6), 1763 (1975).

494. Morgan, D. J., in D. Dollimore, Ed., *Proc. Eur. Symp. Therm. Anal., 1st,* Heyden, London, 1976, p. 355.

495. Morgan, P., and W. W. Wright, in D. Dollimore, Ed., *Proc. Eur. Symp. Therm. Anal., 1st,* Heyden, London, 1976, p. 164.

496. Morisaki, S., *Thermochim. Acta*, **9**, 157 (1974).

497. Morisaki, S. in H. Chihara, Ed., *Thermal Analysis*, Heyden, London, 1977, p. 297.

498. Morris, C. J. O. R., *Lab Pract.*, **23**, 513 (1974).

499. Muenow, D. W., and J. L. Margrave, presented at the 17th Annual Conference on Mass Spectrometry and Allied Topics, Dallas, 1969, p. 217.

500. Müller-Vonmoos, M., and R. Bach, in R. F. Schwenker, Jr., and P. D. Garn, Eds., *Thermal Analysis*, Vol. 2, Academic, New York, 1969, p. 1229.

501. Müller-Vonmoos, M., and R. Müller, in I. Buzás, Ed., *Thermal Analysis*, Vol. 2, Akadémiai Kiadó, Budapest, 1975, p. 521.

502. Munson, M. S. B., *Anal. Chem.*, **43**, 29A, (1971).

503. Murat, M., A. Fevre, and C. Comel, in D. Dollimore, Ed., *Proc. Eur. Symp. Therm. Anal., 1st,* Heyden, London, 1976, p. 98.

504. Murata, T., S. Takahashi, T. Takeda, A. Tsuyama, and K. Kageyama, *Shimadzu Hyoron*, **29**, 191 (1972); *Chem. Abstr.*, **81**, 78412 (1974).

505. Murata, T., and S. Takahashi, *Shitsuryo Bunseki*, **22**, 87 (1974); *Chem. Abstr.*, **82**, 4831 (1975).

506. Murdoch, I. A., and L. J. Rigby, in D. Price, Ed., *Dynamic Mass Spectrometry*, Vol. 3, Heyden & Sons, London, 1972, p. 255.

507. Murphy, C. B., J. A. Hill, and G. P. Schacher, *Anal. Chem.*, **32**, 1374 (1960).

508. Murphy, C. B., and C. D. Doyle, *Appl. Polym. Symp.*, **2**, 77 (1966).

509. Murphy, C. B., "Thermoparticulate Analysis," in W. Lodding, Ed., *Gas Effluent Analysis,* Marcel Dekker, New York, 1967, p. 195.

510. Murphy, C. B., *Instr. Control Systems*, **38**, (8), 101 (1965).

511. Murphy, C. B., F. W. VanLuik, Jr., and A. C. Pitsas, *Plastic Design Process.*, **4**, (7), 16 (1964).

512. Murphy, C. B., F. W. VanLuik, Jr., and G. F. Skala, American Industrial Hygienist Association Meeting, Houston, Texas, May 1965.

513. Nagase, K., *Bull. Chem. Soc. Jap.*, **46**, 144 (1973); *Chem. Abstr.*, **78**, 91987 (1973).

514. Nanjo, M., and K. Taniuchi in H. Chihara, Ed., *Thermal Analysis*, Heyden, London, 1977, p. 329.

515. Nelson, L. S., and N. A. Kuebler, "Heterogeneous Flash Pyrolysis of Hydrocarbon Polymers," in A. L. Myerson and A. C. Harrison, Eds., *Physical Chemistry in Aerodynamics and Space Flight*, Pergamon, Oxford, 1961, p. 61.

516. Nemirovskaya, I. B., V. G. Berezkin, and B. M. Kovarskaya, *Vysokomol. Soedin Ser. A*, **17**(3), 675 (1975).

517. Nerheim, A. G., *Anal. Chem.*, **35**, 1640 (1963).

518. Nesbitt, L. E., and W. W. Wendlandt, *Thermochim. Acta*, **10**, 85 (1974).

519. Nesbitt, L. E., and W. W. Wendlandt, in I. Buzás, Ed., *Thermal Analysis*, Vol. 3, Akadémiai Kiadó, Budapest, 1975, p. 693.

520. Nesmelova, Z. N., and Yu. I. Sokolov, *Tr. Vses. Neft. Nauchno-Issled. Geologarazved. Inst.*, **355**, 137 (1974); *Chem. Abstr.* **84**, 7280 (1976).

521. Newman, R. H., D. F. Glenn, R. W. Jenkins, Jr., and G. E. Lester in H. G. Wiedemann, Ed., *Thermal Analysis*, Vol. 3, Birkhäuser Verlag, Basel, 1972, p. 57.

522. Newton, J. C., R. W. Crawford, and R. K. Stump, Report, UCID-16823, 1975, 17 pp. Avail. NTIS.

523. Nikolaev, A. V., V. A. Logvinenko, V. M. Gorbatchov, L. I. Myachina, and N. N. Knyazeva, in H. G. Wiedemann, Ed., *Thermal Analysis*, Vol. 2, Birkhäuser Verlag, Basel, 1972, p. 667.

524. Nikolaev, A. V., V. A. Logvinenko, V. M. Gorbatchov, and L. I. Myachina, in I. Buzás, Ed., *Thermal Analysis*, Vol. 1, Akadémiai Kiadó, Budapest, 1975, p. 47.

525. Nikolaev, A. V., and V. A. Logvinenko, in H. Chihara, Ed., *Thermal Analysis*, Heyden, London, 1977, p. 364.

526. Nishi, S., and R. Kobayashi, *Raba Daijesuto*, **28**, 2 (1976); *Chem. Abstr.*, **86**, 190803 (1977).

527. Noble, W., B. B. Wheals, and M. M. Whitehouse, *Forensic Sci.*, **3**(2), 163 (1974).

528. Norris, A. C., M. I. Pope, M. Selwood, and M. D. Judd, in I. Buzás, Ed., *Thermal Analysis*, Vol. 1, Akadémiai Kiadó, Budapest, 1975, p. 65.

529. Norris, A. C., M. I. Pope, and M. Selwood, in D. Dollimore, Ed., *Proc. Eur. Symp. Therm. Anal., 1st*, Heyden, London, 1976, p. 79.

530. Norris, A. C., M. I. Pope, and M. Selwood, in D. Dollimore, Ed., *Proc. Eur. Symp. Therm. Anal., 1st*, Heyden, London, 1976, p. 83.

531. Notz, K. J., Jr., U.S. At. Energy Comm. Res. Develop. Rept., NLCO-814, 1960.

532. Oguri, T., in *Proc. Int. Vac. Congr., 7th*, R. Dobrozemski, F. Ruedenauer, and F. P. Viehboek, Eds., R. Dobrozemsky, Vienna, 1977; 1, 149–52; *Chem. Abstr.*, **88**, 122965 (1978).

533. O'Mara, M. M., *Pure Appl. Chem.*, **49**, 649 (1977).

534. Onchi, M., and E. Ma, *J. Phys. Chem.*, **67**, 2240 (1963).

535. Onchi, M., *Nippon Kagaku Zasshi*, **85**, 612 (1964).

536. Onchi, M., and I. Kusunoki, *Nippon Kagaku Zasshi*, **85**, 617 (1964).

537. Onodera, S., *Bull. Chem. Soc. Jap.*, **50**, 123 (1977); *Chem. Abstr.*, **86**, 114747 (1977).

538. Oppermann, H., W. Reichelt, and E. Wolf, in I. Buzás, Ed., *Thermal Analysis*, Vol. 1, Akadémiai Kiadó, Budapest, 1975, p. 403.

539. Orcel, J., and S. Caillere, *Bull. Soc. Franc. Mineral.*, **50**, 75 (1927).

540. Osada, H., and E. Sakamato, *Kogyo Kayaku Kyokaishi*, **24**, 236 (1963); *Chem. Abstr.*, **60**, 7669 (1964).

541. Ota, K., *Yuki Gosei Kagaku Kyokai Shi*, **29**, 796 (1971); *Chem. Abstr.*, **76**, 16021 (1972).

542. Oyama, V. I., *Nature*, **200**, 1058 (1963).

543. Ozawa, T., *Bull. Chem. Soc. Japan*, **38**, 1881 (1965).

544. Ozawa, T., *J. Thermal. Anal.*, **2**, 301 (1970).

545. Pacewska, B., and J. Pysiak, in I. Buzás, Ed., *Thermal Analysis*, Vol. 1, Akadémiai Kiadó, Budapest, 1975, p. 695.

546. Paciorek, K. L., R. H. Kratzer, J. Kaufman, J. Nakahara, and A. M. Hartstein, *J. Appl. Polym. Sci.*, **18**, 3723 (1974).

547. Pal, K., and M. Fodor, *Muanyag Gumi*, **14**, 104 (1977); *Chem. Abstr.*, **87**, 152779 (1977).

548. Papazian, H. A., P. J. Pizzolato, and R. R. Orrell, *Thermochim. Acta*, **4**, 97 (1972).

549. Parrish, J. R., *Anal. Chem.*, **45**, 1659 (1973).

550. Parrish, J. R., *Anal. Chem.*, **47**, 1999 (1975).

551. Pascual, V. L., *Anal. Chem.*, **47**, 2067 (1975).

552. Paul, D. G., D. J. Brindle, and J. A. Wegener, *Pittsburgh Conf. Anal. Chem. Appl. Spectr.*, Cleveland, Ohio, 1972.

553. Paulik, F., J. Paulik, and L. Erdey, *Z. Anal. Chem.*, **160**, 241 (1958).

554. Paulik, F., J. Paulik, and L. Erdey, *Talanta*, **13**, 1405 (1966).

555. Paulik, F., and J. Paulik, *Thermochim. Acta*, **3**, 17 (1971).

556. Paulik, F., and J. Paulik, *Anal. Chim. Acta*, **60**, 127 (1972).

557. Paulik, F., and J. Paulik, *J. Thermal Anal.*, **5**, 253 (1973).

558. Paulik, F., and J. Paulik, *J. Thermal Anal.*, **8**, 557 (1975).

559. Paulik, J., and F. Paulik, *J. Thermal Anal.*, **8**, 567 (1975).

560. Paulik, J., F. Paulik, and L. Erdey, *Microchim. Acta*, 866 (1966).

561. Paulik, J., F. Paulik, and L. Erdey, *Anal. Chim. Acta*, **44**, 153 (1969).

562. Paulik, J., and F. Paulik, in H. G. Wiedemann, Ed., *Thermal Analysis*, Vol. 1, Birkhäuser Verlag, Basel, 1972, p. 489.

563. Paulik, J., and F. Paulik, *Talanta*, **17**, 1224 (1970).

564. Paulik, J. and F. Paulik, *Anal. Chim. Acta*, **56**, 328 (1971).

565. Paulik, J., and F. Paulik, *Thermochim. Acta*, **3**, 13 (1971).

566. Paulik, J., and F. Paulik, *Thermochim. Acta*, **4**, 189 (1972).

567. Paulik, J., and F. Paulik, in I. Buzás, Ed., *Thermal Analysis*, Vol. 3, Akadémiai Kiadó, Budapest, 1975, p. 789.

568. Payer, K., J. Heiszman, S. Békássy, and J. Petró, in I. Buzás, Ed., *Thermal Analysis*, Vol. 3, Akadémiai Kiadó, Budapest, 1975, p. 847.

569. Permyakov, V. M., *Radioaktivnye Emanatsii*, Izvestiya Akademii Nauk SSR, Moscow, 1963.

570. Petersen, D. R., H. W. Rinn, and S. T. Sutton, *J. Phys. Chem.*, **68**, 3057 (1964).

571. Petrella, R. V., "Factors Affecting Combustion of Polystyrene and Styrene" in M. Lewin, S. M. Atlas, and E. M. Pearce, Eds., *Flame-Retardant Polymeric Materials*, Vol. 2, Plenum, New York and London, 1978, p. 185.

572. Pfeil, R. W. in H. G. McAdie, Ed., *Proc. 3rd Toronto Symp., Therm. Anal.*, Chem. Inst. Can., 1969, p. 187.

573. Phillips, D. C., and J. D. B. Smith, *Chem. Instrum.*, **7**, 261 (1976).

574. Pignataro, S., and F. P. Lossing, *J. Organomet. Chem.*, **11**, 571 (1968).

575. Piloyan, G. O., and O. S. Novikova, *Zh. Neorg. Khim.*, **12**, 602 (1967).

576. Pirani, M. von, *Verhandl. Deut. Physik Ges.*, **4**, 686 (1906).

577. Polyakova, A. A., L. O. Koean, G. E. Tsigel'man, A. F. Kuz'min, and A. E. Rafal'son, *Khim. Tekhnol. Topl. Masel.*, **3**, 58 (1974).

578. Poppl. L., in D. Dollimore, Ed., *Proc. Eur. Symp. Therm. Anal., 1st,* Heyden, London, 1976, p. 237.

579. Poppl, L., M. Gabor, J. Wajand, and Z. G. Szabo in D. Dollimore, Ed., *Proc. Eur. Symp. Therm. Anal., 1st,* Heyden, London, 1976, p. 332.

580. Posadov. I. A., N. V. Sirotinkin, Y. V. Pokonova, and V. A. Proskuryanov, *Zh. Prikl. Khim.*, **48**, 2055 (1975).

581. Poulis, J. A., J. M. Thomas, and C. H. Massen, in H. G. Wiedemann, Ed., *Thermal Analysis,* Vol. 1, Birkhäuser Verlag, Basel, 1972, p. 573.

582. Price, D., *Chem. Brit.,* 1968, p. 255.

583. Price, C. C., and T. Yukuta, *J. Org. Chem.,* **34**, 2503 (1969).

584. Probsthain, K., *Kunststoffe,* **66**(6), 379 (1976).

585. Puchkov, V. A., N. S. Wulfson, B. V. Rozinov, Yu. V. Denisov, M. M. Shemyakin, Yu. A. Ovchinnikov, and V. T. Ivanov, *Tetrahedron Lett.,* **10**, 543 (1965).

586. Quet, C., Dissertation, University Claude Bernard, Lyon, France, 1974.

587. Quet, C., and P. Bussiere, *Radiochem. Radioanal. Lett.,* **22**, 91 (1975).

588. Quet, C., P. Vergnon, and P. Bussiere, *J. Chim. Phys.,* **74**, 783 (1977).

589. Quinn, R. K., *Mat. Res. Bull.,* **9**, 803 (1974); *Chem. Abstr.,* **81**, 55256 (1974).

590. Rabbering, G., J. Wanrooy, and A. Schuijff, *Thermochim. Acta,* **12**, 57 (1975).

591. Raskó, J., and F. Solymosi, in I. Buzás, Ed., *Thermal Analysis,* Vol. 1, Akadémiai Kiadó, Budapest, 1975, p. 505.

592. Ray, N. H., *J. Appl. Chem.,* (London), **4**, 21 (1954).

593. Redfern, J. P., B. L. Treherne, M. L. Aspinal, and W. A. Wolstenholme, presented at the 17th Annual Conference on Mass Spectrometry and Allied Topics, Dallas, 1969, p. 158.

594. Reed, C., *Br. Polym. J.,* **6**(1), 1 (1974).

595. Reed, P. R., Jr., and P. L. Warren, *Quart. Colo. Sch. Mines.,* **69**, 221 (1974); *Chem. Abstr.,* **82**, 33117 (1975).

596. Rees, L. V. C., in D. Dollimore, Ed., *Proc. Eur. Symp. Therm. Anal., 1st,* Heyden, London, New York, Rheine, 1976, p. 310.

597. Regel, V. R., and O. F. Posnyakov, *Plaste Kaut.,* **19**, 99 (1972); *Chem. Abstr.,* **76**, 141468 (1972).

598. Reiner, E., *Nature,* **206**, 1272 (1965).

599. Reinhold, K., and K. J. Hauffe, *J. Electrochem. Soc.,* **124**, 875 (1977).

600. Riquier, Y., M. Vant Hourhout, J. M. Levert, and C. Vandael, *Metallurgie (Mons. Belg.),* **16**, 1 (1976); *Chem. Abstr.,* **85**, 146741 (1976).

601. Risby, T. H., and A. L. Yergey, *J. Phys. Chem.,* **80**, 2839 (1976).

602. Risby, T. H., and A. L. Yergey, *Anal. Chem.,* **50**, 326A (1978).

603. Roberts, R. W., *Brit. J. Appl. Phys.,* **14**, 485 (1963).

604. Roche, R. S., *J. Appl. Polym. Sci.,* **18**, 3555 (1974).

605. Robertson, A. J. B., *J. Soc. Chem. Ind.* (London), **67**, 221 (1948).

606. Robertson, A. J. B., *Trans. Faraday Soc.,* **45**, 85 (1949).

607. Rogers, R. N., S. K. Yasuda, and J. Zinn, *Anal. Chem.,* **32**, 672 (1960).

608. Rosen, C. L., and A. J. Melveger, *J. Phys. Chem.,* **68**, 1079 (1964).

609. Rosenthal, K., and R. J. Bambeck, *Air Qual. Instrum.*, **2**, 179–83 (1974).

610. Roth, J., Z. Halmos, and T. Meisel, in I. Busás, Ed., *Thermal Analysis*, Vol. 2, Akadémiai Kiadó, Budapest, 1975, p. 343.

611. Rouquerol, J., in R. F. Schwenker, Jr., and P. D. Garn, Eds., *Thermal Analysis*, Vol. 1, Academic, New York, 1969, p. 281.

612. Rowland, R. A., and E. C. Jones, *Am. Mineralogist*, **34**, 550 (1949).

613. Rudloff, W. K., A. D. O'Donnell, R. G. Scholz, and A. Valaitis, in H. G. Wiedemann, Ed., *Thermal Analysis*, Vol. 3, Birkhäuser Verlag, Basel, 1972, p. 205.

614. Rye, R. R., and R. S. Hansen, *J. Chem. Phys.*, **50**, 3585 (1969).

615. Ryhage, R., *Anal. Chem.*, **36**, 759 (1964).

616. Ryland, A. L., "Mass Spectrometry," in *Encyclopedia of Chemical Technology*, Kirk-Othmer, 2nd ed., Vol. 13, Interscience Publishers, New York, 1967, p. 87.

617. Ryska, M., H. D. R. Schueddemage, and D. O. Hummel, *Makromol. Chem.*, **126**, 32 (1969); *Chem. Abstr.*, **71**, 71126 (1969).

618. Saferstein, R., and J. J. Manura, presented at the 25th Annual Conference on Mass Spectrometry and Allied Topics, Washington, D.C., 1977, p. 111.

619. Saito, Y., Y. Shinata, K. Yokota, and K. Miura, *J. Jap. Inst. Met.*, **38**, 997 (1974).

620. Sakamoto, R., T. Ozawa, and M. Kanazashi, *Thermochim. Acta*, **3**, 291 (1972).

621. Sakamoto, R., Y. Takahashi, and T. Ozawa, *J. Appl. Polym. Sci*, **16**, 1047 (1972).

622. Sano, M., Y. Yotsui, H. Abe, S. Sasaki, *Biomed. Mass Spec.*, **3**, 1 (1976).

623. Saunders, H. L., and V. Giedroyce, *Trans. Brit. Ceram. Soc.*, **49**, 365 (1950).

624. Sayles, D. A., U.S. Pat. 3,865,707 (Cl. 204-195S, GO ln), (Feb. 11, 1975).

625. Sazanov, Yu. N., and L. A. Shibaev, in I. Buzás, Ed., *Thermal Analysis*, Vol. 2, Akadémiai Kiadó, Budapest, 1975, p. 117.

626. Sazanov, Yu. N., and V. A. Sysoev, *Eur. Polym. J.*, **10**, 867 (1974).

627. Sceney, C. G., J. F. Smith, J. O. Hill, and R. J. Magee, *J. Therm. Anal.*, **9**, 415 (1976).

628. Scheve, J., and K. Heise, in H. G. Wiedemann, Ed., *Thermal Analysis*, Vol. 3, Birkhäuser Verlag, Basel, Stuttgart, 1972, p. 71.

629. Schueddemage, H. D. R., and D. O. Hummel, *Kolloid-Z.Z. Polym.*, **220**, 133 (1967); *Chem. Abstr.*, **67**, 117490 (1967).

630. Schueddemage, H. D. R., and D. O. Hummel, *Advan. Mass Spectrom.*, **4**, 857 (1968); *Chem. Abstr.*, **74**, 54265 (1971).

631. Schulten, H. R., H. D. Beckey, H. L. C. Meuzelaar, and A. J. H. Boerboom, *Anal. Chem.*, **45**, 191 (1973).

632. Sedlovitch, L. S., and V. M. Neimark, in I. Buzás, Ed., *Thermal Analysis*, Vol. 3, Akadémiai Kiadó, Budapest, 1975, p. 703.

633. Seeger, M., and E. M. Barrall, *J. Polym. Sci., Polym. Chem. Ed.*, **13**(7), 1515 (1975).

634. Seeger, M., E. M. Barrall, and M. Shen, *J. Polym. Sci., Polym. Chem. Ed.*, **13**(7), 1541 (1975).

635. Seeger, M., H.- J. Cantow, and S. Marti, *Z. Anal. Chem.*, **276**, 267 (1975).

636. Selig, H., and P. K. Gallagher, *Inorg. Nucl. Chem. Lett.*, **13**, 427 (1977).

637. Selig, H., M. J. Vasile, F. A. Stevie, and W. A. Sunder, presented at the 25th Annual Conference on Mass Spectrometry and Allied Topics, Washington, D.C., 1977, p. 221.

638. Selig, H., W. A. Sunder, M. J. Vasile, F. A. Stevie, and P. K. Gallagher, *J. Fluorine Chem.* (In press).

639. Sestak, J., *Talanta*, **13**, 567 (1966).

640. Sestak, J., in H. G. Wiedemann, Ed., *Thermal Analysis*, Vol. 2, Birkhäuser Verlag, Basel, 1972, p. 3.

641. Sevčik, J., *Detectors in Gas Chromatography*, Elsevier, Amsterdam, New York, 1975.

642. Seybold, K., J. Roth, Z. Halmos, Cs. Melykuti, and T. Meisel, in D. Dollimore, Ed., *Proc. Eur. Symp. Therm. Anal., 1st,* Heyden, London, 1976, p. 246.

643. Seymour, R. B., D. R. Owen, G. A. Stahl, H. Wood, and W. N. Tinnerman, *J. Appl. Polym. Sci., Symp.,* **25,** 69 (1974).

644. Shechkov, G. T., V. A. Kaplin, Yu. A. Zakharov, and E. N. Svobodin, *Zh. Fiz. Khim.,* **44,** 529 (1970); *Chem. Abstr., 72,* 136905 (1970).

645. Shen, T. T., and W. N. Stasiuk, *J. Air Pollut. Control Assoc.,* **25**(1), 44 (1975).

646. Shen, J., and M. P. T. Bradley, *Anal. Chem.,* **48,** 2291 (1976).

647. Shulman, G. P., *Polym. Letters, 3,* 911 (1965).

648. Shulman, G. P., and H. W. Lochte, *Am. Chem. Soc., Polymer Preprints, 6,* 36 (1965).

649. Shulman, G. P., and H. W. Lochte, *J. Appl. Polymer. Sci.,* **10,** 619 (1966).

650. Shulman, G. P., *J. Macromol. Sci. (Chem.).,* **A1,** 107 (1967).

651. Shulman, G. P., and H. W. Lochte, *J. Macromol. Sci. (Chem.),* **A1,** 413 (1967).

652. Shulman, G. P., and H. W. Lochte, *J. Macromol. Sci. (Chem.),* **A2,** 411, (1968).

653. Sickfield, J., and B. Heinze, *Double-Liason-Chim. Peint,* **22**(224), 595 (1975).

654. Sigmond, T., *Vacuum,* **25**(6), 239 (1975).

655. Simon, J., and S. Gal, in D. Dollimore, Ed., *Proc. Eur. Symp. Therm. Anal., 1st,* Heyden, London, 1976, p. 438.

656. Sinclair, W. R., J. B. MacChesney, P. K. Gallagher, and P. B. O'Connor, *Thermochim. Acta,* **30,** 225 (1979).

657. Skala, G. F., *Anal. Chem.,* **35,** 702 (1963).

658. Škvara, F., and V. Šatava, *J. Thermal Anal., 2,* 325 (1970).

659. Smith, J. D. B., and D. C. Phillips, *Microchem. J.,* **21,** 27 (1976).

660. Smith, J. D. B., D. C. Phillips, and T. D. Kaczmarek, *Microchem. J.,* **21,** 424 (1976).

661. Smith, J. P., and W. W. Wendlandt, *J. Inorg. Nucl. Chem.,* **26,** 1157 (1964).

662. Smith, J. W., and D. R. Johnson, in R. F. Schwenker, Jr., and P. D. Garn, Eds., *Thermal Analysis,* Vol. 2, Academic, New York, 1969, p. 1251.

663. Smith, J. W., and D. R. Johnson, *American Laboratory, 3,* 8 (1971).

664. Smith, J. W., in H. G. Wiedemann, Ed., *Thermal Analysis,* Vol. 3, Birkhäuser Verlag, Basel, 1972, p. 605.

665. Solymosi, F., and T. Bánsági, in H. G. Wiedemann, Ed., *Thermal Analysis,* Vol. 2, Birkhäuser Verlag, Basel, 1972, p. 289.

666. Sopkova, A., V. Jesenák, J. Chomič, M. Dzurillová, and E. Matejčíková, in I. Buzás, Ed., *Thermal Analysis,* Vol. 1, Akadémiai Kiadó, Budapest, 1975, p. 863.

667. Stammler, M., in R. F. Schwenker, Jr., and P. D. Garn, Eds., *Thermal Analysis,* Vol. 2, Academic, New York, 1969, p. 1127.

668. Stapleton, W. G., presented at the 17th Annual Conference on Mass Spectrometry and Allied Topics, Dallas, 1969.

669. Stapp, A. C., and D. W. Carle, Pittsburgh Conf. Anal. Chem. Appl. Spectr., Cleveland, Ohio, March, 1969.

670. Stern, K. H., *J. Phys. Chem. Ref. Data, 1,* 747 (1972).

671. Stewart, G. H., "Chromatography," in A. Standen, Ed., *Kirk-Othmer, Encyclopedia of Chemical Technology,* 2nd ed., Interscience, New York, 1964, p. 413.

672. Stolle, R., and K. Hoffman, *Ber.*, **37**, 4523 (1904).

673. Stone, R. L., *J. Am. Ceram. Soc.*, **35**, 76 (1952).

674. Stone, R. L., *Anal. Chem.*, **32**, 1582 (1960).

675. Sugimae, A., *Anal. Chem.*, **47**, 1840 (1975).

676. Svec, H. J., and G. A. Junk, *Inorg. Chem.*, **7**, 1688 (1968).

677. Swaminathan, R., and J. C. Kuriacose, *J. Catal.*, **16**, 357 (1971).

678. Sykes, G. F. and P. R. Young, *J. Appl. Polym. Sci.*, **21**, 2393 (1977).

679. Szabó, Z. G., M. Gábor, E. Körös, L. Pöppl, J. Wajand, and N. Varga, in I. Buzás, Ed., *Thermal Analysis*, Vol. 2, Akadémiai Kiadó, Budapest 1975, p. 569.

680. Szabo, Z. G., J. Tompler, E. Hollos, and E. E. Zapp, in D. Dollimore, Ed., *Proc. Eur. Symp. Therm. Anal., 1st,* Heyden, London, 1976, p. 272.

681. Szalkowics, M., in I. Buzás, Ed., *Thermal Analysis*, Vol. 1, Akadémiai Kiadó, Budapest, 1975, p. 445.

682. Szalkowics, M., and B. Gawliczek, in D. Dollimore, Ed., *Proc. Eur. Symp. Therm. Anal., 1st,* Heyden, London, 1976, p. 402.

683. Szekely, T., and F. Till, in I. Buzás, Ed., *Thermal Analysis*, Vol. 3, Akadémiai Kiadó, Budapest, 1975, p. 917.

684. Szekely, T., F. Till, and G. Várhegyi, in D. Dollimore, Ed., *Proc. Eur. Symp. Therm. Anal., 1st,* Heyden, London, 1976, p. 33.

685. Tanaka, M., F. Nishimura, and T. Shono, *Anal. Chim. Acta*, **74**(1), 119 (1975).

686. Thomas, J. Jr., G. M. Hieftje, and D. E. Orlopp, *Anal. Chem.*, **37**, 762 (1965).

687. Thomas, T. J., V. N. Krishnamurthy, and V. R. Gowariker, in D. Dollimore, Ed., *Proc. Eur. Symp. Therm. Anal., 1st,* Heyden, London, 1976, p. 455.

688. Tobola, K., *Termanal '76, Celostatna Konf. Term. Anal.,* ±PR.1 7th, 1976, p. 15; *Chem. Abstr.*, **86**, 149861 (1977).

689. Truex, T. J., R. H. Hammerle, and R. A. Armstrong, *Thermochim. Acta*, **19**, 301 (1977).

690. Tsuge, S., S. Hiramitsu, T. Horibe, M. Yamaoka, and T. Takeuchi, *Macromolecules*, **8**(6), 721 (1975).

691. Tsuge, S., K. Murakami, M. Esaki, and T. Takeuchi, in H. Chihara, Ed., *Thermal Analysis*, Heyden, London, 1977, p. 289.

692. Tutorskii, I. A., E. G. Boikacheva, E. F. Bukanova, T. V. Guseva, and I. G. Bukanov, *Isv. Vyssh. Uchebn. Zaved., Khim. Khim. Tekhnol.*, **18**(3), 460 (1975); *Chem. Abstr.* **83**, 28957 (1975).

693. Uden, P. C., D. E. Henderson, and R. J. Lloyd, in D. Dollimore, Ed., *Proc. First Europ. Symposium on Thermal Analysis*, p. 29, Heyden, London/New York/Rheine, 1976.

694. Uegaki, Y., and T. Nakagawa, *J. Appl. Polym. Sci.*, **20**, 1661 (1976).

695. Valentin, N. D., *Rev. Roum. Phys.*, **17**, 1151 (1972); *Chem. Abstr.*, **78**, 113159 (1973).

696. Valetskaya, N. Ya., M. S. Akutin, B. M. Kovarskaya, M. L. Kerber, L. B. Sokolov, G. A. Kuznetsov, A. B. Blyumenfel'd, and A. N. Belyaeva, *Plast. Massy*, 57 (1970); *Chem. Abstr.*, **74**, 54263 (1971).

697. Vámos, E., Z. Adonyi, G. Körösi, and I. Valasek, in I. Buzás, Ed., *Thermal Analysis*, Vol. 2, Akadémiai Kiadó, Budapest, 1975, p. 287.

698. Vancsó-Smercsányi, I., and Á. Szilágyi, in I. Buzás, Ed., *Thermal Analysis*, Vol. 2, Akadémiai Kiadó, Budapest, 1975, p. 359.

699. Vanderborgh, N. E., R. L. Hanson, and C. Brower, *Anal. Chem.*, **47**, 2277 (1975).

700. Vaniš, M., Š. Varga, and J. Tölgyessy, in H. G. Wiedemann, Ed., *Thermal Analysis*, Vol. 2, Birkhäuser Verlag, Basel, 1972, p. 515.

701. VanLuik, F. W. Jr., and R. E. Rippere, *Anal. Chem.*, **34**, 1617 (1962).

702. Varhegyi, G., *Information Processing Letters*, **2**, 24 (1973).

703. Vassallo, O. A., *Anal. Chem.*, **33**, 1823 (1961).

704. Vastola, F. J., A. J. Pirone, and B. E. Knox, presented at 14th Annual Conf. on Mass Spec. and Allied Topics, Dallas, 1966.

705. Vaughan, H. P., presented at the 17th Annual Conference on Mass Spectrometry and Allied Topics, Dallas, 1969, p. 223.

706. Vaughan, H. P., *Amer. Lab.*, Jan. 10, 1970.

707. Verneker, P. V. R., and J. N. Maycock, *J. Chem. Phys.*, **47**, 3618 (1967).

708. Verneker, P. V. R., M. McCarty, and J. N. Maycock, *Thermochim. Acta*, **3**, 37 (1971).

709. Vidale, G. L., *J. Phys. Chem.*, **64**, 314 (1960).

710. Voboril, M., and V. Balek, in I. Buzás, Ed., *Thermal Analysis*, Vol. 3, Akadémiai Kiadó, Budapest, 1975, p. 555.

711. Volke, K. in I. Buzás, Ed., *Thermal Analysis*, Vol. 1, Akadémiai Kiadó, Budapest, 1975, p. 167.

712. Volmer, M., *Z. Phys. Chem.*, 863 (1931).

713. Von Raven, A., and H. Heusinger, *Angew. Makromol. Chem.*, **42**(1), 183 (1975).

714. Voorhoeve, R. J. H., C. K. N. Patel, L. E. Trimble, R. J. Kerl, and P. K. Gallagher, *J. Catal.*, **45**, 297 (1976).

715. Vymazal, Z., E. Czako, B. Meissner, and J. Stepek, *J. Appl. Polym. Sci.*, **18**, 2861 (1974).

716. Wacks, M. E., *J. Phys. Chem.*, **68**, 2725 (1964).

717. Wahl, A. C., and N. A. Bonner, *Radioactivity Applied to Chemistry*, John Wiley, New York, 1951, p. 284.

718. Walker, W. E., L. A. Cosby, and S. T. Martin, Ger. Offen., 2,426,494 (Cl. GOln) 16 Jan., 1975; *Chem. Abstr.*, **83**, 52908 (1975).

719. Wall, L. A., and S. Strauss, *Polymer Reprints*, **5** (2), 325 (1965).

720. Ware, R. K., *Ind. Res.*, **13**(6), 58 (1971); *Chem. Abstr.*, **75**, 80962 (1971).

721. Ware, R. K., *Thermochim. Acta*, **3**, 49 (1972).

722. Warner, N. A., and T. R. Ingraham, *Can. J. Chem. Eng.*, **40**, 263 (1962).

723. Watson, J. T., and K. Biemann, *Anal. Chem.*, **36**, 1135 (1964).

724. Watson, J. T., and K. Biemann, *Anal. Chem.*, **37**, 844 (1965).

725. Waysman, C., *Double-Liaison-Chim. Peint.*, **23**(246), 51 (1976); *Chem. Abstr.* **85**, 6450 (1976).

726. Wendlandt, W. W., *Anal. Chim. Acta*, 27, 309 (1962).

727. Wendlandt, W. W., and E. Sturm, *J. Inorg. Nucl. Chem.*, **25**, 535 (1963).

728. Wendlandt, W. W., and C. Y. Chou, *J. Inorg. Nucl. Chem.*, **26**, 943 (1964).

729. Wendlandt, W. W. in H. G. McAdie, Ed., *Proc. 1st Toronto Symp., Therm. Anal.*, Chem. Inst. Can., 1965, p. 101.

730. Wendlandt, W. W., and T. M. Southern, *Anal. Chim. Acta*, **32**, 405 (1965).

731. Wendlandt, W. W., T. M. Southern, and J. R. Williams, *Anal. Chim. Acta*, **35**, 254 (1966).

732. Wendlandt, W. W., *Thermochim. Acta*, **9**, 7 (1974).

733. Wendlandt, W. W., *Thermochim. Acta*, **9**, 95 (1974).

734. White, G., R. Fisher, and G. Bradshaw, "The Thermal Analysis of Nitride Phases in Steel," PB 230235, 16 pp, British Steel Corp., Sheffield, England, Feb., 1974.

735. White, G., R. Fisher, and G. Bradshaw, "The Thermal Analysis of Nitride Phases in Steel," Corp. Dev. Lab, British Steel Corp., Sheffield, England, 1974.

736. Wieckowska, J., *Nafte (Katowice, Pol.)*, **31**, 366 (1975).

737. Wiedemann, H. G., in R. F. Schwenker and P. D. Garn, Eds., *Thermal Analysis*, Vol. 1, Academic, New York, 1969, p. 229.

738. Wiedemann, H. G., and H. P. Vaughan, in H. G. McAdie, Ed., *Proc. 3rd Toronto Symp., Therm. Anal.*, Chem. Inst. Can., 1969, p. 233.

739. Wiedemann, H. G., and G. Bayer, in H. Chihara, Ed., *Thermal Analysis*, Heyden, London, Bellmawr, N.J., Rheine, 1977, p. 333.

740. Wiley, R. H., and L. H. Smithon, Jr., *J. Macromol. Sci., Part A*, **2**, (3), 589 (1968).

741. Wilkes, G. L., S. Bagrodia, W. Humphries, and R. Wildnauer, *J. Polym. Sci., Polym. Lett. Ed.*, **13**, 321 (1975).

742. Wilkins, P. E., *Air Qual. Instrum.*, **2**, 246–55 (1974).

743. Willard, H. H., L. L. Merritt, Jr., and J. A. Dean, *Instrumental Methods of Analysis*, 5th ed., Van Nostrand, New York, 1974, p. 455.

744. Willard, H. H., L. L. Merritt, Jr., and J. A. Dean, *Instrumental Methods of Analysis*, 5th ed., Van Nostrand, New York, 1974, p. 522.

745. Wilson, D. E., and F. M. Hamaker, in R. F. Schwenker, Jr., and P. D. Garn, Eds., *Thermal Analysis*, Vol. 1, Academic, New York, 1969, p. 517.

746. Wist, A. O., in R. F. Schwenker, Jr., and P. D. Garn, Eds., *Thermal Analysis*, Vol. 2, Academic, New York, 1969, p. 1095.

747. Wist, A., and J. Frohliger, *Proc., 2nd Internat. Air Congr., Washington, D.C.*, 1971.

748. Wist, A. O., in H. G. Wiedemann, Ed., *Thermal Analysis*, Vol. 1, Birkhäuser Verlag, Basel, 1972, p. 515.

749. Wist, A., J. Funt, and J. Magill, in I. Buzás, Ed., *Thermal Analysis*, Vol. 2, Akadémiai Kiadó, Budapest, 1975, p. 259.

750. Wolk, B., *J. Electrochem. Soc.*, **105**, 89 (1958).

751. Woodward, P. W., G. P. Sturm, Jr., and J. E. Dooley, *Proc., 25th Annual Conference on Mass Spectrometry and Allied Topics*, Washington, D.C., 1977, p. 586.

752. Wright, W. W., *Br. Polym. J.*, **6**, 147 (1974).

753. Wyden, H., H. M. Muller, and H. Z. Dokuzoguz, in I. Buzás, Ed., *Thermal Analysis*, Vol. 1, Akadémiai Kiadó, Budapest, 1975, p. 931.

754. Yamada, K., S. Oura, and M. Maruta, *Shimadzu Hyoron*, **28**, 157 (1971); *Chem. Abstr.*, **77**, 20286 (1972).

755. Yamada, K., T. Okino, and S. Ohura, *Shimadzu Hyoron*, **29**, 157 (1972); *Chem. Abstr.*, **78**, 161245 (1973).

756. Yamada, K., S. Oura, and T. Haruki, in I. Buzás, Ed., *Thermal Analysis*, Vol. 3, Akadémiai Kiadó, Budapest, 1975, p. 1029.

757. Yang, R. T., and M. Steinberg, *J. Phys. Chem.*, **81**, 1117 (1977).

758. Yatabe, T., *Kagaku Kojo*, **9**, 14 (1975); *Chem. Abstr.* **84**, 168838 (1976).

759. Yates, J. T., *Chem. Eng. News*, **52**(34), 19 (1974).

760. Yergey, Al. L., T. H. Risby and H. M. Golomb, *Biomed. Mass Spec.*, **5** (In press).

761. Yoshimura, M., and E. Tajima, in H. Chihara, Ed., *Thermal Analysis*, Heyden, London, 1977, p. 71.

762. Zavitsanos, P. D., *Carbon*, **6**, 731 (1968).

763. Zeeman, P. B., *Can. J. Phys.*, **32**, 9 (1954).

764. Zeman, A., in H. G. Wiedemann, Ed., *Thermal Analysis*, Vol. 3, Birkhäuser Verlag, Basel, 1972, p. 219.

765. Zeman, A., *Angew. Makromol. Chem.*, **31**, 1 (1973); *Chem. Abstr.*, **79**, 42988 (1973).

766. Zeman, I., L. Novak, L. Mitter, J. Stekla, and O. Holendova, *J. Chromatogr.*, **119**, 581 (1976).

767. Zemany, P. D., *Anal. Chem.*, **24**, 1709 (1952).

768. Zemany, P. D., *Nature*, **171**, 391 (1953).

769. Zerilli, L. F., *Proc., Chromatography and Electrophoreses, Symp. Int. 9th*, May 1978.

770. Zitomer, F., *Anal. Chem.*, **40**, 1091 (1968).

771. Zitomer, F., and A. H. DiEdwardo, in J. Chiu, Ed., *Polymer Characterization by Thermal Methods of Analysis*, Marcel Dekker, New York, 1974, p. 119.

772. Zurakowska-Orszàgh, J., and S. Kobiela, in I. Buzás, Ed., *Thermal Analysis*, Vol. 2, Akadémiai Kiadó, Budapest, 1975, p. 147.

ADDENDUM

Since completion of the original manuscript in early 1980, a number of publications have appeared and they are summarized in this addendum.

In contrast to the earlier years, lately more attention is being paid to applications of EGA rather than to developing and testing new methods and equipment. In the instrumental category, the trend is both towards simplification and towards constructing universal apparatus. Mass spectrometers alone, or in combination with other analytical instruments, become more and more dominant in evolved gas analysis. Needless to say, data-handling with some type of computer has become the rule rather than the exception.

Typical examples for simple yet sensitive EGA methods are use of hydrometers for water release down to ppm levels or for kinetic studies (230c,324a,722a), and detection of carbonate minerals at the 100 ppm level by DTA and a nondispersive CO_2 detector (487a).

The majority of reports on methodology involve mass spectrometry, covering aspects such as "limited budget approach" (121a), or the combination of a special furnace combined with modulated-beam mass spectrometry for analysis of condensibles above 500°C as from borosilicate glasses at 1250°C and 1500°C (176a). Other reports deal with various TA-MS interfaces which allow simultaneous operation, trapping, or analysis *in vacuo* (115a,115b, 115c,115d,332a,761a,761b). The most interesting mass spectrometric method, however, results in the detection of volatile compound (eg., H_2O, O_2, Ar) from polymeric materials as a function of depth. This is accomplished by a rotary mill located near the ion source of a time-of-flight instrument (257a).

An important finding was based on kinetic data obtained by mass spectrometry for the decomposition of calcium oxalate monohydrate (583a). The authors concluded that the mechanism is not simple but dependent on experimental conditions. Therefore, the compound should not be used to demonstrate the applicability of a particular method for obtaining kinetic data *in vacuo*.

Turning to applications, release of water is still of considerable interest, be it thermally from hydrates (155a,230c,230d) or by hydrogen reduction of metal oxides (230e). Reduction with graphite has been followed by plotting CO_2 and CO from a quadrupole mass spectrometer interfaced to a thermal analyzer via capillary and molecular leak (684b). For $Ba(NO_2)_2 \cdot H_2O$ thermal decomposition produced mainly NO, some NO_2 and N_2, and less O_2 at 500–750°C (230d). Kinetics of $Ni(NO_3)_2 \cdot 6\,H_2O$ decomposition were followed by mass spectrometric plotting of NO and NO_2 release (155a). Several papers dealt with detection of impurities in materials for electronic applications (18a,230a,230b). Mixed metal oxide refractories were found to release alkaline earth oxide at high temperatures (284a). Pyrolysis products from foundry core binders were detected by EGA (486a), emanation thermal analysis was applied to study the hydration of cement from 20°C to 85°C (26a), and the presence of chrysotile (an asbestos mineral) could be determined with combined TG-GC by quantitative analysis of water evolved between 560°C and 650°C (626a). In this manner, vermiculite, calcium carbonate, or calcium sulfate hydrate did not interfere. Supported metal catalysts were characterized by observing the release of H_2O, CO, and CO_2 with temperature-programmed direct-probe mass spectrometry (750a), while temperature-programmed desorption with mass spectrometric identification of evolved species produced an understanding of the conversion of CH_3OH to CH_2O over an oxide catalyst (197b). The history of natural and artificial glasses was determined at high vacuum by a TA-MS combination for O_2, CO, CO_2, and HF (282a), while emanation thermal analysis was carried out on a semiconductor glass $Ge_{.25}Te_{.60}Se_{.15}$ (26b).

Products from boron–oxygen complexes with organic acids were identified by GC after a combination of DTA, TG, and DTG (631a), and GC also identified CO and H_2O as the main decomposition products from nickel squarate under nitrogen (90a). For tetrakis(dithiocarbamato)tin(IV) complexes, the decomposition involves reduction of tin(IV) to tin(II) as shown by temperature-programmed direct-probe mass spectrometry (84a). As an example of high-temperature chemistry involving heavy metals, Knudsen cell effusion for sublimation enthalpy studies of tungsten and molybdenum complexes should be mentioned (101a).

In the field of organic chemistry, a major effort is still continuing with polymer-related problems. Studies on polymer degradation and pyrolysis involved thermal volatilization analysis at ambient and subambient temperatures (472a,472b), modulated molecular-beam mass spectrometry (433b), and a TG-GC-MS combination (296a). Computer-aided pyrolysis-GC-MS was also used for trace analysis of organic contaminants in gyroscopes (122a). Similarly, types and amounts of silane coupling agents bonded to glass were determined (373a). Fire retardancy (230b,386a,433a) and flammability, especially of cellulosics (197a,551a) are receiving more attention.

The largest increase for publications on EGA, however, has been observed on energy-related topics. Separation and quantitation of C_1–C_8 hydrocarbons and oxides of carbon in a single run have been achieved by flash pyrolysis-GC (296b); pyrolysis and oxidation products from DSC of coal and peat at 1–50 atm pressure have been identified by time-of-flight mass spectrometry; atomic absorption and infrared spectroscopy (7a), and fossil fuels have been characterized by computer-aided TG-MS (684a). During an investigation of trace pyrolysis products from air oxydation and chlorination of bituminous coal by three-stage trapping, TLC, HPLC, and finally quantitative GC-MS analysis, it was found that presence of dibenzo-p-dioxins and similar toxic species is associated with coal oxidation and presence of chlorine sources (735a). Retorting conditions for fuel production from oil shales have been simulated by combined pyrolysis-GC-MS, which caused different compositions to be recognized for different shales (122b). The effect of oxidation and cracking on product distribution was studied by pyrolysis-EGA (via flame ionization or thermal conductivity detector), vapor-phase IR, and pyrolysis-GC-MS under helium, air, and carbon dioxide (693a).

Using TG-GC, it was detected that iron compounds—especially hematite—catalyze the hydrogasification of coal, resulting in production of 85% methane (546a).

Finally, there has been an environmentally significant result. Fuel-bound nitrogen from several forest fuel sources was found to convert to NO and NO_x under conditions of TG-EGA by chemiluminescence analysis (121b).

REFERENCES

7a. Amey, R. L., and C. D. West, in *Proc., 11th NATAS Conf.*, New Orleans, Oct. 18–21, 1981, p. 141.

18a. Bagley, B. G., and P. K. Gallagher, in *Proc., 10th NATAS Conf.*, Boston, Oct. 26–29, 1980, p. 193.

26a. Balek, V., J. Dohnálek, and W. D. Emmerich, in H. G. Wiedemann, Ed., *Thermal Analysis*, Vol. 1, Birkhäuser Verlag, Basel, 1980, p. 375.

26b. Balek, V., and M. Vobořil, in H. G. Wiedemann, Ed., *Thermal Analysis*, Vol. 1, Birkhäuser Verlag, Basel, 1980, p. 403.

84a. Bratspies, G. K., J. F. Smith, and J. O. Hill, in W. Hemminger, Ed., *Thermal Analysis*, Vol. 2, Birkhäuser Verlag, Basel, 1980, p. 147.

90a. Brown, M. E., A. K. Galwey, and M. LePatourel, in W. Hemminger, Ed., *Thermal Analysis*, Vol. 2, Birkhäuser Verlag, Basel, 1980, p. 153.

101a. Cavell, K. J., J. M. Ernsting, and D. J. Stufkens, *Thermochim. Acta*, **42**, 343 (1980).

115a. Chiu, J., and A. J. Beattie, *Thermochim. Acta*, **40**, 251 (1980).

115b. Chiu, J., and A. J. Beattie, in H. G. Wiedemann, Ed., *Thermal Analysis*, Vol. 1, Birkhäuser Verlag, Basel, 1980, p. 245.

115c. Chiu, J., and A. J. Beattie, in *Proc., 10th NATAS Conf.*, Boston, Oct. 26–29, 1980, p. 221.

115d. Chiu, J., and A. J. Beattie, *Thermochim. Acta*, **50**, 49 (1981).

121a. Clarke, E., *Thermochim. Acta*, **51**, 7 (1981).

121b. Clements, H. B., and C. K. McMahon, *Thermochim. Acta,* **35,** 133 (1980).

122a. Coldiron, S. J., J. Garrett, M. L. Taylor, and T. O. Tiernan, in *Proc., 10th NATAS Conf.,* Boston, Oct. 26–29, 1980, p. 225.

122b. Coldiron, S. J., M. L. Taylor, and T. O. Tiernan, in *Proc., 10th NATAS Conf.,* Boston, Oct. 26–29, 1980, p. 229.

155a. Dollimore, D., G. A. Gamlen, and T. J. Taylor, *Thermochim. Acta,* **51,** 269 (1981).

176a. Fabricant, B. L., in *Proc., 10th NATAS Conf.,* Boston, Oct. 26–29, 1980, p. 153.

197a. Franklin, W. E., in *Proc., 11th NATAS Conf.,* New Orleans, Oct. 18–21, 1981, p. 471.

197b. Frederick, C. G., A. W. Sleight, and V. Chowdry, in *Proc., 11th NATAS Conf.,* New Orleans, Oct. 18–21, 1981, p. 421.

230a. Gallagher, P. K., *Thermochim. Acta,* **41,** 323 (1980).

230b. Gallagher, P. K., in H. G. Wiedemann, Ed., *Thermal Analysis,* Vol. 1, Birkhäuser Verlag, Basel, 1980, p. 13.

230c. Gallagher, P. K., and E. M. Gyorgy, in H. G. Wiedemann, Ed., *Thermal Analysis,* Vol. 1, Birkhäuser Verlag, Basel, 1980, p. 113.

230d. Gallagher, P. K., *Thermochim. Acta,* **51,** 233 (1981).

230e. Gallagher, P. K., E. M. Gyorgy and W. R. Jones, in *Proc., 11th NATAS Conf.,* New Orleans, Oct. 18–21, 1981, p. 431.

257a. Grayson, M. A., and C. J. Wolf, in *Proc., 11th NATAS Conf.,* New Orleans, Oct. 18–21, 1981, p. 193.

282a. Heide, K., in W. Hemminger, Ed., *Thermal Analysis,* Vol. 2, Birkhäuser Verlag, Basel, 1980, p. 341.

284a. Hirayama, C., R. L. Kleinosky, and R. S. Bhalla, *Thermochim. Acta,* **39,** 187 (1980).

296a. Hosaka, Y., T. Kojima, and S. Kudo, in W. Hemminger, Ed., *Thermal Analysis,* Vol. 2, Birkhäuser Verlag, Basel, 1980, p. 393.

296b. Hovsepian, B. K., in *Proc., 11th NATAS Conf.,* New Orleans, Oct. 18–21, 1981, p. 301.

324a. Jones, W. R., *Thermochim. Acta,* **52,** 305 (1982).

332a. Kaisersberger, E., in H. G. Wiedemann, Ed., *Thermal Analysis,* Vol. 1, Birkhäuser Verlag, Basel, 1980, p. 251.

373a. Lahr, S. K., and A. M. C. Walker, in *Proc., 11th NATAS Conf.,* New Orleans, Oct. 18–21, 1981, p. 725.

386a. Langer, H. G., and J. D. Fellmann, in *Proc., 11th NATAS Conf.,* New Orleans, Oct. 18–21, 1981, p. 327.

433a. Lum, R. M., in *Proc., 10th NATAS Conf.,* Boston, Oct. 26–29, 1980, p. 119.

433b. Lum, R. M., in *Proc., 11th NATAS Conf.,* New Orleans, Oct. 18–21, 1981, p. 459.

472a. McNeill, I. C., in H. G. Wiedemann, Ed., *Thermal Analysis,* Vol. 1, Birkhäuser Verlag, Basel, 1980, p. 319.

472b. McNeill, I. C., and A. Hamoudi, in *Proc., 11th NATAS Conf.,* New Orleans, Oct. 18–21, 1981, p. 751.

486a. Miller, R. L., and M. A. Oebser, *Thermochim. Acta,* **36,** 133 (1980).

487a. Milodowski, A. E., in W. Hemminger, Ed., *Thermal Analysis,* Vol. 2, Birkhäuser Verlag, Basel, 1980, p. 289.

546a. Padrick, T. D., D. D. Dees, and T. M. Massis, in *Proc., 11th NATAS Conf.,* New Orleans, Oct. 18–21, 1981, p. 401.

551a. Patil, K. C., J. P. Vittal, and C. C. Patel, *Thermochim. Acta,* **43,** 213 (1981).

583a. Price, D., D. Dollimore, N. S. Fatemi, and R. Whitehead, *Thermochim. Acta,* **42,** 323 (1980).

626a. Scalera, J. V., in *Proc., 11th NATAS Conf.*, New Orleans, Oct. 18–21, 1981, p. 303.

631a. Schwartz, J., I. Vitol, G. Sergeeva, A. Bernane, and A. Terauda, in W. Hemminger, Ed., *Thermal Analysis*, Vol. 2, Birkhäuser Verlag, Basel, 1980, p. 169.

684a. Székely, T., F. Till, and G. Várhegyi, in W. Hemminger, Ed., *Thermal Analysis*, Vol. 2, Birkhäuser Verlag, Basel, 1980, p. 365.

684b. Szendrei, T., and P. C. VanBerge, *Thermochim. Acta*, **44**, 11 (1981).

693a. Uden, P. C., in *Proc., 10th NATAS Conf.*, Boston, Oct. 26–29, 1980, p. 233.

722a. Warrington, S. B., and P. A. Barnes, in H. G. Wiedemann, Ed., *Thermal Analysis*, Vol. 1, Birkhäuser Verlag, Basel, 1980, p. 327.

735a. Whiting, L. F., and N. H. Mahle, in *Proc., 10th NATAS Conf.*, Boston, Oct. 26–29, 1980, p. 243.

750a. Wood, C. D., and H. G. Langer, in *Proc., 10th NATAS Conf.*, Boston, Oct. 26–29, 1980, p. 211.

761a. Yuen, H. K., G. W. Mappes, and W. A. Grote, in *Proc., 11th NATAS Conf.*, New Orleans, Oct. 18–21, 1981, p. 321.

761b. Yuen, H. K., G. W. Mappes, and W. A. Grote, *Thermochim. Acta*, **52**, 143 (1982).

THERMODILATOMETRY

By Michel Murat,
Laboratoire de Chimie Appliquée,
Institut National des Sciences Appliquées de Lyon,
France

Contents

I. THERMAL EXPANSION—THEORETICAL ANALYSIS

A. ORIGIN OF THERMAL EXPANSION

All solid substances, when heated or cooled, undergo a reversible change in dimensions which is called thermal dilatation or thermal expansion. This dilatation is a result of the thermal motion of the atoms or groups of atoms of which the matter is composed. In the case of material such as an ideal gas, in which there are no interactions among the atoms or molecules, the energy of the motion of these molecules or atoms is directly proportional to the absolute temperature so that as the temperature increases, the energy of impact on the container increases and the gas tends to expand. The expansion coefficient of a gas is relatively large. It is smaller in the case of liquid because the effects of molecular interaction begin to appear.

In the case of crystalline solid, the atoms do not move about freely but are bound to certain sites in the lattice structure. They can however vibrate about these sites. It is well-known that the binding forces between the atoms of a solid are the result of an attractive force and a repulsive force. When two atoms of a solid are a distance a apart, the expression of binding forces can be written:

$$F = F_A + F_R = + \frac{A}{a^M} - \frac{B}{a^N} \tag{1}$$

A, B, M, and N are constants that depend on the nature of the bond. Attractive forces are generally electrostatic forces ($M = 2$). Repulsive forces can be either of metallic type ($7 \leqslant N \leqslant 10$) or of ionic or covalent type ($10 \leqslant N \leqslant 12$). N is thus higher than M.

In a diagram where attractive and repulsive forces are drawn versus the distance between the atoms of simple crystalline solid composed of one kind of atom arranged in a single cubic lattice, the resulting force $F(a)$ is represented by a full line (Fig. 7.1). When the attractive and repulsive forces are in equilibrium, the resulting force is null. The corresponding distance a_0 represents the equilibrium position of the atoms, the thermal agitation being null and corresponding to $T = O~K$.

The distance a_m, corresponding to the minimum value of F, is the maximum distance by which the atoms can be separated under tension. When $a > a_m$, the cohesive force of the material decreases rapidly and the solid breaks.

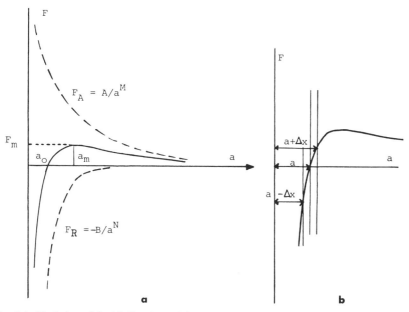

Fig. 7.1. Variation of the binding force F in a solid versus the distance "a" between atoms.

A closer examination of Fig. 7.1b shows that a greater force is required to move two atoms a given distance Δx toward each other than to increase the distance between them by the same amount. When $T > O\ K$, the atoms vibrate about their equilibrium position. If the temperature is increased, the amplitude of the vibrations increases, but because the repulsive and attractive forces are not symmetrical, the atoms are drawn back with greater force from the position $a - \Delta x$ than they are from $a + \Delta x$. Consequently, the center of vibration is displaced towards $+ \Delta x$, the interatomic distance increases, and the solid expands. The same conclusion can be made from the potential energy diagram because binding forces derive from a potential energy.

Some qualitative conclusions can be drawn from this simplified mechanism: for example, since thermal expansion arises from interatomic forces, the force exerted when a substance expands can be very great. Austin (10) expresses this force in terms of the pressure required to prevent expansion. As a specific example, the pressure required to prevent any increase in volume of a piece of M_gO when it is heated from room temperature to 100°C is of the order of 70,000 lb/in^2. That explains the large and often disruptive thermal stresses that can be developed, for instance, in a refractory body subject to a severe temperature gradient, or in certain geological phenomena such as the development of stress during the cooling of an igneous rock.

This simple mechanism, which shows how the anharmonic nature of the vibrations causes thermal expansion, explains that the entire structure should expand by an increase in scale and that is certainly what happens in the expansion of most crystalline solids. However, in addition to this change of scale, it is possible to have an expansion of the lattice resulting from a change of bond angle with temperature without significant change in bond length. Megaw (164) cited, for example, the case of diamond and silicon carbide. For diamond to remain cubic on heating, the tetrahedral angle must be maintained, whereas in silicon carbide the angle may vary without violating the symmetry so that the expansion can be considerable without any great increase of interatomic distance.

The mechanism of expansion becomes more complex in the case of substances containing groups of atoms such as in sulfates, carbonates, tungstates, etc. Also, the forces acting in certain directions in a complex solid may be such that *contraction* may occur in certain directions in the crystal. Examples of such contraction along a particular axis of nonisotropic solids will be given later (see Section I.C.7).

B. DEFINITION OF THERMAL EXPANSION COEFFICIENTS

The following different definitions have to be taken into consideration:

Specific thermal expansion. That is, the relative variations in length L (or in volume V) of the material between a reference temperature Θ_0 and a temperature Θ

$$\Delta L\Big]_{\Theta_0}^{\Theta} = \frac{L_\Theta - L_{\Theta_0}}{L_{\Theta_0}} \tag{2}$$

$$\Delta V\Big]_{\Theta_0}^{\Theta} = \frac{V_\Theta - V_{\Theta_0}}{V_{\Theta_0}} \tag{3}$$

The reference temperature can be chosen in different ways: absolute zero ($\Theta_0 = O\ K$) or ambient temperature (20°C or 25°C).

Average thermal expansion coefficients (linear or bulk) between two temperatures Θ_0 and Θ. The linear coefficient is:

$$\alpha_L\Big]_{\Theta_0}^{\Theta} = \frac{1}{\Theta - \Theta_0} \cdot \frac{L_\Theta - L_{\Theta_0}}{L_{\Theta_0}} \tag{4}$$

The bulk coefficient is:

$$\alpha_V\Big]_{\Theta_0}^{\Theta} = \frac{1}{\Theta - \Theta_0} \cdot \frac{V_\Theta - V_{\Theta_0}}{V_{\Theta_0}} \tag{5}$$

True thermal expansion coefficients (linear or bulk) at a given temperature Θ. The Linear coefficient is:

$$(\alpha_L)_\Theta = \frac{1}{L} \cdot \frac{dL}{d\Theta} \tag{6}$$

TABLE 7.I
True Thermal Expansion Coefficients of MgO and CaO in Different Temperature Ranges

Oxide	Temperature range (°C)	$\alpha_L \times 10^6$
	20–300	13.47
MgO	20–600	13.45
	20–1200	14.45
	20–300	13.12
CaO	20–600	12.80
	20–1200	13.57

The bulk coefficient is:

$$(\alpha_v)_\Theta = \frac{1}{V} \cdot \frac{dV}{d\Theta} \tag{7}$$

They are the derivatives of the specific expansion with respect to the temperature, or the slope of dilatometric curves at the temperature Θ.

In the case of isotropic materials (identical expansion in the three dimensions), the bulk thermal expansion coefficient is related to the linear coefficient by the relation:

$$(\alpha_V)_\Theta = 3 (\alpha_L)_\Theta \tag{8}$$

After Weyl (276), measurements using small increments of ΔT are generally made only if a scientist expects some particular phenomenon such as a phase change or a transition of a higher order. In the range of temperature between 25°C and 100–1000°C, a solid that does not show anisotropy or phase change (or shrinkage and sintering) generally has a true thermal expansion coefficient that stays approximately constant. That is the case, for example, for a number of oxides such as MgO and CaO, as can be seen in the Table 7.I. However, certain solids can show a large variation of α with temperature.

Generally, the thermal expansion coefficient of a substance can be given analytically as a polynomial expression versus temperature:

$$\alpha = A + BT + CT^2 + DT^3 \tag{9}$$

A, B, C, D being negative or positive constants that can be evaluated by fitting the experimental data by methods such as the method of least squares.

In the case of anisotropic crystals or in the case of fibrous materials or fiber-reinforced composite materials, particular expansion coefficient will be defined such as $\alpha_\perp$ or $\alpha_\parallel$ corresponding to coefficients measured in perpendicular or parallel direction either to the principal axis in the case of anisotropic crystals or to the extrusion axis or fiber in the case of composite materials.

C. RELATION BETWEEN THERMAL EXPANSION AND PHYSICOCHEMICAL PROPERTIES OF SOLIDS

1. Thermodynamic Properties and Heat Capacity

The thermal expansion coefficient of a substance is related to the thermodynamic properties. The partial derivatives of the free energy G, with respect to pressure P and temperature T are:

$$\left(\frac{\delta G}{\delta P}\right)_T = V \text{ and } \left(\frac{\delta G}{\delta T}\right)_P = -S \tag{10}$$

V and S are molar volume and molar entropy, respectively. Differentiating again gives:

$$\left(\frac{\delta^2 G}{\delta P \delta T}\right) = \left(\frac{\delta V}{\delta T}\right)_P \text{ and } \left(\frac{\delta^2 G}{\delta T \delta P}\right) = -\left(\frac{\delta S}{\delta P}\right)_T \tag{11}$$

then

$$\left(\frac{\delta V}{\delta T}\right)_P = -\left(\frac{\delta S}{\delta P}\right)_T \tag{12}$$

which gives the relation between thermal expansion coefficient and entropy. From the definition of the volumetric thermal expansion coefficient, we can write:

$$\alpha_v = \frac{1}{V}\left(\frac{\delta V}{\delta T}\right)_P = -\frac{1}{V}\left(\frac{\delta S}{\delta P}\right)_T \tag{13}$$

If ΔS represents the increase in entropy that accompanies an isothermal change in a solid, then, provided that the internal stability has not been influenced by the change, we may write:

$$\lim_{T \to 0} \Delta S = 0 \tag{14}$$

On this basis, we should expect that

$$\lim_{T \to 0} \left(\frac{\delta S}{\delta P}\right)_T = 0 \tag{15}$$

and from equation 13, it follows that $\alpha_v \to 0$ as $T \to 0$.

As for the thermal coefficient of expansion, the magnitude of the heat capacity is related to the amplitude of thermal vibrations, which, in turn, is proportional to absolute temperature. Consequently, one may expect to find proportionality between the thermal expansion coefficient and the specific heat. If β is the isothermal compressibility coefficient, given by

$$\beta = -\frac{1}{V}\left(\frac{\delta V}{\delta P}\right)_T \tag{16}$$

a simple calculation shows, for example, that α and β are related to molar specific heat by the expression

$$C_p - C_v = \frac{T \alpha_v^2 V}{\beta} \tag{17}$$

where C_p and C_v are specific heat at constant pressure and constant volume, respectively. But the more interesting correlation between α and specific heat was developed by Grüneisen (100), who expressed α as follows:

$$\alpha_L(T) = \gamma(T) \frac{\beta C_v}{3 V} = \gamma(T) \frac{C_v}{3BV} \tag{18}$$

with

$$B = \frac{1}{\beta} = - V \left(\frac{\delta P}{\delta V}\right)_T \tag{19}$$

$$\gamma = 3 \alpha_L \frac{BV}{C_v} = \alpha_v \frac{BV}{C_v} \tag{20}$$

B is the isothermal bulk modulus of elasticity, and $\gamma(T)$ is the macroscopic Grüneisen constant.

At the time at which these equations were developed, B was taken to be independent of temperature, and this led directly to the conclusion that $\alpha_v \propto C_v$ to a good approximation, which was known as Grüneisen's rule or law. Experiments show that Grüneisen's law is only true for the majority of familiar solids at intermediate and high temperatures, and to this extent it is to be regarded as a first approximation.

At low temperatures, it was observed according to theories of Debye and of Einstein that a good approximation for a large number of substances is:

$$C_v \approx f(T/\Theta_D) \tag{21}$$

where f is the same function for a wide range of substances, and Θ_D is the Debye temperature, the value of which is characteristic of each substance.

When $T < \Theta_D/12$ (temperatures near $O K$), $C_v \simeq k(T/\Theta_D)^3$ to an accuracy of approximately 1%.

When $T \geqslant \Theta_D$, $C_v \to 3 R$, which agrees with the rule of Dulong and Petit formulated in 1819 and stating that the atomic heats of a large number of elements were equal to approximately 6 cal (gram $\cdot$ atom)$^{-1} \cdot {}^{\circ}C^{-1}$.

From these thermodynamic considerations, the following conclusions dealing with thermal expansion coefficient can be pointed out:

1. At $T = O K$, C_v and $\alpha = 0$.

2. At low temperature α and C_v vary proportionally to $(T/\Theta_D)^3$.

3. At higher temperatures, α and C_v tend towards a limit. Consequently, the thermal expansion coefficient of a solid becomes constant at high temperature.

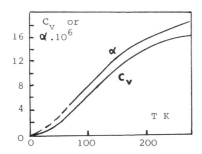

Fig. 7.2. Variation with temperature of the coefficient of thermal expansion and the specific heat of calcium fluoride.

Such a proportionality between α and C_v was observed many years ago for a number of substances (diamond, fluorite, and pyrite). A typical example is given in Fig. 7.2 in the case of fluorite: the course of the specific heat curve is closely parallel to that of the curve of the expansion coefficient. An interesting compilation and analysis of results concerning specific heat and thermal expansion was made by Borelius in 1963 (37).

The proportionality between α and C_v seems to be less valid at high temperature, except for particular solids such as MgO at 800°C. For complex substances, it appears to have only limited validity.

From elementary thermodynamics, we may show that the Grüneisen constant γ may be written in an alternative form:

$$\gamma = \alpha_v \, VB_s/C_p \qquad (22)$$

where B_s is the adiabatic bulk modulus of elasticity.

Another expression of γ is also given:

$$\gamma = -\left(\frac{V}{\Theta}\right)\left(\frac{d\Theta}{dV}\right)_T = -\,d(\ln \Theta)/d(\ln V) \qquad (23)$$

A lot of theoretical data about the Grüneisen constant γ can be found in the Proceedings of the Thermal Expansion Symposia (255–260) and in Yates (283). We will not discuss these data here because they have only limited interest in this chapter.

2. Chemical Bond and Heat of Formation

It seems reasonable to expect that, as temperature is raised, atoms which are held together by a strong chemical bond should not separate as much as would two atoms joined by a weak bond. Consequently, *the thermal expansion should be inversely proportional to the strength of the bond.*

So, it is interesting to compare the expansion coefficient with the heat of formation, which is a criterion of bond energy. More specifically, Henglein (109) studied the product of the expansion coefficient and the heat of formation ($\alpha_x \Delta H$). He found that this quantity was substantially constant in the case of alkali halides (Table 7.II), but a comparison of this kind on more complex molecules is difficult to make because of the presence of several

TABLE 7.II

Relation between Thermal Expansion Coefficient, Heat of Formation, ΔH, and Melting
Temperature, T_m, in the Case of Alkali Halides (109–110)

Halide	$\alpha \times 10^6$ (0–79°C)	$\alpha \times \Delta H$ (arbitrary units)	$\alpha_x T_m \times 10^4$
LiF	92	11.0	8.0
LiCl	122	11.8	7.5
LiBr	140	12.2	7.6
LiI	167	11.9	7.5
NaF	98	10.9	9.6
NaCl	110	10.7	8.7
NaBr	119	10.8	9.0
NaI	135	10.3	8.8
KF	100	10.9	8.8
KCl	101	10.5	7.9
KBr	110	10.7	8.0
KI	125	10.6	9.1
RbCl	98	10.4	7.5
RbBr	104	10.3	7.5
RbI	119	10.4	7.6

different kinds of bonds. For instance, scattered data in the case of oxides of relatively simple structure show that the correlation between α and ΔH has but a limited validity.

In other hand, Klemm (141) noted that in the case of alkali halides, α_L for a given alkali increases from that of the fluoride to that of the iodide, but no such relationship can be found for a series ranging from the sodium halide to the corresponding rubidium halide (see Table 7.II).

3. Melting Temperature and Entropy

Lindeman (154) and Grüneisen (99) were the first scientists who correlated the melting and the expansion of metals with thermodynamic quantities and with the amplitudes of thermal vibrations. Another criterion of bond energy is the melting temperatures. To see to what extent melting points correlate with thermal expansion coefficients, the product of the two was computed for the series of alkali halides (Table 7.II). It appears that the product is sensibly constant, but the extension of this correlation to oxides or more complex substances does not hold.

On the other hand, Straumenis (253) showed that the linear coefficients of expansion of cubic elements plotted against their melting point gave a smooth curve that started with the high-expansion, low-melting cesium and ended with the low-expansion, high-melting tungsten and diamond. Only slight deviations from a smooth curve were observed, but there were no exceptions to the rule that *the expansivity of cubic elements decreases as their melting points increase*. The cubic elements are the most striking ex-

TABLE 7.III
Thermal Expansion Coefficient, α_L, Values of the Field Strength of the Cations (Z/r^2),
Lattice Energy $(-U)$, and Melting Points (°C) of some Alkaline Earth Oxides (276)

Oxide	$\alpha_L \Big]_0^{300} \times 10^6$	Z/r^2	$-U$ kcal · mole^{-1}	T_m (°C)
MgO	13.5	0.45	939	2800
CaO	13.1	0.33	831	2570
SrO	13.7	0.27	766	2430

ample for a fairly rigorous relationship between melting point and coefficient of thermal expansion.

On the other hand Weyl showed, for example, that even the ionic crystal of NaCl structure does not show a rigorous relationship between thermal expansivity, lattice energy, and melting point, as can be seen from the alkaline earth oxides (Table 7.III).

After Weyl (276), one cannot generalize that low-melting ionic crystals have high expansivity and vice versa. Austin (10) remarked that the observations made on alkali halides do not apply to oxides and wrote that "there appears to be a tendency for the compounds with higher melting point to have a greater expansion."

Based on the fact that the driving forces that cause thermal expansion are the same as those of melting, Weyl (276) suggests the use of the Gibbs function as a basis for explaining the widely different thermal expansivities of crystals.

$$\Delta G = \Delta E - T (\Delta S_{\text{vibr}} + \Delta S_{\text{config}}) \qquad (24)$$

This function offers the advantage that we do not need to know the absolute values of the internal energy E of the system nor its absolute vibrational entropy S_{Vibr} and configurational entropy S_{Config}. These parameters have to be considered only with respect to their changes if the temperature T is raised. To keep the value of ΔG at a minimum, the crystal increases its entropy in a way that its internal energy is kept as low as possible. Thermal expansion is one of several possible mechanisms by which a crystal can increase its entropy and, in order to interpret the widely different expansivities of ionic crystals, one has to examine which mechanism the crystal had available for *increasing its entropy*. The analysis of Weyl on this subject will not be developed here, but it is certain that the work of this author has given a clearer understanding of this problem, particularly about the role of the atomic structures of crystals in their response to temperature changes.

4. Valency and Coordination Number

Some attempts have been made to derive a quantitative expression for the expansion coefficient in terms of specific formulations for the forces acting

TABLE 7.IV

Relation between Thermal Expansion Coefficients α_L and Valency Z (164)

Substance	Structure type	Valency Z of the cation	$\alpha_L \times 10^6$	$\alpha \cdot Z^2 \times 10^6$
Na Cl	NaCl	1	40	40
MgO	(cubic)	2	10	40
CaF$_2$	Fluorite	2	19	76
ZrO$_2$	(cubic)	4	4.5	80
CuBr	Zincblende	1	19	19
ZnS	(cubic)	2	6.7	27
MgF$_2$	Rutile[a]	2	11	44
SnO$_2$	(tetragonal)	4	3.4	54
NaNO$_3$	Calcite[a]	1	4.7	47
MgCO$_3$	(rhombohedral)	2	11	44
KNO$_3$	Aragonite[a]	1	60	60
CaCO$_3$	(orthorhombic)	2	21	84

[a] For substances of structure other than cubic, α_L is taken as one-third of α_v.

between atoms. The simplifying assumptions made in these theoretical treatments are more nearly fulfilled in the alkali halides that have been chosen to test the theories.

According to Megaw (164), for crystals having the same structure, the expansion coefficient is inversely proportional to the square of the valency Z

$$\alpha_L = k/Z^2 \qquad (25)$$

(Table 7.IV). This would mean that the lattice energy of a crystal is a major factor that controls its volume expansion on heating. But after Weyl (275), this rule is not always obeyed because for many crystals the observed volume expansion results from two mechanisms: (1) an increase of all internuclear distances, and (2) a variation of the bond angles.

Another property that has been taken into account is the coordination number, that is, the number of neighboring ions of opposite sign. Megaw (164) discovered that for the same valency, *the average expansion is directly proportional to the square of the coordination number.* Making use of Pauling's concept of electrostatic share, which is defined as the charge of one ion divided by its coordination number, Megaw expresses this as

$$\alpha_L = c/q^2 \qquad (26)$$

C is a constant of the order of 1.10^{-6} and q is the electrostatic share (valency divided by the coordination number). Attempts to apply the rule to complex compounds have yielded somewhat poor results, but this relation achieved a considerable degree of success in the case of certain simple substances. (Table 7.V).

TABLE 7.V
Relation between Thermal Expansion Coefficient α and Electrostatic Share q (164)

Substance	q	$\alpha_L \times 10^6$	$\alpha \cdot q^2 \times 10^6$
CsCl	$\frac{1}{8}$	53	0.87
NaCl	$\frac{1}{6}$	40	1.11
CaF$_2$	$\frac{2}{8} = \frac{1}{4}$	19	1.19
CuBr	$\frac{1}{4}$	19	1.19
MgO	$\frac{2}{6} = \frac{1}{3}$	10	1.11
ZrO$_2$	$\frac{4}{8} = \frac{1}{2}$	4.5	1.12
ZnS	$\frac{2}{4} = \frac{1}{2}$	6.7	1.67

5. Refractive Index

In relatively simple compounds such as the isometric alkali halides, there is a very direct relation between thermal expansion coefficient and refractive index. This relation is linear for the halides of potassium and rubidium, as shown by Austin (10).

The situation is not clear, but in a rough way there appears to be a correlation in that those compounds *whose mean refractive index is high tend to have high expansion coefficients.* Austin compared the refractive indexes and α_L values of some more complex compounds and classed these substances in the way presented in Table 7.VI.

In a general way, the order of the decreasing index of refraction parallels the order of decreasing expansion. There are, however, some significant exceptions, such as zircon, ZrSiO$_4$, which has a high refractive index but a relatively low expansion. It is interesting to note that Jeppesen (126) used change of the refractive indices of a sapphire crystal with temperature to calculate its thermal expansivity.

TABLE 7.VI
Coefficient of Thermal Expansion and Refractive Index of Some Substances

Substances	Average refractive index	$\alpha_L \times 10^6$
MgO	1.736	13.5
CaO	1.838	13.1
BeO	1.719–1.733	5.0
Al$_2$O$_3$	1.76	7.2–8.1
MgAl$_2$O$_4$ (Spinel)	1.72–1.75	7.6–9.0
BeAl$_2$O$_4$ (Chrysoberyl)	1.74	7.0–8.0
Be$_2$SiO$_4$ (Phenacite)	1.65	6.4
Be$_2$Al$_2$Si$_6$O$_{18}$ (Beryl)	1.580–1.547	1.3
Na$_2$O $\cdot$ Al$_2$O$_3$ $\cdot$ 6 SiO$_2$ (Albite)	1.5–1.6	7.0–9.7
CaO $\cdot$ Al$_2$O$_3$ $\cdot$ 2 SiO$_2$ (Anorthit)	1.583	4.5–6.4
ZrSiO$_4$ (Zircon)	1.96	$\alpha_a = 3.2 \; \alpha_c = 5.4$

6. Binding Forces and Crystal Structure

Thermal expansion of solids is one of the principal properties that depends on the binding forces between the constituent atoms. As the theoretical evaluation of these forces in terms of the interionic distances is complicated, even in the case of simple substances, an experimental study of the thermal expansion of crystals is of considerable importance.

Various important points arise from the literature data published in the past, for example:

1. The stronger the structure of the crystal is, the lower is the expansivity

Megaw (164) attributes the low expansion of ZnS ($\alpha = 6.7 \times 10^{-6}$) as compared with that of MgO (13.5×10^{-6}) to the high bond strength resulting from the low coordination. According to Weyl (276), the rule of Megaw that links expansivity with the strength of the binding forces seems to have a wide range of application. The conclusions of Megaw were, for example, confirmed by Rigby (224) in the case of silicates: for comparable structures, the compound with the greatest bond strength has the lowest expansivity. This relationship applies also within a crystal such as graphite or brucite where the forces are different in different directions.

Another aspect of this correlation of structure-expansivity was given in 1950 by Hummel (120): the expansivity is related to the complexity of the structure; crystals that have simple close-packed structures such as MgO or CaO have, as a rule, high expansivity; crystals that have more open structure and are more complex and less symmetrical, have, as a rule, a lower thermal expansion.

2. Compounds having same crystal geometry (case of $AlPO_4$ and of cristobalite) have essentially the same thermal expansivity and undergo a similar structural change at nearly the same temperature (119).

3. Although certain minerals are isostructural, their expansivities may differ greatly. That is the case, for example, for willemite, Zn_2SiO_4 ($\alpha = 3.2 \times 10^{-6}$), and phenacite, Be_2SiO_4 ($\alpha = 6.4 \times 10^{-6}$). Using the field strength of the cations or their radii as a guide, one would expect that the binding forces between the small Be^{2+} ion (0.34 Å) and the SiO_4 group would be much stronger than those between the larger Zn^{2+} ions (0.83 Å) and the SiO_4 groups. Nevertheless, according to Weyl (276) the difference between expansivities of these two minerals is very important. The importance of the strength of the binding forces must not be underestimated, but according to Weyl (276) we have to treat them on a par with geometrical considerations as they were emphasized primarily by Hummel. His work on the crystals that are isostructural with silica (119) revealed the paramount importance of the geometry. A perusal of the literature shows that data on the thermal expansion of alkali halides was, for a long time, the only example of data on a number of crystals belonging to the same crystal type. As seen in this chapter, these data have been used to correlate the thermal expansion with

interatomic distances, melting point, and so on. In the same way, it appeared that the cubic metals were the only solids for which thermal expansivity can be correlated with their melting point or with other thermodynamic quantities. In 1952, Rigby (224) examined the thermal expansion coefficients of a number of refractory materials belonging either to the *cubic* structure (oxides, spinels) or to more complex structures (silicates). He explained the different values of expansion coefficients on the basis of structure, distortion of the lattice, coordination number, nature of the packing of oxygen in the lattice, and so on. In 1958, Gibbons (96) gave information about thermal expansion of semiconductors of diamond structure (*cubic*) and showed that silicon and indium antimonide have negative values for γ (Grüneisen constant) at low temperature and some of the requirements for a structure to behave in this manner were suggested, namely, fourfold coordination in the lattice, covalent bonding, and openness of structure.

More recently, many authors have reported new information about thermal expansion and crystal structures of solids (other than alkali halides). For example:

1. Krishna Rao (146) studied the case of a number of crystals belonging to the *rutile type* or to the *calcite type*. In the case of rutile-type structure, in general, the coefficient of thermal expansion increases with the increase in the percentage ionic character and can be correlated to the interionic distance and the electronic configuration of the ions. In the case of calcite-type structure, the thermal behavior shows large variations in the value of the coefficients of thermal expansion, but α decreases linearly with increasing cation radius, the coefficients of nitrates being higher than the coefficients of carbonates.

2. Bayer studied in 1971 (27) the thermal expansion of various oxide compounds crystallizing with the *orthorhombic pseudobrookite* structure and showed that the usual expansion anisotropy appears to be a structure-related property. All pseudobrookite compounds are mechanically very weak due to their strong expansion anisotropy.

Numerous anomalies appear when one tries to establish correlations between crystal structure, melting temperature, and thermal expansion coefficient. A typical example is that of the compound $3\ CaO \cdot Al_2O_3$ which forms in the phase diagram $CaO - Al_2O_3$ (10,225). This compound, which has a fairly high melting point and a tetragonal structure, shows an anomalously low expansion.

Another interesting case is that of the variation of thermal expansion coefficient of solid solutions. This problem was examined by Austin (10) but it would take too long to discuss that in this chapter. The particular problem of the thermal expansion of nonisotropic crystals will be discussed in the next paragraph (Section I.C.7.).

7. Crystal Symmetry—Anisotropy of Thermal Expansion

In a compound of complex structure, the forces associated with different bonds may differ greatly, resulting in a marked anisotropy in expansion. This anisotropy is associated with crystalline symmetry in a very definite way, analogous to anisotropy in the conduction of heat or the refraction of light in crystals. Two cases can be considered:

1. Isotropic crystals. In crystals belonging to the cubic system the expansion is equal in all directions and the following relation holds:

$$\alpha_v = 3\,\alpha_L \tag{27}$$

2. Anisotropic crystals. Crystals that do not have cubic symmetry expand differently in different directions so that the angles between faces change with temperature.

In crystals with rotational symmetry, which belong to one of the optically uniaxial systems (tetragonal, hexagonal, or trigonal symmetry), the maximum and minimum expansions occur in the direction of the axis of symmetry and in any direction perpendicular thereto. Two measurements, one in the direction of the principal axis and another perpendicular to it (equatorial direction), are required to derive the cubic expansion

$$\alpha_v = (\alpha_L)_{\parallel} + 2\,(\alpha_L)_{\perp} \tag{28}$$

The expansion coefficient in any direction making an angle φ with the axis of rotation varies with the square of $\cos\varphi$ according to the relation

$$(\alpha_L)_{\varphi} = (\alpha_L)_{\perp} + [(\alpha_L)_{\parallel} - (\alpha_L)_{\perp}]\cos^2\varphi \tag{29}$$

$(\alpha_L)_{\parallel}$ and $(\alpha_L)_{\perp}$ being coefficients parallel to and perpendicular to principal axis, respectively.

The plot of $(\alpha_L)_{\varphi}$ against $\cos^2\varphi$ yields a straight line, which provides a useful method for obtaining the value of $(\alpha_L)_{\perp}$ and of $(\alpha_L)_{\parallel}$.

In other crystals, which belong to one of the optically biaxal systems (rhombic, monoclinic, or triclinic symmetry), there are two directions perpendicular to each other along which the expansion is respectively maximum or minimum whereas the expansion in a direction perpendicular to both is intermediate. The three mutually perpendicular directions form the axes of a triaxial ellipsoid analogous to the optical ellipsoid.

This general case of nonisotropic crystal deformation was developed in the past century by Fizeau. The theory of "homogeneous deformation" of Fizeau (89), shows that a sphere of unit radius at a given temperature in any crystal changes at a neighboring temperature into an ellipsoid, whose axes serve, in length and direction, to determine the expansion of the crystal over this temperature range. Let the expansion coefficients in the direction of the axes be α_1, α_2, and α_3, then the semi-axes of the ellipsoid corresponding to a temperature change of 1°C are $1 + \alpha_1$, $1 + \alpha_2$, $1 + \alpha_3$; and the expansion coefficient α_L in any other direction, making angles φ_1, φ_2, φ_3 with the axis of the ellipsoid, is given by the equation

$$\frac{1}{(1 + \alpha_L)^2} = \frac{\cos^2\varphi_1}{(1 + \alpha_1)^2} + \frac{\cos^2\varphi_2}{(1 + \alpha_2)^2} + \frac{\cos^2\varphi_3}{(1 + \alpha_3)^2} \tag{30}$$

In the case of an uniaxial crystal, the ellipsoid becomes an ellipsoid of revolution with its axis coincident with the crystallographic axis; its orientation is thus fixed and $\alpha_1 = \alpha_2 = (\alpha_L)_\perp$. In the case of the cubic symmetry $\alpha_1 = \alpha_2 = \alpha_3$, the ellipsoid becomes a circle.

To determine the cubical (or volumetric) thermal expansion coefficient from linear measurements, it is necessary to have data for three mutually perpendicular directions and the cubical coefficient is the sum of these three linear coefficients.

$$\alpha_v = \alpha_1 + \alpha_2 + \alpha_3 \tag{31}$$

If we refer to the works of Fizeau (89), we can consider that the knowledge of the volume change or of the cubic dilatation can be obtained in fact in two ways: (1) either by the determination of one, two, or three distinct linear coefficients according as the crystal belongs to the categories (a) or (b) discussed before or (2) by making a measurement of the expansion along a direction which is equally inclined at 54° 44' to the three axes of elasticity of the crystal. In this case the measured dilatation is equal to the average linear dilatation, whatever the crystalline system of the substance and whatever the values (positive or negative) of the three principal expansions of the crystal.

That leads to the following geometrical considerations: for a small temperature increase and whatever the crystalline system to which the crystal belongs, the expression which gives the increase in unit length along any direction making the angles φ_1, φ_2, and φ_3 with the three dilatation rectangular axis (axis which coincide with the crystallographic axis) in terms of the three principal expansion coefficient α_1 α_2 and α_3, is

$$D = \alpha_1 \cos^2\varphi_1 + \alpha_2 \cos^2\varphi_2 + \alpha_3 \cos^2\varphi_3 \tag{32}$$

The relation between the three angles φ_1, φ_2, and φ_3 is

$$\cos^2\varphi_1 + \cos^2\varphi_2 + \cos^2\varphi_3 = 1 \tag{33}$$

If $\varphi_1 = \varphi_2 = \varphi_3$ (direction equally inclined on the three axes), the relation in equation 33 becomes

$$\cos^2\varphi = \tfrac{1}{3} \quad \text{and} \quad \varphi = 54°44'$$

Equation 32 then becomes

$$D = \frac{\alpha_1 + \alpha_2 + \alpha_3}{3} \tag{34}$$

That is the average dilatation of a crystal in the general case.

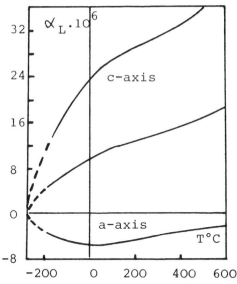

Fig. 7.3. Anisotropic expansion in a single crystal of calcite (10).

Fizeau (90–92) obtained interesting experimental results on many minerals such as quartz, diamond, rutile, gypsum, and periclase, by this method.

A typical example of expansion anisotropy is that of calcite (Fig. 7.3). On heating, an expansion occurs along the c axis but along the a axis a contraction is observed at temperatures below 0°C.

In explaining the mechanism of thermal expansion, it has been assumed that the anharmonic vibration of atoms is such that a crystal always expands on heating. This is generally true for the volume expansion but in a complex compound it is quite possible that the forces acting in certain directions may be such that contraction may occur in other directions in a crystal. That is the case for calcite: the low-temperature negative coefficient along the a axis has been ascribed to the tendency of the carbonate ions to rotate out of their own plane so that the distance between carbon and oxygen atoms as projected on the plane decreases with increasing temperature (10).

Two other examples are given on Fig. 7.4: that is, the case of the anisotropy of dilatation of TiO_2 and of GeO_2 along the a and c axes. A polycrystalline aggregate of the material gives an intermediate expansion.

The anisotropy of thermal expansion can lead to the formation of mechanical stresses between adjacent crystals of different orientations. That promotes a nonreversibility of the thermal expansion (hysteresis in dilatometric curves) in the case of a low-expansion anisotropy. In numerous cases, the stress level resulting from the expansion anisotropy can exceed the elastic limit of the material. For metals and alloys, a wide relaxation of these stresses occurs by plastic deformation of the material which promotes dimension change of the pieces after several heating and cooling cycles. In the case of polycrystalline ceramics with anisotropic expansion, stresses lead to forma-

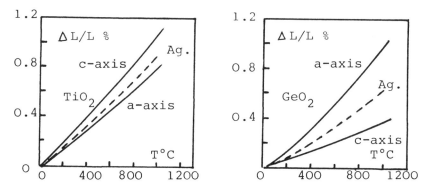

Fig. 7.4. Axial (———) and aggregate (- - - - - -) thermal expansion curves for TiO_2 and GeO_2 (Reprinted, with permission, from Ref. 232.)

tion of microcracks and the propagation of these through the material can promote the breaking of the piece.

8. Defects in Solid State

a. VACANCIES AND INTERSTITIALS

The presence, in a crystal, of *vacancies* or *interstitials,* defects which can be created by thermal agitation can yield differences between the expansion curves obtained either by X-ray methods or by dilatometer methods. At high temperature and just before melting point, the $\Delta L/L$ values obtained by the dilatometer method are slightly higher than the $\Delta a/a$ values (a = lattice parameter) otained by X-ray method. This difference can be calculated theoretically and is a consequence of the variation of the density of the solid. This variation of density affects the macroscopic thermal expansion but has only very slight effect on the lattice parameters.

If N is the number of atoms contained in a metallic crystal and V_a the volume of one atom, the relaxation effect of atoms near a vacancy leads to a slight variation of the volume of this vacancy which becomes smaller than the volume of the atom. The volume of the vacancy is fV_a, where f is a coefficient slightly less than 1. To remove n_1 atoms out of the lattice has the effect of replacing their volume $n_1 V_a$ by the smaller volume $n_1 f V_a$ of vacancies. The relative volume variation (contraction) of the lattice becomes

$$n_1 V_a (f - 1)/NV_a$$

and affects the X-ray measurements. The n_1 atoms, the removal of which has created vacancies inside the solid, are located on the surface of the crystal. The relative variation of volume of the crystal (which affects the macroscopic expansion measured by the dilatometer method) is

$$n_1 f V_a / NV_a$$

So, the difference of expansion, measured by the two methods, is

$$3\left(\frac{\Delta L}{L} - \frac{\Delta a}{a}\right) = \frac{n_1}{N} \tag{35}$$

This difference is the concentration of vacancies.

Experimentally, very accurate dilatometer methods (optical methods) have to be chosen for observing this difference. Some examples have been published in the literature, such as those of lead, aluminum, silver, gold, and copper (236). In the case of aluminum, the difference between X-ray and dilatometer methods is of an order of 0.03% at 650°C and corresponds to a vacancy concentration of 9.10^{-4} (246). The difference between X-ray and optical dilatometer method appears above 900°C for gold and 1100°C for platinum (86).

A similar effect is observed in the case of solids containing interstitials, but here the $\Delta L/L$ curve lies below the $\Delta a/a$ curve.

b. Nonstoichiometry

Another important cause of variation of the expansion coefficient is the *noinstoichiometry* of solids. For instance, the measurements by X-ray methods of the expansion coefficient of non stoichiometric titanium dioxide show that the variation of α_L must be very important (160,169).

For pure TiO_2:

$$\alpha_a = 7.86 \times 10^{-6} \qquad \alpha_c = 9.92 \times 10^{-6} \qquad \bar{\alpha}_L = 8.55 \times 10^{-6}$$

for $TiO_{1.97}$:

$$\alpha_a = 6.8 \ \times 10^{-6} \qquad \alpha_c = 10.45 \times 10^{-6} \qquad \bar{\alpha}_L = 8.00 \times 10^{-6}$$

The nonstoichiometry of TiO_2 (tetragonal) is due to oxygen vacancies which form Ti^{3+} ions, the ionic radius of which (0.76 Å) is higher than for Ti^{4+} ions ($r = 0.68$ Å).

c. Doping of Solids

The low-temperature thermal expansion of solids can also be disturbed by defects created by the *doping of solids*. That is the case for solids such as alkali halides containing small quantities of OH^- impurities or for CN^--doped NaCl and Li^+-doped KCl. These anomalies (Fig. 7.5) have been ascribed to tunnel-splitting of the impurity ground states and can be expressed in term of Grüneisen parameters which range from 40 for OH^- in NaCl to 300 for Li^+ in KCl (51).

d. Dislocations

Another aspect of the influence of defects on the thermal expansion of solids is seen in the role played by dislocations. The hammer-hardening of a metal increases the dislocation content which places the atoms of the material in a stress field and promotes local relative deformations of the lattice.

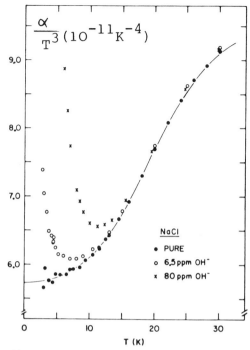

Fig. 7.5. Influence of impurities on the low-temperature expansion of NaCl. (Reprinted, with permission, from Ref. 51.)

The thermal expansion coefficient is then modified and dilatometric curves during the first heating will differ from the curve of non-hammer-hardened metal; a small curvature towards temperature axis appears in the curve and an hysteresis takes place during the first cycle. But this first heating treatment, if realized at a sufficiently high temperature, "cures" defects, and a normal dilatometric curve is obtained again during a second cycle.

9. Textural Properties—Expansion of Aggregates

It is well known that thermal expansion of aggregates may differ from that of a single crystal. The exact role of porosity was not clear in the past. For example, Rigby (224) found experimentally that increasing the porosity of a fireclay material from 15.3% to 43.5% increased the coefficient of expansion from 5.4×10^{-6} to 5.9×10^{-6}; earlier, however, Houldsworth and Cobb (117) had concluded that expansion decreased with an increase in porosity.

According to Austin (10), who studied the way in which the thermal expansion can be influenced in the case of aggregates, two extreme types of porosity can be taken into account: (a) one in which there is a continuous matrix of solid material which contains holes, and (b) a bonded group of solid grains surrounded in part by air spaces, an extreme case being a mate-

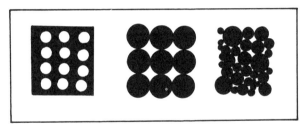

Fig. 7.6. Schematic illustration of different types of porosity (10).

rial composed of sand grains sintered to each other at a few points. The schematic illustration of different types of porosity is shown in Fig. 7.6 (10).

In case (a), the observed expansion of the solid is simply that of the matrix material, the holes having no effect. A typical example is that of a series of alumina bricks of different porosity in which there was essentially a solid matrix with the pores scattered through it: the thermal expansion of such materials is exactly the same as that of the matrix material, as determined on a smaller sample free from holes.

Case (b) is more complex. The thermal expansion of an aggregate of nonisotropic crystals is difficult to predict and often shows evidence of hysteresis which is not observed in the case of the dilatation of a single crystal.

Three cases were discussed by Sosman (248). They are illustrated schematically in Fig. 7.7a:

1. Behavior of an aggregate of isotropic crystals. If the heating rate is slow enough to avoid a large temperature gradient, each grain expands equally in all directions and there is no differential movement among them. On cooling, the aggregate returns to its original size and shape. The thermal expansion curve does not show hysteresis (Fig. 7.7a). However, some differential movement may arise and may cause the formation of intergranular spaces when the heating rate is too high.

2. Behavior of an aggregate of nonisotropic grains. During heating, the grains expand differently in different directions and push each other apart with the resultant formation of open spaces. During cooling, these openings tend to close, but never quite retrace their first movement, leaving some spaces unfilled. So, the dilatometer curve presents an hysteresis (Fig. 7.7a). When heated a second time, there is a tendency to retrace the first cooling curve, but there is, in addition, usually a small amount of new differential motion resulting in an additional permanent expansion not so large as the first one. The dilatometric curve of the second heating–cooling cycle does not coincide with the first one. It is displaced towards expansion and presents an hysteresis. The attainment of a reproducible expansion curve sometimes requires a number of heating–cooling cycles and, therefore, is not a

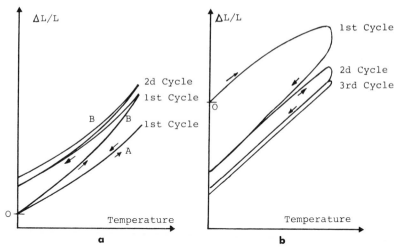

Fig. 7.7. Thermal expansion curves of aggregates. (*a*) Curve A, Reversible curve obtained with an aggregate of isotropic substance; curve B, Nonreversible curves obtained with an aggregate of aeolotropic substance (248). (*b*) Evolution of dilatometric curves obtained with a compacted powder during different heating and cooling cycles.

proof that the final curve represents the true dilatation of the material. The dilatometric curves of rocks obey this general rule of expansion during heating cycles, as shown by Thirumalai and Demou (261).

In conclusion, the expansion of an aggregate may vary, depending on the orientation of its components and on the extent to which they are bonded. According to Austin (10) in many cases, the expansion of the aggregates appears to be less than that of a single crystal, but this is by no means generally true. On the other hand, the dilatometric curve is not always displaced towards expansion when the sample is submitted to several heating cycles. That is the case, for example, for some composite materials heated up to 150–200°C. The shrinkage observed during the first heating cycle is associated with a small weight loss from 0.1% to 0.3% (wt) and is related to absorbed moisture (195).

Another interesting case is that of the thermal expansion of *compacted powders*. Generally, during heating, a creep of variable intensity can appear inside the sample because grains are not well bonded. That leads to a shrinkage that can be enhanced by the beginning of a sintering process at high temperature. So, the dilatometric curves do not present the same aspect as described by Austin (Fig. 7.7*a*) but are of the type shown in Fig. 7.7*b*. It is necessary to operate several heating cycles (3–5) to obtain a stable reversible dilatometric curve.

The form of the first dilatometric curve can be influenced by many textural factors directly or indirectly related to porosity: these factors are grain size, specific surface, compacting pressure (in the case of compacted pow-

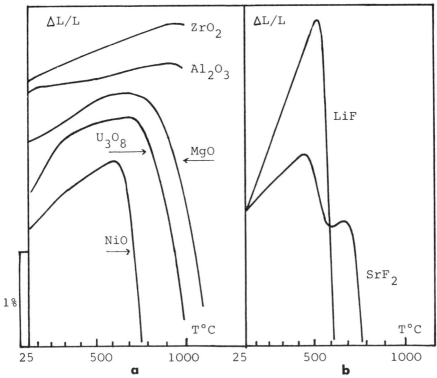

Fig. 7.8. Dilatometric curves of some metallic oxides (*a*) and fluorides (*b*) compacted powders [(Author's results) (181).]

ders), crystallization state, etc. The influence of such factors varies with the nature of the solids investigated and, for a given solid, with its mineralogical or textural properties.

For example, in general the contraction observed in dilatometric curves of refractory oxides such as ZrO_2 or Al_2O_3 is very small in the case of compacted powder and does not affect greatly the curve up to 1000°C (Fig. 7.8). It is more important in the case of other oxides, such as U_3O_8 or NiO, the sintering of which appears between 700°C and 800°C (Fig. 7.8*a*). Metallic salts such as fluorides (181–190), which do not have very high melting temperatures, may show very different dilatometric curves, for example high grain-size ($\phi > 1$ μm) and low-specific-surface ($< 5m^2 \cdot g^{-1}$) fluorides contract above 400°C in one step (case of LiF) or in two steps (case of SrF_2) (Fig. 7.8*b*).

For a given solid (for instance MgF_2), the variation of the temperature θ_F (above which the thermal expansion coefficient becomes negative) versus the compacting pressure P_c of the powder is not very large. On the other hand, the amplitude of the contraction that occurs above θ_F and θ_F itself depend on the textural characteristics of the powder (porosity, specific sur-

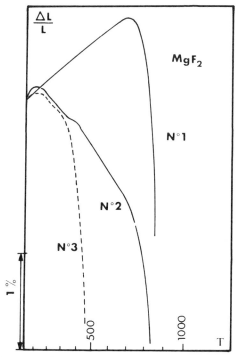

Fig. 7.9. Dilatometric curves of three samples of MgF₂ (compacted powder) [(Author's results) (181).]

face). The study of three MgF_2 samples of different texturals properties, showed that (Fig. 7.9): (a) a fine grain (0.03–1 μm) and high-specific-surface (44 m² · g⁻¹) sample (MgF_2 number 3) contracts from 80°C and the contraction is large (4.5% in length at 450°C); (b) a large grain (> 1 μm) and low-specific-surface (< 5m² · g⁻¹) sample (MgF_2 number 1) contracts only at high temperature (700°C); (c) a sample with large grain (0.3–1 μm) and very high specific surface (110 m² · g⁻¹) (MgF_2 number 2) contracts continuously and slowly above 100°C and the amplitude of contraction lies between that observed with the samples numbers 1 and 3. This phenomenon can be due to the high internal porosity of the grains of this sample.

This example shows the complexity of the influence of many textural characteristics of the solid. The interpretation of the phenomena observed by dilatometry needs the use of many other techniques such as microscopy, grain size, porosity and specific surface determination, etc.

From a more general point of view, variations in thermal expansion coefficients from point to point in a polycrystalline ceramic body or single crystal give rise to localized or large-scale residual stresses when the temperature is changed.

Deviation of the coefficient of thermal expansion between the heating and cooling curve of a polycrystalline body can be explained by Bussem's basic approach (43) to the creation of internal ruptures and recombinations in anisotropic polycrystalline materials. A polycrystalline body will develop internal tensile stresses, voids, and ruptures in its microstructure during cooling from its fabrication temperature; these stresses, voids, and ruptures are caused by the *anisotropy* of the thermal expansion of individual crystallites within the body. The only mechanism for the relief of tensile stress set up in the body is the creation of microcracks. The tendency to form these cracks increases with increasing grain size.

The problem of the thermal expansion of aggregates and of compacted powders becomes more complicated when the sample is subjected to chemical or structural transformations when heated. So, in an aggregate, some transformations, such as that of quartz or cristobalite can be entirely or partially masked as discussed later.

D. EXPANSION OF MIXTURES AND OF COMPOSITE MATERIALS

On cooling a material or a body consisting of two or more intimately mixed phases (crystal–crystal, glass–crystal, metal–ceramic, etc.) from its firing temperature, each phase is restrained to the same over-all contraction. In the case of two or more phases having different expansion coefficients, this restraint leads to substantial stresses on each phase and to an intermediate over-all expansion coefficient. In some cases, differences in expansion coefficients may produce stresses sufficient to cause plastic deformation or cracks which decrease strength, elasticity, and thermal conductivity in the material and lead to thermal expansion hysteresis.

This problem of correlations between thermal expansion and microstresses in composite materials was studied many years ago, particularly by Turner (269) and Kingery (136) and later by many other authors. But it seems even today the question is not entirely resolved.

In Turner's method for calculating coefficients of thermal expansion of mixtures, Turner (269) considers that an internal stress system in a mixture is such that the stresses are nowhere sufficient to disrupt the material, the sum of the internal forces can be equated to zero, and an expression of the thermal expansion coefficient of the mixture is obtained. When small particles or fine filaments are incorporated into a mixture, the small dimensions appear to permit combinations of materials that would be incompatible on a larger scale.

If it is assumed that each component in the mixture is constrained to change dimensions with temperature changes at the same rate as the aggregate and that shear deformation is negligible, the stress acting on the particles of the various components can be written (product of volume strain and bulk modulus):

$$S_i = [(\alpha_v)_r - (\alpha_v)_i] \Delta T \, B_i \tag{36}$$

where $\Delta T = \Theta_0 - \Theta$, with Θ_0 being temperature of zero stress (near the firing temperature), S the stress, α_v the volume expansion coefficient, and B the bulk modulus. Subscripts i and r refer to the property of the ith component and of the resultant mixture, respectively.

The resultant of the forces acting on any cross section of the mixture must vanish. Therefore,

$$\Sigma S_i A_i = 0 \tag{37}$$

where the A refers to the part of the cross-sectional area formed by the various components. However, in a homogeneous mixture, the relative areas formed in the cross section by the different components are proportional to their relative volume. Therefore, it follows from equation 37 that

$$\sum S_i V_i = 0 \tag{38}$$

Substitution for S from equation 36 in equation 38 then yields:

$$\sum [(\alpha_v)_r - (\alpha_v)_i] \Delta T V_i B_i = 0 \tag{39}$$

As

$$\sum V_i = V, \qquad \text{then } V_i = \frac{P_i d_r V_r}{d_i}$$

(where P is the fraction or percent by weight, V the volume, and d the density) which can be substituted for V in equation 39. Also, as ΔT, dr, and Vr are common factors, they can be eliminated from each term of the expression. Solving for $(\alpha_v)_r$, the following expression is obtained:

$$(\alpha_v)_r = \frac{\displaystyle\sum \frac{(\alpha_v)_i P_i B_i}{d_i}}{\displaystyle\sum \frac{P_i B_i}{d_i}} \tag{40}$$

As to the coefficient of linear thermal expansion (α_L) is directly proportional to the cubical coefficient, α_L can be substituted wherever α_v appears with the following

$$(\alpha_L)_r = \frac{\displaystyle\sum \frac{(\alpha_L)_i P_i B_i}{d_i}}{\displaystyle\sum \frac{P_i B_i}{d_i}} = \frac{\displaystyle\sum \frac{(\alpha_L)_i P_i}{d_i \beta_i}}{\displaystyle\sum \frac{P_i}{d_i \beta_i}} \tag{41}$$

with $\beta = 1/B = $ value of the isothermal compressibility coefficient. It is apparent by inspection that equation 41, based on stress equilibrium, reduces to a percentage by volume calculation if the ingredients have the same

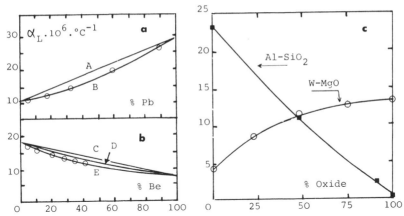

Fig. 7.10. Thermal expansion coefficient of mixtures or composite materials. (a) Lead-antimony mixture. Curve A, Percentage by weight calculation; curve B, percentage by volume or stress equilibrium calculation (269). (b) Beryllium-aluminum alloy. Curve C: Percentage by weight calculation; curve D, percentage by volume calculation; curve E, percentage by stress equilibrium calculation (269). (c) Tungsten-magnesia system and aluminum-SiO₂ glass mixture. (Reprinted, with permission, from Ref. 136.) (○, □) Experimental points.

bulk moduli. If the ingredients have the same modulus-to-weight ratios, the calculation amounts to a percentage by weight interpolation.

According to Kingery (136), for mixtures having the same value for Poisson's ratio, B can be replaced by modulus of elasticity, E, giving

$$(\alpha_L)_r = \frac{\sum \dfrac{(\alpha_L)_i P_i E_i}{d_i}}{\sum \dfrac{P_i E_i}{d_i}} \tag{42}$$

which can also be used to a good approximation if bulk moduli are not available.

Relation 42 has been verified experimentally by Turner for certain two-phase compositions such as mixtures of lead and antimony (Fig. 7.10a), beryllium and aluminum (Fig. 7.10b), and for some plastic compositions. This calculation was also verified later by Kingery (136) for W-MgO and Al-SiO₂ compositions (Fig. 7.10c). It was shown that the thermal expansion data are not simply averages of end-member values but agree quantitatively with the calculations of Turner, based on the assumption of substantial residual microstresses resulting from the restraint of each phase on cooling. However, although the calculation of Turner is relatively simple, it does not seem to be general and cannot be applied for all composite systems.

Other equations for predicting thermal expansion of composites were proposed later for two-phase systems (subscripts d and m pertain, respectively, to dispersed phase and to matrix phase; G, shear modulus; v, volume fraction; μ, Poisson's ratio).

Turner's equation is:

$$(\alpha_L)_r = \frac{[(\alpha_L)_d v_d B_d + (\alpha_L)_m v_m B_m]}{v_d B_d + v_m B_m} \qquad (43)$$

Kerner's equation (129) is:

$$(\alpha_L)_r = \frac{[(\alpha_L)_d v_d B_d/(3B_d + 4G_m)] + [(\alpha_L)_m v_m B_m/(3B_m + 4G_m)]}{[v_d B_d/(3B_d + 4G_m)] + [v_m B_m/(3B_m + 4G_m)]} \qquad (44)$$

Blackburn's equation (7) is:

$$(\alpha_L)_r = \frac{[(\alpha_L)_d + v_m((\alpha_L)_m - (\alpha_L)_d)]\,\tfrac{3}{2}\,(1 - \mu)}{\tfrac{1}{2}\,(1 - \mu_d) + v_m(1 - 2\mu_d) + (1 - v_m)(1 - 2\mu_m)} \times \frac{E_d}{E_m} \qquad (45)$$

Thomas's equation (264) is:

$$(\alpha_L)_r = (\alpha_L)_d^{v_d} \times (\alpha_L)_m^{V_m} \qquad (46)$$

Tummala and Friedberg's equation (268) is:

$$(\alpha_L)_r = (\alpha_L)_m - \frac{(1 + \mu_m)/2\, E_m}{[(1 + \mu_m)/2\, E_m] + [(1 - 2\,\mu_d)/E_d]}$$
$$\times V_d[(\alpha_L)_m - (\alpha_L)_d] \qquad (47)$$

(This equation is valid only for isotropic dilute binary composites.) Fahmi and Ragai's equation (80) is:

$$(\alpha_L)_r = (\alpha_L)_m - \frac{3[(\alpha_L)_m - (\alpha_L)_i][1 - V_m]\,\chi_i}{(2E_m/E_i)(1 - 2\mu_i)\chi_m + 2\chi_i(1 - 2\mu_m) + (1 + \mu_m)} \qquad (48)$$

This equation concerns a matrix, or shell (subscript i) containing spherical inclusions of the second substance (subscript m); x_m and x_i are the volume fraction of the matrix and of the inclusion, respectively; E is the Young modulus.

The application of these equations to different composites was made in 1970 by Tummala and Friedberg (268). It shows that if some equation can be verified experimentally for certain practical systems, poor agreement was found between theoretical analysis and experimental results with certain other systems.

Fahmi and Ragai (80) who studied the thermal expansion behaviour of the two duplex systems lead-fused silica and aluminum–silicon, found that the expansion coefficients always fell below those predicted by the simple rule of mixtures. Equation 48 proposed by these authors fits their experimental results better than Turner's formula.

Some particular cases were also studied, for example:

1. The case of *unidirectional fiber composites*. The thermal expansion of such materials was predicted by Fahmy and Ragai-Ellozy (82) by the use of a discrete element method of analysis taking into account a representative "cell" which is selected and subdivided into triangular elements. Matrix equations relating the forces and displacements at each vertix, with allowance for thermal expansion together with equilibrium, compatibility, and boundary conditions, provide the longitudinal and transverse thermal expansion coefficients. It was found that in the case of examples studied (boron–aluminum composites, graphite–epoxy composites), Poisson's ratios of the fiber and its transverse Young's modulus have very little effect on the thermal expansion behavior of the composite.

2. Nakamura and Larsen (195) studied the thermal expansion of unidirectional and balanced symmetrical unidirectional-angle ply-laminated boron and graphite fiber-reinforced resin matrix composites. Based on stress equilibrium and strain compatibility considerations, as discussed by Schapery (234) and Ashton et al. (8), the composite thermal expansion coefficient in the longitudinal direction (0°) (parallel to reinforcing fibers) is given analytically by an expression analogous to Turner's equation:

$$\alpha(0°) = (\alpha_M V_M F_M + \alpha_F V_F E_F)/(V_M E_M + V_F E_F) \tag{49}$$

where α_M, α_F = expansion coefficients of matrix and fiber, respectively; V_M, V_F = matrix and fiber fractions; E_M, E_F = Young's moduli of matrix and fiber. In the transverse direction (90° fiber orientation), the composite expansion coefficient (90°) is given analytically by Kreider and Patarini (145) as:

$$\alpha_{(90°)} = (1 + \mu_m)\alpha_M V_M + (1 + \mu_F)\alpha_F V_F - \alpha_{(0°)}\mu_{(0°)} \tag{50}$$

where μ_M, μ_F = Poisson's ratio of matrix and fiber, respectively.

3. The case of *laminated plates and cylinders* (103). The application of plane stress and plane strain equations leads to exact expressions for the composite expansion of laminated plates and cylinders. Under the assumption of equal Poisson's ratio for all layers (which introduces negligible error), these expressions simplify considerably and provide a conceptually simple relationship between the composite expansion and constituent properties. Calculations are too long for development here.

So, although a number of authors have proposed calculation methods for otaining the value of α in the case of mixtures or composite materials, it appears that the experimental data do not always fit the theoretical calculations; this is due to the complexity of the materials studied. It must be kept in mind that each composite or mixture shows its own textural properties, consequently it is difficult to propose a general equation α that is representative of all materials investigated.

Some other points should be discussed, such as the influence of *external stresses* upon the thermal expansion of materials, or the nature of *atmosphere* in which experiments are made (148). It appears that either the application of external stresses or the creep of high-plasticity materials promoted by their own weight can modify the thermal expansion coefficient values at elevated temperatures by deformation of the material. At high temperature, thermal expansion of materials can be modified by interactions between either the specimen and the surrounding atmosphere or the specimen and the material with which they are in contact (sample holder, push-rod). For example, measurements with oxide specimens when made in a neutral atmosphere or in a vacuum can lead to stoichiometric deviations by formation of defects such as anionic vacancies. When measurements are taken in an oxidizing atmosphere, cationic vacancies and intersticial oxygen ions can be formed. In the same way, sublimation phenomenon can take place at high temperature, particularly when experiments are made in a vacuum.

II. EXPERIMENTAL METHODS FOR MEASURING THERMAL EXPANSION

Three types of measurement can be made when length variations must be obtained with a sample XY placed in horizontal or vertical position along a *xx'* direction (Fig. 7.11): (a) to observe and to note these variations in two directions A and B at right angle of *xx'*, which leads to absolute measurements; (b) to fix the end X of the sample and transmit the change of place of the other end Y, via a push-rod placed in *xx'* direction, to a detection device that amplifies the length variations, which leads to relative measurements; (c) to fix the end X of the sample and to modify some physical properties (optical or electrical property) of an arrangement directly applied on the end Y of the specimen.

Case (a) corresponds to optical (direct optical sighting) or mechanical (measurement with lever system) dilatometers. Case (b) corresponds to pushrod dilatometers with mechanical, electrical, or optical detection systems. Case (c) corresponds to optical (interferometry) or electrical dilatometers. It is more adapted to low-temperature studies while cases (a) and (b) are more suited to high-temperature investigation. These types of measurements can suggest a first, but not too clear, way to classify dilatometers.

There are other classification ways such as low-temperature, middle-temperature, or high-temperature dilatometers, differential or absolute dilatometers, optical, electrical, or mechanical-dilatometers, or ways based on the nature of atmosphere around the sample—in air, in low or high pressure, in vacuum, in particular atmosphere (nitrogen, argon, water vapor).

We will adopt in this work a classification that is based on physical methods that characterizes method of measurement, i.e.: optical methods, electrical methods, mechanical methods, and X-ray diffraction methods.

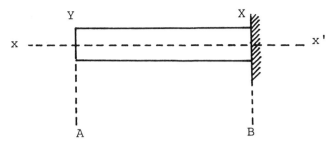

Fig. 7.11. Schematic way for measuring thermal expansion of a solid specimen.

As literature is particularly plentiful on this subject of techniques, all the articles or papers published cannot be cited here. We will just try to cite the better-known publications. A bibliography about this subject (152 references from 1952 to 1972) can be found in the work of Mignot (173) and another in the paper of Cizeron (61). Much information about dilatometers is also related in Yates (283), Rubens and Skochdopole (229) and in the Proceedings of Thermal Expansion Symposia (255–260).

A. OPTICAL METHODS

1. Direct Optical Measurement

The direct measurement of the length variation of the sample by observation perpendicular to the sample axis can be made with two optical field glasses or telescopes supplied with graduated eyepieces for direct reading or by use of optical comparator. Measurements can be made to temperature up to 2500–2800°C, as was described in 1953 by Chalmin (52) who heated the sample with a graphite resistor furnace or with a solar furnace.

The apparatus described by Miccioli and Shaffer (171) and used by Naum et al. (196–197) to study graphite and carbon–carbon composites, permits measurements up to 2843°C. Reproducibility of $\pm 12.7 \times 10^{-4}$ cm in length determinations yields accuracy of $\pm 0.02\%$. Changes in sample length are measured directly by simultaneous sighting on the ends of the sample by means of two 8-in. focal Gaertner optical micrometers. The limit of measurement of each optical micrometer is 2.54×10^{-5} cm with a specified detectable displacement of $\pm 5.8 \times 10^{-5}$ cm at a focal length of 8 in. Temperature corrections for the second surface mirror and quartz window can be determined by calibration against a NBS tungsten ribbon lamp. A period of 15 min is employed to allow the temperature to stabilize after each power adjustment. At temperatures above 1300°C, the temperature variations along the length of the sample and reported temperatures are within approximately ± 20°C.

Nielsen and Liepold (203) make use of telemicroscopes for measuring the thermal expansion of materials (MgO, CaO, Al_2O_3, ZrO_2) at temperatures up

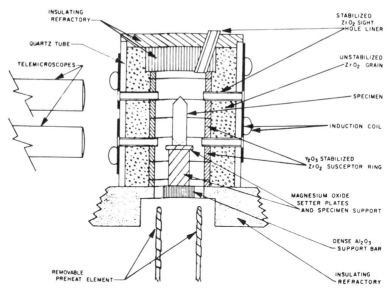

Fig. 7.12. Schematic cross-section of ultrahighfrequency induction furnace and the optical dilatometer of Nielsen and Liepold. (Reprinted, with permission, from Ref. 203.)

to 2200°C (Fig. 7.12), but their furnace is particular and original: heating is produced by induction (5–6 megacycles frequency) with an yttrium-stabilized zirconia suceptor ring (ZrO_2 + 8 mol % Y_2O_3). A preheating of the furnace up to 1200°C is necessary and it is realized with two resistance heaters. Above 1200°C, the electrical conductivity of the Y_2O_3–ZrO_2 system becomes sufficient for the establishment of current flow.

A twin-microscope technique was used also by Hahn and Kirby (107) to measure thermal expansion of platinum from 1000 K to 1900 K. Previously described by Rothrock and Kirby (228), it makes use of a pair of telemicroscopes, 50× magnification, rigidly attached to an invar bar. Filar micrometer eyepieces are used to compare the length of the specimen at elevated temperatures to that of a fused silica scale kept at room temperature. The fiducial marks that define the length of the specimen (nominally 10 cm) are machined into it in such a way that they form sharp images when viewed against either a bright or a dark background. A five-zone vacuum furnace is used to obtain a constant temperature over the length of the specimen under equilibrium conditions. The precision of the expansion is 24 μm · m^{-1}. The precision on the temperature (determined with a W–Re and a Pt–Rh 30-6 thermocouple) is 0.7 K.

Gaal described a specially built system capable of testing ten specimens concurrently by use of motor-driven microtelescopes (94).

2. Interferometry

It seems that the most outstanding early method of any notable precision which was suitable for use with all specimens was that due to Fizeau (87,88).

The method involved preparing blocks from crystals and polishing one of the faces. The crystal under investigation was placed on a table, above which was supported a glass plate. A beam of approximately monochromatic light was directed down onto the system and reflections occurred from the lower surface of the glass plate and the upper face of the specimen. When these two faces were adjusted to be at small angle to each other, the two reflected beams interfered upon being combined and movement of the fringe pattern accompanying subsequent temperature changes allowed the coefficient of thermal expansion to be measured relative to the expansion of the supports. Repeating the observations with the specimen absent facilitated allowance for the expansion of the support.

In the Fizeau's experiments, the fringes, or Newton's rings, were produced with a NaCl-colored flame light, and the apparatus was placed in a heating chamber allowing isothermal experiments at a temperature up to 80°C. With the raising of temperature, the displacement of fringes was measured with regard to fixed black points used as reference.

The length variation ΔL of the crystals is obtained by the determination of the number n of displaced fringes

$$n \text{ fringes} \Rightarrow \frac{n\lambda}{2}$$

$$\Delta L(\text{mm}) = \frac{n\lambda}{2} \tag{51}$$

The precision is $\Delta L(\text{mm}) = \dfrac{\lambda}{2}$ and depends on the wavelength of the light used.

This method has been employed by a number of later workers, particularly by Waterhouse and Yates (273), who make use of quartz or glass optical flats between adjacent faces, the interference of which is arranged to take place when these are separated by the specimen under investigation. The sensitivity was improved greatly and, for instance, Rubin et al. (230) produced circular fringes by directing a nonparallel beam of light on to parallel optical flats, producing a system capable of detecting values of $\Delta L/L$ of approximately 3×10^{-7}. Meincke and Graham (165) used a Fabry-Perot standard with a sensitivity $\Delta L/L$ of approximately 4×10^{-9} when using specimens 2 in. long.

Waterhouse and Yates (273) have described improvements to earlier versions of a Fizeau system with which a sensitivity of approximately 10^{-7} has been achieved. This apparatus represents an improvement over earlier verions of the Fizeau arrangement employed by Yates and Panter (284) and by James and Yates (124). The specimen chamber is shown in Fig. 7.13 and the optical system in Fig. 7.14. The experimental chamber is supported by three adjustable 2-mm-bore German silver tubes from the top of the brass jacket L, which is immersed in liquid nitrogen or liquid hydrogen. The space between C and L was normally evacuated and observations were made at temperatures maintained steady to within 0.01 K with the aid of the heater P

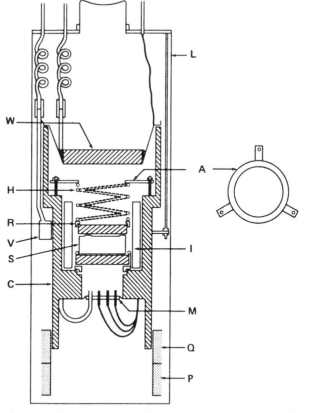

Fig. 7.13. Specimen chamber assembly used in the interferometric method of measuring thermal expansion at low temperature (124). L, outer jacket; A, spring compression ring; I, Indium resistance thermometer; M, metal-glass seal; Q, copper-sensing element; P, heating coal; W, window; H, helical brass spring; R, grooved collar; V, vapor pressure bulb; S, specimen; C, specimen chamber.

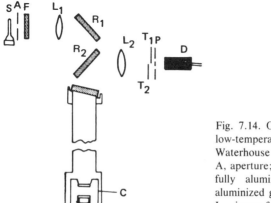

Fig. 7.14. Optical system employed in the low-temperature thermal expansion work of Waterhouse and Yates (273). S, mercury lamp; A, aperture; F, filter, L_1, collimating lens; R_1, fully aluminized glass plate; R_2 partially aluminized glass plate; C, specimen chamber; L_2, image-focusing lens; T_1, T_2, shutters; P, slits; D, photomultiplier tube detector.

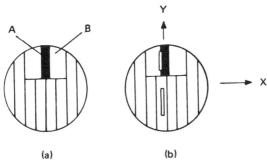

Fig. 7.15. Appearance of the interferometer used by Waterhouse and Yates (273). (a) Plan view of interferometer showing fully aluminized strip A on lower surface of upper flat over blackened region B of upper surface of lower flat. (b) Plan view showing slits in relation to fiduciary and fringe systems.

and the copper sensing element Q, which formed part of an automatic temperature controller.

The following fringe movement detection system is used: a fully aluminized strip A (Fig. 7.15a) was deposited on the partially aluminized lower surface of the upper flat. With the interferometer assembled, this lay above the region B of the fully aluminized upper surface of the lower flat, on which a piece of black tape had been attached, presenting the appearance of Fig. 7.15b, in which Fizeau fringes are shown in the rest of the field of view. The slits were lined up on the fiduciary and fringe systems as shown, through which light passed to a photomultiplier arrangement, the detector of the system being a galvanometer. Each slit was masked in the appropriate sequence, and the galvanometer was calibrated in terms of fringe displacements resulting from alterations of the pressure of the exchange gas between the interferometer plates by a carefully measured amount. Fringe movement accompanying a dimension change in sample resulting from temperature changes can be measured and is used to calculate α_L of the specimen. The vapor lamp which forms the original source of illumination, can be replaced by a laser possessing high-amplitude stability.

Instead of counting the passage of the fringes by eye and measuring the fractional parts with a filar micrometer eyepiece, a photographic technique was developed by Saunders (233) who utilizes a slowly moving 35-mm film strip positioned behind a slit. The recorded fringes on the film are called "interferograms." Once the interferogram has been developed, the change in the order of interference is determined by using a measuring densitometer (106).

The Fizeau interferometer is particularly well-suited for measurements with materials that can be employed as *reference standards*, such as copper, tungsten, etc. (see Section II.G.3); Hahn (104) has developed and reported the apparatus he made use of for the study of copper standard, either in low or in high temperature. The specimen length is 1 cm, and the green spectral

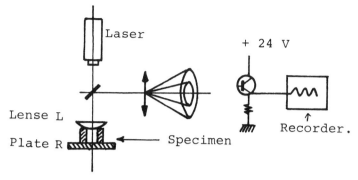

Fig. 7.16. Principle of the laser Fizeau interferometry applied to dilatometry (167).

line of mercury light source was used to produce the interference fringes. Fringe motion was measured with a filar-micrometer eyepiece. The expansion measurements were made from 20 K to 800 K.

Finally we must signalize the fast interferometric dilatometer developed by Ruffino et al. (231). The technique utilizes an image follower and a Michelson interferometer. The sample may be heated to temperatures close to the melting point in times of less than 1 sec.

A different laser application to dilatometric measurement was proposed some 10 years ago. The intrinsic coherence properties of laser beams, and particularly of gas laser beams, allow easy measurement of important length variations of optical paths with great sensitivity and with not too expensive an apparatus.

Laser Fizeau interferometry was developed by Plummer and Hagy (216) in 1968 for obtaining precise thermal expansion measurements on low-expansion optical materials. Plummer (214), with a 10-cm-long sample and a helium–neon laser, found the precision is of 0.1×10^{-6} cm. cm^{-1}, corresponding to a precision of 0.001×10^{-6} °C^{-1} (in ΔL) for an interval of temperature of 100°C. For example, measurements have been made by Plummer on two samples of Corning Code 7971 ULE titanium silicate glass and the values of α obtained were:

$$\alpha_L\Big]_{25°C}^{125°C} = 0.038 \times 10^{-6} \text{ °C}^{-1} \quad \text{and} \quad 0.047 \times 10^{-6} \text{ °C}^{-1}$$

In the laser Fizeau interferometry, an improved Fizeau arrangement is used (Fig. 7.16): a reflecting plate R upon which lies the specimen of tubular form; a plane-convex lens L placed upon the specimen; and a monochromatic light source (for example a helium–neon laser), the coherence length of which is suitable with respect to the optical path of the specimen.

When the laser beam falls upon the optical arrangement, reflection upon R and L gives rise to a system of interference fringes which can be projected upon a screen fitted out (or not) with phototransistor cells.

The expansion or shrinkage of the specimen promotes a change in the interference effect. The sensing of past-marching of the fringes and their number account for the nature (expansion or shrinkage) and the value of the length change of the specimen. For example, an expansion by a quarter of a wavelength ($\simeq 0.15$ μm in the case of He–Ne laser) converts an interference maximum to a minimum.

In the case of laser interferometry, the precision is not related to the variation in optical path length but to other parameters such as temperature gradient, lens motion due to length variation of the specimen, and variation of the gas refractive index. This precision can be improved by using large-focal-distance quartz lenses and by working in helium atmosphere.

Two laser Fizeau interferometric dilatometers (Models Sirius 1A and Sirius 2A), based on this principle and working in the temperature range of 10 K to 1300 K, were developed in France recently by Merard (167) in the C.E.A.,* to study for example the behavior of certain concretes during their short-term hydration.

In Europe, a He–Ne laser dilatometer has been commercialized by Linseis (155). Its construction is similar to a Michelson double-beam interferometer. The beam emerging from the laser is divided by a beam splitter (semitransparent mirror) into two beams—the measurement and the reference beam. After reflection at the mirrors, the measurement and the reference beams are recombined in the beam splitter to produce interference effects which depend on the difference in the optical path lengths of the two beams. The periodic interference fluctuations produced by an expansion of the specimen are detected by a photoelectric system.

Likewise, Jacobs et al. (123), of the University of Arizona, developed an ultraprecise laser dilatometer that yields absolute determinations of α which in turn can be used to calibrate other faster methods. The method relies on multiple-beam optical interference, with the sample used as a Fabry-Perot standard spacer. Extreme precision is made possible by monitoring in the electrical domain the thermally induced changes in optical interference. A frequency stable laser is used to probe the resonance of a Fabry-Perot standard whose mirrors are optically contacted to the end of the sample. The incident beam is electro-optically modulated to impress sidebands on it which can be moved in frequency until one coincides with a Fabry-Perot transmission peak. When the sample temperature is changed by an amount ΔT, the subsequent length change ΔL causes the Fabry-Perot resonances to shift an amount $\Delta \nu = (\nu/L)\Delta L$. This change in modulation frequency is related to the thermal expansion coefficient.

$$\alpha = (1/\Delta T)(\Delta L/L) = (1/\Delta T)(\Delta \nu/\nu) \qquad (52)$$

For small values of α, the precision of measurement is limited by the laser's frequency stability, which is approximately one part on 10^9. Measurements have been made in the temperature range 0–300°C.

---

* French Commissariat à L'Energie Atomique.

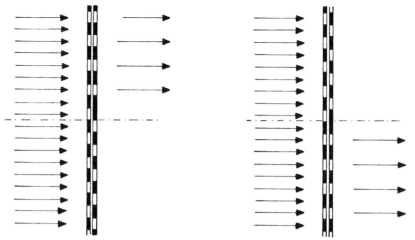

Fig. 7.17. Optical grids used by Andres (4).

3. Other Optical Methods

These are essentially the photometer method and the optical lever method. The Photometer method was proposed by Andres (4). It consists of two grids constructed from optically flat pieces of glass on which chromium had been deposited and upon which lines had been scratched with a discontinuity in the spacing as shown in Fig. 7.17a. The principle of operation was to maintain one grating fixed, while dilatational changes in the specimen were communicated to the second grid. If the disposition of the grids resembled the arrangement shown in Fig. 7.17a, it is clear that light might pass freely through the upper half of the assembly, while none would pass through the lower half. The reverse is true in the case shown in Fig. 7.17b, while in general a limited amount of light would be passed by both halves, the relative amounts being a periodic function of the relative displacement. This method is applied to measurement at low temperature, the arrangement being housed in a cryostat. The sensitivity $\Delta L/L$ claimed for this system is approximately 10^{-9}.

The optical lever method can be applied for measuring small displacements and a review of it was made by Jones in 1961 (127).

An image of the source S (Fig. 7.18) is focused onto a mirror M by a lens L_1. A lens L_2, which is identical to L_1, transmits this image to the split photocell D. In a subsidiary system, an image of a grid G_1 is cast on a grid G_2 with the aid of a lens L_3. The spacing of the gratings G_1 and G_2 resemble those of the Andres setup in that when light from the upper half of G_1 can pass through G_2, light passing through the lower half of G_1 will be stopped by G_2. A prism P deflects the light so that the light from half of the system falls on one photocell while the light from the other half falls on the other photocell. A slight rotation of the mirror M thus results in the image of G_1 moving

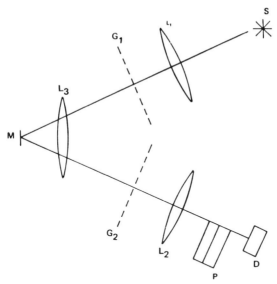

Fig. 7.18. Optical lever method. A typical optical level system: S, source; L_1, L_2, L_3, lenses; G_1, G_2, grids; M, concave mirror; D, split photocell; P, prism. (Reprinted, with permission, from Ref. 283.)

over G_2, which in turn results in a change in the difference of the response of the two photocells.

Huzan et al. (122) made use of an alternative form of the optical lever for the measure of low-temperature thermal expansion of aluminum. The sensitivity obtained was $\Delta L/L = 6 \times 10^{-8}$. Shapiro et al. (243) with the same type of method, obtained a sensitivity of $\Delta L/L = 4 \times 10^{-11}$ in the case of the expansion of copper at low temperatures. Bunton and Weintroub (44) reach a sensitivity of 2×10^{-9} by developing a combined adaptation of a lever–grid assembly.

Pereira et al. (210) described also a dilatometer which consists of a doubly twisted strip of beryllium copper with a mirror attached to its central region. This system is immersed in liquid helium. Dilatations of the sample are sensed by this system via a thin diaphragm, causing rotations of the mirror, which is detected by an external optical level. This dilatometer was employed for low-temperature thermal expansion study of copper. The sensitivity is of 10^{-11} in $\Delta L/L$.

B. ELECTRICAL METHODS

1. Capacity Measurements

It is well known that a small movement of one of the plates of a capacitor results in a change in capacity that may be measured electronically. This change may be related to the corresponding linear displacement.

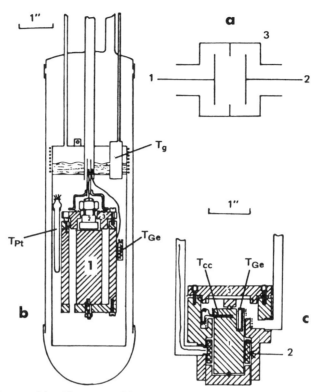

Fig. 7.19. The capacitive dilatometer of Carr et al. (50). Electrical capacitance between 1 and 2 is measured. Parts marked 3 form a guard ring and grounding shield. (*a*) Principle of measurement. (*b*) Differential cell. (*c*) Absolute cell. T_{Pt}, T_{Ge}, T_g, and T_{cc} are platinum-resistance thermometer, germanium thermometer, helium gas thermometer, and copper-constantan thermocouple, respectively. (Reprinted, with permission, from Ref. 50.)

In 1955, Bijl and Pullan (34) applied this technique to the measurement of thermal expansion at low temperature and the sensitivity obtained was 6×10^{-7}. Other authors, e.g., White (277) and Carr et al. (50), constructed capacitor dilatometers based on the principle described above.

The principle of the capacitor dilatometer of Carr, McCammon, and White (50) is to compare the capacitances of the three-terminal capacitor to parts in 10^8 with the aid of a bridge circuit based upon that of Thompson (264*b*). The specimen, represented by 1 on Fig. 7.19*b*, and its upper face, is polished flat, forming one electrode. The components, labeled 2 and 3 in the cryostat diagram, correspond to those bearing the same number in the capacitor diagram (Fig. 7.19*a*), the capacitance between 1 and 2 being measured, while 3 forms the guard ring and grounding shield. This dilatometer is a comparative one, giving the expansion of the specimen relative to that of the containing chamber. An absolute measurement of the linear coefficient of thermal expansion can be obtained by the use of an absolute cell (Fig. 7.19*c*).

The dilatometer is employed for low-temperature measurements. The sensitivity is $\Delta L/L \simeq 2 \times 10^{-10}$.

This apparatus has been used by White et al. to investigate thermal expansion of a wide variety of solids at low temperature [for example, samples of copper (278)] and the results of the researches represent certainly one of the major experimental contribution to low-temperature lattice dynamics.

The three-terminal capacitance dilatometer was also employed by Willemsen et al. (281) for the measurement of low-temperature thermal expansion of pure zinc and a dilute Zn–Mn alloy.

2. Transformer System

Another method based upon an electrical principle and having a high sensitivity is the variable transformer system described by Carr and Swenson (49) and Sparks and Swenson (250).

In this method, windings of the secondary coil of a transformer are wound in opposite directions. When this coil is centered within the primary coil, there is no net flux linkage between the two coils. Upon moving the secondary with respect to the primary, the mutual inductance changes, and this change may be used as a measure of the linear displacement causing it. Thermal expansion of the sample is communicated to the secondary coil, thermal isolation of the specimen being achieved with the aid of sapphire spacers at each end. The mutual inductance bridge used for detecting changes in mutual inductance, allows one to obtain a sensitivity of $\Delta L/L \simeq 3 \times 10^{-10}$. Case et al. (51), for instance, have made use of such a technique at the Iowa State University for measuring the low-temperature thermal expansion of alkali halides containing defects.

3. Electrical Resistance Measurements

Such an electrical method was used by Shrivastava and Joshi (245). Application of longitudinal stress on wires results in typical lattice distortions. Measured adiabatic and isothermal stress coefficients of electrical resistance, Poisson's ratio, Young's modulus, along with standard values of density and specific heat of the material, allow one to calculate its desired linear coefficient of expansion to quite a good accuracy. This method has been applied to the measurement of linear thermal expansion of Cu, Ag, and Au wires at 300 K.

C. MECHANICAL METHODS—PUSH-ROD DILATOMETERS

There are essentially two methods of determining the thermal expansion of a solid by mechanical means: (1) to measure the variation of length of the sample by use of amplifier lever systems; (2) to transmit the length variation to an amplifier measurement arrangement via a mechanical equipment, generally a push rod. The amplifier device is outside the furnace.

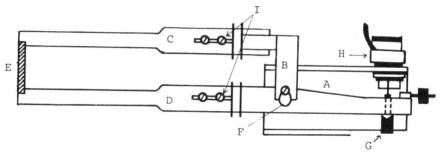

Fig. 7.20. Dilatometer of Baudran (25,26). A, Frame of the apparatus; B, invar body fixed to the apparatus frame, and protected from calorific flow by water-cooling I; C, sintered alumina bar fixed to the frame A; D, sintered alumina bar, movable around vertical axis F; E, specimen with 60-mm length; F, vertical axis; G, part of the movable bar D against which rests the movable part of the displacement transducer H; H, displacement transducer fixed to the apparatus frame A; I, water-cooling arrangement.

1. Mechanical Lever Systems

The simplest arrangement is that of Miehr et al. (172). The specimen is placed between two vertical fused quartz cylinders; the lower cylinder is fixed and serves as support, and the upper cylinder transmits the expansion of the specimen to a needle which moves on a graduated dial.

A more sophisticated and precise arrangement is that developed by Baudran (24–26) for studies of clays and ceramics. The specimen, placed inside a furnace is disposed at a right angle between the ends of two bars C and D, the first is fixed and the second can move around a fixed point F (Fig. 7.20). The expansion of the specimen promotes the displacement of the bar D around the point F. The end of this bar D is placed against the movable part of an inductive Philips displacement transducer, itself rigidly fixed on the apparatus frame. The voltage which is produced in the displacement transducer is measured by the use of an extensometric bridge and can be correlated directly to the expansion of the specimen. The furnaces developed by Baudran allow one to work in air or in controlled atmosphere up to 1650°C, the specimen dimensions being 50 mm in length and about 6–8 mm in diameter.

2. Push-Rod Dilatometers

a. Technical Aspects

Different arrangements can be used to measure the length variation of the sample. The push-rod can transmit the expansion to the following systems:

1. Mechanical lever system. Such a system was used by Kiefer (131), who obtained an amplification of 100 for temperature measurements up to 1400°C. The push-rod transmits the expansion to an amplifying lever, the end of which is provided with a pen which allows the direct recording of the expansion on a roller.

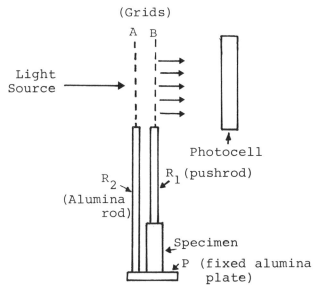

Fig. 7.21. Scheme of the photometer process used by Liebermann and Crandall (153).

2. Mechanical comparator. The push-rod transmits the expansion to a mechanical comparator which gives, on a dial, the length variation of the sample. This is a "point-to-point" method which needs the presence of the operator near the apparatus for the observation of micrometer indications.

Such a system was employed by many authors who developed push-rod dilatometers, for example Baron (18), Mark and Emmanuelson (159) (apparatus employed in the NBS), Thormann and Buchmayer (265) [high-temperature (1600°C)] dilatometer developed at the Technische Hochschule Clausthal, West Germany) and Shaffer and Mark (242) for measurements in inert atmosphere and up to 2000°C, of thermal expansion of metallic carbides.

3. Liquid micrometer. The push-rod is connected to a flexible membrane or diaphragm which can modify the hydrostatic pressure that exists inside an enclosure or vessel filled with a suitable liquid. The enclosure is terminated by a capillary in which the liquid level varies when the specimen length changes (5).

4. Interferometric process. The sample is placed between two vertical push-rods; the upper push-rod is connected to an optical system (prism + lens) which allows one to obtain Newton rings. This interferometric device was used by Terpstra (254).

5. Photometer process. Such a process was developed by Liebermann and Crandall in 1952 (153) (Fig. 7.21). The push rod R_1, which maintains the specimen against a fixed alumina plate P, is rigidly locked with a grid B. Four

vertical sintered alumina rods R_2 ixed 、ɔ the plate P are joined to a grid A. An expansion of the specimen pronʌotes a motion of the grid B with respect to the grid A. That modifies the light flow given by a light source situated at right angles to the grids. This variation of the light flow is detected by a photocell and, after amplifying and recording, allows determination of the expansion (or shrinkage) of the specimen. For a grating with 500 lines/in., a full deflection is obtained for each 0.001 in. of movement.

The dilatometer of Liebermann and Crandall, which allows measurements up to 1400°C, was perfected in 1958 by Beals and Lauchner (28) who replaced the grids by special glass diffraction gratings and used two photocells in opposition. This optical and photometer process must be compared with photometer method described in Section II.A.3.

6. Dynamometer gauge. The push-rod is connected to an elastic strip which is fixed a strain gauge itself connected to a bridge arrangement. Such a system was used by Sinha et al. (247).

7. Displacement transducer system and differential transformer. The expansion or shrinkage of the specimen promotes, via the push-rod, the displacement of a ferrite core in a differential transformer (LVDT). The resultant voltage can be rectified and recorded. It is proportional to the length variation of the sample. The LVDT is generally thermostated.

This system is increasingly used for the measurements of length change in modern apparatus. It is also used in differential dilatometry. One rod supports the coil of the differential transformer with the core suspended from the other rod by means of an adjustable Invar holder as in the Plummer's differential dilatometer (213,215). A scheme of the complete Differential Dilaflex System with core and coil mounting is given Fig. 7.22.

8. Electronic micrometer. It consists of an accurate machine screw that is motor driven and electrically controlled. This system was used by Kollie et al. (142) who interfaced a fused quartz differential dilatometer with a minicomputer to automatically measure the thermal expansion of solids from 300 K to 1000 K. The electronic micrometer readable to 2.5×10^{-6} cm was employed for length change measurements accurate to 2.5×10^{-5} cm.

It is difficult to include in this chapter functional diagrams and detailed description of the numerous push-rod dilatometers described in the literature. Table 7.VII gives some information about the best-known apparatus, commercialized or not, but the list given is certainly not an exhaustive one.

An illustration on push-rod dilatometers, we will give some information about two classical arrangements: the Heraeus apparatus and the Chevenard dilatometer.

The TMA 500 Heraeus dilatometer (Fig. 7.23) is a vertical dilatometer provided with a calibrated linear, variable differential transformer (LVDT). This apparatus allows direct length variation measurements as a function of temperature from −190°C to 1000°C. The sample holder is of transparent

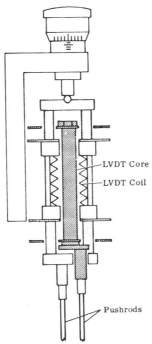

LVDT Core

LVDT Coil

Pushrods

Fig. 7.22. Complete differential Dilaflex system with core and coil mounting. (Reprinted, with permission, from Ref. 66.)

fused silica. The sample dimensions are up to 9 mm in diameter and 50 mm in length. Measurements can be made under vacuum or in an inert atmosphere. Recording of both length variations and derivative is obtained by a two-pen X–Y recorder. The probe and sample holder can be easily exchanged individually for the determination of linear and volumetric expansion coefficients, penetration, and stress behavior (240). Resolution of 0.1 μm per centimeter of chart is obtained.

The same type of push-rod dilatometer was commercialized by Netschz and Linseis (horizontal dilatometer) and by Rigaku Denki (vertical dilatometer), but complementary penetration or stress behavior cannot be studied as in the Heraeus dilatometer.

The Chevenard's dilatometer (31,54,56) was developed at the beginning of this century and was applied essentially to metallurgy. Initially, it was a differential dilatometer. The specimen S and a reference sample P are placed side by side and symmetrically in a sample holder consisting of two silica tubes, themselves placed in the center of a furnace. Expansion of S and P are transmitted to a 3-ft amplifier piece via two silica rods R1 and R2 which are extended by two metallic slides C1 and C2 (Fig. 7.24). The length raise which acts on the slides is the expansion excess of each sample (reference and specimen) over the expansion of an equal silica length L_0 (Fig. 7.24b).

TABLE 7.VII
Information about Some Push-rod Dilatometers

Authors or companies	a	b	Temperature range	Atmosphere	Amplifyer system	Recording	Material
Plummer (213,215) (Corning Design)	D	V	25/1200°C		Diff. transf.	Electron.	Glass-ceramic
Theta Industries U.S.A.[c]	D	H			Diff. transf.	Electron.	Vitreous silica
Netzsch[c]	A	H	25/1320–1550–1700°C	Control. atmosph. vacuum	Displ. transd.	Electron.	SiO_2 or Al_2O_3
	A	V	−160/+420°C		Displ. transd.	Electron.	Fused-silica
Linseis[c]	A	H	25/1000°C	Control. atmosph. vacuum	Mech. transd.	Dial comp.	Alumina
	D	V	25/2200°C		Diff. transf.	Electron.	Fused-silica
Rigaku Denki Ltd.[c]	D	V	−200/+300°C	Air or vacuum	Diff. transf.	Electron.	Fused-silica
	A	V	25/1000°C	Vacuum	Displ. transd. + diff. transf.	Electron.	Fused-silica
Kollie et al. (142) (Oak Ridge Nat. Lab. Des.)	D	V	27/700°C	Inert	Electron. Micrometer	Direct read. or electr. count.	Fused-silica
Evans and Winstanley (79)	A	V	−196/+25°C		Displ. transd.	Electron.	Fused-silica
Shaffer and Mark (242)	A	V	25/2000°C	Inert	Mech. transd. or LVDT	Microm. dial or electron.	Silicon carbide
Thormann and Buchmayer (265)	A	V	25/1600°C		Mech. transd.	Microm. dial	Ceramics
Mark and Emmanuelson (NBS design) (159)	A	V	25/1500°C		Mech. transd.	Dial. microm.	Silicon carbide
Heraeus[c]	A	V	−190/1000°C	Vacuum or inert	Displ. transd.	Electron.	Fused-silica
Dupont Instruments[c]	A	V	−190/1500°C	Inert	LVDT	Electron.	Fused-silica
Adamel Lhomargy[c]	D	H	25/1100–1500°C	Air, inert, or vacuum	Mechanical or displ. transd.	Graph., photogr. or electron.	Fused-silica or alumina
	A	V	−196/1400°C	Air, inert, or vacuum	Displ. transd.	Electron.	Fused-silica

[a] Differential (D) or absolute (A).
[b] Horizontal (H) or vertical (V).
[c] Commercialized companies.

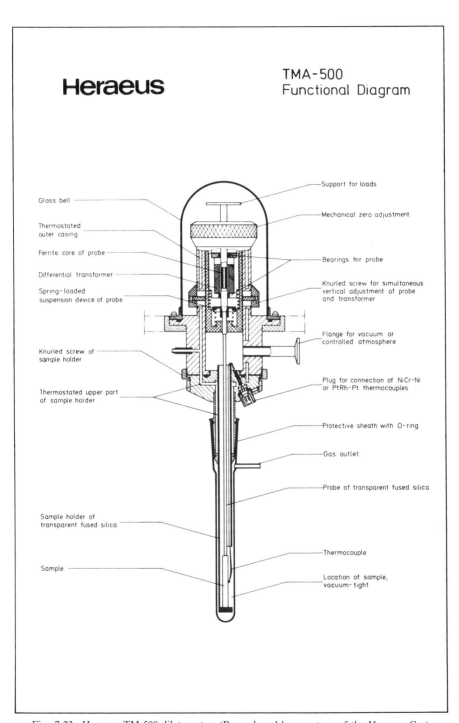

Heraeus

TMA-500
Functional Diagram

Support for loads

Glass bell

Mechanical zero adjustment

Thermostated
outer casing

Ferrite core of probe

Bearings for probe

Differential transformer

Spring-loaded
suspension device of probe

Knurled screw for simultaneous
vertical adjustment of probe
and transformer

Flange for vacuum or
controlled atmosphere

Knurled screw of
sample holder

Plug for connection of NiCr-Ni
or PtRh-Pt thermocouples

Thermostated upper part
of sample holder

Protective sheath with O-ring

Gas outlet

Probe of transparent fused silica

Sample holder of
transparent fused silica

Thermocouple

Sample

Location of sample,
vacuum-tight

Fig. 7.23. Heraeus TM 500 dilatometer. (Reproduced by courtesy of the Heraeus Co.)

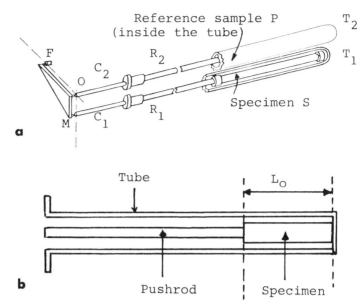

Fig. 7.24. Principle of the Chevenard's dilatometer. For differential measurement, the fixed point F is at the summit of the acute angle (such as in the above scheme). For obtaining the true expansion curve, the fixed point F must be at the summit of the right angle.

The 3-ft amplifier piece is of rectangular triangle form. It allows one to obtain the dilatometric curve on a cartesian coordinate diagram: the thermal expansion of the reference P is such that the temperature scale is directly obtained in the Ox direction of the diagram, and the expansion between the sample and the reference is directly obtained in the Oy direction. The dilatometric curve can be recorded either by a mechanical process or photographically. In the first case, a long metallic stem ended by an inked pen gives the dilatometric curve point by point on a paper fixed on a moving support allowing the paper to come to touch the pen alternatively. In the photographic recording, the 3-ft amplifier piece is provided in its center with a small concave mirror which sends towards a photographic plate the small light spot it receives from a light source. This photographic recording system is more precise than the mechanical one. The reference sample is an air oxidation-stable alloy (composition: Cr 8%, W 4%, Mn 3%, Fe 3%, Ni 82%) and is characterized by a reversible and well-known thermal expansion. Reference samples other than pyros can be chosen for certain particular studies.

If Θ is temperature and ΔL the difference in length changes between the specimen S and the reference P, the differential dilatometer records the curve $\Delta L = \Delta_{Sp.} - \Delta_{Ref.} = f(\Theta)$.

Chevenard in 1950 (58) brought certain improvements to its dilatometer and built an apparatus with higher performances, obtaining true dilatation

Fig. 7.25. Adamel-Lhomargy LK 02 dilatometer with furnace open. (Reproduced by courtesy of the Adamel-Lhomargy Co.)

curves and the possibility of recording isothermal expansion curves, extending of the experimental temperature range, particularly towards the high temperatures, working in vacuum or in controlled atmosphere such as nitrogen or argon. Possibility exists of studying the quenching of the sample by use of a vertical sample holder.

The French Company Adamel has largely constructed and commercialized a series of Chevenard-type dilatometers and today certain older apparatus such as the simple mechanical DM50 or more recent DHT 60 (63) are always used in research laboratories. In recent years, more sophisticated apparatus were developed by the Adamel-Lhomargy Company, such as the DI-10/2 apparatus which allows to work up to 1500°C, in air, vacuum, or controlled atmosphere, and the amplifying system of which is a displacement transducer. The more up-to-date and particularly interesting apparatus, recently developed by Adamel-Lhomargy Company, is the vertical LK-02 dilatometer (Fig. 7.25). This apparatus allows one to study phase transformations simultaneously by dilatometry and thermal analysis in a temperature range from -196°C and 1400°C with heating rate up to 300°C $\cdot$ sec^{-1} and cooling rate of 700°C $\cdot$ sec^{-1}. The heating system is a radiation furnace which focuses on a small size sample (12 mm in length and 2 mm in diameter), the power radiated by two tungsten filament lamps. The quenching is obtained by blowing helium onto the sample. Such a rapid-quenching dilatometer finds many interesting applications for metallurgical research.

Besides these more sophisticated arrangements, the Adamel-Lhomargy Company continues to manufacture simpler apparatus such as lever-mechanical recording or optical recording dilatometers for studies and controls on metals and alloys, coals, glasses, ceramics, minerals and rocks, plastics, etc. The length amplification of dilatometers constructed by Adamel-Lhomargy Company varies from 75 to 10,000 according to the type of apparatus considered.

A mechanical dilatometer of the same type as that developed by Chevenard is the Leitz-Bollenrath apparatus (36) which uses a photographic recording. Also the IRSID absolute dilatometer must be cited; finds the same applications as the Ditirc Adamel dilatometer for particular studies of the transformation curves of steels and alloys.

b. Corrections for Calculation of Expansion Coefficient from Push-Rod Dilatometer Data

In the case of push-rod dilatometers, the rod is partially situated inside the furnace and presents its own expansion during heating. This expansion adds to the expansion of the specimen, which introduces an error in the measurements. This disadvantage can be minimized by making use of the same material for both the rod and the sample holder as seen in Fig. 7.24b. So, the rod expansion is approximately compensated by that of the sample holder *except for the length of the specimen*. For precise measurement, the value of the sample holder- or rod-material expansion coefficient must be known.

So, the curve obtained with the nondifferential push-rod dilatometer is

$$\Delta_{Sp.} - \Delta_{Mat.} = f(\Theta) \tag{53}$$

where $\Delta_{Sp.}$ = thermal expansion of the specimen of length L_0 and $\Delta_{Mat.}$ = thermal expansion of a length L_0 of the material that composes the push rod and the sample holder (SiO_2 or Al_2O_3, for example). Once the thermal expansion of the material is known, it is easy to obtain the thermal expansion coefficient of the specimen.

To provide correction data, a plot can be made of the expansion of the quartz (or alumina) assembly versus temperature with no specimen in place and the measuring tube resting directly on the end of the sample holder, or by replacing the specimen by a rod sample of the same length and of the exactly same material as the assembly (quartz or alumina). In the case of quartz assembly, it was shown by Griffin and Demou (98) that this expansion turned out to be nearly linear with temperature, and this relation was assumed consistent to simplify the correction procedure.

In the case of differential dilatometry, which measures the expansion of a specimen relative to that of a reference sample, the recorded curve gives

$$(\Delta_{Sp.} - \Delta_{Mat.}) - (\Delta_{Ref.} - \Delta_{Mat.}) = \Delta_{Sp.} - \Delta_{Ref.} \tag{54}$$

We can assume that, with exactly identical push-rods, the expansion of the assembly does not have to be taken into account. When the differential system is not ideal, due to primary differences in expansion between the two push-rods, there is a contribution to the measured expansion which can be termed a baseline correction Δ_{BL} (215), so we can write

$$\Delta_{\text{Sp.}} - \Delta_{\text{Ref.}} = A \times \Delta_{\text{meas.}} - \Delta_{BL} \qquad (55)$$

where A is a calibration constant and $\Delta_{\text{meas.}}$ is the recorded value.

The baseline correction can be easily made by a pair of measurements on two samples a and b: a is placed at the specimen place and b at the reference place, so the measurement gives (first experiment)

$$\Delta_a - \Delta_b = A\Delta_1 - \Delta_{BL} \qquad (56)$$

The position of a and b are reversed in the dilatometer (second experiment), which leads to

$$\Delta_b - \Delta_a = A\Delta_2 - \Delta_{BL} \qquad (57)$$

Adding these two expressions eliminates the two sample expansions and gives

$$\Delta_{BL} = \tfrac{1}{2} A (\Delta_1 + \Delta_2) \qquad (58)$$

So, the baseline can be determined as a function of temperature over the entire range of the dilatometer.

c. Calculation of Thermal Expansion Coefficient in Differential Dilatometry

The calculation of expansion coefficients from differential curves is not difficult because the expansion of the reference material is generally given in tables. The knowledge of the true expansion curve of the specimen must be first obtained graphically as follows: after having determined the scale in the experimental chart, values of $\Delta_{\text{Ref.}}$ (obtained from tables) must be added to the experimental $(\Delta_{\text{Sp.}} - \Delta_{\text{Ref.}})$ curve at different temperatures in the temperature range investigated. This leads to the graphically obtaining the $\Delta_{\text{Sp.}}$ curve. The true expansion coefficient $\alpha_{\text{Sp.}}$ can be obtained by drawing, at the desired temperature Θ_1, the tangent AB on the $\Delta_{\text{Sp.}}$ curve. The determination of the average thermal expansion coefficient $\alpha]_{\Theta_1}^{\Theta_2}$ can also be easily calculated from the $\Delta_{\text{Sp.}}$ curve by drawing the line AC and measuring its slope (Fig. 7.26).

3. Particular Devices

Some particular dilatometers were developed by different authors for specific measurement of specific materials. Clusener (66) described a multisample dilatometer head which allows one to save instrument time by measuring several samples at the same time. This multisample head is made up of six transducer push-rod assemblies. A recommended procedure is to

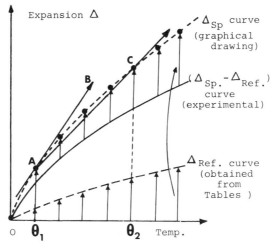

Fig. 7.26. Determination of thermal expansion coefficients from differential dilatometry results.

measure five unknown materials and one reference material. A multichannel recorder has to be used.

Prime et al. (217) described a method for the measurement of linear thermal expansion in the plane of thin materials, such as thin fiber-reinforced composites. The measurements are performed by a thermomechanical analysis (TMA) technique developed for polymer films, in which samples are held in a specially designed set of Invar chucks.

Liquid and pasty substances are generally studied by volumetric dilatometry (see Section II.D). Linseis Messgeräte GMBH has developed another method which uses an Invar steel cylinder provided with a movable plunger. A fine needle allows the volume of the liquid or pasty substance to be constant from one measurement to another because the plunger is brought in contact by its point. Such an arrangement is placed in the normal Linseis dilatometer and is heated.

D. VOLUMETRIC DILATOMETRY

Dilatometric measurements on liquid or paste are generally made by use of simple apparatus consisting of a glass or stainless steel vessel provided with a very fine capillary. A smooth heating of the vessel inside which the substance for investigation is placed leads to expansion that promotes the rise of the substance in the capillary. The substance to be investigated can be a solid which is placed in a liquid (for example mercury) in the vessel. Expansion or contraction of the solid modifies the level of the liquid in the capillary. This motion can be directly read with an optical system (cathetometer), an electric design, or a photoelectric follower system.

Volumetric dilatometry was widely used in polymer science, either to

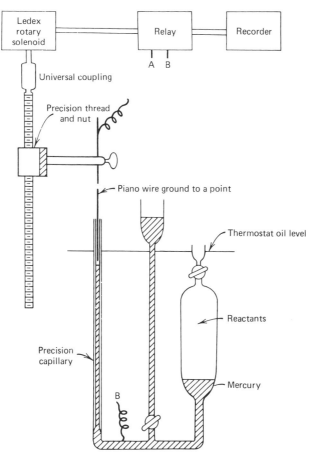

Fig. 7.27. Dilatometer and recording device with mercury sensing fluid. [After Bell (30), reported by Rubens and Skochdopole (229).]

study the kinetics of polymerization or to investigate the thermal behavior of polymers, plastics, or rubbers. A description in this chapter of the numerous volumetric dilatometers developed by many authors during the last 30 or 40 years would be too lengthy. An interesting paper about this subject was published by Rubens and Skochdopole in 1966 (229). The general design of this apparatus is the same as that used for study on liquid and pasty substances. But some peculiarities are encountered in the dilatometer arrangement according to the nature of the reactant being studied and the type of data required. Consequently, the type of dilatometer developed varies whether as the polymerization reaction in realized in homogeneous bulk, in solution, or in emulsion, and whether studies are made with highly viscous systems or in presence of heterogeneous catalysts.

Fig. 7.27 gives as an example the volumetric mercury dilatometer with the recording device developed by Bell (30): the wire, which is in direct contact with the mercury in the capillary, is driven up or down by a screw connected to a rotary solenoid. As the mercury level falls and breaks contact with the wire, a relay is energized which closes a circuit causing the solenoid to rotate, until contact between the wire and mercury is reestablished. The position of the screw is recorded versus time.

Bekkedahl (29) reviewed the design and use of volume dilatometers both for measuring volume coefficients of thermal expansion of solids and liquids and for the study of phase changes. We agree with Rubens and Skochdopole (229) recognizing that although the paper of Bekkedahl was published in 1949, there have been relatively few changes in the experimental apparatus that he described, except for the use of a linear differential transformer to convert a linear displacement into a measurable electrical signal.

E. X-RAY DIFFRACTION METHODS

1. Principle of the Method

The X-ray diffraction method allows one to determine directly the change in lattice spacing. In accordance with Bragg's law

$$n\lambda = 2d \sin \Theta \tag{59}$$

where λ is the wave length of the incident X-ray, d the lattice spacing, and Θ the incident angle of X-ray. A change Δd in the spacing corresponds to a change $\Delta\Theta$ in the glancing angle, given by

$$\frac{\Delta d}{d} = cotg \ \Theta\Delta\Theta \tag{60}$$

The thermal expansion corresponds generally to an increase in the lattice spacing d and consequently to a decrease of the Bragg angle Θ. So the thermal expansion promotes a displacement of diffraction lines to small Bragg's angles. The higher the variation of $\Delta\Theta$ (coming from the thermal expansion $\Delta d/d$), the higher will be the Bragg angle; thus, the higher sensitivity will be obtained for Bragg angles near $90°$.

On the other hand, the lattice spacing is directly related to the lattice parameters and the connection formulae depend on the crystal system. So, it is possible to know the variation of these parameters versus temperature and consequently to determine the thermal expansion along each crystallographic axis of the solid, the simplest case being that of cubic symmetry for which the relation

$$d = a/\sqrt{h^2 + k^2 + l^2} \tag{61}$$

leads to

$$\frac{\Delta d}{d} = \frac{\Delta a}{a} \tag{62}$$

by derivation.

According to Megaw (163), the X-ray method was used as early as about 1920 to study thermal expansion of solids such as diamond, graphite, bismuth crystals, zinc, cadmium, long-chain paraffins, etc. Megaw studied thermal expansion of some anisotropic crystals after developing an X-ray method based on the considerations seen above.

By a suitable choice of X-ray wavelength, a reflection from some order of the plane to be investigated is obtained at a glancing angle of nearly at right angle and recorded on a film in a back camera. Since Θ is nearly $\frac{\pi}{2}$, a small change of spacing thus gives a large change of angles, making possible accurate determination of the spacing. From measurements of the position of the spot produced by the ray reflected from the crystal at the different temperatures, the thermal expansion normal to the reflecting plane is calculated.

2. High-Temperature Techniques

Great progress has been made in the technology since the first works in this subject and today high-temperature camera techniques are well developed (238) and allow the determination of thermal expansion up to 900–1000°C in air or in controlled atmosphere, or up to 2000°–3000°C in vacuum. Many authors have proposed and developed high-temperature diffractometer furnaces since Westgren in 1921 (275). The characteristics and uses of these instruments were summarized by McKinstry in 1970 (175). One important problem for realizing a high-temperature furnace for X-ray diffraction measurements is the gradient of temperature around the sample and the determination of temperature. McKinstry (175), by considerations of the optimum conditions for high-temperature X-ray diffractometer, described in some detail a two-hemisphere-shaped furnace with a minimum of compromise with these optimum conditions. This furnace is adaptable for use with either the GE, Norelco, Siemens, or Picker diffractometers and is capable of continuous operations at any temperature from room temperature to 1400°C, in air or in gas. A commercial furnace based on this design has been marketed by Tem-Pres Research Inc., State College, Pennsylvania.

Other commercial high-temperature cameras are:

1. The Raymax-60 demountable X-ray unit with a Unicam 19-cm high-temperature powder camera. Temperature control to within 1°C is obtained with the use of a voltage stabilizer and a variac. The temperature range investigated is 30–700°C. This apparatus was employed for example, by Krishna Rao to correlate thermal expansion and crystal structure (146).

2. The Guinier-Lenne camera, commercialized by Enraf-Nonius. Measurement of thermal expansion can be made on powder samples up to 1200°C in a Guinier-Sehmann-Bohlin principle.

3. The high-temperature Guinier camera 631/632. This camera works with a strictly monochromatic X-ray radiation. Diffraction patterns can be

made of single crystals as well as of powder specimens, planar specimens of a maximum size $2 \times 3 \times 0.2$ mm. The specimen can be heated up to a maximum of 1000°C. Film motion can take place continuously or in steps. The camera is commercialized by Huber.

4. The Rigaku X-ray diffraction cameras (Japan) allowing work either in high vacuum (maximum temperature 1500°C or 2500°C; heating element, Pt-Rh, or W) or in air or inert gas (maximum temperature 1400°C or 1700°C; heating element, Pt-Rh). This apparatus uses the Rigaku horizontal goniometer to study crystal structure, changes of metals, ceramics, high polymers, organic chemicals, etc. A special attachment was developed also by Rigaku for simultaneous high-temperature X-ray diffractometry and differential thermal analysis up to 1200°C.

5. The *C.G.R.* (Compagnie Générale de Radiologie) camera developed in France by Barret and Gerard (19), patented by the C.N.R.S. and commercialized by C.G.R. The diffraction pattern is recorded by use of a Geiger counter, or proportional counter, or scintillator counter. The heating of the sample is realized by thermal radiation from two hemispheric cups. The measurement can be made in air or in a controlled atmosphere up to 900°C, or in vacuum up to 1500°C. The apparatus of Barret et al. allows simultaneous DTA measurement in air or in controlled atmosphere.

Some other experimental instruments were developed recently in France, but they are not commercialized. For example:

1. The high-temperature camera developed by Revcolevschi et al. (223). This is a cylindrical Debye-Scherrer camera of high diameter (320 mm). The sample is in the center of the camera and the X-ray diagram is recorded photographically on a standard 35-mm film. The heating process of the sample is original: it is obtained by a image furnace, and allows measurements up to 3000°C. This furnace is a biellipsoid one and contains two mirrors of 300-mm diameter. The source of light, situated in the focus of one mirror, is a xenon arc lamp of 6.5 kW power. With this apparatus, measurements can be made in oxidizing or reducing atmosphere, or in vacuum.

2. The heating apparatus developed in I.N.S.A. of Lyons (France) by Mentzen for the study of thermal behavior of some metallic formates (Th, U, Sr, and Ca formates) and particularly the polymorphism and volume change of the unit cell of calcium formate during heating (166). The X-ray generator is a Siemens Kristalloflex II model equipped with a temperature-regulating moving-film camera (TRMF camera), recording the diffraction line against temperature. Different heating cells were built by Mentzen.

3. Low-Temperature Techniques

Some X-ray diffraction cameras are commercialized for low-temperature measurements, for example the low-temperature X-ray diffractometer attachment of Rigaku (Japan) designed for use on the Rigaku horizontal

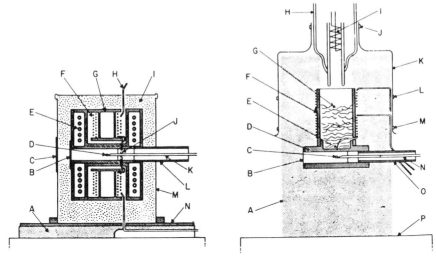

Fig. 7.28. X-ray method design of Mauer and Hahn (161). (*Left*) Vertical section through the furnace showing: A, asbestos base; B, aluminum foil window; C, aluminized Mylar window; D, crystal; E, nichrome heater; F, thermocouple leads; G, copper block; H, thermocouple leads; I, refractory brick; J, thermocouple junctions (in copper block); K, spindle; L, copper sleeve; M, outer shell; N, copper base plate with cooling fins. (*Right*) Vertical section through the cryostat showing: A, styrofoam; B, aluminized Mylar window; C, crystal; D, copper block; E, F, thermocouple junctions; G, copper turnings; H, Dewar tube; I, gas stream heater; J, L, M, rubber bands; K, plastic sleeve; N, spindle; O, copper sleeve; P, base. (Reprinted, with permission, from Ref. 161.)

goniometer. When liquid nitrogen is used, the specimen temperature can be lowered to $-190°C$. For temperature measurement, it employs a unique structure with the thermocouple welded directly to the specimen holder, ensuring high accuracy of temperature measurement and small gradient.

4. Apparatus for Both Low and Middle Temperatures

Another interesting X-ray apparatus was used in 1971 by Mauer and Hahn (161) to measure the thermal expansion of some azides by a single crystal X-ray method. The originality of this apparatus is that it allows measurements between $-180°C$ and $325°C$. The diffractometer is a commercial instrument that has been extensively modified by Mauer and Hahn. The furnace (Fig. 7.28*a*) can be installed after the crystal is in place. A cryostat (Fig. 7.28*b*) is interchangeable with the furnace. It is cooled by a stream of cold nitrogen gas delivered at a constant rate by a commercial evaporator with an automatic refill device. The temperature measurements are made with a copper–constantan thermocouple, the limit of error being estimated to be $\pm 0.4°C$ from $-60°C$ to $100°C$, increasing to $\pm 1.8°C$ at $-180°C$ and to $\pm 1.2°C$ at $325°C$. The sample is an elongated crystal approximately $0.5 \times 0.5 \times 1$ mm, mounted in a trough formed at the end of the thin walled Pyrex capillary.

Remark: The variation of lattice parameter versus temperature can also be measured by high-temperature neutron diffraction. For example, Bowman et al. made use of this technique to measure the thermal expansion of UC–ZrC solid solution (40), to study the anisotropic thermal expansion of refractory carbides (39) or to determine the thermal expansion of manganese diboride (MnB_2) at low temperatures (5–298 K) (41).

It would be too long to detail this method here, but according to after Bowman, the neutron diffraction technique must be preferable to X-ray diffraction. However the apparatus used is expensive and there are only few laboratories where they are available.

Finally, the coefficient of thermal expansion can be calculated from the determination of the force constants in the substance.

5. Calculation of Thermal Expansion Parameters from X-Ray Diffraction Data

The method used by Mauer and Hahn (161) is taken as example:

The value of each lattice parameter as a function of temperature is represented by a polynomial, the coefficients of which are determined by a least-squares fit of the experimental points. For example:

$$a_T = k_0 + k_1 T + k_2 T^2 + k_3 T^3 + \ldots \qquad (63)$$

where a_T is the value of any lattice parameter at temperature T, in °C. Polynomial expressions for all lattice parameters and the volume of the unit cell are given.

The linear expansivity may be obtained by differentiation of the polynomial form:

$$\alpha_a = \frac{1}{a_{20}} \times \frac{da}{dT} = \frac{k_1}{a_{20}} + \frac{2k_2}{a_{20}} T + \frac{3k_3}{a_{20}} T^2 + \ldots \qquad (64)$$

where a_{20} is the value of the parameter at 20°C. The volume expansivity may be obtained in like manner from the expression for the volume. Values are given for a temperature of 20°C.

The linear expansion may be defined by the expression

$$\frac{a_T - a_{20}}{a_{20}} = \frac{k_0 - a_{20}}{a_{20}} + \frac{k_1}{a_{20}} T + \frac{k_2}{a_{20}} T^2 + \frac{k_3}{a_{20}} T^3 + \ldots \qquad (65)$$

Again, the volume expansion is obtained from the same expression using the volume.

F. SIMULTANEOUS METHODS

1. Combined Thermal Analysis and Dilatometry

In the first arrangement combining DTA. and dilatometry, where the two methods are carried out simultaneously in a single piece of apparatus, data were recorded visually and proved time consuming (209). An interesting automatic apparatus was constructed by Pearce and Mardon (209) for studies on small metallic samples at temperature up to 1200°C in a vacuum. They used a dial gauge push-rod dilatometer which is provided with two parallel and vertical silica tubes, the first containing the specimen for thermal expansion measurements. The dial gauge was modified to make it record automatically. The differential thermal analysis trace is obtained by amplifying the difference between a thermocouple the junction of which is sheathed in tantalium or alumina and inserted into a hole drilled in the center of the specimen investigated by dilatometry, and a similar couple immersed to the same depth in a reference block in the second silica tube of the dilatometer. The apparatus was fitted into a glove box because expansion measurements were made on plutonium alloy system and neptunium.

Another more complete arrangement is that of Paulik and Paulik (208) who have incorporated a very simple dilatometric device in the Derivatograph. The sample, in most cases a microcrystalline powder, is compressed into a hollow cylindrical shape to enable the measuring of its thermal expansion. This block rests on a silica tube on one arm of the derivatograph balance. The tube is encased by another silica tube, the upper end of which is formed into a stirrup which rests on the test piece. If the sample expands, the two tubes move, changing their relative position, together with that of two diaphragms which are mounted one on each tube. The diaphragms are illuminated by a parallel light beam, the intensity of which varies with the length-change of the specimen. This variation generates an electric signal in a photocell, as in the photometer system described in Section II C.1. The signal measured by a galvanometer is recorded on a photographic paper that gives the thermal expansion curve which can be compared directly with the other data of the derivatograph. (Temperature, T; thermogravimetry, TG and derivative DTG; differential thermal analysis, DTA; thermogas titrimetry, TGT and derivative, DTGT). This apparatus gives also the derivative curve, DTD, of the dilatometric curve. It was applied to the study of the thermal decomposition of salts.

We must also recall the apparatus developed by Barret et al. (19) which allows simultaneous thermal analysis (DTA) and high-temperature X-ray diffraction studies. In the same way the Rigaku high-temperature X-ray camera can be provided with a special attachment for DTA investigations.

An original apparatus was developed by Balek (16) to study simultaneously the DTA., the thermal expansion, and the emanation thermal analysis (ETA). This latter method consists of incorporating an inert radioactive gas in a solid and measuring the rate of release of this gas when the sample is submitted to a linear heating rate. The comparison of the three types of results obtained gives interesting information about structure or surface state change of the solid investigated. Numerous applications of this technique were made in the case of salts, oxide gels, etc. In this technique, experiments are made on three different samples placed in the same heating cell, the sample for dilatometry is a compacted powder whereas the samples for DTA and ETA are not compacted powders.

2. Combined Dilatometry and Electrical Properties Measurements

Thermal expansion and electrical conductivity are two properties which are particularly interesting to be studied for a more precise knowledge of solid state phase transition such as allotropic– or order–disorder transformations.

The development of a combined electrical conductivity–thermal expansion apparatus was described by Duclot and Deportes (77) who obtained simultaneous measurements by adding an electrical arrangement in the Netzsch 1550°C push-rod dilatometer. The specimen is placed between two thin plates of platinum (Fig. 7.29a). One of these plates is fixed on the end part of the sample holder, the other on the end part of the rod. Thermal expansion coefficients of platinum ($\alpha = 9.6 \times 10^{-6}$) and of alumina ($9.0 \times 10^{-6}$) are approximately of the same order. The influence of adding the two platinum plates to the dilatometric arrangement is therefore negligible. The variation of electrical conductivity is recorded on one pen of the six-pen recorder used for the dilatometric and temperature measurements. Such an arrangement was employed by Duclot and Desportes to study the monoclinic-quadratic transformation of a calcium-stabilized zirconia sample and the order–disorder transformation of an yttrium-double oxide.

Another arrangement (Fig. 7.29b) was used by Deportes and Gauthier for the determination of cationic transport number in solid oxides (MgO) by dilatometer with dc flow inside the sample. It differs from the precedent by the specimen form which must be adapted to the type of measurement desired which is here the electrolysis of a solid (73).

3. Other Combined Devices

Shelley (244) has described an apparatus that allows simultaneous measurement of the ratio of complex specific heat to thermal conductivity and the linear coefficient of thermal expansion. This method was applied to the

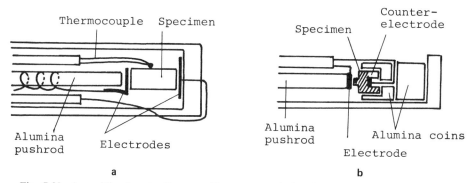

Fig. 7.29. Assemblies for simultaneous dilatometry and electrical measurements (reprinted, with permission, from Ref. 77) or electrolysis in the solid state (reprinted, with permission, from Ref. 73).

study of the volume and thermal relaxation of chlorinated polyethylenes near their glass–rubber transition.

The Dupont Instruments dilatometer can work either for expansion measurements or as a penetrometer. The recording of creep or shrinkage can be obtained by loading a small disc plate on the top of the apparatus. The probe and the sample holder of the TMA 500 Heraeus dilatometer can be easily exchanged individually for the determination of expansion coefficients, penetration, and stress behavior of the specimen.

G. COMPLEMENTARY TECHNICAL DATA

Some complementary data about the nature of the material chosen for the assembly, the temperature measurement, the heating and cooling processes, etc. will be given.

1. Generalities on Experimentals

a. MATERIAL FOR SAMPLE HOLDER AND EXTENSION ROD

The following materials can be used depending on the temperature range investigated: up to 1000°C, fused silica; up to 1200°C, porcelain; 1700°C, sintered alumina; 1800°C, sapphire; 2000°C, graphite, molybdenum, tungsten (in reducing atmosphere), and silicon carbide.

b. TEMPERATURE MEASUREMENTS

The nature of thermocouples and the type of thermometer differ according to the temperature range investigated. In Table 7.VIII lists some recommendations for measuring temperature in dilatometry.

TABLE 7.VIII
Methods of Temperature Measurement Versus Temperature Range Investigated

Temperature range	Nature of thermocouple and type of thermometer used
Low-temperature	Indium resistance thermometer
$T > 11$K	Platinum resistance thermometer
$4 < T < 11$K	Helium gas thermometer
	Germanium thermometer
4 K $- 300$K	Au (Fe)–NiCr thermocouple
$-230°C + 380°C$	Fe–constantan, Cu–constantan
$-170°C + 380°C$	Ni–NiCr thermocouples
$-50°C + 800°C$	Pt 100
$+20°C + 1000°C$	Ni–NiCr thermocouple
$+20°C + 1550°C$	Pt–Pt · 10 Rh thermocouple
$+20°C + 1700°C$	EL 18
$+20°C + 2200°C$	W · 5 Re–W · 26 Re Mo–MoRe W–graphite thermocouple
Up to 2500°C	Optical pyrometer

c. Heating and Cooling Equipment

For low temperature (79,162), the specimen can be placed on the sealed end of a fused-silica tube suspended in a copper block, which, in turn, is suspended from the main frame by means of brass rods in such a manner that it can be immersed in a Dewar vessel which contains liquid nitrogen or helium and whose position is controlled by a system of counterweights. The heating rate is controlled by varying the position of the Dewar vessel and heating rates of < 0.5 deg $\cdot$ min^{-1} can be obtained. The sample can also be cooled by the addition of liquid gas (nitrogen) to the space within the push rod around the specimen. Liquid gas is added as needed to keep the specimen covered until it reaches its minimum temperature as indicated by the recorder. At this time, the tube furnace of the dilatometer is carefully closed around the specimen. When the liquid gas boils away, the specimen temperature starts to rise and as the temperature approaches 0°C, a small amount of current is applied to the furnace windings to maintain the rate of temperature rise. Then, as needed, more current is applied to the furnace to raise the temperature to the desired higher limit at a rate of about 5°C min. In some arrangements, the cryostatic enclosure can be constituted of two Dewar vessels, D_1 and D_2, the first placed just around the dilatometer tube, the second D_2 around the first D_1.

For middle and high temperatures, different types of furnaces can be used versus the temperature range investigated. Some information about this subject is collected in Table 7.IX.

d. Different Ways of Improving Push-rod-Type Dilatometer Efficiency

This problem was studied in detail by Clusener in 1971 (66). The following parameters have to be taken into account for obtaining valuable results:

1. Sample length. International standards have been checked and it was found that the sample length should be 5×10^4 times the accuracy of the dilatometer. Based on the fact that an accurate dilatometer allows measurement to 5×10^{-6} in., one arrives at a ¼-in. sample length, that is to say .635 cm. All dilatometers do not have such an accuracy, so a simple length between 1 and 5 cm could be considered sufficient. Higher sample lengths are used (for example 10 cm), which raises the accuracy of the measurement but needs a longer furnace for providing an uniform temperature zone around the sample.

2. Sample diameter. A reasonable diameter is one-quarter to one-eighth of the sample length.

3. Heating or cooling rate. These factors depend on the specimen dimensions; the larger is the specimen size, then the slower must be the heating rate. Table 7.X gives some information about this point.

The calibration standard cost (use of platinum as standard) diminishes with the sample dimensions.

TABLE 7.IX

Different Furnace Types Used in High-Temperature Dilatometers

Type of furnace	Maximum temperature investigated	Authors
Graphite resistance furnace	1750°C	Baron (18)
Pt–Rh resistor furnace	1500°C	Adamel Co.
Induction furnace	2000°C	Shaffer and Mark (242)
	1600°C	Thormann and Buchmayer (265)
Kanthal-rod furnace	1700°C (air)	Baudran (24–26)
Kanthal Super 1 (Mo/Si) furnace	1650°C (air)	Baudran (24–26)
Two-resistor furnace		
Kanthal	1300°C	Kiefer (131)
Pt/Rh	1400–1450°C	
Oxide induction furnace (Y_2O_3-stabilized ZrO_2 suceptor ring)	2200°C	Nielsen and Leipold (203)
Radiation furnace[a]	1400°C	Adamel-Lhomargy
Solar furnace	1950°C	Chalmin (52)
Pulse heating furnace[a]	Melting point	Ruffino et al. (231)

[a] These furnaces allows one to heat the sample at high temperature in a very short time (some seconds).

4. Attachment of thermocouple. A good illustration of the type of thermocouple attachment is given in Fig. 7.30 versus the heating rate. In cooling or quenching experiments, two recommendations are given by Clusener. The best method when using forced cooling is to spot weld the thermocouple onto the specimen; when simulating quenching, a hollow specimen with inside thermocouple attachment is preferable.

TABLE 7.X

Relationship between Dilatometer Sample and Required Total Heating and Cooling Cycle (66)

Size of dilatometer sample	Material	Grain-size	Required time to heat to 1600°C (in hours)	Heating rate (in $°C \cdot h^{-1}$)
25 mm, 1 in.	Brick, carbon	3 mm, (⅛ in.)	17	100
25 mm, 1 in.	Coarse ceramic	3 mm, (⅛ in.)	12	130
6 mm, ¼ in.	Ceramic composites	Homogeneous	7	225
3 mm, ⅛ in.	Glass, metal	Homogeneous	3	500
1.5 mm, 1/16 in.	Plastic, crystal	Homogeneous	0.5	3000

Reprinted, with permission, from Ref. 66.

Rate °C.min.$^{-1}$	Thermocouple Attachment	
0.1	Touching under spring load	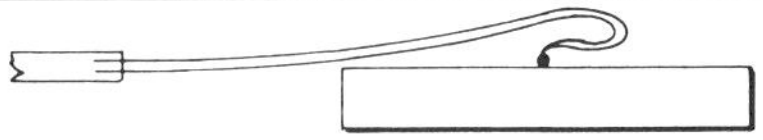
1	Inserted into center hole	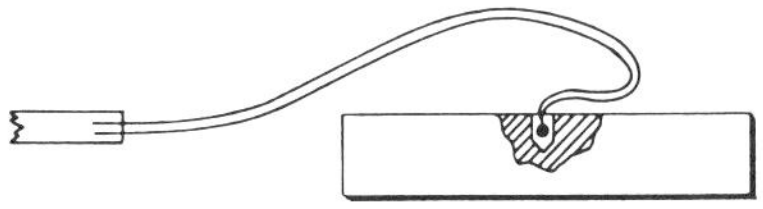
5	Inserted into center hole	
10	Spotwelded onto specimen	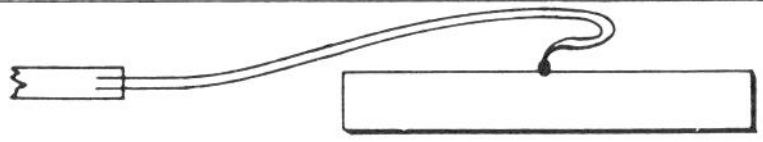
100	Welded into a Hollow specimen	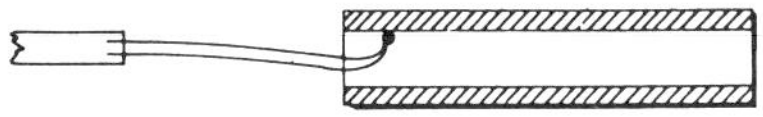

Fig. 7.30. Effect of sample thermocouple attachment on the recommended heating rate. (Reprinted, with permission, from Ref. 66.)

5. Choice of heating versus thermal conductivity of the sample materials. Table 7.XI gives information about this subject, as recommended by Clusener. The higher the thermal conductivity of the sample, the higher can be the heating rate.

6. Recording of thermal expansion. The most interesting method is to obtain the direct recording of the length variations (or $\Delta L/L$ curves) but the use of a Multi-pen Line Recorder or Multi-point Recorder is recommended, because, utilizing a zero suppression, it is possible to have the expansion signal recorded with various magnification factors, such as $10\times$, $1000\times$, and $10,000\times$. The lower magnification signal guarantees that the expansion signal of an unknown material does not run off the chart, and amplification of particular phenomena can be obtained with the higher magnification. With such recorders, the temperature of the sample can be obtained continuously.

TABLE 7.XI

Thermal Conductivity or Better Thermal Diffusivity of the Sample Materials in Relationship to Applicable Heating Rate (66)

Heating rate		Materials	Thermal Conductivity (Watt $\cdot$ meter$^{-1}\cdot$ K^{-1})
°C $\cdot$ min^{-1}	°C $\cdot$ hr^{-1}		
0.1	6	Thermal insulators	< 1
1	60	Pyrex glass	> 1
5	300	Quartz, single crystal	10
10	600	Metal, aluminum	> 100

Reprinted, with permission, from Ref. 66.

We can add to these considerations of Clusener that Multi-pen Line Recorders or Multi-point Recorders are generally more expensive than simple recorders. But they also allow one to obtain the recording of the $\Delta L/L$ first derivative (that is to say the expansion coefficient curve) when the dilatometer is provided with a derivator.

A digital recording can be also recommended for calculation and punched tape can be used for data reduction with a computer.

In high-speed dilatometry, oscillographic recording can be used. It allows one to obtain an image on a screen. This image can be photographed or cinematographed (38).

2. Preparation of Specimens for Dilatometry

Different processes can be used to prepare specimens for dilatometry:

1. Cutting of the sample from blocks or plates of the material to be investigated. This cutting is made with a metallic or diamond saw according to the hardness of the material.

In the case of anisotropic solids, samples of required size (for example $\frac{1}{8} \times \frac{1}{8} \times (\frac{2}{8} - \frac{7}{8})$ in.) can be cut either with the sample axis perpendicular to the fabric planes, or with the sample axis parallel to one fiber direction of the fabric. These were designated perpendicular ($\perp$) and parallel ($\parallel$), respectively. After careful grinding to square, parallel surfaces, the length of each sample is measured exactly. Such a sample preparation was used for example by Naum et al. (197) for the study of carbon–carbon composites.

Bars can also be cut from a cylinder of material obtained by sedimentation. That allows one to obtain specimens with required orientation, for example in perpendicular or parallel direction of an oriented deposit in the case of kaolinite (26) when the role of particle orientation inside the specimen must be analyzed.

2. Pressing or compacting of powders. Powders are pressed or compacted in a matrix. The compression can be made isostatically or nonisostatically, at normal temperature or at elevated temperature (hot-pressing). The compaction can also be made on dry samples or in presence of liquids such as ethanol. In certain cases of difficult pressing, some solids such as camphor, starch, gum arabic, etc., can be mixed with the powder before compacting. But this process can modify the first heating curve by thermal decomposition of the binder.

3. Fashioning or molding of the solid previously wetted in a pasty state. This leads to formation of small bars of an adequate dimension. These bars can be dried in a drying-oven (case of clays for example) before the dilatometric investigation. The molded sample can also be prepared by casting molten material in a mold.

Certain apparatus such as the Dupont Instruments dilatometer allows work with samples in the form of film, or fiber, or with powder previously mixed with an inert solid such as alumina, but not compacted (case of polymers).

3. Expansion Standards and Calibration

As a general rule, dilatometers require calibration before use. The National Bureau of Standards has been working for the last 10 years to establish standard reference materials for thermal expansion. At the present time four standard reference materials (SRM) are proposed by the NBS: fused silica (SRM 739), borosilicate glass (SRM 731), copper (SRM 736), and Tungsten (SRM 737). The apparatus used for the expansion measurements was a Fizeau interferometer described by Hahn (104). For tungsten, experiments were also made above 970 K by the twin-microscope technique of Rothrock and Kirby (228).

Tables 7.XII to 7.XV give thermal expansion and expansivity as a function of temperature for the four SRMs cited. These materials can be purchased at the NBS for price from \$81 to \$177, for lengths of 2, 4, and 6 in. (prices in June 1977).

TABLE 7.XII

Thermal Expansion, As a Function of Temperature, of SRM 739
(Fused-Silica) (137)

T	Expansion $\Delta L/L_{293}$	Expansivity α	T	Expansion $\Delta L/L_{293}$	Expansivity α
80 K	-1×10^{-6}	$-0.70 \times 10^{-6}/K$	320 K	$13._5 \times 10^{-6}$	$+0.53 \times 10^{-6}/K$
90	$-7._5$	-0.61	340	$24._5$	0.56
100	-13	-0.53	360	36	0.58
110	-18	-0.46	380	$47._5$	0.60
120	$-22._5$	-0.38	400	$59._5$	0.61
130	-26	-0.31	420	72	0.62
140	$-28._5$	-0.24	440	85	0.63
150	$-30._5$	-0.17	460	97	0.63
160	-32	-0.10	480	110	0.63
170	$-32._5$	-0.04	500	122	0.63
180	$-32._5$	$+0.02$	520	135	0.62
190	-32	0.08	560	159	0.61
200	-31	0.13	600	183	0.59
210	$-29._5$	0.19	640	206	0.56
220	$-27._5$	0.23	680	228	0.54
230	-25	0.28	720	249	0.51
240	-22	0.32	760	269	0.49
250	$-18._5$	0.36	800	288	0.47
260	$-14._5$	0.39	840	307	0.44
273	-9	0.43	880	324	0.42
280	-6	0.45	920	340	0.40
293	0	0.48	960	356	0.38
298	$+2._5$	0.49	1000	371	0.37

Reprinted, with permission, from Ref. 137.

TABLE 7.XIII
Thermal Expansion, As a Function of Temperature, of SRM 731
(Borosilicate Glass) (105)

T	Expansion $\Delta L/L_{293}$	Expansivity α	T	Expansion $\Delta L/L_{293}$	Expansivity α
80 K	-819×10^{-6}		293 K	0×10^{-6}	$4.78 \times 10^{-6}/K$
90	-797		300	34	4.82
100	-771	$2.64 \times 10^{-6}/K$	320	131	4.91
110	-744	2.86	340	230	4.99
120	-714	3.07	360	330	5.06
130	-683	3.25	380	432	5.11
140	-649	3.43	400	535	5.15
150	-614	3.58	420	638	5.19
160	-578	3.72	440	742	5.21
170	-540	3.85	460	847	5.23
180	-501	3.97	480	952	5.25
190	-460	4.08	500	1057	5.26
200	-419	4.17	520	1162	5.26
210	-377	4.26	540	1267	5.27
220	-334	4.34	560	1372	5.27
230	-290	4.41	580	1478	5.27
240	-246	4.48	600	1583	5.27
250	-201	4.54	620	1689	5.28
260	-155	4.60	640	1794	5.29
270	-109	4.66	660	1900	
280	-62	4.71	680	2007	

Reprinted, with permission, from Ref. 105.

There is also a recommended set of values for the thermal expansion of platinum, and a new SRM of single crystal sapphire is being prepared. Hahn and Kirby (106) cited also a 431 stainless steel and an aluminum alloy.

Platinum has been widely used as a reference material in thermal expansion work in spite of its relatively high cost and softness because it has a high melting point (2045 K) and its expansion is expected to be reproducible. Platinum is recommended by an ASTM method of test (9) for calibrating fused-silica push-rod dilatometer. Hahn and Kirby (107) have determined the linear thermal expansion (from 293 to 1900 K) of three samples.

$$\frac{L_T - L_{293}}{L_{293}} \times 10^6 = -2279 + 6.117\,T + 8.251 \times 10^{-3}T^2 - 1.1187 \times 10^{-5}T^3$$

$$+ 9.1523 \times 10^{-9}T^4 - 3.6754 \times 10^{-12}T^5 + 5.893 \times 10^{-16}T^6$$

$$\alpha \times 10^6\,K = 6.117 + 1.6503 \times 10^{-2}T - 3.3561 \times 10^{-5}T^2 + 3.6609$$

$$\times\ 10^{-8}T^3 - 1.83772 \times 10^{-11}T^4 + 3.5357 \times 10^{-15}T^5$$

The problem of standard reference materials was widely studied by many authors as can be seen in the Thermal Expansion Symposia (255–260). For

TABLE 7.XIV
Thermal Expansion, As a Function of Temperature, of SRM 736
(Copper) (138)

T	$\dfrac{L\text{-}L_{293}}{L_{293}}$	$\dfrac{1}{L_{293}}\dfrac{dL}{dT}$	T	$\dfrac{L\text{-}L_{293}}{L_{293}}$	$\dfrac{1}{L_{293}}\dfrac{dL}{dT}$
20 K	-3250×10^{-6}	$0.27 \times 10^{-6}/\text{K}$	340 K	793×10^{-6}	$17.07 \times 10^{-6}/\text{K}$
30	-3245	0.98	360	1135	17.22
40	-3229	2.29	380	1481	17.38
50	-3198	3.87	400	1831	17.53
60	-3151	5.48	420	2183	17.68
70	-3089	6.98	440	2538	17.82
80	-3012	8.30	460	2896	17.97
90	-2923	9.46	480	3256	18.11
100	-2823	10.46	500	3620	18.25
110	-2714	11.32	520	3986	18.39
120	-2597	12.05	540	4356	18.53
130	-2474	12.67	560	4728	18.67
140	-2344	13.20	580	5102	18.81
150	-2210	13.64	600	5480	18.95
160	-2072	14.01	620	5860	19.09
180	-1785	14.63	640	6244	19.24
200	-1487	15.14	660	6630	19.38
220	-1180	15.57	680	7019	19.53
240	-865	15.94	700	7411	19.69
260	-543	16.24	720	7807	19.84
280	-215	16.50	740	8205	20.00
293	0	16.64	760	8607	20.16
300	117	16.71	780	9012	20.33
320	453	16.90	800	9420	20.51

Reprinted, with permission, from Ref. 138.

example experiments made at the National Standards Laboratory (CSIRO, Sydney, Australia) (279) and at Iowa State University (USA) of the linear expansion and for silicon, copper, silver, aluminum, and germanium below 35 K agree within limits of experimental uncertainty and should constitute reliable low-temperature reference data.

Graphite was recommended by Gaal (95) to cover the temperature range above 1400°C. The equation describing the behavior of heat-treated CS 1000 series stock is:

$$\text{Percent expansion} = -5.27 \times 10^{-2} + 6.98T \times 10^{-4} + 7.76T^2 \times 10^{-8}$$

where T is the temperature in degrees Celsius.

Calibration can also be made by use of the change of a physical property. An interesting method was used by Periera and Graham (211) to calibrate the low-temperature optical dilatometer they developed in 1970 (210). They incorporated a calibration signal using the inverse piezoelectric effect of a quartz x-cut crystal. The slab used for measurements can be cut into a set of

TABLE 7.XV

Thermal Expansion, As a Function of Temperature, of SRM 737
(Tungsten) (139)

T	$\dfrac{L\text{-}L_{293}}{L_{293}}$	$\dfrac{1}{L_{293}}\dfrac{dL}{dT}$	T	$\dfrac{L\text{-}L_{293}}{L_{293}}$	$\dfrac{1}{L_{293}}\dfrac{dL}{dT}$
80 K	-814×10^{-6}	$2.30 \times 10^{-6}/\text{K}$	520 K	1040×10^{-6}	$4.70 \times 10^{-6}/\text{K}$
90	-790	2.61	560	1229	4.73
100	-762	2.88	600	1418	4.76
110	-732	3.11	650	1657	4.79
120	-700	3.30	700	1898	4.83
130	-666	3.46	750	2140	4.86
140	-631	3.59	800	2384	4.90
150	-595	3.71	850	2630	4.93
160	-557	3.81	900	2878	4.97
180	-479	3.97	950	3127	5.01
200	-398	4.10	1000	3378	5.04
220	-315	4.20	1100	3887	5.12
240	-231	4.27	1200	4404	5.21
270	101	4.36	1300	4930	5.31
293	0	4.42	1400	5467	5.43
320	120	4.47	1500	6016	5.56
360	300	4.53	1600	6578	5.70
400	483	4.59	1700	7157	5.87
440	667	4.63	1800	7754	6.07
480	853	4.66			

Reprinted, with permission, from Ref. 139.

disks ~1 cm in diameter, one or two of which are mounted in line with the thermal expansion specimen in the dilatometer (quartz disk and specimen glued together or quartz disk glued to the frame assembly).

The signal obtained when the quartz crystals were excited by dc voltages (of from 10 V to 200 V) showed very sharp response, no asymmetry with sign, and no hysteresis. The effect, thus, affords an excellent calibrating signal: the measurement of d_{11} (relevant piezoelectric coefficient) gives $d_{11} = 2.32 \times 10^{-10}$ cm. volt^{-1}.

H. COMPARISON OF ADVANTAGES AND OF DISADVANTAGES OF EACH METHOD

It is well known that different methods of thermal expansion measurements yield different results. For example, the dilatometer method measures the large-scale change in length or in volume while X-ray methods measure the change in volume or length of a unit cell.

Austin (10) reports as particular example the case of calcite crystal, used as a standard for precise X-ray work. It is essential to know whether all

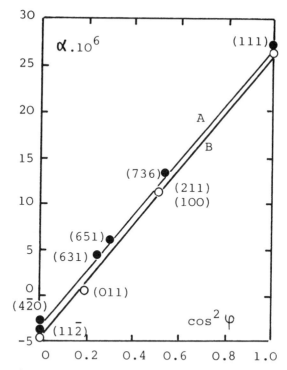

Fig. 7.31. Comparison of coefficient of thermal expansion of the same crystal of calcite as measured by the X-ray method (curve A) and by the dilatometric method (curve B) (14).

crystals are alike and whether macromeasurements by means of a dilatometer yield the same results as X-ray methods. Weigle and Saini (274), using an X-ray method, had determined accurately the thermal expansion of calcite and obtained results which differed greatly from those reported early by Benoit who had used a precision dilatometer. For example, the expansion coefficient parallel to the principal axis was found to be 25.7×10^{-6} by Benoit but only 21.0×10^{-6} by Weigle and Saini. According to Austin, it was not certain whether this difference, which was much greater than the experimental error, was due to the use of different methods of measurements or was, in fact, a real difference in the specimens of calcite. Austin and Pierce measured the thermal expansion of a single crystal of calcite in several different directions and sent their specimens to Weigle and Sani for measurement by their X-ray method. The results (14) reported in Fig. 7.31 show that within the error measurement, the expansion measured by the two methods is the same and agrees with the measurements of Benoit. It can be concluded from this example that different specimens of the same mineral may differ significantly in thermal expansion.

We have discussed in Section I.C.8. the difference given by dilatometric measurement and by X-ray methods in the case of solids containing defects such as vacancies or interstitials. This difference is small and appears only near the melting point for metals. It is more important, as shown by Goetz and Hergenrother (163), in the case of bismuth: these authors measured the thermal expansion of bismuth along its trigonal axis up to the melting point, and showed that while the expansion coefficient as measured by X-ray increased with temperature, the expansion coefficient measured optically remained constant until near the melting point, when it decreased rapidly. The explanation of such a difference, suggested by Goetz and Hergenrother, was that crystals possess a mosaic structure, the layers between the microcrystals being occupied by amorphous or "decrystallized" atoms with a small expansion coefficients.

These typical examples and that dealing with the large shrinkage which appears in dilatometric curves of a solid during sintering (phenomenon which does not affect greatly the X-ray lattice parameter evolution) show that the more important differences will appear if the dilatometric and X-ray data are compared.

Independently of these considerations, each method possesses its own advantages and disadvantages, which can be summarized as follows: True expansion curves are very useful when it is not necessary to point out a discrete or unknown physicochemical transformation. Their interpretation is more difficult when the reference and the specimen have very different expansion coefficients.

Differential dilatometry finds numerous interesting applications such as: establishing the expansion of new reference materials; quality control of materials such as ceramics for turbine heat exchangers, etc.; rapid determination of the difference of expansion between two materials (such as a glass and a metal) which have to be sealed (glass-metal seals); phase-transformation studies.

It allows one to measure both differential and absolute expansion of materials. This technique is a powerful tool and was detailed by many authors, such as Keyser (130) and Plummer (215), but partiularly by Chevenard (57). However, the heating rate must be limited and not too high when the expansion coefficients of the reference and of the specimen differ.

Push-rod dilatometers need a correction of the expansion curves because there is an additive expansion of the push rod and sample holder. This problem is rapidly resolved if the expansion coefficient of the arrangement material is well-known. They are well-suited for middle and high-temperature studies.

Mechanical and push-rod dilatometers cannot be employed to investigate rubbers and elastomers because these substances are highly deformable when placed in the sample holder. The pressure promoted on them by the push-rod, even small, can either modify the dilatometric curve or prevent any measurement when the elasticity of the sample is too high. Volumetric dilatometry or apparatus based on optical method (interferometry) are certainly better suited to their investigation.

Interferometric and optical methods allow experiments with massive samples without specifications on their structural state (glass, crystals, etc.). The temperature range of interferometric methods is limited (low and middle temperature). The sensitivity is high, particularly in the case of laser-interferometric methods. They give absolute measurement of expansion coefficients and are well-suited for low-temperature experimentation and for the determination of the expansion of standard reference materials.

X-ray methods can be used only with crystalline materials but not with amorphous or glassy solids. They need only small quantity of matter. They give measurement of the intrinsic expansion of crystal-lattice but not the macroscopic expansion. Some other problems can arise with X-ray methods, for example, possible deformation of the specimen when small-radius film cameras are employed and difficulty in the exact temperature measurement.

Simultaneous methods such as dilatometry + DTA or dilatometry + electrical conductivity measurements, give interesting complementary information insofar as the sample form and state are adapted to the technique.

III. USE OF THE TECHNIQUE

A. DETERMINATION OF THERMAL EXPANSION COEFFICIENTS

The problem that can be resolved in the first place by dilatometry is the determination of thermal expansion coefficients of solids (classical dilatometry) or eventually of liquids (volumetric dilatometry). Such determinations may furnish either technical data or controls (metallurgy, ceramics, and, more generally, material technology), or information when correlations must be obtained between thermal expansion and physicochemical or physical and structural characteristics of solids. Literature is particularly plentiful about this subject and it would be too long to cite in this chapter the list of thermal expansion coefficients of all the substances studied. We give in Table 7.XVI thermal expansion coefficients of elements at 298 K as reported by Gschneidner (101). This table allows interesting comparisons between the α values and the atomic structure of these substances as we will see later (Section IV.B.) in the case of metals. Other values of coefficients of expansion are given in this chapter but only as informative data in relation with particular structure or materials.

B. CRYSTALLINE STRUCTURE CHANGES AND TRANSITIONS IN THE SOLID STATE

Transitions may be described as being of different "orders" according to the way in which properties such as free energy, entropy, volume, and heat capacity vary in the vicinity of the transition point.

Normal polymorphic change, which concerns an element or compound

TABLE 7.XVI
Linear Coefficient of Thermal Expansion of Elements at 298 K (101)

Element	$\alpha \times 10^{-6}$ $(°C^{-1})$	Element	$\alpha \times 10^{-6}$ $(°C^{-1})$
3 Li	45	50/Sn(g)	5.3[d]
4 Be	11.5	50 Sn(w)	21.2[a]
5 B	8.3	51 Sb	10.9
6 C(g)	3.8 ± 3.1	52 Te	16.77 ± 0.03
6 C(d)	1.19 ± 0.01	55 Cs	97
11 Na	70.6 ± 0.6	56 Ba	18.8 ± 0.8
12 Mg	25.7 ± 0.7	57 La	10.4[a]
13 Al	23.1 ± 0.5	58 Ce(γ)	8.5
14 Si	3.07 ± 0.07[a]	59 Pr	6.79[a]
15 P(w)	124.5 ± 0.5	60 Nd	9.98[a]
15 P(r)	(66.5)[b]	61 Pm	(9.0)[b]
16 S(r)	64.1 ± 0.1	62 Sm	10.4
16 S(m)	(63)[b]	63 Eu	33.1[a]
19 K	83.0	64 Gd	8.28[a,f]
20 Ca	22.4 ± 0.1	65 Tb	10.3[a]
21 Sc	10.0[a]	66 Dy	10.0[a]
22 Ti	8.35 ± 0.15	67 Ho	10.7[a]
23 V	8.3	68 Er	12.3[a]
24 Cr	8.4[c]	69 Tm	13.3[a]
25 Mn	22.6 ± 0.3	70 Yb	24.96 ± 0.04
26 Fe	11.7	71 Lu	8.12[a]
27 Co	12.4	72 Hf	6.01 ± 0.16
28 Ni	12.7 ± 0.2	73 Ta	6.55 ± 0.05
29 Cu	16.7 ± 0.3	74 W	4.59 ± 0.03
30 Zn	29.7	75 Re	6.63 ± 0.06
31 Ga	18.1 ± 0.2	76 Os	4.7 ± 0.1
32 Ge	5.75	77 Ir	6.63 ± 0.12
33 As	4.28 ± 0.42	78 Pt	8.95 ± 0.05
34 Se	36.9 ± 0.1	79 Au	14.1 ± 0.1
37 Rb	88.1 ± 1.9	80 Hg	61[f]
38 Sr	20	81 Tl	29.4 ± 1.0
39 Y	12.0[a]	82 Pb	29.0 ± 0.3
40 Zr	5.78 ± 0.07	83 Bi	13.41 ± 0.09
41 Nb	7.07 ± 0.05	84 Po	23.0 ± 1.5
42 Mo	4.98 ± 0.15	87 Fr	(102.)[b]
43 Tc	(8.06)[b]	88 Ra	(20.2)[b]
44 Ru	9.36 ± 0.27	89 Ac	(14.9)[b]
45 Rh	8.40 ± 0.10	90 Th	11.2 ± 0.4
46 Pd	11.5 ± 0.4	91 Pa	(7.3)[b]
47 Ag	19.2 ± 0.4	92 U	12.6 ± 0.4
48 Cd	30.6 ± 1.3	93 Np	27.5
49 In	31.4 ± 1.4	94 Pu	55

[a] X-ray data.

[b] Estimated value.

[c] See the original paper (101).

[d] Value at 215 K.

[e] Value at 361 K.

[f] Value for solid mercury at its melting point, 234 K.

Reprinted, with permission, from Ref. 101.

which forms two or more crystalline solid phases differing in atomic arrangement, are first-order transitions (as phase changes, e.g., fusion or vaporization). They correspond to a discontinuity in the first derivative of G

$$\left(\frac{\delta G}{\delta T}\right)_P = -S \quad \text{and} \quad \left(\frac{\delta G}{\delta P}\right)_T = +V \tag{66}$$

and are associated with a latent heat, a volume change, and an entropy change.

A transition for which ΔS and ΔV are both zero is called a "second-order" transition. It corresponds to a discontinuity of the second derivative of G:

$$\left(\frac{\delta^2 G}{\delta T^2}\right)_P = -\frac{dS}{dT} = -\frac{Cp}{T} \quad \text{(specific heat anomaly)} \tag{67}$$

and

$$\left(\frac{\delta^2 G}{\delta P \delta T}\right)_P = \frac{dV}{dT} = \alpha_V \quad \text{(expansion coefficient anomaly)} \tag{68}$$

Transitions of higher order are possible.

Although these distinctions seem clear, it is sometimes difficult experimentally to define exactly the type of a transition because many of them that are supposed to be of first order occur as second-order transformations.

The above considerations show that dilatometry is of great interest for the study of polymorphism and second-order transitions because they are associated with volume or thermal expansion coefficient anomalies.

Among other solid-state transformations that give rise to a volume or length change, the transition may be considered from one phase to another (or several others) in phase diagrams and particular phenomena such as electrolysis in the solid state.

1. Polymorphic Transformations

Utilization of dilatometry to this field is particularly well suited because polymorphic transformation, which corresponds to a change from one atomic structure to another, is accompanied by a discontinuous change in volume. According to the nature of the crystalline change, the dilatometric anomaly may be either an expansion or a contraction. The magnitude of volume (or length) change related to a polymorphic transition may differ greatly according to the structural change of the solid investigated. Table 7.XVII gives some information about several polymorphic transitions, studied by dilatometry.

Different manifestations of polymorphic transformations can be observed

TABLE 7.XVII
Some Polymorphic Transitions Studied by Dilatometry

Compound	Transition	Temperature of transition (°C)	$\frac{\Delta V}{V}(\%)$	References
NaNO₃	(Rhomb.) → (rhomb.)	150–280		12
Na₂WO₄	(Rhomb.) → (rhomb.?)	585	17.0	13
K₂SO₄	β (Orth.) → α (hex.)	586 ± 2		206
	Other	300 ± 2		32
		350 ± 2		
		449 ± 2		
KNO₃	II (Orth.) → I (rhomb.)	128	1.0	144
	III (Rhomb.?) ← I (rhomb.)		−2.8	
	II (Orth.) ← III (rhomb.?)		1.87	
NH₄Br	(Cubic) → (cfc)	139–144	22–30	271
	(Cubic) ← (cfc)	117–124		
NH₄Cl	(Cubic) → (cfc)	181–184	23	271
	(Cubic) ← (cfc)	170–175		
CsCl	(Cubic) → (cfc)	468–474	23	271
	(Cubic) ← (cfc)	453–459		
	Aragonite → calcite	> 400	8.3	
CaCO₃	(orth) (rhomb)			218
	Vaterite → calcite	> 360	−24.0	177
	(hex) (rhomb)			
Sn	α (Grey) → β (white)	13	>20	45
	(Cubic) ← (tetrag.)	−40ᵃ		
ZrO₂	Monocl. → tetrag.	1100	−7.7	128
	←	900		
	Tetrag. → cub.	2300		223
	or tetrag.			
SiO₂	α-quartz → β-quartz	573	0.86	128
	(Rhomb.) ← (hex.)	570 (?)		
AlPO₄ (quartz structure)	α → β	586		119
SiO₂	α-Cristobalite → β-cristobalite	220–27	3–7	
	(Quadr.) ← (cub.)	198–240		
AlPO₄ (cristobalite struct.)	α → β	210		119
Na₂O·Al₂O₃·2 SiO₂ (cristobalite struct.)	α → β	670		180
SiO₂	α-Tridimite → β₁-tridymite	117	0.15	128
	(orth.) (hex.)			
	β₁-Tridimite → β₂-tridymite	163	0.20	
	(hex.) (hex.)			
AlPO₄ (tridymite struct.)	α → β₁	93		119
	β₁ → β₂	150		

ᵃ Temperature of maximum rate.

a. Manifestation versus Temperature

Polymorphic transitions can take place at a given temperature or be spread over a temperature range. This is illustrated Fig. 7.32 (12,13) corresponding to two extreme cases of sodium tungstate and sodium nitrate, respectively. In the case of sodium nitrate, investigated as a single crystal, the solid is subjected to a subtle type of transformation with an unusual gradual transition over the temperature range 150–280°C: there is a discontinuity in volume or length but a point of inflexion at which the coefficient of expansion passes through a maximum. The transition appears more clearly along the c axis than along the a axis which is related to the particular motion of structural elements in sodium nitrate.

b. Apparent Nonreversibility of Polymorphic Transitions on Cooling

Generally a polymorphic transition is reversible on cooling. But apparent nonreversibility can be observed. A typical example is given by potassium nitrate. This solid exists at 128°C and 82 mb pressure in three varieties I, II, and III (144). On heating $KNO_3(II)$ inverts directly to $KNO_3(I)$; on cooling dry $KNO_3(I)$ first changes metastably to $KNO_3(III)$ which inverts to $KNO_3(II)$. That leads to nonreversible dilatometric curve with the particularities that the transition II → I corresponds to a small expansion while the transitions I → III and III → II correspond to a contraction and an expansion, respectively (Fig. 7.33a).

Hysteresis often occurs in polymorphic transitions on cooling and certain transitions on cooling must appear as "jerks" or "leaps" (see the case of

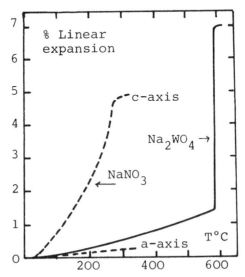

Fig. 7.32. Linear thermal expansion of an aggregate of sodium tungstate (———) (13) and of a single crystal of sodium nitrate (---------) (12).

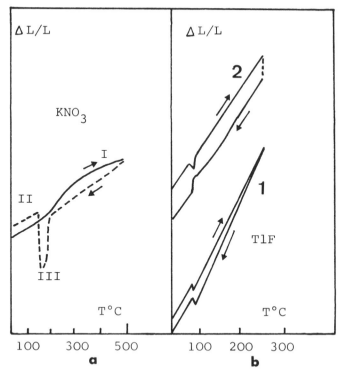

Fig. 7.33. Thermal expansion of potassium nitrate (*a*) (compacted powder) and thallium fluoride (*b*) (compacted powder, curve 1, and cooled cast sample from the melt, curve 2). [Author's results.]

transition of $2CaO \cdot SiO_2$ in Section IV.E.1., and of plutonium in Section IV.B.4.).

c. POLYMORPHISM OF ISOSTRUCTURAL COMPOUNDS

Certain inorganic compounds having an atomic arrangement similar to that of cristobalite have an expansion curve which is remarkably similar to that of cristobalite, even to the extent of having the same type of transformation at about the same temperature. As typical example can be cited the case of aluminum orthophosphate (Fig. 7.34) (119).

d. PARTICULAR CASE OF SILVER IODIDE POLYMORPHISM (156)

Silver iodide presents a polymorphic transition at about 146°C. This transition appears in dilatometry as a large contraction from the low-temperature to the high-temperature phase. This phenomenon was ascribed to the partial melting of the high-temperature AgI in which the cations have a very high mobility while the anions form a rigid network. That is a typical example of a solid which, during heating, retains the lowest possible value of *G* (Gibbs

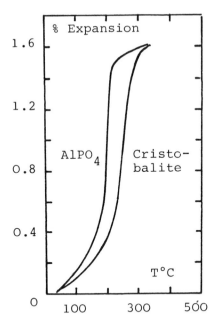

Fig. 7.34. Comparison of thermal expansion of aluminum orthophosphate and that of cristobalite (119).

function) and does not expand during the transition, but presents a "pre-melting" to increase its entropy (276).

e. MATRIX EFFECT (181)

Certain polymorphic transformations are not always easy to investigate by dilatometry (other than X-ray method) when the sample is a compacted powder. For example, in the case of calcium formate, the volume change related to the $\alpha \to \gamma$ transition (expansion) is so large that the specimen collapses before the end of the transformation. Then it is necessary to make the measurements on a sample previously mixed with a diluent (e.g., kaolinite) which does not present dilatometric anomalies in the temperature range investigated. But this method leads to a matrix effect, that is to say a hardly detectable transition on cooling or during the second heating of the sample.

Polymorphic transformations of calcium formate can be represented by the following scheme (65).

$$
\begin{array}{c}
\alpha \\
\uparrow \\
25°C \\
+ \epsilon\,H_2O \\
\downarrow \\
\beta
\end{array}
\qquad
\begin{array}{c}
\underset{180°C}{\nearrow}\;\epsilon\,H_2O \\
\\
\underset{\underset{100°C}{\longleftarrow}}{\overset{150°C}{\longrightarrow}}
\end{array}
\quad \gamma \;
\underset{300°C}{\overset{300°C}{\rightleftarrows}}
\; \delta
$$

During the first heating of α formate (diluted with kaolinite), there occurs successively expansions due to $\alpha \to \gamma$ and $\gamma \to \delta$ transitions (Fig. 7.35A).

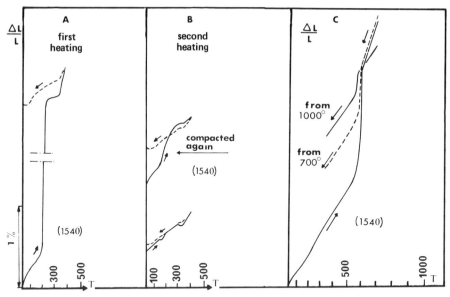

Fig. 7.35. Matrix effect on the polymorphic transitions of calcium formate mixed with kaolinite (compacted powder) during the first heating (A) and second heating (B), and of potassium sulfate mixed with alumina (compacted powder) (C). [Author's results (181).]

During cooling, there takes place $\delta \to \gamma$ and $\gamma \to \beta$ transitions. During a second heating, length changes related to $\beta \to \gamma$ and $\gamma \to \delta$ transitions can be measured with accuracy only if the specimen is smoothly ground and compacted again after the first heating (Fig. 7.35B) (181).

Matrix effects occur also in other mixtures of an inert substance and a specimen the polymorphism of which must be studied. Fig. 7.35C gives the example of potassium sulfate-alumina mixture. The higher the temperature attained during heating, the smaller is the length change associated with the polymorphic transition of K_2SO_4 in the cooling.

f. INFLUENCE OF SAMPLE PREPARATION

Dilatometric anomaly associated with a polymorphic transformation can be different according to the preparation of the specimen. In the case of thallium fluoride, the orthorhombic-tetragonal transition (82°C) appears as a contraction on heating and as an expansion of cooling (266). When the specimen is obtained by cooling the liquid-fused salt in a mold, the dilatometric curve presents on heating a small contraction immediately followed by an expansion, and on cooling essentially a contraction (Fig. 7.33b) (190), that is to say, an inverse effect with respect to that observed with a compacted powder. An explanation of such a difference may be found in the possibility of a certain orientation of anisotropic crystals during the specimen preparation by casting.

g. COMPARISON BETWEEN DILATOMETRY AND X-RAY DATA

Potassium sulfate presents in dilatometry an expansion at about 570–590°C related to its β (orthorhombic) → α (hexagonal) transition which was detected at 586 ± 2°C by X-ray diffraction (206). Besides this unambiguous transition, Bernard and Jaffray (32) pointed out by DTA and dilatometry three new transitions at 300 ±2°C, 350 ±2°C, and 449 ±2°C, respectively. These transitions do not appear clearly in X-ray diffractometry as shown by Pannetier and Gaultier (206) who observed only very low anomalies in X-ray lines of the orthorhombic pattern at 260°C for a and c parameters, at 360°C for a and b parameters, and at 460°C for a, b, and c parameters.

h. DILATOMETRIC STUDY OF THE KINETICS OF POLYMORPHIC TRANSITIONS

Isothermal dilatometry is well-suited to study the kinetics of polymorphism. In dilatometry the degree of reaction (or degree of transformation) x is proportional to the length change of the specimen.

$$x = \frac{L - L_0}{L_\infty - L_0} \tag{69}$$

where L_0 is the length of the specimen at the initial time t_0; L is the length at a time t; and L_∞ is the length when the transformation is ended.

We will summarize three typical examples of this application:

1. Aragonite–calcite and vaterite–calcite transitions. These transitions are nonreversible on cooling. The transformation of aragonite into calcite leads to an expansion at temperatures >400°C and is followed by a sintering of the calcite formed. The transformation of vaterite into calcite gives rise to a contraction at θ >360°C. Kinetic investigation of these transitions was made by Pruna et al. (218) and by Mondange-Dufy (177) who both used a Chevenard dilatometer. The recording of isothermal expansion curves and the graphical drawing of the derivative curves, which gives the reaction rate versus time, led to interesting conclusions: (a) all isothermal dilatometric curves show an induction period which is a period of nucleation; (b) many factors affect the kinetics of the transitions (aging of the sample, use of natural macrocrystals or of compacted powders, presence of impurities in the solid, nature of the gas in the dilatometer sample holder, etc.).

The S-shaped form of kinetics results obtained by Mondange-Dufy (177) suggest that aragonite → calcite transition is controlled by the Avrami equation

$$1 - x = \exp(-At^k) \tag{70}$$

where t is time and A and k are constants.

2. Polymorphic transition of tin. The mechanism and kinetics of the allotropic transformation of tin were studied by Burgers and Groen (45) who determined the rate of transformation by volumetric dilatometry (glass di-

latometer containing both solid tin and a liquid, the level of which is measured in a capillary tube). For the transformation white → grey, the results comply with the Avrami equation for the fraction transformed x with $k = 3$, which corresponds to three-dimensional growth of nuclei formed at the beginning of the transformation process. For grey → white transformation, the kinetics are also controlled by the law (equation 70) but with $k = 1$ to 2.5. For values of k about 1, the transformation is of unimolecular type. k Values greater than 1 point out an autocatalytic reaction, and spontaneously formed nuclei (which can be seen by direct observation) give rise to induction of further nuclei in their neighborhood.

3. Polymorphic transitions of alkaline halides. Kinetics of alkaline halides were studied by Vernay (271) with a Netzsch 1550°C dilatometer. Specimens were investigated as compacted powder.

For ammonium bromide, the cubic → cfc transition occurs between 139°C and 144°C. Activation energy varies with the degree of transformation from 62 kcal · mol^{-1} at $x = 0.1$ to 135 kcal · mol^{-1} at $x = 0.6$ and to 118 kcal · mol^{-1} at $x = 0.8$.

For ammonium chloride, the cubic → cfc transition occurs between 139°C and 144°C. Activation energy (average value) is of 95 ± 7 kcal · mol^{-1}.

For cesium chloride, the cubic → cfc transition occurs between 468°C and 474°C and the activation energy varies with x from 98 kcal · mol^{-1} at $x = 0.1$ to 196 kcal · mol^{-1} at $x = 0.6$.

But from these experiments, it was pointed out that it is difficult to obtain accurate data either on the temperature at which the transition begins or on the variation of volume change, whatever the technique used (dilatometry, DTA, or X-ray crystallography). That explains the divergences of literature data about this subject. On the other hand dilatometric data show that kinetics are not controlled only by thermal processes and it was shown that defects in the solid state lower the temperature of the transition.

The method for determining the exact temperature T_t of a polymorphic transition (31) varies according to the form of the curve around this point. When the curve is linear before the transition T_t is the point where the curve changes its slope (Fig. 7.36a). When no linear portion is observed, the point T_t can be determined by drawing the tangent at the curve, this tangent making an angle of 30° with the temperature axis (Fig. 7.36b). If the transition occurs at a minimum or a maximum in the dilatometric curve, T_t is the point of horizontal tangent intersection at the minimum or maximum of the curve (Fig. 7.36c). T_t can also be determined as the intersection of two tangents when the dilatometric curve presents a characteristic slope change with a curvature, (Fig. 7.36d), but the accurate determination of T_t is difficult when the transition is spread over a large temperature range.

Another approach of the determination of T_t is to submit the specimen to successive heating and cooling cycles while increasing the limit temperature

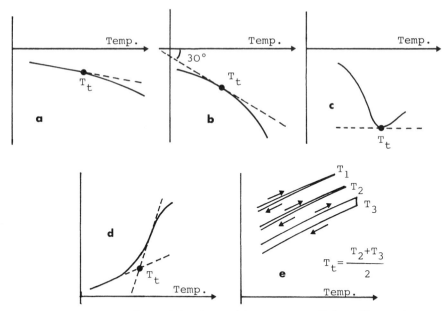

Fig. 7.36. Different ways for obtaining the transition temperature T_t from dilatometric curve.

on heating. When nonreversibility occurs (appearance of a vertical shrinkage on the curve), the temperature T_t can be fixed as the average temperature between the last cycle and the preceding one (Fig. 7.36e).

2. Order–Disorder and Second-Order Transitions

An order–disorder transition is characterized by the fact that below a given temperature, some ions that were statistically distributed in the structure, are subject to an ordering which modifies properties such as reactivity, the micrographic aspect, and especially the X-ray diffraction patterns. Order–disorder transitions occur in the case of certain alloys and this phenomenon may be detected by dilatometry (see Section IV.B.4.). They occur also in numerous mineral compounds and particularly solids with spinel structure. A typical example of the volume anomaly that occurs in the case of lithium aluminate $LiAl_5O_8$ was given by Lejus and Collongues (151) (Fig. 7.37).

In considering transitions of silica and related compounds (84,248), silica exists as numerous phases such as quartz, cristobalite, and trydimite (crystalline phases) and as fused silica (amorphous phase). Crystalline phases show different polymorphic or second-order transitions that are easily detected by dilatometry (see Table 7.XVII). The $\alpha \rightarrow \beta$ transition of quartz at 573°C is a typical second-order transition. Fig. 7.38 gives the typical curve of quartz, which has a negative value of the thermal expansion coefficient above 573°C.

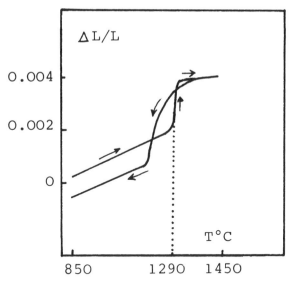

Fig. 7.37. Thermal expansion curve of LiAl$_5$O$_8$ with the second-order transition at 1290°C. (Reprinted, with permission, from Ref. 151.)

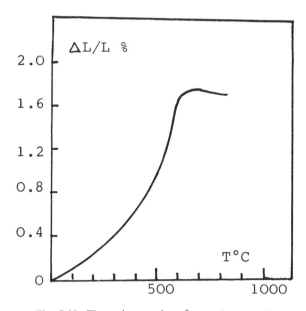

Fig. 7.38. Thermal expansion of a quartz aggregate.

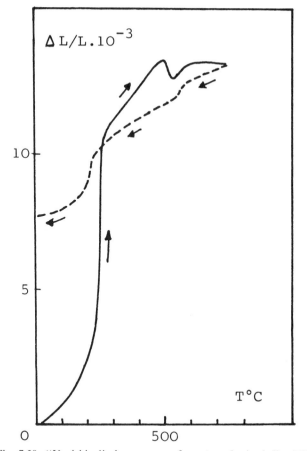

Fig. 7.39. "Vanishing" phenomenon of quartz and cristobalite (178).

Quartz is encountered in many minerals such as clays (see Section IV). Quartz and cristobalite are components of many ceramic bodies. Transitions that are observed on heating do not present the same length-change during cooling when quartz and cristobalite are as components of a mineral, a ceramic body, or a ceramic raw paste. This effect, mentioned already in the case of polymorphic transitions, was called "vanishing effect" by Munier (178) (Fig. 7.39). This phenomenon brings some difficulties when dilatometric curves are used from an analytical point of view (184).

Carnegieite, $Na_2O \cdot Al_2O_3 \cdot 2 SiO_2$, has a structure close to that of cristobalite and presents a reversible $\alpha - \beta$ second-order transition at 670°C. This alumino-silicate appears as a transitory phase during the thermal decomposition of 4A-zeolites above 800°C. The author (180) used the appearance of the second-order transition to study by DTA and dilatometry the kinetics of the formation of carnegieite and its transformation in nepheline at higher temperature (Fig. 7.40).

Dilatometry was also used to study other solids isostructural with silica,

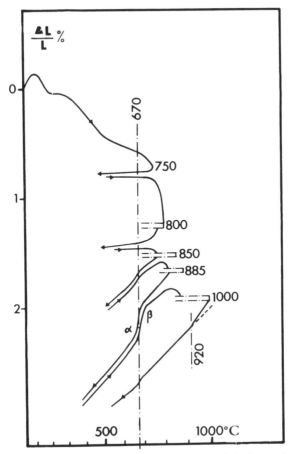

Fig. 7.40. Dilatometric curve of a 4A zeolite (compacted powder). Successive heating cycles up to 750°, 800°, 850°, 885°, and 1000°C. The $\alpha \to \beta$ transition of carnegieite is visible at 670°C. [Author's result (180).]

and particularly aluminum orthophosphate (119), which presents transitions in temperature range close to that of silica phases (see Table 7.XVII and Fig. 7.34).

3. Magnetic Transitions

The volume of a magnetic solid can change when the magnetization value is modified. That can occur either by variation of temperature or by application of a magnetic field. The first phenomenon is called "volume anomaly of ferromagnetic compounds," the second "volume magnetostriction."

a. VOLUME ANOMALY OF FERROMAGNETIC SOLIDS

Thermal expansion curves of ferromagnetic materials present a slope change when they are heated through the Curie point. The coefficient of

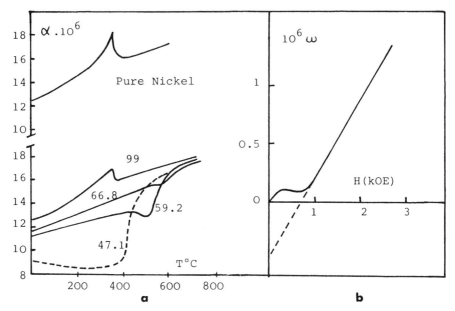

Fig. 7.41. Variation of thermal expansion coefficient of pure nickel and nickel-iron alloys (Ni content is written on each curve) (*a*) (112) and volume variation of an iron sample versus magnetic field (*b*) (112).

thermal expansion presents an optimum at this point. That is the case, for example, for magnetite (576°C), iron (770°C), gadolinum (16°C), and nickel (360°C) (Fig. 7.41*a*, for the case of nickel). This anomaly comes from the variation of the molecular field coefficient with distances between magnetic atoms or atomic volume. The theory was developed by Neel (198) and reported by Herpin (112). It will not be detailed here, but is based on the fact that in magnetic solids, it is necessary to take into account the magnetic free energy F_m which adds to the elastic free energy F_{el} and thermic free energy F_{th}. Calculations show that the principal term of the anomaly of expansion is expressed by

$$\Delta\alpha = \frac{aW'}{18(1/\beta)} \cdot \frac{dM^2}{dT} \tag{71}$$

where W' is the derivative of the function $W(\Delta V/V)$ with respect to the interatomic distance a, M the magnetization, β the coefficient of isothermal compressibility, and T the temperature.

The anomaly of expansion may be positive or negative. It is positive for nickel and nickel-rich alloys and negative for alloys with low nickel contents (Fig. 7.41*a*). The negative anomaly can even be so important to compensate the thermal expansion so that α becomes very small. That is the case for invar alloy (36% Ni), the thermal coefficient of which is about 2×10^{-6}. In the same way, for certain Fe–30% Pt alloys, the magnetic anomaly is higher in absolute value than the normal expansion coefficient so that a negative thermal expansion occurs in a certain temperature range. Dilatometric study of ferromagnetic materials is more accurate when the derivative curve (which represents the variations of α versus temperature) are recorded because the change of slope in expansion curve is not always evident.

In the particular case of antiferromagnetic substances, a similar phenomenon may occur, that is to say, an expansion anomaly when the material is heated through the Neel temperature. A recent dilatometric study on this subject was published by Woolfrey in the case of sintered- or monocrystalline-nickel oxide (282).

b. VOLUME MAGNETOSTRICTION

When a ferromagnetic substance is submitted to a magnetic field, a small variation of volume (about 10^{-6} for an applied field of some thousands oersteds) is observed. This phenomenon, which is different from the dimension variations of a sample when it is magnetized, occurs in high fields. It can be detected by liquid (or volumetric) dilatometry, the sample having an ellipsoid form. An example of volume change in the case of iron is given Fig. 7.41b.

Change in sample length was also studied with a three-terminal capacitor dilatometer by Schlosser et al. (235) who studied the temperature and magnetic field dependence of the thermal expansion and forced magnetostriction of worked and annealed Fe-Ni-Co alloys, at low temperature. Working and annealing modify greatly the forced magnetostriction and produce a strong invar thermal expansion anomaly.

4. Vitreous Transition

It is well-known that certain liquids, when cooled, do not crystallize but keep a rigid state with a certain short-distance order. Such cooled metastable liquids constitute glasses or amorphous materials, the thermodynamic properties of which are different of the thermodynamic properties of the same material in a crystallized state. That is the case for the specific volume as shown in Fig. 7.42. The same phenomenon appears in polymers or rubbers which can adopt an amorphous or glassy state on cooling.

A particular temperature T_g (vitreous transition temperature) is defined for both glasses and polymers. This temperature corresponds to the slope change of specific volume curve (Fig. 7.42). It can be easily detected and measured by dilatometry, either directly when the slope change is abrupt, or graphically when the slope change occurs with a certain curvature in the chart. In this case, T_g is the contact point of the extrapoled straight lines corresponding to the expansion before and after the transition.

Vitreous transitions in polymers and glasses will be discussed in more detail in Sections IV.C and IV.D, respectively.

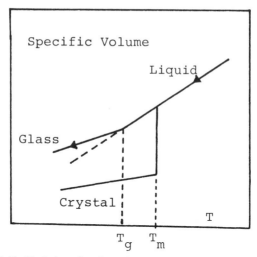

Fig. 7.42. Variation of molar volume during the cooling of a liquid.

5. Miscellaneous

Many applications of dilatometry for the study of phase change in the solid state are related to metallurgy (see Section IV.B), but some examples deal with inorganic chemistry such as the study of the different varieties of wüstite (nonstoichiometric iron oxide FeO_x) with demonstration of a metastable triple point between the three phases (48) and dilatometric study of wüstite inside its existency domain (22). It was shown that $W_2 \rightleftharpoons W_3$ allotropic transition of wüstite is of second order (47).

Dilatometry was also applied to investigate electrolysis in the solid state. When a direct current passing through a system such as

$$O_2 \text{ (1 atm.), Pt / solid oxide / Pt, } O_2 \text{ (1 atm.)}$$

it is partially transported by cationic defects, and it is accompanied by a matter transport from anode to cathode. That promotes a motion of platinum–oxide interface which can be detected and measured by dilatometry. The cationic transport number is calculated by the equation

$$t_{\text{cat.}} = \frac{\Delta L S a Z F}{M I t} \tag{72}$$

where ΔL is the Pt-oxide interface displacement, S the area of Pt-oxide interface, a the density of the oxide, Z the charge of cationic carriers, F the Faraday, M the molecular weight of the oxide, I the current intensity, and t the electrolysis time.

Deportes and Gauthier (73) employed this technique (dilatometry with direct current flow inside the specimen) to determine cationic transport number in magnesium oxide monocrystal in the temperature range 1000–1400°C.

C. CHEMICAL REACTIONS

Many chemical reactions such as the thermal decomposition of solids, oxidation, or formation of compounds by solid–solid reactions were studied by classical dilatometry. Particular solid–liquid reactions may be investigated by volumetric dilatometry.

1. Thermal Decomposition of Solids

Reactions of type "solid → solid + gas" modify more or less the structure and the texture of the solid studied and many applications of dilatometry were made in this field, especially in the case of dehydration reactions. So, in 1936, Rencker and Dubois (222) made use of dilatometry to study the dehydration of manganese sulfate pentahydrate as compacted microcrystalline powder and pointed out shrinkages related to the formation of the tetra-, tri- and monohydrate, and subsequently the formation of anhydrous phase.

The dilatometric anomalies related to the thermal decomposition of hydrated solids depends on the nature of the water bonds, the structure of the initial hydrate, the nature of intermediate phases, and final compound. The shrinkage is slight in the case of the departure of "zeolitic water" [case of zeolites (179), fibrous clays, and calcium sulfate hemihydrate (20)]. It is more important in the case of true hydrates such as hydrated Portland cement (70), gypsum (21), and metallic hydroxides, while in the latter case certain hydrated solids (boehmite) show an expansion during their decomposition (188). (Differences between dilatometric curves of both boehmite and gibbsite may serve to characterize bauxites as it will be seen in Section IV.A.3).

Fig. 7.43 compares dilatometric curves of copper sulfate pentahydrate and calcium sulfate dihydrate (181). Thermogravimetry curves obtained in the same experimental conditions are given in dotted lines.

For copper sulfate the following reactions occur:

$CuSO_4 5 H_2O$ (tricl.)   → $CuSO_4 3 H_2O$ (monocl.)   part "bc"

$CuSO_4 3 H_2O$ (monocl.) → $CuSO_4 1 H_2O$ (monocl.)   part "cde"

$CuSO_4 1 H_2O$ (monocl.) → anhydrous $CuSO_4$ (orth.)   part "ef"

The second reaction, which corresponds to monoclinic → monoclinic transition (no important structural change), does not greatly affect the dilatometric curve.

For calcium sulfate dihydrate the following reaction occurs

$CaSO_4 2 H_2O$ (monocl.)   → $CaSO_4 \frac{1}{2} H_2O$ (hex.)   part "bc"

$CaSO_4 \frac{1}{2} H_2O$ (hex.)   → $CaSO_4 \epsilon H_2O$ (hex.)   part "cd"

$CaSO_4 \epsilon H_2O$ (hex.)   →   anhydrous $CaSO_4$ (orth.)   part "fg"

(part "de" corresponds probably to sintering of $CaSO_4 \epsilon H_2O$).

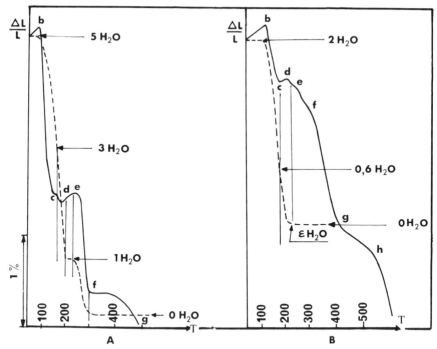

Fig. 7.43. Dilatometric and TG curves during thermal decomposition of copper sulfate pentahydrate (A) and calcium sulfate dihydrate (B) (compacted powders). [Author's results (181).]

As in the case of $CuSO_4$, the second reaction, which does not correspond to an important structural change, does not lead to an apparent shrinkage but to a small expansion on the dilatometric curve.

This information is interesting to point out the nature of particular transformations or the existence of a solid solution such as in the case of $CaSO_4 \frac{1}{2}$ H_2O − $CaSO_4 \epsilon H_2O$. Faivre and Chaudron (83) had shown many years ago, that the transition $CaSO_4 \epsilon H_2O$ (soluble anhydrite) → anhydrous $CaSO_4$ (insoluble anhydrite) promotes a shrinkage on the dilatometric curve: that was not a polymorphic transition as suggested by Faivre and Chaudron but a sintering of $CaSO_4 \epsilon H_2O$ followed by a precipitation of orthorhombic $CaSO_4$ from the solid solution of water in hexagonal $CaSO_4 \epsilon H_2O$.

A similar phenomenon was studied by Murat and Charbonnier (189) in the case of hydrated cobalt molybdate $CoMoO_4$, $0.9 H_2O$. Dehydration between 25°C and 320°C does not affect greatly the crystal structure of the hydrate (small shrinkage AB in the dilatometric curve, Fig. 7.44), but when the solid is almost entirely dehydrated there occurs an important shrinkage BC in the dilatometric curve: DTA shows two peaks at 320°C and 336°C, respectively, when this shrinkage occurs. We do not think that the BC anomaly is a true polyporphic transition but a precipitation of an anhydrous phase, the structure of which is entirely different from that of the hydrate. At higher temper-

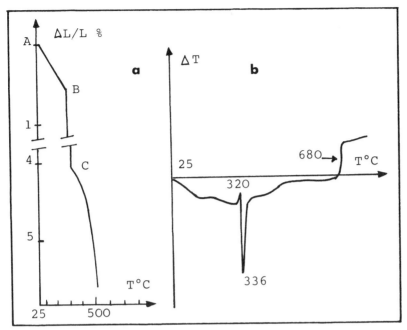

Fig. 7.44. Dilatometric curve (*a*) and DTA curve (*b*) of hydrated cobalt molybdate. [Author's results (189).]

ature, there occurs a large shrinkage related to sintering. At 650–700°C, a crystal-growth phenomenon occurs and on cooling the specimen breaks as a consequence of the large volume change related to polymorphism of anhydrous cobalt molybdate that occurs reversibly at about 400°C.

Many other examples of use of dilatometry to study thermal decomposition reactions will be given in Section IV (characterization of clays, bauxites, minerals, and study of ceramic raw-material and ceramic pastes).

2. Kinetics of Polymerization Reactions

Generally, the contraction which occurs during polymerization of a monomer is the result of the shortening of the intermolecular distance between monomer units as they enter the chains. This contraction can be very large, up to 20–30%, according to the type of polymerization and the nature of the monomer.

The major problem in interpretation of dilatometric data is establishment of the precise relationship between the density of monomer–polymer solution and the polymer concentration.

In the ideal case of monomer and polymer volume additivity, the following equations are verified

$$V_{ps.} = (w_p V_p + w_m V_m)/100 \tag{73}$$

$$d_{ps} = 100/[(w_p/d_p) + (w_m/d_m)] \tag{74}$$

where V, d, and w are specific volume, density, and weight percent, respectively, of the polymer solution (index ps), monomer (index m) and polymer (index p).

When there are deviations from these relations, an empirical equation can be used (229)

$$d_{sp} = [w_p d_p + w_m d_m]/100 \tag{75}$$

If we suppose the additivity of monomer and polymer volume according to equation 73, then the percent conversion x calculated from dilatometric data can be expressed versus initial height Ho (at time $t = 0$), H_t (at time t) and H_f (at final time t_f) of mercury in capillary, and x_f percent of conversion of monomer at final time t_f.

$$x_t = [(H_0 - H_t)/(H_0 - H_f)]\chi_f \tag{76}$$

When additivity of monomer and polymer densities is assumed in equation 75, then a similar equation giving x_t can be expressed

$$\frac{x_t}{x_f} = \frac{V_m V_p}{(V_m - V_p)\left[\dfrac{(H_0 - H_t)}{(H_0 - H_f)}(V_p - V_m) + V_m\right]} - \frac{V_p}{V_m - V_p} \tag{77}$$

The use of parallel absolute methods, such as the titration of residual monomer, is necessary because a polymerization reaction often stops short of completion.

Many examples of dilatometric rate studies were given by Rubens and Skochdopole (229); it appears from these data that if volume change can be measured with good precision by the dilatometric methods, it is very necessary to have accurate conversion factors to relate the data to conversion.

3. Other Types of Reaction

Dilatometry could be used for study of oxidation reactions, but it seems that literature is poor about this subject. It should be interesting to employ this technique to study, for example, the oxidization of metal powders that leads to expansion and could give interesting data compared with that obtained by other techniques.

Volume changes that may happen by reactions in solid–liquid or in liquid systems can be investigated by volumetric dilatometry (detection of volume change by motion of a liquid in a capillary). There must be cited as examples of physicochemical studies: (a) determination of phase diagrams and transitions between phases. An interesting example is that of the transition gypsum-hemihydrate around 100°C (249), and (b) crystallization processes, the study of precipitation of beryllium hydroxide from metastable sodium beryllate solutions (116).

D. SINTERING AND CHANGE IN SURFACE STATE OF SOLIDS

Sintering is a phenomenon which occurs in a compacted powder when it is heated at a temperature below its melting point. The processes involved are first formation of joining zones between grains and later the elimination of residual porosity. The first step leads to a densification of the specimen; it can be detected as length change (shrinkage) easily detectable and measurable by dilatometry either in nonisothermal way for a rapid shrinkage evaluation and characterization of the beginning temperature of sintering, or in isothermal way for kinetics studies and determination of sintering process. A lot of work has been done on this application of dilatometry from the study of, e.g., Nicol and Domine-Berges (202) on the sintering of silica, alumina, platinum, and silver.

As we saw before, sintering is a phenomenon that always affects the thermal decomposition products of a given sample, e.g., hydrated salt, and occurs very often when the dilatometric specimen is a compacted powder but more or less according to its chemical nature and the mineralogical properties of grains.

Sintering of solids such as metals or metallic oxides has been widely investigated. The case of metallic powders will be discussed in Section IV.B. We will give here two examples of the sintering of oxide powders (see also Fig. 7.12)

Uranium dioxide compacted powder. This oxide is of great interest as nuclear fuel. Fig. 7.45 shows the typical dilatometric curve of a UO_2 compact during nonisothermal heating (part $ABCDO_2$) and during an isothermal heating (part after the point O_2). Parts AB and BC are not related to sintering, they correspond to a desorption of water vapor and to a reduction of an overstoichiometric oxide, respectively (15). Shrinkage after point O_2 corresponds to the first step of sintering.

The experimental recorded dilatometric curves during isothermal sintering of compacted powder obey to the equation

$$\frac{\Delta L}{L_0} = kt^n \tag{78}$$

k is a constant that depends on temperature, grain-size, and characteristics of the specimen and t is time.

The quantitative determination of experimental results by using specific methods (e.g., method of Dorn, method of the tangents, method of Running) is improved by recording the derivative of the dilatometric curve and obtaining experimentally a rapid approach of the isothermal sintering temperature. Dilatometry brings a good accuracy in kinetics of sintering, for example, the activation energy obtained from dilatometric data for UO_2 sintering (100 kcal $\cdot$ mol^{-1}) agrees with the values determined from

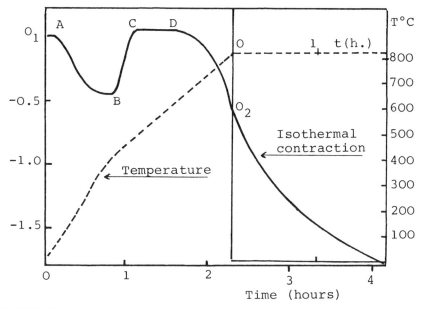

Fig. 7.45. Dilatometric and temperature curves during heat treatment of an uranium dioxide compacted powder (hydrogen atmosphere, DHT 50 Chevenard dilatometer). (Reprinted, with permission, from Ref. 15.)

analysis of isothermal shrinkage curves in the same temperature interval (15).

Alumina. Generally, sintering of alumina-compacted powders does not occur before about 1250°C. Vergnon et al. (270) employed the "retractometry" (method close to optical dilatometry by direct optical measurement of the length change of the sample) to point out that the "flash sintering" or very high-speed heating of the specimen greatly increases the shrinkage rate of alumina.

Zeolites. Another interesting example is that of surface evolution of microporous grains such as zeolites. When heated in air, these materials lose their water vapor absorption capacity at about 700°C. No particular thermal effect is detectable in the DTA curve in this temperature range, the macroscopic structure break taking place only at about 800°C. Dilatometry allows one to detect an important shrinkage precisely at the temperature at which the absorption capacity falls (Fig. 7.46) (179). This phenomenon is due to a partial, if not total occlusion of the structural microporosity (channels) of the sample by a phenomenon which can be compared with a first step of a sintering.

Another particular example of a phenomenon affecting volume change is the study of the *rheology* of certain mixtures such as mixtures of lithium sulfate Li_2SO_4 with potassium sulfate or sodium sulfate (125).

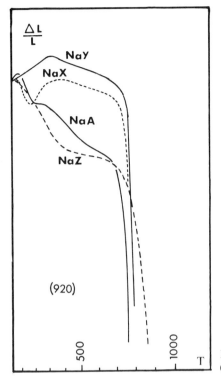

Fig. 7.46. Thermal expansion curves of zeolites (compacted powder). [Author's results (181).]

E. ADSORPTION OF GAS ON SOLIDS

The adsorption of a gas on a solid gives rise to a length change of the sample. This problem was widely discussed for example by Sereda and Feldman (241). This length change can serve to study the mechanism of adsorption. It was shown by Bangham et al. (17) that "unactivated" wood charcoal samples change in length (or volume) when gas such as pyridine, benzene, and the lower alcohols adsorb on them. The relation between the linear expansion and the surface pressure π is of the form

$$\frac{\Delta L}{L} = A\pi \tag{79}$$

where A is a constant.

A typical example of the length change during adsorption or desorption of a gas (water vapor) on a solid (charcoal) is shown in Fig. 7.47. A slight contraction, which becomes greater at higher humidities and reaches a maximum at 65% R.H., is observed during adsorption. It is followed by an expansion. On desorption, a larger contraction is observed and can be explained by assuming that the desorption branch and thus the hysteresis of the isotherm are caused by evaporation from concave menisci.

Sereda and Feldman have discussed the relationship between length

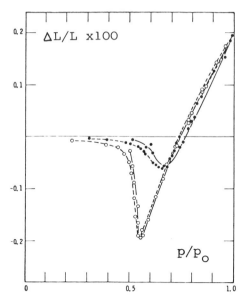

Fig. 7.47. Linear expansion $\Delta L/L$ of charcoal as a function of the relative vapor pressure of adsorbed water. (●) Adsorption points; (○) desorption points; (———) first adsorption and desorption; (--------) second adsorption and desorption (280).

changes and either capillary condensation or properties of the adsorbent but this point cannot be detailed in this chapter.

IV. PRINCIPAL FIELDS OF APPLICATION

Dilatometry has been employed in practically all the science fields dealing with geology, mineralogy, and material sciences (metal and alloys, ceramics, glass, polymers, hydraulic binders, refractories, and so on).

A. CHARACTERIZATION OF ROCKS, MINERALS, AND NATURAL RAW MATERIALS

1. Dilatometry of Rocks and Minerals

a. THERMAL BEHAVIOR OF ROCKS

Rock is a complex composite solid generally consisting of heterogeneous granular aggregates of polycrystalline mineral constituents which are characteristically anisotropic. Thermal stresses induced during the heating or cooling of rocks cause fracture and fragmentation. Thermal expansion is a significant parameter for the creation of thermal stresses to fragment rocks by heat.

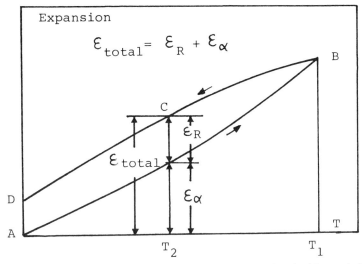

Fig. 7.48. Ideal thermal expansion cycling for rock material. (Reprinted, with permission, from Ref. 261.)

The physical properties of individual grains of rock material are insufficiently known to permit quantitative analysis of induced thermal stresses during heating. Expansion constraints of individual grains during uniform heating or cooling store thermoelastic strain energy which tends to release itself by inducing inelastic response of the rock material. Inelasticity pertains to the behavior of material beyond the elastic range. The inelastic response is immediately irrecoverable and constitutes the instantaneous reaction of rock material to induced thermal stresses. The response will represent immediately irrecoverable structural damage including the formation and/or extension of microcracks between the granular constituents of the rock material.

After Thirumalai and Demou (261), the example of ideal thermal expansion of a rock material for cycling heating from temperature T_0 to T_1 and cooling to T_0 is that presented in Fig. 7.48. The total thermal dilatation ϵ (total) at any point C for temperature T_2 can be considered as sum of observed expansion ϵ_α during heating and immediately irrecoverable thermal dilation ϵ_R. The magnitude of ϵ_R will be proportional to the structural damage of the rock material induced by internal thermal stresses and serves to measure the response of rock material to induce thermal stresses. The magnitude of immediately irrecoverable dilation falls off with successive thermal expansion cycling and reaches a steady state after about three to five consecutive cycles.

b. Some Dilatometric Data on Rocks and Minerals

Many papers were published in this field and we must give only certain typical examples.

(1) Lunar and Volcanic Rocks

During past years, some thermal expansion studies were made on rocks chosen as likely to be representative of those to be found on the lunar surface. Through a program of multidisciplinary research of the Bureau of Mines and NASA, basic scientific and engineering knowledge was provided that will be needed to utilize extraterrestrial mineral resources for support of future space missions.

Based on the fact that thermal stresses induced during the heating or cooling of rocks cause fracture and fragmentation, and that thermal expansion is a significant parameter for the creation of thermal stresses to fragment rocks by heat, Thirumalai and Demou (261) reported in 1970 on some thermal expansion measurements of rock such as granadiorite (plagioclase + quartz), Sioux quartzite, and obsidian (glass + microlites) in vacuum, using surface-strain gauges as transducers. It was shown that the response of these rock materials to induced thermal stresses was independent of reduced environmental pressure down to 10^{-5} torr. Later, Griffin et al. (98) made thermal expansion measurements on fourteen kinds of rocks, such as basalts, obsidian, pumice, tuff, and rhyolite over a temperature range of approximately $-140°C$ to $950°C$ with a modified version of a commercial vertical quartz tube dilatometer. The curves of percentage expansion and coefficient of thermal expansion as a function of temperature, shown in Fig. 7.49, are representative of the curves found among the 14 rock types. Among the distinct slope changes in the case of granodiorite and altered rhyolite, the most important is near the $\alpha - \beta$ quartz inversion point ($\approx 570°$), this seems logical since these rocks have about 39% and 2% quartz, respectively. The percentage expansion is small for practically all the rocks, but the structure is altered permanently during heating and these materials do not return to their original dimensions on cooling.

(2) Granite

Thirumalai and Demou (262) studied also granite and charcoal granites to estimate the rock damage induced by cyclic temperature changes. This is important for planning utilization of underground space for storing energy. Predominant damage takes place during initial exposure of intact rock to temperature cycling between 25°C and 400°C, but it reaches a steady state after three successive thermal cycles.

(3) Slates

These minerals, used as building materials (roof material) are micaschists and contain quartz, oligoclase, muscovite, biotite, chlorite, plus some accessory minerals. Dilatometric curves vary with the cutting orientations of the specimen (53), which is a consequence of the anisotropy of the mineral. An anomaly of expansion, observed at about 320°C, and which is independent of the hydration of the rock, appears in certain slates such as that of Travassac

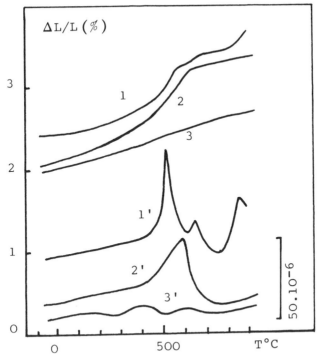

Fig. 7.49. Thermal expansion % (1,2,3) and coefficient of thermal expansion (1', 2', 3') of granadiorite (curve 1), altered rhyolite (curve 2), and theiiolitic basalt (curve 3). (Reprinted, with permission, from Ref. 98.) Scale ½ for curves 1 and 1'.

(Correze, France). This anomaly was related to chlorite expansion which promotes the formation of microcracks inside the sample and an hysteresis on the dilatometric curve. Murat (181) studied an Anjou (France) slate as compacted powder (after grinding), the dilatometric curve shows an important expansion to 800°C followed by a shrinkage above 800°C. The cooling curve allows one to detect the β − α transition of quartz present in the mineral (Fig. 7.50). The anomaly at 320°C, pointed out by Chaye D'Albissin and Morlier (53) on Travassac slate, does not appear, but that can arise from the different state and origin of the sample studied.

(4) Gypsum

Thermal expansion of natural gypsums, studied either as sawed specimen in natural microcrystals or as compacted powder, was studied by Barriac et al. (21). It was shown that the higher the impurity content is, the smaller the shrinkage related to the thermal decomposition reaction. The shrinkage related to the sintering of hexagonal anhydrite can be almost entirely hidden when the sample contains quartz (sand-rose specimen, Fig. 7.50a). Results obtained with specimens prepared by direct sawing in gypsum blocks show

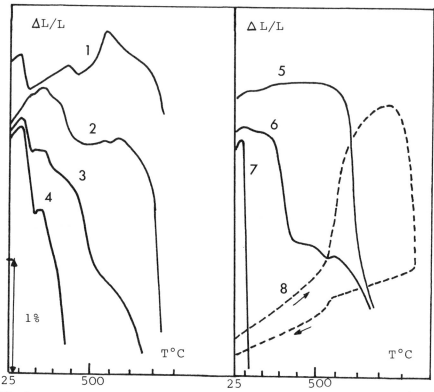

Fig. 7.50. Dilatometric curves of some minerals: compacted powders after grinding. 1, sand-rose gypsum; 2, clayish gypsum; 3, porous gypsum; 4, alabaster gypsum; 5, serpentine; 6, oolitic limonite; 7, allophan; 8, micaschist (slate). [Author's results.]

that the texture of the rock plays an important role and great differences are observed in the dilatometric curves.

(5) Feldspars

Feldspars begin to contract at about 1100°C. When quartz is present in potassic feldspar, an expansion occurs between 1330°C and 1380°C; this is due to a solid-state reaction between quartz and feldspar (285). Melting occurs at about 1400°C and gives rise to an important shrinkage as all melting phenomena when studied by methods other than volumetric dilatometry.

Dilatometric curves obtained with compacted powders and related to allophane, serpentine, and oolithic limonite are given in Fig. 7.50b. They are very different according to the nature of the mineral. Cases of clays and bauxites will be more detailed later.

Many authors have determined thermal expansion coefficients of minerals and the first more important papers about this subject are those of Fizeau (90,91,92) who used interferometric dilatometry. The values obtained by this author are certainly very old but accurate enoughto be taken into account.

2. Clays and Clay Minerals

Clays are phyllitic silico-aluminates (layer structure) that find many uses as raw material in ceramic industry. According to their di- or trioctahedral character and to the interlayer distances (7 Å, 10 Å, 14 Å), the commonest pure clays can be classified as in Table 7.XVIII. There must be added to this classification, the fibrous clays (Sepiolite, Attapulgite) and mixed-layer clays such as Rectorite (Pyrophyllite-Vermiculite), Allewardite (Illite-Montmorillonite), and Corrensite (Chlorite-Montmorillonite).

During heating, clays are subjected to numerous transformations such as decomposition reactions (dehydration and dehydroxylation), structure changes (recrystallization of amorphous phases), recombination reactions, sintering, melting, vitrification, etc. Consequently, the interpretation of thermal expansion curves of clays can be difficult.

Numerous works were published about dilatometric analysis and characterization of clays, e.g., those of Caillere and Henin (46), Kiefer (133), Steger (252), Baudran (26), Paquin (207), Pampuchova and Szmal (205), and the recent technical Note of Adamel-Lhomargy Co. (1) The intensity of expansions and contractions of clays during heating can vary widely from one sample to another for a given mineral, but it is actually recognized that the dilatometric curve of a given type of clay (such as kaolinite or montmorillonite, for example) is characteristic enough of this type. When compared with other thermal methods (TG, DTA) and with X-ray diffraction results, thermal expansion data can be used for a rapid control and qualitative characterization of clays. Typical dilatometric curves of clays can be seen in Fig. 7.51 in the temperature range 25–1000°C.

As pointed out by Kiefer (133), the shape of dilatometric curves of clays may be related to the structure of the mineral but also to the texture and

TABLE 7.XVIII
Classification of Clays

7 Å		10 Å		14 Å	
Dioct.	Trioct.	Dioct.	Trioct.	Dioct.	Trioct.
Kaolinite Halloysite (Dombassite)	Serpentine	Pyrophyllite	Talc (Stevensite) (Hectorite)	Sudoites	Chlorites
	Antigorite Berthierine	Montmorillonite (Beidellite)	Saponite		Di-Tri
		Vermiculite	Vermiculite		
		⎰Muscovite[a]	⎰Biotite	Dombassite	
		⎱Illite (Glauconite)	⎱Phlogopite (Ledikite)		
		Margarite[b]	Clintonite		

[a] Micas.
[b] Brittle micas.

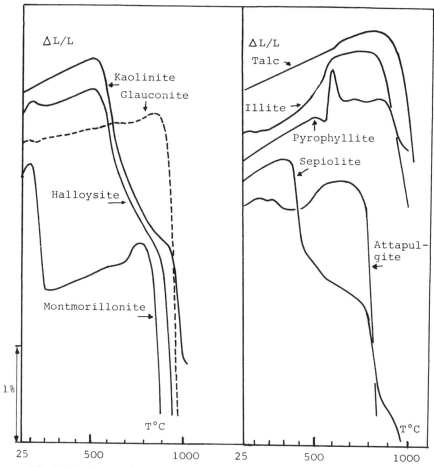

Fig. 7.51. Dilatometric curves of clays (compacted powder). [Author's results.]

mechanical changes in structure on a macroscopic scale. Consequently the following different factors have to be taken into account: structure, crystallization state, orientation, grain size, and impurities.

a. STRUCTURE

Dehydration, dehydroxylation, and thermal recrystallization of anhydrous phases formed at 600–700°C and at higher temperatures lead systematically to contractions in the case of 7 Å clays, and the magnitude of phenomena is more important with antigorites than with kaolinite.

In the case of 10 Å clays, an analogy of thermal expansion was observed by Kiefer (133) for minerals with saturated tetrahedral layers and minerals with nonsaturated tetrahedral layers (micas). This analogy is clear for homologous minerals, for example for pyrophyllite and muscovite or for talc and phlogopite, but the magnitude of the length change is higher in the case of mica-type minerals. However, there is not necessarily an analogy between the curves of two minerals having the same type of layer but different octahedral organization, for example for pyrophyllite (dioctahedral) and talc (trioctahedral), respectively.

Fibrous clays have a structure similar to zeolites: the loss of water at low temperature does not greatly affect the crystal structure, which leads in dilatometry to a limited contraction for attapulgite and an expansion for sepiolite (179).

A peculiarity that appears on dilatometric curves is that for certain minerals (kaolinite) dehydroxylation leads to a contraction while for other minerals (illite) this reaction leads to a large expansion. This last phenomenon depends on the texture of the clay as we will see later.

b. Orientations

Phyllite clay particles are generally oriented and form an anisotropic body. This is due to their layer and laminar structure. So, clays can be considered to have two preferential directions: one, or "parallel" (∥), located in some manner in the basal plane (001); the other, or "perpendicular" (⊥), is at right angle of the basal plane. Consequently, all the phyllitic clays have two different dilatometric curves, depending on whether the measurement is made along the ⊥ or ∥ direction. The curves obtained without taking orientations into account, are always contained between the two extreme graphs obtained by measurement in the ⊥ and (∥) direction.

Kiefer showed in 1966 (134) that expansion anomalies following each direction defined above are similar for a given clay but anomalies measured in the ⊥ direction always begin at lower temperature and their intensity is always greater than in the ∥ direction. Fibrous clays present the same phenomenon, except for the nature of the anomalies (expansion or shrinkage) which are not always the same in the dilatometric curve in the ⊥ or ∥ direction. Typical example of micas is given Fig. 7.52.

c. Grain Size

According to Kiefer (133), the following can be observed with the decreasing grain-size:

1. a decreasing of the mineral stability and a lowering of transformation temperatures;

2. a decreasing of expansion phenomenon (when expansion occurs), the importance of which can be so large that expansion can be replaced by a contraction;

3. an increase of contraction amplitude.

A typical example is given in Fig. 7.53a in the case of phlogopites. A similar phenomenon is found in the case of vermiculite, as shown by Caillere

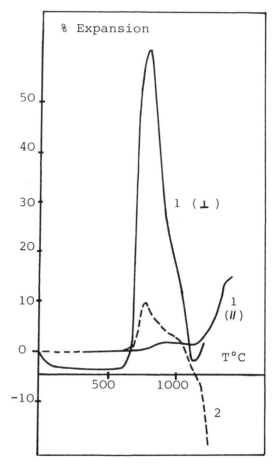

Fig. 7.52. Dilatometric curves of muscovites (133). (1) α-Muscovite in large crystals, measure-ments in (⊥) and (⫽) directions to the cleavage plane: (2) muscovite in flakes, measurements in (⊥) direction to the cleavage plane of oriented sample. (Reprinted, with permission, from Ref. 133.)

and Henin (Fig. 7.53*b*). Expansion observed at about 500°C, which disap-pears in the case of small crystals, is due to the swelling of all the layers as a consequence of hydration water vaporization.

d. Crystallization State

This factor must be connected to change in grain size when samples are ground before experiments, which leads to an amorphization of the mineral. But natural clay minerals are also more or less amorphous according to their geological origin. A study of this problem in the case of kaolinite was studied by Murat and Negro (194) who tried to correlate intensity of certain X-ray lines [(001) and (002) for the characterization of amorphous state, and ($1\bar{1}1$),

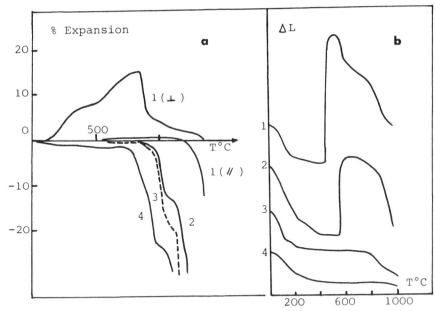

Fig. 7.53. Dilatometric curves of phlogopites (*a*) (133) and vermiculite (*b*) (46). (*a*) 1, α-Phlogopite in large crystal, measurement along (⊥) and (//) directions to the cleavage plane (scale 1/10); 2, β-phlogopite obtained by grinding α-phlogopite for 30 hr in water; 3, β-phlogopite obtained by grinding α-phlogopite in H_2SO_4 1 *N*; 4, altered phlogopite obtained by grinding α-phlogopite in KOH 1 *N*. (*b*) Vermiculite in large crystal (curve 1), measurement in direction (⊥) to the cleavage plane. Crystal with φ <0.2 mm (curve 2) with φ ≈100 μm (curve 3), and with φ <2 μm (curve 4). (Reprinted, with permission, from Refs. 133 and 46).

(1$\bar{1}$0) and (020) for the characterization of crystal disorder] and the shrinkage associated to thermal decomposition on dilatometric curves. It appears that a good correlation was made with (001) and (002) lines which characterize the amorphization state of the mineral; the higher the amorphous state of the clay, the larger is the shrinkage during dehydroxylation.

e. IMPURITIES

Dilatometric curves of clays can be modified by the presence of impurities in the sample. These impurities are mainly quartz, calcareous minerals, organic matters, or small quantities of other phyllosilicates (such as micas) in a given clay (kaolinite for example). Another case is that of mixed clays which lead to more complex dilatometric curves.

f. QUANTITATIVE AND SEMIQUANTITATIVE DETERMINATION OF IMPURITIES IN KAOLINITE

An attempt was made in France by Paquin (207) to determine the content of impurities such as quartz or muscovite in kaolinites of different origins. It was shown by Paquin that for a given geological bed, a linear correlation can

be found between the impurity content and the value of the expansion at a
given temperature above 200°C (700°C was chosen by Paquin). But according to the geological bedplace, the straight line can be different and placed on
both sides of the theoretical correlation line. Consequently dilatometric
curves are specific of each kaolinite bed. However, they allow a rapid and
sufficiently accurate control to detect qualitative or eventually quantitative
variation in a given kaolinite bed.

The differentiation of the nature of the impurity (quartz or muscovite) can
be made by recording the cooling curve, quartz only giving a transition at
573°C. Recording both heating and cooling curves allows one to obtain both
the total (quartz + muscovite) impurity content, and quartz content only.
What is true for a given geological bed does not lead systematically to
generalization.

3. Alumina Hydrates and Bauxites (182,183,192,193)

The DTA curves, recorded during heating of alumina hydrates at atmospheric pressure, show generally only endothermic effects due to dehydration reactions, such as:

Gibbsite $Al(OH)_3$ $\xrightarrow{280-350°C}$ $Al_2O_3 \; \epsilon \; H_2O + AlOOH + H_2O$ (vap)

Boehmite $AlOOH$ $\xrightarrow{500°C}$ $Al_2O_3 \; \epsilon' \; H_2O + H_2O$ (vap)

Diaspore AlO_2H $\xrightarrow{500°C}$ $Al_2O_3 \; \epsilon'' \; H_2O + H_2O$ (vap)

In dilatometry, the dehydration reactions appear either as large "bc"
(case of gibbsite) or small "bc" (case of diaspore) contractions, or as very
marked expansion "de" (case of boehmite). Fig 7.54 shows these differences in the case of gibbsite and of boehmite, and Fig. 7.55 in the case of
diaspore.

These effects are found again when samples are not pure alumina hydrates but natural bauxites. Impurities such as kaolinite or quartz can modify the thermal expansion curves as shown in Fig. 7.55 where either the
contraction associated with the dehydration of kaolinite (curve N°2) or the
expansion associated with the $\alpha - \beta$ transition of quartz (curve N°4), respectively, can be detected in the case of bauxites containing boehmite and
kaolinite or gibbsite and quartz. In the case of a bauxite containing gibbsite
+ boehmite, there can be seen one after another a contraction and an
expansion (curve N°3).

It is difficult to determine the content of each mineral in impure bauxites
or in bauxites containing both gibbsite and boehmite. This is due to a matrix
effect or to the crystallization state of the solid. For example, no quantitative
correlations can be obtained between the intensity of both expansion and

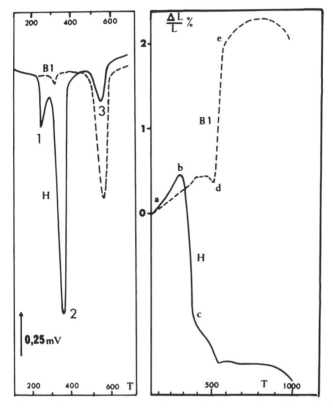

Fig. 7.54. DTA and dilatometric curves of crystallized boehmite (B₁) and gibbsite (H). [Author's results (188).]

contraction in the case of bauxites containing both boehmite + kaolinite (Fig. 7.56). Only the nature of the two minerals can be pointed out, as in the case of bauxites containing both gibbsite + quartz (quartz appears also on the cooling curves). However, it is possible to detect low content of boehmite in bauxites containing both diaspore and boehmite as shown in Fig. 7.55. That is more difficult by X-ray diffraction.

Finally it can be considered that DTA and dilatometry give good complementary data in this subject, the first being particularly well suited for the detection of gibbsites, the second for the detection of boehmite.

B. METALS AND ALLOYS

Independent of the first experiments about thermal expansion, e.g., that of Fizeau (interferometric method), it appears that the large development of dilatometry, and particularly of mechanical dilatometry, is due to a metallurgist (Chevenard) for studies on metals and alloys. This field was largely investigated in France during the last 20 years, particularly by Cizeron and

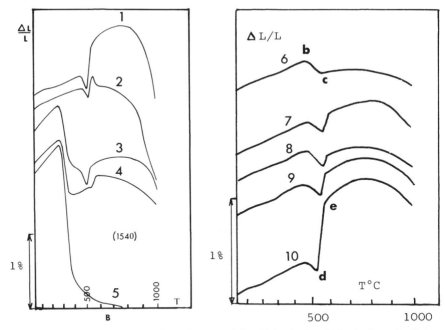

Fig. 7.55. Dilatometric curves of bauxites containing (1) boehmite; (2) boehmite + kaolinite; (3) gibbsite + boehmite; (4) gibbsite + quartz; (5) gibbsite; (6) diaspore; (7) diaspore + boehmite; (8,9,10) diaspore + 5%, 10%, 25% of boehmite, respectively (synthetic mixtures). [Author's results (182,192).]

co-workers at the University of Paris (Orsay) and in Mines Engineering School, in Center of Metallurgical Chemistry Studies (CNRS) and in the French CEA (Saclay).

Apart from some particular problems dealing with the influence of defects on thermal expansion of metals (detailed in Section I.C.8), of magnetic transitions (see Section III.B.3), and of change of expansion coefficients in the absence of crystalline structure changes, it appears, according to Cizeron (62) that two points were particularly studied by dilatometry: (1) length anomalies related to crystalline structure changes and (2) length changes during sintering of compacted metallic powders.

1. Thermal Expansion Coefficients of Metals and Alloys

Dilatometry was widely employed to determine expansion curves of metals to obtain the thermal expansion coefficients. Thermal expansion coefficients of metals vary greatly from one metal to another and are related to the electronic structure of constitutive atoms. Table 7.XVI gives the values of linear coefficient of thermal expansion of metals at 298 K (25°C) as reported by Gschneider (101) from different literature data. It can be noted that the coefficients of thermal expansion when plotted as a function of

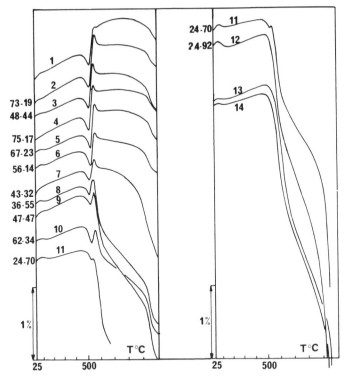

Fig. 7.56. Dilatometric curves of bauxites containing boehmite + kaolinite (percentages indicated by the first and the second number, respectively, in front of each curve), and of two kaolinites (curves 13 and 14). [Author's results (193).]

atomic number show a behavior which is approximately the opposite of that shown by Young's modulus, the shear, and the bulk moduli, the melting point and the boiling point, and the cohesive energy. The coefficient α is very large for the alkali metals but it decreases rapidly as one proceeds through the alkaline-earth and group III-A metals. It can be observed that the minimum value for each period is attained in the element that has the $s^2 d^4$ configuration (Cr, Mo, W) or when the p level is half filled (Sb, Bi). The maximum value is obtained at approximately the configuration $s^2 d^{10}$ (Zn, Cd). Manganese presents its anomalous behavior ($\alpha = 22.6 \times 10^{-6}$) in the transition metals series.

The variation of expansion coefficients of alloys can vary approximately linearly for certain alloys, e.g., Fe-Cr alloys, but can present anomalies in other cases. A typical example is that of the conventional Fe $-$ 35 at. pct. Ni alloy known commercially as Invar. Its expansion coefficient ($\alpha \simeq 2.0 \times 10^{-6}$) is very different from the expansion coefficient of both iron ($\alpha = 11.7 \times 10^{-6}$) and nickel ($\alpha = 13.0 \times 10^{-6}$).

In the case of binary systems and supposing that α varies linearly with composition, Berger (31) has proposed an analytical method to determine the composition of a binary system by measuring the slope of the dilatometric curve. The coefficient of thermal expansion α_x of an alloy containing $x\%$ of a compound having a coefficient of expansion α_1, and $(100 - x)\%$ of a compound having a coefficient of expansion α_2, is given by the expression

$$\alpha_x = x \, \frac{\alpha_1 - \alpha_2}{100} + \alpha_2 \tag{80}$$

Consequently, the value of x is

$$x = 100 \, \frac{\alpha_x - \alpha_2}{\alpha_1 - \alpha_2} \tag{81}$$

For example, a quenched 18/10 Mo alloy (composition after quenching, austenite ferrite) has a thermal expansion coefficient (measured from dilatometric curve) of 18.7×10^{-6}. Coefficients of thermal expansion of austenite ($\alpha = 19.6 \times 10^{-6}$) and ferrite ($\alpha = 13 \times 10^{-6}$) being known, the calculation of x percent of ferrite given by equation 81 leads to $x = 13.6\%$, which is confirmed by linear analysis on microphotographs.

2. Thermal Expansion and Magnetic Transitions

Dilatometry allows one to point out graphically the magnetic transitions that occur in metals and alloys. In the case of α-iron the ferromagnetic $\rightarrow$ paramagnetic transition (Curie point) occurs at 768°C and is revealed by a small reversible anomaly on the dilatometric curve (Fig. 7.58). The same detection can be made on alloys that can also show magnetic transitions, e.g., invar alloys at about 200°C. In this case, the transition occurs as a slope change in the dilatometric curve.

Magnetic transitions in certain steels are interesting to study by dilatometry because they allow one to determine the carbon content of the steel. The addition of a sufficient content of carbon in steel leads to a biphasic structure (α Fe + cementite Fe_3C). The magnetic transition of cementite takes place at 210°C and leads to a slope change, the importance of which is related to the carbon content which influences the quantity of cementite formed (Fig. 7.57) (55). In the same way, the curve point of Fe-Cr alloys varies linearly versus Cr content which can serve analytical purposes.

3. Particular Textural Changes in Metals

Dilatometry can also be employed to point out particular textural changes of nonisotropic metals, which results either in preferential orientation and internal stresses and dislocations produced by the working of the metal, or in thermal recrystallization which occurs in the material previously submitted to an isostatic high pressure. These textural changes lead to dilatometric

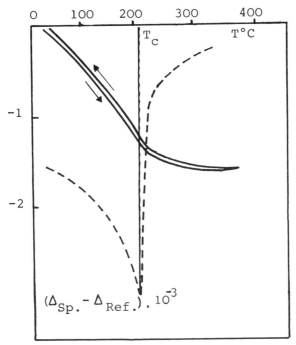

Fig. 7.57. Dilatometric curve of cementite (———) and derivative of the dilatometric curve
(---------) (55).

curves very different from that observed with a nonworked material, and
essentially to hysteresis and variation in expansion coefficients.

The case of sintering will be detailed later in this section.

4. Length Anomalies Related to Crystalline Structure
Changes of Metals and Alloys

The following cases can be considered:

a. POLYMORPHISM

Dilatometry was used widely for the characterization of the polymorphism of metals and alloys and for measurement of the associated length or
volume change. A typical example is that of iron which is the principal
constituent of steels. α-Cubic centered iron transforms firstly in cfc γ form at
910°C (contraction) and in δ form at 1530°C (expansion). Dilatometry reveals
those transformations that are isothermal but present an hysteresis and
whose length change on cooling is larger than on heating (Fig. 7.58). That
was explained by the formation of internal stress in the metal as a consequence of the volume change. The value of this stress is of 4 kbar for pure
iron, which has an elasticity modulus of about 1200 kbar and which presents
a volume change of about 1% during the $\alpha \rightarrow \gamma$ transition (149). In the case of

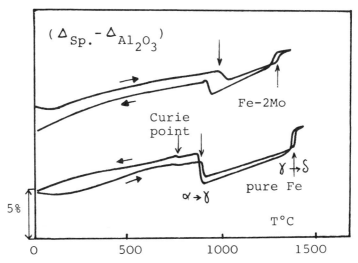

Fig. 7.58. Dilatometric curves of pure iron and of Fe-2 Mo alloy. The Curie point of iron is clearly visible on the curves (31,62).

metal, these stresses lead to an irreversible dimension change when polymorphism occurs.

In iron-containing alloys, the temperature at which polymorphism occurs and associated length change is modified, as can be seen on the dilatometric curve of an Fe-2 Mo alloy (Fig. 7.58).

Certain polymorphic transformations do not occur isothermally, as in the case for zirconium and titanium (64) or cobalt (33) for which the transition is spread over a certain temperature range; this could be due to the presence of interstitials such as oxygen atoms, of iron as impurity, or to the martensitic nature of the transition (62).

In the case of uranium (150), the $\beta \rightarrow \alpha$ transformation on cooling is associated with a wide textural change which leads to a large change in the value of the expansion coefficient of the α phase.

Plutonium presents different polymorphic transformations between 20°C and 600°C (Fig. 7.59). They are isothermal except the $\delta \rightarrow \gamma \rightarrow \delta$, which occurs as jerks on the dilatometric curve on cooling, as shown by Hocheid et al. (114).

b. Structural Transformations Associated with Structural Changes Predicted by Equilibrium Phase Diagrams

In a binary diagram, these structural changes can be:

1. the passage through a boundary curve between a monophasic and a biphasic range. The slow diffusional process related to this phenomenon leads to its spreading over a temperature range;

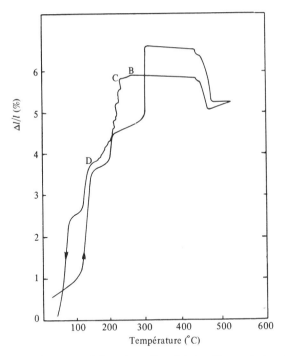

Fig. 7.59. Dilatometric curve of plutonium. (Reprinted, with permission, from Ref. 114.)

2. the passage through a level corresponding to an invariant transformation, e.g., eutectoïd and monotectoïd;

3. an order–disorder transformation (e.g., Au-Cu and Cu-Zn equiatomic alloys);

4. a congruent transformation (case of Fe-Co approximately equiatomic system).

As an example, we give in Fig 7.60 the dilatometric curves of Au-Cu and Cu-Zn equiatomic alloys and of Fe-24.8%Al alloy. These curves point out the progressive appearance of disorder which becomes complete above point P.

c. Transformations Associated with Strong Hysteresis Structural Changes

These happen when homogeneous high-temperature solid solutions are cooled, and can be governed either by a diffusional mechanism or by a nondiffusional mechanism. The behavior of alloys can be represented by two types of dilatometric curves: (1) continuous cooling curves and (2) isothermal quenching or temper curves.

We give in Fig. 7.61 an example of continuous cooling curves during

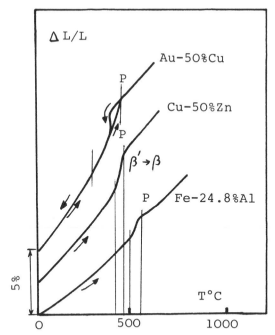

Fig. 7.60. Dilatometric study of order–disorder in alloys (62).

cooling of type 100 C6 steel, at different cooling rates. The transformation of austenite occurs in only one step either as martensitic step when cooling rate is $35° \cdot s^{-1}$ or as perlitic step when cooling rate is $2° \cdot s^{-1}$. For intermediate cooling rates (e.g., $10° \cdot s^{-1}$), the transformation occurs in several steps. Dilatometric curves can change widely if quenching is stopped at a given intermediate temperature and the isothermal length change of the sample is recorded versus time. The aspect of the curves is very different versus the temperature for a given sample.

d. Temper Drawing

Temper drawing, which consists of reheating at more or less high heating rates a quenched specimen, gives rise in dilatometry to curves in which the form depends on the heating rate and on the nature of metastable phases which appeared during quenching.

Figure 7.62 gives an example of dilatometric curves recorded during temper drawing of a rapidly quenched Zr-8 Nb alloy. This quenching from the β domain leads to an unstable martensitic structure ω in type. It can be seen that the $\omega \rightarrow \beta$ takes place directly when the temper drawing is realized with high heating rate ($15° \cdot s^{-1}$), but when heating rate is decreased one can observe a radical change in the thermal evolution of the alloy with formation of intermediate phases ($\omega \rightarrow \omega'$, $\omega' \rightarrow \alpha$, $\alpha \rightarrow \beta$).

Another interesting example is that of the graphitization phenomenon, which is observed in pig irons (Fig. 7.63). During heating, a rapidly

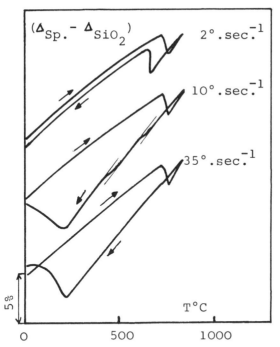

Fig. 7.61. Influence of cooling rate on the transition which occurs in a 100 C6 steel (62).

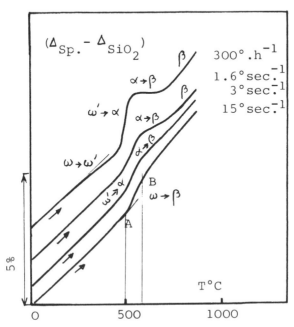

Fig. 7.62. Dilatometric curves of temper-drawing of a Zr-8 Nb alloy, at different heating rates (62).

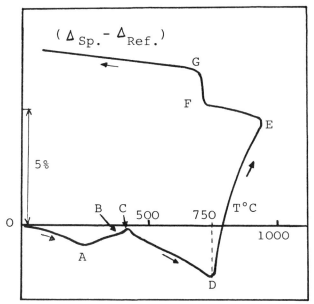

Fig. 7.63. Dilatometric curve of a quenched white pig iron (graphitization phenomenon) (31,62).

quenched white pig iron shows the slope change (point A) related to the magnetic transition of cementite, an expansion AB due to the presence of residual austenite which transforms (part BC). The part CD is related to the decomposition of martensite formed during quenching. The decomposition of cementite Fe_3C in iron and graphite occurs at point D. This reaction leads to an important expansion DE. During cooling, an expansion FG appears which is related to the formation of ferrite.

The study of temper drawing can also be made by dilatometry in isothermal conditions. Many other problems also can be studied by dilatometry, e.g., the stabilization of a given phase (e.g., austenite), the influence of several high-rate cycles on the stabilization of particular phases in alloys, or the influence of external stresses applied to the sample, etc.

Recently in France many researchers have been concerned with metallurgical problems and the use of particular dilatometers [e.g., Dirtic dilatometer, Irsid dilatometer, Adamel-Lhomargy LK-02 dilatometer, or apparatus developed for special applications such as the low-inertia dilatometer of Mignot (173)] associated with other techniques (thermomagnetometry, X-ray diffraction, microscopy, etc.), which leads to a better knowledge of metals and alloys.

5. Sintering of Metallic Compacted Powders

Dilatometry can provide interesting information about the sintering process of metallic powders and has been applied to different studies in this field, e.g., sintering of silver and copper (152,121), nickel (102,121), cobalt

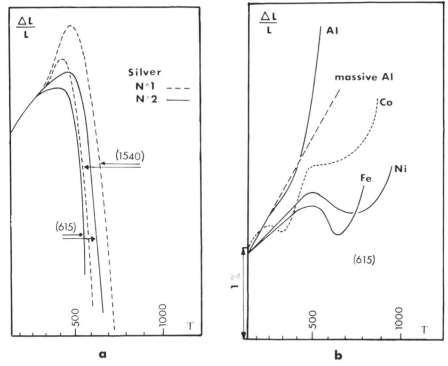

Fig. 7.64. Dilatometric curve, in air, of metal powders compacted at different compacting pressures in bar (number between brackets). [Author's results (181).]

(170), iron (15,60,102), etc.; metal powders often are prepared by thermal decomposition of metal carbonyls, $Me(CO)_n$.

Dilatometric experiments must be performed with specimens placed in an atmosphere free of oxygen (hydrogen or argon atmosphere or vacuum). Except in some cases, such as in study of noble metals, e.g., silver (Fig. 7.64a), thermal shrinkage is hidden by the large expansion related to oxidation reactions when samples are heated in air or in oxygen-containing atmospheres. According to the reactivity of the sample and its mineralogical properties, one can detect either an expansion associated with oxidation reaction (case of aluminum powder), or a beginning of shrinkage (sintering) followed by an expansion related to oxidation (Fig. 7.64b; case of cobalt, nickel, iron).

It appears also that the lower the compacting pressure, the lower is the starting temperature of sintering and the higher is the associated shrinkage. This observation is general enough for the sintering process.

An interesting observation made by dilatometry is that sintering is closely related to the nature of polymorphic phases. The dilatometric curve of nonisothermal sintering of a compacted calcio-thermic uranium powder (Fig. 7.65) shows very accurately that sintering occurs when the γ phase appears at about 900°C. In fact, sintering begins suddenly when the β → γ

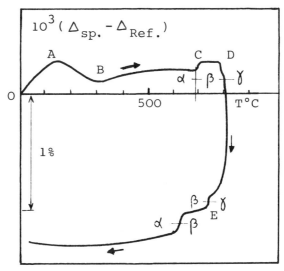

Fig. 7.65. Dilatometric curve, in vacuum, of a calciothermic uranium compacted powder. Part AB of the curve is related to the departure of the binder (1% camphor) (62).

transition occurs and the expansion which is associated with this transition in the case of a massive sample, is replaced by a shrinkage with compacted powder. According to Cizeron (62), this particular capacity of the γ phase for sintering is directly related to the high values of its volume autodiffusion coefficient, which is 300–500 times higher than for the β phase. After sintering, the cooling dilatometric curve presents the anomalies of uranium polymorphism ($\gamma \rightarrow \beta$ and $\beta \rightarrow \alpha$).

In the case of powder mixtures, the thermal behavior and sintering process can be widely changed because interdiffusion phenomena superpose upon the sintering processes which leads either to solid solution or definite compound formation, or smelting of certain constitutive components, thus giving dilatometric curves different from those obtained with the pure metal. The formation of certain phases can lead to expansion in a temperature range where the shrinkage of each metal powder occurs.

Dilatometry allows one to show the influence of certain additives on the shrinkage associated with sintering. A typical example is that of the addition of sulfur in nickel powders. Sulfur, when added at a low content ($< 0.4\%$), promotes an acceleration of sintering (Fig. 7.66) by continued growing of grains and appearance of a transitory liquid phase. This process seems to be lowered when sulfur content is greater than 0.4%, the total shrinkage becoming limited by formation of Ni_3S_2 which promotes a swelling of the compact (74).

C. POLYMERS, PLASTICS, AND RUBBERS

Dilatometry has been widely used for characterization and study of plastics and rubbers and this technique was largely described in the *Encyclope-*

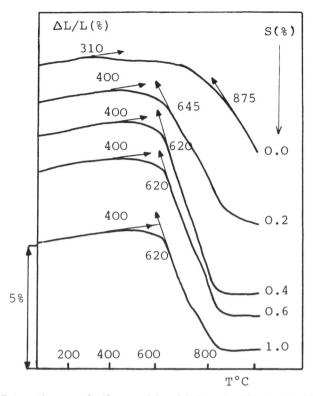

Fig. 7.66. Dilatometric curve of sulfur-containing nickel powder. (Reprinted, with permission, from Ref. 74.)

dia of Polymer and Technology (3,174,229). Many papers have been published on this subject, e.g. (23,42,72,143). Two types of data can be obtained from dilatometric investigations: (1) kinetic data about the polymerization reactions, which requires volumetric dilatometers, and (2) thermomechanical and structural data. Volumetric or more classical dilatometers may be used to determine the glass transition point T_g, the melting point, the thermal expansion coefficient, and the percent crystallinity. Some other investigations can be made about the influence of aging or previous thermal history on thermal properties of polymers.

As in the case of other composite materials (glass–metal, ceramic–metal, and cermets), the knowledge of the dilatometric curve of a polymer is necessary when plastic composite materials have to be prepared.

1. Kinetics Data about the Polymerization Reaction

As discussed in Section III.C.2, many investigations were made on this subject. A typical example which is studied in the INSA of Lyons as experimental work for students, is that of styrene polymerization. The applied dilatometric equation is:

$$x = \frac{S}{V_0} \cdot \frac{dr}{(dp - dm)} \cdot (H - H_0) \tag{82}$$

where x is the degree of reaction, S the capillary section, $(H - H_0)$ the level change of the liquid in the capillary, and d densities of polymer (dp) and monomer (dm), respectively, and V_0 the volume of the dilatometer.

It is shown that in the case of styrene the curve $x = f(t)$ is strictly linear at the beginning of the polymerization leading to a very simple kinetic law.

2. Thermomechanical and Structural Data

Before detailing this point, it is necessary to recall the representation of how the specific volume of a polymer varies with temperature (Fig. 7.67); depending on whether the polymer is entirely in glassy state, partially crystallized or entirely crystallized, three variation curves can be encountered, ABCD, A' B' C D, and A" B" C" C D, respectively. The liquid above B is "stable." It is "metastable" between B (or B') and C.

a. Determination of Thermal Expansion Coefficients

Thermal expansion coefficients of polymers are very much higher than expansion coefficients of metals or the most common inorganic materials. These coefficients are not the same on both sides of the glass transition temperature, T_g. Table 7.XIX gives some data about thermal expansion

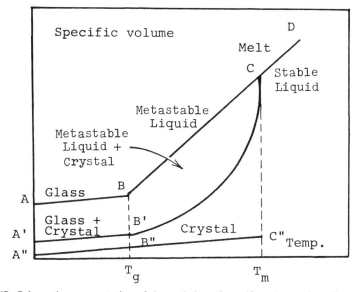

Fig. 7.67. Schematic representation of the variation of specific volume of a polymer versus temperature. Mass crystallinity is given by the position of the "metastable liquid + crystal"-like relative to the "metastable liquid" and "crystal" lines (174).

TABLE 7.XIX
Thermal Expansion Coefficients of Some Polymers (3)

Polymers	$\alpha_L \times 10^6$
Polymethyl methacrylate (bulk)	45
Polyvinyl acetate	230
Polyvinyl chloride (rigid)	50–185
Polyvinyl chloride (flexible)	70–250
Polyethylene (low density)	180–200
Polyethylene (high density)	110–130
Polyurethane (cast)	100–200
Polyurethane (elastomer)	100–200
Polystyrene (general purpose)	60–80
Polystyrene (high impact)	10–30
Epoxy-resin	56–65
Polytetrafluorethylene	100
Polypropylene	58–102
Nylon (-6; $-6,6$; $-6,10$)	59– 90

coefficients of polymers. The relationship between expansion coefficients and structure of polymers was studied by Wakelin and White (272).

b. Determination of T_G and Other Transitions

As shown in Section III.B.4, polymers are partially or entirely vitreous. Consequently their dilatometric curves allow one to determine the glass transition temperature T_g. Figure 7.68 shows typical examples of the slope change which gives the values of T_g for polymers that are in an amorphous or partially crystalline state. It can be observed that the values of T_g for the partially crystalline polymer can be either lower or higher than the T_g point of the substance in an amorphous state.

Some polymers may exhibit more than one transition. This is the case for polymers with a very large amount of branching when measured at low temperature. Different second-order transitions were observed for example by Quenum et al. (219) who studied chlorinated polyethylene with the Dupont 942 TMA apparatus. These different second-order transitions were assigned to the molecular structures which can belong either to the same polymeric chain or to different polymeric chains.

Irregular nonrepeatable breaks near or at the T_g point can be attributed to stresses frozen into the polymer by too-rapid cooling. In the same way, the polymer may suddenly contract near or at the glass-transition temperature or the melting temperature, or both. This is also due to mechanical stresses during cooling (Fig. 7.68 c). The glass transition region may be larger and the position of the total dilatometric curve may be changed when the rate of heating of a correctly annealed polymer is higher than the cooling rate.

The T_g value must depend on the time scale of the volumetric measurement. It has, in fact been demonstrated that a slower measurement pattern

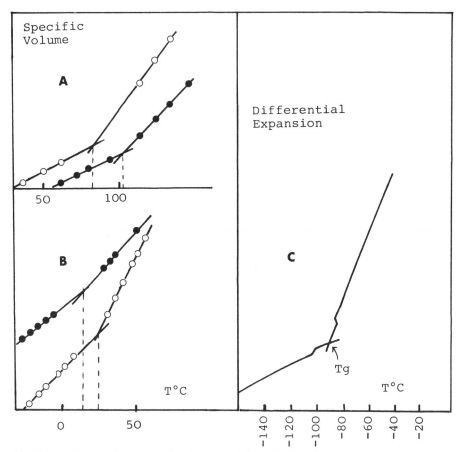

Fig. 7.68. Dilatometric curves of polystyrene (*A*), poly(methyl-4-pentene-2) (*B*), and pale crepe rubber (*C*) (220,3). (○) Crystalline polymer (35% for polystyrene, 78% for poly(methyl-4-pentene-2); (●) amorphous polymer.

leads to a lower value of T_g. Kovacs (143) studied the variation of the specific volume versus time of a sample well equilibrated above T_g and cooled to a temperature T_1 near T_g. These isothermal volume change measurements showed that first contraction occurs gradually over many hours and second if T_1 is not very far below T_g a stable equilibrium volume v_∞ is eventually reached. That leads to the definition of a $T_{g\infty}$ value which differs from T_g. For example, in the case of polyvinyl acetate, $T_g = 32.5°C$ at 0.02 hr while at 100 hr, the value of T_g is 24.5°C. In principle, one can picture a value of $T_{g\infty}$ obtained from an experiment of infinite duration.

From a technological point of view, some caution has to be taken for obtaining good accuracy in the dilatometric determination of T_g. In mechanical dilatometers equipped with push-rod, and spring arrangement, a pressure is exerted on the specimen: this can lead to nonexact values of T_g

TABLE 7.XX
Values of T_g For the Most Common Polymers (85)

Polymers	T_g (°C)
Polyisobutylene	$-$(68–70)
Polyvinylacetate	$+$ (32–35)
Polystyrene	$+$ 100
Poly-α-methylstyrene	$+$ 172
Polymethyl acrylate	$+$ (0–3)
Polymethylmethacrylate	$+$ (100–115)
Polyhexene-1	$-$55
Polydimethyl siloxane	$-$ 123
Polypropylene oxide	$-$ 75
Polyvynylchloride	$+$ 85
Hevea Rubber	$-$ (70–73)
Poly-1, 4-butadiene	$-$ 100
Poly-1, 2-butadiene	$-$ 12
Polyurethane	$-$ 35
Polyethylene	$-$125

because the glass transition is modified when the sample is subjected to an external stress.

T_g values for the most common polymers and plastics are given in Table 7.XX. The glass-transition temperature for elastomers are always below 0°C.

c. Determination of Softening and Melting Point

In the case of entirely amorphous polymers, volumetric dilatometry cannot detect the melting point because there is no discontinuity in the volume change along the line BCD (Fig. 7.67). With crystalline or partly crystalline polymers, the melting point (C) is easily detectable because it corresponds to a transition in the dilatometric curve.

In the case of dilatometers other than volumetric dilatometers, a contraction occurs between T_g and the melting point. It corresponds to the softening temperature of the specimen, that is to say, the temperature at which the polymer begins to soften. This contraction becomes more and more important when temperature rises. In this case, it is impossible to determine the melting point when the sample is not in a totally crystalline state. But true crystalline polymers are very uncommon and generally no polymer is totally crystalline.

Crystallinity can cause the dilatometric curve to contain a discontinuity at higher temperature than the T_g point. This discontinuity is S-shaped and encompasses a larger temperature range than the glass-transition. It is due to the melting of the crystallites in the polymer. That was observed for example by Dannis (72) in the dilatometric curve of a flexibilized epoxy resin.

d. Determination of Crystallinity (174)

Based on the fact that volume and mass are additive properties for a two-phase system, both total volume V and mass M are the sum of amorphous volume V_a or mass M_a and crystalline volume V_c or mass M_c.

$$V = V_a + V_c$$

$$M = M_a + M_c$$

Taking into account the sample, crystalline, and amorphous specific volumes $\bar{v}$, $\bar{v}_c$, and $\bar{v}_a$, respectively, and the densities ρ, ρ_a, and ρ_c, it is possible by a single calculation, to give equations of mass crystallinity X_m and volume crystallinity X_v

$$X_m = (\bar{v}_a - \bar{v})/(\bar{v}_a - \bar{v}_c)$$

$$X_v = (\rho - \rho_a)/(\rho_c - \rho_a)$$

and

$$X_m = (\rho_c/\rho) X_v$$

When $\bar{v}$, $\bar{v}_c$, and $\bar{v}_a$ are known as a function of temperature, from volume dilatometric data, the experimental crystallinity at any temperature may be obtained. Fig. 7.67 illustrates a typical observation of the variation of specific volume with temperature. The mass crystallinity is given by the position of any point on the "metastable liquid + crystal" line ($\bar{v}$) relative to the "metastable liquid" ($\bar{v}_a$) and "crystal" ($\bar{v}_c$) curves at the same temperature. The crystalline specific volume ($\bar{v}_c$) is determined from the X-ray diffraction data about the variation of unit cell dimensions with temperature. The specific volume of the amorphous phase ($\bar{v}_a$) is obtained by extrapolation of the dilatometric curve of the liquid or melt. Crystallinity is very different according to the nature and the structure of the polymer and the cooling rate. Natural rubber, for example, does not crystallize entirely when cooled and tends to an equilibrium corresponding to about 25–30% crystallinity (226).

e. Other Applications

Many other applications of dilatometry can be found in literature for the investigation of polymeric materials, for example:

1. Study of the degree of cure of thermosetting composites, e.g., both epoxy and phenolic resins, which are used in the electronics and other related industries (157). TMA was used to determine expansion coefficients and to show the increase of the glass-transition temperature T_g versus cure time.

2. Study of the thermal expansion and softening temperature of polymer samples coated with lacquer, for example cellophane coated with PVDC lacquer (240).

3. More generally, study of plastic-containing composites. Study of compatibility between materials and of stresses induced therein needs to have the knowledge of the expansion coefficient of the substrate. Many investigations about this subject were published since the paper of Turner (269) who studied with a fused-quartz dilatometer the expansion of many plastic compositions and mixtures of polystyrene and aluminum oxide, phenolformaldehyde and glass fibers, etc. More recently, Fahmy and Ragat Ellozy (82) and Nakamura and Larsen (195) investigated by dilatometry the thermal behavior of reinforced materials such as epoxy-resin matrix composites.

4. Study of the linear length change and determination of expansion coefficient of silicone polymers used as dental materials. An interesting work about this subject was made recently by Pirel (212) who used a particular method, holography, to study the dimensional stability of silicon elastomers.

D. MINERAL GLASSES AND GLASSY MATERIALS

1. Glasses

Dilatometry can be used for resolving different problems dealing with properties of glasses, such as:

1. Influence of the composition of the glass upon the expansion coefficient.

2. Influence of temperature on the structure or texture change of glasses (softening, shrinkage, recrystallization)

3. Influence of the cooling process on concomitant bulk density changes.

4. Determination of the process of certain modifications by isothermal measurements.

Consequently this method is largely used in glass technology.

During heating, the dilatometric curve of an annealed glass presents the following phenomena (2,227):

1. A monotonic expansion characterized by a thermal expansion coefficient α, and that from the origin 0 to a point T_C (Fig. 7.69a).

2. From the point T_C, the expansion increases and α is maximum at the point P corresponding to an inflection point in the dilatometric curve.

3. Above the point T_D, a slowing of the expansion is observed and above T_R the curve falls down very quickly. T_R is called the "softening temperature" of the glass.

Between the temperatures of T_C and P, the glass is subjected to the "glassy transition," which characterizes the fact that certain properties such as viscosity, specific heat, and expansion coefficient are greatly changed (this appears in DTA as an endothermic peak). This glassy transition is

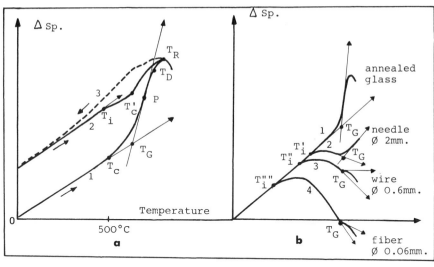

Fig. 7.69. Dilatometric curves of a sodiocalcic glass (2).

located by the temperature of the point T_G which is obtained graphically by the intersection of the tangent at the inflexion point P and the portion OT_C of the dilatometric curve. The viscosity value corresponding to the point T_G is generally about 10^{13} poises. The higher the heating rate, then the higher is the T_G temperature, but this variation is not important (T_G passes from 530°C to 550°C when the heating rate passes from $1°hr^{-1}$ to $6°hr^{-1}$). According to Heide et al. (108), who made use of the derivatograph, the break point in DTA curves of glass blocks does not coincide with the T_G point, but under identical conditions they can serve for a rapid characterization of the transformation range.

The rapid cooling or quenching of the sample heated at a temperature just below T_R gives the dilatometric curve N° 2 in Fig. 7.69a. This curve does not return to the origin 0, which indicates that the volume of the quenched glass is higher than the volume of the original glass. When the sample is heated again (slow heating), the thermal expansion coefficient is not greatly changed, that is to say the thermal history of the glass before the softening temperature does not affect the value of α. But a small contraction appears at the point T_i, and corresponds to the relaxation of the stresses created by the quenching. The determination of T_i is important because it determines the lower limit of annealing allowing the relaxation of stresses in a quenched glass.

The value of T_i is greatly influenced by the quenching process or quenching rate. The quenching rate can be raised by decreasing the sample dimensions. It will be higher with fine fiber than with needles or bulkier samples. Fig. 7.69b shows the variations of T_i as a function of the sample form, which is directly related to the quenching rate. The position of the point T_G is not

greatly changed but it becomes difficult to determine the softening temperature with the raising of quenching rate.

The coefficient of thermal expansion of glasses, which can be determined
by dilatometry, increases slightly with temperature before the point T_C. A
typical example is shown by the following values of α (measured between
20°C and a temperature θ) reported by Scholze (237) in the case of a
sodacalcic glass:

$$20\text{--}100°C \quad \alpha = 8.9 \ \times \ 10^{-6}$$
$$20\text{--}200°C \quad \alpha = 9.1 \ \times \ 10^{-6}$$
$$20\text{--}300°C \quad \alpha = 9.35 \times \ 10^{-6}$$
$$20\text{--}400°C \quad \alpha = 9.6 \ \times \ 10^{-6}$$
$$20\text{--}500°C \quad \alpha = 9.85 \times \ 10^{-6}$$

This coefficient is close to that of crystalline solid, whereas the coefficient
of the liquid is nearly three times greater. But the fact that silica glass, as
well as boric glass, has lower expansion than the same material in the crystalline state led to the suggestion many years ago that glass always has a
lower expansion than a crystal of the same composition. That is not generally true and, for example, in the case of certain lithium silicates, the glass
has a larger expansion than the crystalline substance.

The expansion coefficient α and the dilatometric curves change with the
composition of a glass. As an example, Fig. 7.70 shows the influence of the
Li_2O content in glasses of the system $SiO_2 \cdot Al_2O_3 \cdot TiO_2 \cdot K_2O$ (168). It is
clearly seen that expansion rises with the content of Li_2O. The addition to

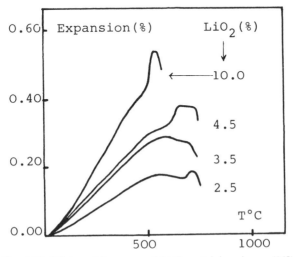

Fig. 7.70. Dilatometric curves of Li_2O-containing glasses (168).

silica of modifying oxides depolymerizes the network and increases the asymmetry of vibrations, which results in a substantially increased thermal expansion. This effect is more pronounced for cations with low field strength and increases in the order $Li^+ < Na^+ < K^+$.

A minimum value of α can be found in certain binary system such as in binary alkali borates (minimum of α for about 20% mol of alkali oxide) as shown by Biscoe and Warren (35). This phenomenon is due to a change of the coordination of B^{3+} ions from 3 to 4.

As discussed above, the determination of expansion coefficient is important for a better knowledge of the structure and properties of glasses.

An important technological problem is the sealing of glass with a metal. This needs very close matching of the thermal coefficients of the two materials in the temperature range where the glass is not plastic. If we consider the two types of glass generally employed—smooth glasses with $\alpha \simeq 9 \times 10^{-6}$ and hard glasses with $\alpha \simeq 5 \times 10^{-6}$—it appears that special alloys have to be developed for realizing a good glass–metal sealing. Dilatometry can be used to compare the thermal expansion of the two materials that have to be sealed, either absolute dilatometry, which allows one to compare directly the expansion curves of the metal and of the glass, or differential dilatometry which can give both the two thermal expansion curves of the materials and the differential curve between these materials.

2. Glass-Ceramics (176)

Glass-ceramics are polycrystalline solids prepared by the controlled crystallization of glasses. These materials are of great interest in various technological fields. Dilatometry is one of the experimental methods used for the study of glass-ceramics to determine the expansion coefficient the softening temperature, or particular transitions of crystalline phases which constitute the material.

The knowledge of dimensional changes that occur with change of temperature is of great interest from a number of points of view, in connection with the uses of these materials. The thermal expansion should be as low as possible to minimize strains resulting from temperature gradient within the material for glass-ceramics required to have a high thermal shock resistance, or to prevent the generation of high stresses when the glass-ceramic is to be sealed or otherwise rigidly joined to another material such as a metal. Some applications need glass-ceramics of near zero thermal expansion coefficient.

Glass-ceramics are remarkable for the very wide range of thermal expansion coefficients which can be obtained, the extremes being ceramics with negative coefficients such as those containing β-eucryptite, $LiO_2 \cdot Al_2O_3 \cdot SiO_2$ ($\alpha = -64 \times 10^{-7}$ between 20°C and 1000°C), or aluminum titanate, $Al_2O_3 \cdot TiO_2$ ($\alpha = -19 \times 10^{-7}$ between 25°C and 1000°C), or with high coefficients such as those containing different forms of crystallized silica (quartz, cristobalite, trydimite).

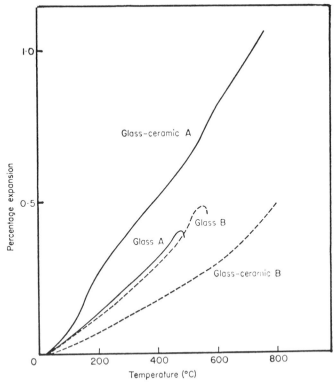

Fig. 7.71. Thermal expansion curve of glass-ceramics and parent glasses. (Reprinted, with permission, from Ref. 176.)

On the other hand, the thermal expansion coefficients of glass-ceramics are generally markedly different from those of the parent glasses. Devitrification of the glass may result in raising or lowering of the thermal expansion coefficient depending upon the types of crystal that are formed. The thermal expansion curves for glasses and glass-ceramics show that parent glasses having very similar thermal expansion coefficients can give rise to glass-ceramics having markedly different coefficients of expansion due to the formation of different crystal phases (Fig. 7.71). Another typical example is that given by Mercier (168) for Li_2O-containing glasses.

Dilatometry is also a convenient method of comparing the softening temperature of glasses and corresponding glass-ceramics, this softening temperature being very considerably increased by the crystallization process. Similarly dilatometry provides information about the nature of some particular crystallized phases which undergo phase inversions reflected in the shape of the expansion curves versus temperature. A typical example of such curves showing the influence of phase inversions is given Fig. 7.72. The material A is a glass-ceramic containing neither quartz nor cristobalite. Materials B and C contain cristobalite and quartz, respectively. The occurrence of phase

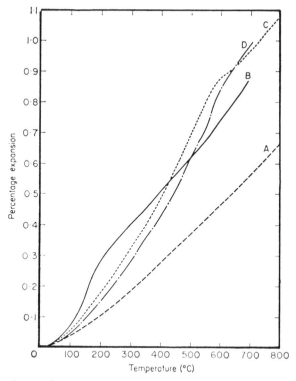

Fig. 7.72. Thermal expansion curves of glass-ceramics showing the influence of phase inversions. (Reprinted, with permission, from Ref. 176.)

inversions of this type can have important bearing on the properties of the glass-ceramic. For example, the important volume change (about 5%) associated with the cristobalite inversion, can generate high stresses within the material and, even if actual fracture of the glass-ceramic does not take place, there may be weakening due to the production of microcracks. Thus, it is clearly of importance to limit the proportion of cristobalite present in the material. The problem is less grave in the case of glass-ceramics containing quartz rather than cristobalite because these materials have higher mechanical strengths.

The thermal expansion of a glass-ceramic can also be markedly affected by the heat-treatment schedule which determines the proportions and nature of the cystal phases present. The curves B and D in Fig. 7.72 concern two glass-ceramics prepared from the same parent glass composition, so that chemically they are identical. However, in the case of curve B, all of the crystalline silica is in the form of cristobalite, while in the case of curve D, the silica is entirely in the form of quartz. These differences come from variation of the heat-treatment process; glass-ceramics containing mixtures of quartz and cristobalite in various proportions can also be obtained from the same original glass.

3. Vitreous Slags

According to how they are cooled or quenched, blast-furnace slags can be more or less in a vitreous state. These by-products can be used as substitution raw materials in chemistry of cement, in glass or glass-ceramic manufacture, and as materials for road building. During heating in air, different phenomena take place in slags, such as sintering and thermal recrystallization. Dilatometry was used initially by Nicol (201) who showed that when heated as compacted powder with compacting pressure about 1000 bar, quenched blast-furnace slags present a large shrinkage at about 600–700°C associated with sintering.

Dilatometric curves of slags can give interesting information on the recrystallization process when these materials have to be used for vitro-ceramic manufacture. This point of view recently led Negro and Murat (185,200) to compare the data obtained by dilatometry, DTA, and X-ray diffraction on a series of slags from different origins and having different compositions. The DTA curves show generally, just before the recrystallization exothermic peak, a small endothermic wave which can be related either to the nucleation process or to sintering of the powder. Dilatometric curves in the same temperature range [700–850°C present a shrinkage, the intensity of which is very different depending on the sample studied (Fig. 7.73)]. Parallel scanning election microscopy studies on samples heated either up to point A or up to point B (Fig. 7.73) gave the following conclusions:

1. When the shrinkage is low ($\Delta L/L \simeq 1\%$), no morphological change is observed on the grains of which the sample is made. X-ray diffraction shows that thermal recrystallization is just beginning.

2. When the shrinkage is important ($\Delta L/L \geqslant 15\%$), two cases can be encountered: either a beginning of the sintering of the powder with appearance of a crystallized phase, or no formation of crystallized phase but important sintering of the material what leading to very large shrinkage up to 25%.

Dilatometry was also recently used by Negro and Bachiorrini (199) to study the devitrification of silica sand-slags mixtures.

E. HYDRAULIC BINDERS

1. Chemistry of Cement

Marchese et al. (158) have made use of the Leitz UBD dilatometer to characterize polymorphic modifications of tricalcium silicate 3 CaO · SiO$_2$

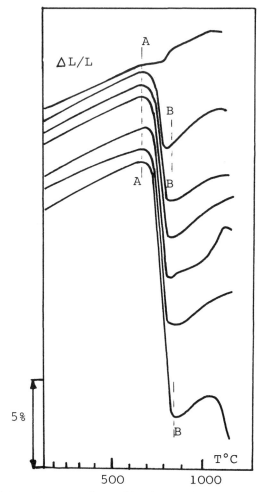

Fig. 7.73. Dilatometric curves of some blast-furnace slags. [Author's results (200).]

(or C₃S) and some of its solid solutions. Between ambient temperature and 1100°C, tricalcium silicate crystallizes in six forms (221)

$$T_I \underset{}{\overset{600}{\rightleftharpoons}} T_{II} \underset{}{\overset{920}{\rightleftharpoons}} T_{III} \underset{}{\overset{980}{\rightleftharpoons}} M_I \underset{}{\overset{990}{\rightleftharpoons}} M_{II} \underset{}{\overset{1050}{\rightleftharpoons}} R$$

where T = triclinic, M = monoclinic, and R = rhombohedral. All these varieties have crystal lattices very close to the trigonal lattice. DTA gives four reversible peaks of low energy at 600, 920, 980, and 990°C, respectively. The dilatometric curve obtained by Marchese et al. shows typical and reproducible transitions: contraction for $T_I \rightarrow T_{II}$, expansion for $T_{II} \rightarrow T_{III}$, and contraction for $T_{III} \rightarrow M_I$. Above this later transition, the slope of the di-

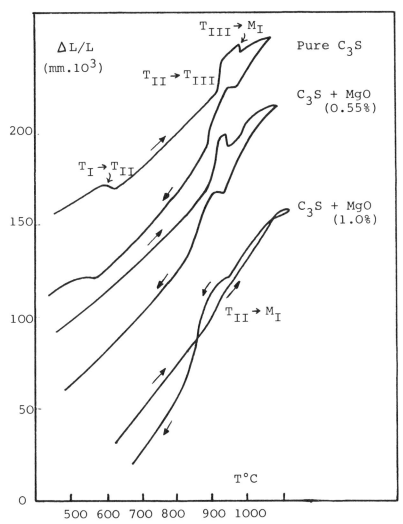

Fig. 7.74. Dilatometric curves of pure tricalcium silicate (curve 1) and of MgO-containing tricalcium silicate (curve 2 and 3) (158).

latometric curve decreases. The cooling does not repeat the heating curve: hysteresis is evident and the final length of the sample is less than at the beginning. These phenomena can be attributed to grain rearrangement in the specimen which was studied as a compacted powder (Fig. 7.74).

Magnesium solid solutions obtained by adding a small amount of magnesium compound to the base oxides before synthesis at 1550°C, give dilatometric curves different from that of the pure product. The 0.55% MgO solid solution stabilized the T_{II} phase after quenching, so the $T_I \rightarrow T_{II}$ transition disappears on both DTA and dilatometric curves (Fig. 7.74). The appearance of transition $T_{II} \rightarrow T_{III}$ is not greatly affected, but the temperature of the $T_{III} \rightarrow M_I$ transition is lowered about 25°C. The dilatometric curve of the 1.0% MgO solid solution shows only one barely outlined transition at about 925°C during heating, which corresponds to $T_{II} \rightarrow M_I$ transition, the stability field of the M_{III} phase being cancelled out by the presence of MgO in the lattice. A similar phenomenon appears with 0.75% ZnO solid solution.

Dilatometry was also used to study some transformations of dicalcium silicate $2CaO \cdot Al_2O_3$ for which five crystalline forms can exist according to the temperature: α (1500°C), α'_H (1250°C), α'_L (1000°C), β (650°C), and γ (20°C) (221). Forest (93) studied some of these transitions by the use of a Chevenard dilatometer. He started with a γ C$_2$S compacted powder sample and showed particularly that the $\alpha'_L \rightarrow \alpha'_H$ transition gives a length change of about 1.5%. Length changes associated with $\beta \rightarrow \alpha'_L$, $\alpha'_H \rightarrow \alpha$, and $\gamma \rightarrow \alpha'_L$ are less important. The $\beta \rightarrow \gamma$ transition, which takes place between 500°C and 450°C on cooling, leads to a large volume change (about 13%).

Thomas and Stephenson (263) have studied, by DTA and dilatometry, the kinetics of the $\beta \rightarrow \gamma$ transition in pure or oxide-doped dicalcium silicate and in air-cooled slags, by-products in which β C$_2$S can be encountered and which can lead to damage during the cooling and storage of slags. The β C$_2$S present in blast-furnace slags (less than 5%) does, very occasionally, invert to the γ form on cooling. The volume expansion associated with the formation of γ can completely disrupt the slag, which then forms a fine powder. Dilatometric recording of the $\beta \rightarrow \gamma$ transformation shows that the transformation is proceeding in a series of discrete bursts even though the cooling rate was fixed at $5° C \cdot min^{-1}$. When γ C$_2$S is formed, the volume expansion disrupts the pellet so that it has not been possible to follow the transformation completely using this technique because the solid pellet crumbles to dust.

An exhaustive study of the dilatometric behavior of hydrated portland cement (fly ash CPAC cement) was made by Cubaud et al. (70,71) in the temperature range 25–1000°C. The hydration conditions and the process of preparing samples greatly affects the appearance of dilatometric curves. For example, Fig. 7.75 shows the results obtained with molded cement pastes (water/cement ratio = 0.30) stored in ambient air (R.H. 60%) for 17 hr after mixing and then stored either in deaerated water (Fig. 7.75A) or in ambient atmosphere (Fig. 7.75B). For the first process of storage, dilatometric curves show different shrinkage portions which can be related to dehydration of calcium sulfoaluminate (portion AB), thermal decomposition of portlandite Ca(OH)$_2$ associated with gradual departure of cement hydration water (part CDE), and thermal decomposition of calcium carbonate (part EF). For samples stored in ambient atmosphere, it appears that the shrinkage between 100°C and 500°C becomes less important, probably by transformation of portlandite into calcium carbonate (a phenomenon which was verified by chemical analysis of the samples at different hydration times) and by the pozzuolanic effect of fly ash. When samples are stored in water-vapor-

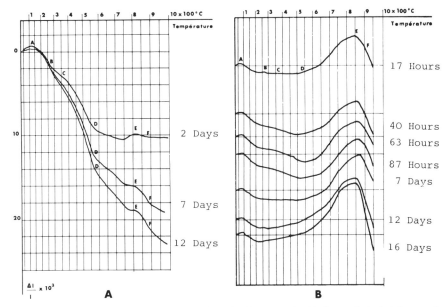

Fig. 7.75. Dilatometric curves of hydrated portland cement (molded samples). (*A*) Storage in water; (*B*) storage in air (70,71).

saturated atmosphere for 17 hr after mixing, the same appearance of dilatometric curves is obtained with, however, a more important shrinkage associated with decomposition of hydration producs, which can be explained by a more important formation of cement hydration products during storage.

When experiments are made not with molded pasty specimens, but with compacted powders obtained by grinding or with samples cut directly from hydrated cement pieces, the same phenomena appear in the dilatometric curves. An interesting case is that of CPA hydrated samples obtained by saw-cutting of cement pieces after 400 hydration days. The curves (Fig. 7.76) show that for normal ambient atmosphere storage, the higher the hydration time then the lower are shrinkages associated with dehydration reaction between 100°C and 500°C; also the higher the expansion appears and then shrinkage at 700–800°C related to carbonation of cement. After 1 year, the CO_2 content, chemically analyzed, is more than 10%.

Dilatometry was also applied recently to the study of thermal expansion of Barya-alumina cements used as binder for superrefractory concretes (75). Thermal expansion coefficient of dry cement is about 5×10^{-6}. A significant shrinkage (about 7% in $\Delta L/L$) was observed between 20°C and 1000°C in the case of hydrated water-cement pastes.

Other dilatometric studies were made on the following subjects related to cement chemistry: (1) formation of calcium dialuminate or hexaaluminate in refractory cements and concretes (69); (2) characteristics of calcium alumi-

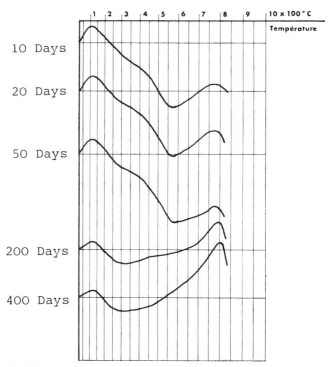

Fig. 7.76. Dilatometric curves of hydrated portland cement (saw-cut samples) (71).

nates (68); and (3) thermal expansion of different phases in the $CaO-Al_2O_3$ system (225).

2. Analysis of Industrial Gypsum Plasters (186,187)

The first work on the application of dilatometry to problems of calcium sulfates was that of Barriac et al. (21).

One important problem of analytical chemistry of gypsum plaster is to determine the α- and β-hemihydrates and orthorhombic anhydrite contents. This problem can be easily resolved by use of dilatometry.

The α- and β-hemihydrates show very different dilatometric curves (Fig. 7.77a). With the β sample, the following can be seen successively: the expansion *ab* of hemihydrate between 25°C and 200°C, the dehydration *bc* between 200°C and 300°C (reaction which is associated a very small shrinkage, the structural change is practically nonexistent), the sintering of the hexagonal anhydrite formed and its transformation into orthorhombic anhydrite [$CaSO_4(II)$] (part *cd* of the curve between 320°C and 450°C), and finally the sintering of $CaSO_4(II)$ above 500°C (part *ef* of the curve). With the α sample, anhydrite $CaSO_4(II)$ appears just after the dehydration (part *b'c'* of the curve) and the important shrinkage *cd* is not observed as in the case of β hemihydrate.

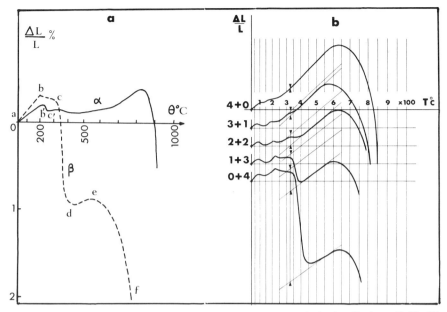

Fig. 7.77. Dilatometric curves of α- and β-hemihydrates (*a*) and of β-hemihydrate-CaSO₄(II) mixtures (*b*). The first number indicates the weight of CaSO₄(II) in the mixture, the second the weight of hemihydrate (186,187).

The dilatometric analysis needs the preliminary preparation of three calcium sulfate samples: (1) Anhydrite(II) by burning gypsum in air at 450°C; (2) α-hemihydrate by warming gypsum in autoclave at 120–130°C; (3) β-hemihydrate by dehydration of gypsum in air at 150°C.

a. Study of β-Hemihydrate–Anhydrite(II) Mixtures

The dilatometric curves of either pure anhydrite(II) or mixtures with β-hemihydrates are given in Fig. 7.77*b*. The contraction due to the presence of β-hemihydrate is measured by vertical distance between two parallel lines which have the same inclination as the dilatometric curve of pure CaSO₄(II), and which pass through the points that mark the beginning and the end of the contraction of β-CaSO₄(III). The standardization curve, N°2 (Fig. 7.78), presents some slight difference in position according to the original gypsum used in the study but it allows determination of β-hemihydrate content with precision of ±2% for a given gypsum.

b. Study of Mixtures of α- and β-Hemihydrates

The same operation as the preceding one is performed—recording of dilatometric curves of mixtures with different contents of α- and β-hemihydrates (total α + β = 100% in weight). But dilatometric curves of pure hemihydrate are disturbed by the grinding of the material, a necessary operation for obtaining α-hemihydrate with grain size lower than 100 μm.

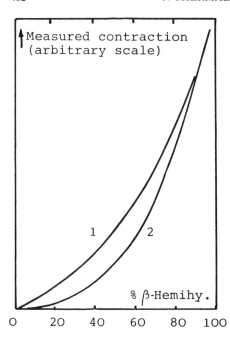

Fig. 7.78. Contraction measured on dilatometric curves in function of the β-hemihydrate content in the mixtures: β-hemihydrate-CaSO₄(II) (curve 1) and β-hemihydrate–α-hemihydrate (curve 2) (186).

The grinding favors the destruction of crystallinity and the greater the grinding, the greater is the contraction between 350°C and 450°C. Consequently, it is necessary to define with accuracy the grinding conditions for obtaining representative results. As described previously, it is possible to obtain a standardization curve (Fig. 7.78) by plotting the distance measured between two parallel lines versus the percentage of β-hemihydrate in the mixture. This standardization curve differs from the curve obtained with β-hemihydrate–CaSO₄(II) mixtures because in the former case CaSO₄(II) acts as an inert matrix, while in the latter it appears and sinters during the warming. The standardization curve of α- and β-hemihydrate mixtures allows the determination of β-hemihydrate content with ±4% only.

F. CERAMICS AND REFRACTORIES

Among the main applications of dilatometry in the field of ceramics and refractories, we will consider the cases of raw-material thermal behavior and thermal expansion characteristics of conventional ceramics, and provide information dealing with oxides and carbides used as refractories. The problem of glass-ceramics is discussed in Section IV.D.2.

1. Conventional Ceramics (128,132)

Phyllitic minerals play a very large part in the ceramic industry. Independently of making use of dilatometry to characterize the raw material (e.g., clays; see Section IV.A.2), the knowledge of the thermal expansion or thermal shrinkage of raw-material mixtures allows one to study easily either reactions occurring in ceramic bodies during firing (and especially sintering and vitrification at high temperatures) or to determine in advance the most favorable heating conditions. Kiefer (132) recorded, for example, the dilatometric curves of a kaolinite–feldspar (microcline) mixture and showed that the intensity of contraction between 25°C and 1400°C increases with the increase of heating rate, time of grinding of the mixture, and grain size of samples.

The dilatometric curve of kaolinite base mixtures containing alkalis (K_2CO_3, muscovite, or feldspar) differs greatly on both sides of about 1100–1500°C. The shrinkage above 1150°C is higher than the shrinkage of kaolinite alone, and it depends on the nature of alkaline product added. In the case of natural kaolinite, always containing some alkaline, the appearance of dilatometric curves allows one to foresee certain properties of ceramics.

The study of dilatometric curves of kaolinite–β-muscovite mixtures (132) shows that the higher is the β-muscovite content, the higher is the sandstone action of this mineral at low temperature. This effect is used in the manufacture of earthenware or sandstone pastes and vitreous bodies. In the same way, the recording of the dilatometric curves of mixtures such as chalk–kaolinite and chalk–β-muscovite (the composition of which corresponds to numerous calcareous clays used in the manufacture of potteries and tiles of sedimentary origin) is also very interesting for obtaining information about high-temperature solid-state reactions, as shown by Kiefer (132).

The control of length changes and the determination of expansion coefficients of ceramics is of great importance from a technological point of view. For example, in the case of porcelain, different kinds of material can be manufactured, such as feldspar porcelains ($\alpha \simeq 5 \times 10^{-6}$ for normally burnt product and $\alpha \simeq 4 \times 10^{-6}$ for overburnt product) which contain residual quartz that leads to the dilatometric anomaly at about 573°C, and soft porcelains ($\alpha \simeq 8 \times 10^{-6}$) which contain residual quartz and cristobalite which lead to length changes at about 200–300°C for cristobalite and 573°C for quartz. Application of dilatometry in this field is detailed in a recent paper of Schueller and Groschwitz (239). Other examples concern silica bricks (11,267), diaspore bricks (10), forsterite bricks (251), stoneware tiles (147), and so on.

Considerable improvements in mechanical properties of ceramics can be achieved through informed manipulation of these variations in thermal expansion properties. This problem was particularly studied by Kirchner (140) in the case of polycrystalline ceramics and oxide single crystals.

In the case of composite materials, such as cermets or ceramic metal composites, dilatometric measurements at high-temperature are necessary to compare the shrinkage of the metal and of the ceramic before manufacturing the cermet. If these shrinkages differ widely, the phase that presents the most important shrinkage promotes the appearance of a nondesirable residual porosity in the cermet.

2. Oxide and Carbide Refractories

Dilatometry is an interesting method to complete electrical, thermal, and mechanical investigations on refractories and ceramic oxides and carbides. Thermal expansion measurements up to 2200°C or more were made by numerous authors such as Nielsen and Liepold (203), Engber and Zehm (78), etc. Results are not always concordant, but divergences are not too great, and it was shown for example by Nielsen and Liepold (203), who worked on polycrystalline single-phase ceramics oxides such as MgO, CaO, magnesium aluminate, and alumina, that the expansion coefficient α is not affected by changes in grain size or by fabrication technique. Fig. 7.79 shows the thermal expansion curves of the most common refractories oxides. It can be seen that magnesia shows the largest expansion coefficient. Except for zirconia and silica, no polymorphism occurs with refractory oxides (113).

Zirconia is an interesting refractory material for applications up to 2200°C, but it shows two polymorphic transitions at 1100°C and 2400°C, respectively. The first one leads to an important volume change ($\Delta V/V \simeq 7\%$, contraction on heating and expansion on cooling). On cooling, this leads to formation of cracks and sometimes the material made of sintered zirconia may break. The knowledge of how this transition occurs and how to prevent it by stabilizing the material (as solid solutions with oxides, e.g., CaO, MgO, Y_2O_3, etc.) is obtained by dilatometric measurements or controls. Anthony gave examples of dilatometric curves obtained with pure and commercial zirconia and with stabilized zirconia, which does not show the polymorphic transition and then becomes a good refractory (6).

Metallic carbides, which present generally low thermal expansion coefficients, are also interesting refractories and were widely investigated by dilatometry up to 2800–3000°C. With their push-rod dilatometer, Shaffer and Mark (242) gave the following values for average thermal expansion of several carbides:

	$10^6\ \alpha\Big]_{20}^{1000}$	$10^6\ \alpha\Big]_{20}^{1500}$	$10^6\ \alpha\Big]_{20}^{2000}$
SiC	4.0	5.0	5.85
TiC	7.9	8.94	10.20
MoC	5.67	8.23	11.5
ZrC	6.93	7.98	9.05
8 TaC · ZrC	6.70	7.53	8.10

Incremental thermal expansion data have been given by Miccioli and Shaffer (171) for the monocarbides of niobium, tantalum, and zirconium, and for binary carbides in the system TaC · ZrC. Dilatometry shows that for the monocarbides of both Zr and Ta, an expansion anomaly occurs in the region of 2000°C which agrees with similar anomalies reported from observations of other properties. Analogous inversions were observed for the binary car-

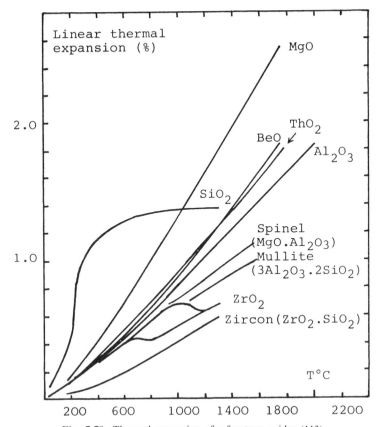

Fig. 7.79. Thermal expansion of refractory oxides (113).

bides in the system TaC · ZrC. These anomalies were not indicated if the thermal expansion studies were made in too limited a temperature range (up to 2000°C only), as may be seen in the paper of Houska (118) who reported investigations of monocarbides of titanium, hafnium, niobium, tantalium, zirconium, and silicon (X-ray method).

It appears from these different data, that silicon carbide, which shows the lowest expansion coefficient, may find the most interesting use as refractory material.

G. MISCELLANEOUS

Among other applications of dilatometry, we will give some examples of the study of graphite and graphite composites, carbon–carbon composites, catalysts and support of catalysts, nuclear materials, compatibility between materials, and about the problem dealing with the determination of quartz and cristobalite contents in minerals or materials.

1. Graphite and Graphite Composites

Graphite finds many uses in the aerospace and high-temperature nuclear reactor industries (as a reentry nose cone material or as neutron moderator), in electrometallurgy (as electrode material), and in electromechanics (for alternator brushes manufacture). Consequently, it is important to know the thermal expansion of the material but it is a complex phenomenon due to anisotropy, material variation, method of preparation and graphitization temperature, particle size binder, molding pressure, additional atomic scale parameters, etc., as shown by different authors.

Naum and Jun (196) studied by dilatometry the thermal expansion up to 2200°C of three samples of extruded graphite. The test samples were taken perpendicular and parallel to the extruded axis and referred to as perpendicular (⊥) and parallel (∥). They observed hysteresis with two "parallel" samples when heated at high temperature and cooled (Fig. 7.80). This phenomenon, which reduces in the second cycle, was explained by the creation of internal ruptures and recombinations in anisotropic polycrystalline materials [Buessem's theory (43)], and by complementary data on the microstructure. The sample which did not present hysteresis exhibited the highest degree of both particle and void homogeneity and certainly contained free carbon. Its degree of crystallinity increases as a result of the heating cycles

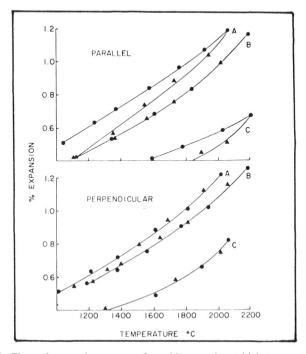

Fig. 7.80. Thermal expansion curves of graphite samples at high temperature (196).

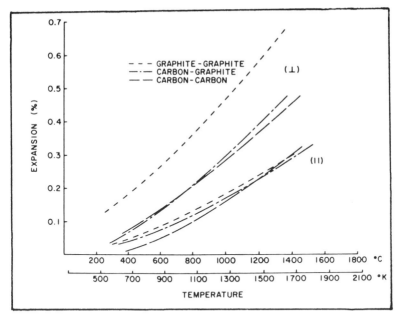

Fig. 7.81. Thermal expansion curves of carbon–carbon composites (197).

associated with thermal expansion measurements. No such effect could be observed in the case of the two other samples which exhibit hysteresis in their thermal expansion curves.

The quantitative model of Buessem (43) agrees also with the results of Hollenberg and Ruh (115) who observed wide variations in the bulk thermal expansion of polycrystalline graphites, from 3.2×10^{-6} for ATJ-S material to 13.1×10^{-6} for isostatically pressed powder. This is due to the presence of internal microcracks or tessellated stresses in the material, and is the consequence of the large anisotropy in the thermal expansion of individual crystallites. The low thermal expansion coefficient of certain specimens was explained by the existence of very small microcracks between the layer planes, the high thermal expansion of isostatically pressed samples, by the presence of significant tessellated stresses or distortion. Addition of boron to isostatically pressed graphite reduced the thermal expansion in the c direction by approximately 30%.

Carbon–carbon composites are a potentially useful high-temperature engineering materials. Thermal expansion of a series of carbon–carbon composites (carbon fibers in a carbon matrix, graphite fibers in a carbon matrix, and graphite fibers in a graphite matrix) has been determined as a function of temperature by Naum et al. (197). Thermal expansion was measured on each composite as a function of orientation relative to the fiber phase, and in an argon atmosphere. The results, summarized in Fig. 7.81, show that all of these composites have, for all practical purposes, the same expansion behavior in the plane of the reinforcing fiber, whether the reinforcing fiber is carbon–carbon or carbon–graphite and graphite–graphite. Expansion perpendicular to the plane of reinforcement is greater in all cases; the expansion of the sample with the graphite matrix being approximately 50% greater than the expansion of those having a carbon matrix. On the other hand, these measurements show that the carbon phases [carbon–carbon ($\|$), carbon–carbon ($\perp$), and carbon–graphite ($\perp$)] show a somewhat lower expansion at lower temperatures, rising somewhat more rapidly with increasing temperature than do the corresponding graphite phases. Complementary X-ray diffraction measurement of the degree of preferred crystal orientation shows that a linear relationship between the expansion anisotropy and the structural anisotropy factor is observed.

Other dilatometric studies were made on graphite-containing materials such as graphite–epoxy composites (80,81) and graphite–ZrC carbide composites (67), but we cannot detail them in this work.

2. Catalysts and Catalyst Supports

An unpublished French work about the study of catalysts and catalyst supports was written some years ago by Duchene (76). This work discussed thermal expansion of refractory oxides, silica, alumina and alumina oxides, and clays, all materials used as support. More specific data were given about particular oxide such as vanadium oxides which present dilatometric anomalies at 70°C (V_2O_4) and 250°C (V_2O_5). Vanadium pentoxide presents a very low thermal expansion coefficient ($\alpha_L = 0.63 \times 10^{-6}$) between 30°C and 450°C (135). Thermal expansion of nickel oxide was studied by Nielsen and Liepold (204) who showed that length changes of the sample are not affected by the grain size when experiments are made on compacted samples. Study of the densification of alumina when heated is of great importance when this substance has to be used as catalyst support. As seen in section III.D, the knowledge of thermal behavior of zeolites (179) gives variable information for determining the upper temperature to which these solids can be heated without suffering surface and adsorption properties damage.

Dilatometry should give valuable data about the aging of catalysts, but it does not seem that data have been published on this subject.

3. Nuclear Materials

The problem of sintering of particular oxide powders (UO_2, ThO_2, ZrO_2, MgO), of polyporphism of certain metals (plutonium, uranium) was discussed in other sections. Another particular application of dilatometry in this field of nuclear material is the determination of thermal expansion of carbon-saturated plutonium carbide used as plutonium-containing fuels for nuclear reactors, which was studied by Green and Learly (97) by X-ray method.

4. Compatibility between Materials

Dilatometry can find application in study compatibility between materials because stresses induced in composite materials made of, e.g., plastic and metal or oxide and metal, require the knowledge of the expansion of the substrate. The case of laminated glass or glass ceramics, used as substrate onto which a second material is fused, bonded, or coated, was discussed some years ago by Gulati and Plummer (103). The case of metal–metal composites, e.g., boron-aluminium composites, was studied by Fahmy and Ragai Ellozy (82). The problems of plastic–metal composite and glass–metal sealing were briefly cited in Sections IV.C and IV.D, respectively.

We must give an example of the compatibility between metallic oxides and metals, which can be deduced from thermal expansion measurement, when metal has to be coated by an oxide to obtain a composite material utilizable at more or less high temperature. In this field, dilatometry was used recently (direct optical sight measurements with microscope) by Henry and Thompson (111) to determine the thermal expansion coefficients of oxides with spinel or perovskite structure. These data were compared with the thermal expansion coefficient of molybdenum and titanium–zirconium–molybdenum alloys, material which had to be coated by oxides cited above. It was shown that spinels are more convenient for coating than oxides with perovskite structure. The most interesting oxides are orthochromites, the expansion coefficients of which between 25°C and 1000°C are only 20% higher than coefficient of expansion of molybdenum and molybdenum–titanium–zirconium alloys.

5. Attempt to Determine Quartz or Cristobalite Contents in Materials and Raw Materials

An attempt was made many years ago by Austin and Pierce (11) to determine, by dilatometry, the relative amount of quartz and cristobalite present in a coke-oven liner by determining the relative change in volume at the transformation points. The scheme failed because the expansion curves showed no indication of the transformation of quartz. This phenomenon is due to the following cause: the outer layers of the quartzite grains used in making the green brick are usually converted to cristobalite during firing whereas the core of the grain remains untransformed. When heated again, the cristobalite shell, in passing through the inversion point, suffers such a large increase in volume that it expands away from the core, leaving the quartzite enclosed in what is equivalent to a hollow shell. So, when the inversion point of quartz is reached, the increase in volume is relatively smaller so that the core does not refill the hollow shell and there is no visible effect externally. This is a typical case of the effect of porosity or matrix effect on the determination of quartz content by dilatometry.

Many other authors studied the possibility of using dilatometric data to determine the quartz (or cristobalite) content in materials or minerals such as clays (59,132,207), but it appears that such determinations are very difficult. Paquin (207) showed that, in the case of quartz-containing kaolinites, the curves [$\Delta(900°C) - \Delta(700°C)$] vary linearly with quartz content, but the slope of the curve depends on the kaolinite bed. Kiefer (132) proposed an analytical method which consists of measuring the difference [$\Delta(600°C) - \Delta(550°C)$] on dilatometric curves of clays previously burned at 950°C. The difference corresponds to the sum of the clay expansion and of quartz expansion if quartz is present in the clay. But such a method leads to large inaccuracies in the determination of quartz content. Murat and Didier (191) studied the dilatometric curves of clays during the second firing cycle, and showed that, if linear correlations between expansion and quartz content can be found in the case of kaolinite, no linear relationship is found in the case of montmorillonites and illites.

Consequently, although dilatometry should be a well-suited experimental technique for the determination of quartz and cristobalite in certain materials or raw materials, it appears that the exploitation of the "expansion–shrinkage" curves, either during the first heating or during cooling or during a second heating cycle, is very difficult. Some years ago, we studied this problem (184) and showed that in the case of quartz, the dilatometric anomaly which occurs at about 580°C on cooling varies very greatly depending on the nature of the matrix (oxides such as MgO or Al_2O_3, hydrates, and clays). We showed that it would be illusory to make use of dilatometry for quantitative analysis; effectively, every matrix promotes its own specific lowering or enhancement of the quartz expansion anomaly during cooling for a given quartz content. Some scanning electron microscopy photographs obtained by observation of the fractures of cooled samples give some information dealing with the textural feature of the composites, but do not permit a rigorous explanation of the observed phenomena. From these data it may be supposed that the same difficulty occurs when other second-order or polymorphic transition anomalies have to be used for quantitative analysis.

In conclusion, we can say that dilatometry is a very interesting method to characterize materials or raw materials, but from an analytical point of view, this technique provides less information than other techniques such as DTA or X-ray radiocrystallography.

V. CONCLUSION

As pointed out in this chapter, dilatometry is a valuable experimental technique that allows one to obtain a lot of interesting data about materials and solids, from the determination of thermal expansion to studies of transitions, control of manufactured products, and characterization of raw materials.

Some more theoretical points of view were not detailed in this chapter, for example the interpretation of low-temperature expansion which can be correlated to thermodynamics, vibration spectra, or defect and impurities state in solids.

We have tried to remain as practical as possible, but the most common examples of use of dilatometry, which were detailed in this chapter, show that this technique is certainly not a specific method of chemical analysis, except in certain cases, such as in the field of metallurgy when bulk samples are investigated. The "true" analysis of a sample, that is to say the quantitative determination of phase content in a mixture or in a composite material, is difficult, particularly in the case of aggregates or compacted powders. On the other hand, dilatometry is a good complementary method, either to characterize a material (mineral, glass, polymer) from its anomalies of expansion, or to study the reactivity and the phase changes that occur in a specimen when heated or cooled. Dilatometry also finds many applications for industrial control of many manufactured products.

REFERENCES

1. Adamel-Lhomargy Co., "Dilatometric Analysis of the Thermal Behavior of Clays," Technical Notice (in French).

2. Adamel-Lhomargy Co., "Dilatometric Analysis of the Thermal Behavior of Glasses," Technical Notice (in French).

3. Anderson, D. R. and R. U. Acton, "Thermal Properties," in *Encyclopedia of Polymer Science and Technology*, John Wiley, New York, 1970, Vol. 13, p. 764–88.

4. Andres, K., *Cryogenics*, **2**, 93 (1961).

5. Andrew, J. H., J. E. Rippon, C. P. Rippon, and A. Miller, *J. Iron Steel Inst.* (London) **101**, 527 (1920).

6. Anthony, A. M., *L'Ind. Ceram.*, **686**, 483 (1975).

7. Arthur, G., and J. A. Coulson, *J. Nucl. Mater.*, **13**, 242 (1964).

8. Ashton, J. E., et al., *Primer on Composite Materials Analysis*, Technomatic, Stamford, Conn., 1969.

9. A. S. T. M., Method of Test, E 228, for Linear Thermal Expansion of Rigid Solids with a Vitreous Silica Dilatometer, in *Book of A. S. T. M. Standards*, 1971, Part 30.

10. Austin, J. B., *J. Am. Ceram. Soc.*, **35**, 243 (1952).

11. Austin, J. B., and R. H. H. Pierce, Jr., *J. Am. Ceram. Soc.*, **16**, 102 (1933).

12. Austin, J. B., and R. H. H. Pierce, Jr., *J. Am. Chem. Soc.*, **55**, 661 (1933).

13. Austin, J. B., and R. H. H. Pierce, Jr., *J. Chem. Phys.*, **3**, 683 (1935).

14. Austin, J. B., H. Saini, J. Weigle, and R. H. H. Pierce, Jr., *Phys. Rev.*, **57**, 931 (1940).

15. Backmann, J. J., P. Cahour, and G. Cizeron, *Mem. Scient. Rev. Metallurgie*, **65**, 481 (1968).

16. Balek, V., *J. Mater. Sci.*, **4**, 919 (1969).

17. Bangham, D. H., and N. Fakhoury, *Proc. Roy Soc.* (London), **A130**, 81 (1930).

18. Baron, J., *Bull. Soc. Fr. Ceram.*, **18**, 4 (1953).

19. Barret, P., N. Gerard, and G. Watelle-Marion, *Bull. Soc. Chim. Fr.*, **8**, 3172 (1968).

20. Barriac, P., and M. Murat, *Bull. Soc. Chim. Fr.*, **12**, 4772 (1968).

21. Barriac, P., M. Murat, and C. Eyraud, *Rev. Mat. Constr.* (Fr.), **606,** 115 (1966).

22. Bars, J. P., and C. Carel, *C. R. Hebd. Seances Acad. Sci.,* Sect. C, **269,** 1152 (1969).

23. Barsamyan, S. T., and K. N. Babayan, *Plast. Massy* (S. S. S. R.), **9,** 54 (1974).

24. Baudran, A., *Bull. Soc. Fr. Ceram.,* **27,** 13 (1955).

25. Baudran, A., *Silic. Ind.,* Belg., **25,** 397 (1960).

26. Baudran, A., Thesis, Paris, 1968.

27. Bayer, G., *J. Less-Common Metals,* **24,** 129 (1971).

28. Beals, R. J., and J. H. Lauchner, *Ceram. Bull.,* **37,** 486 (1958).

29. Bekkedhal, N., *J. Res. NBS,* **43,** 145 (1959).

30. Bell, C. L., *J. Sci. Instrum.,* **38,** 27 (1961).

31. Berger, G., *Differential Dilatometry Applied to the Study of Alloys,* Dunod, Paris, 1965 (in French).

32. Bernard, M., and J. Jaffray, *C. R. Hebd. Seances Acad. Sci.,* **240,** 1078 (1954).

33. Bibring, H., and F. Sebilleau, *Rev. Métallurgie,* **52,** 569 (1955).

34. Bijl, D., and H. Pullan, *Physica, 21,* 285 (1955).

35. Biscoe, J., and B. E. Warren, *J. Am. Ceram. Soc.,* **21,** 287 (1938).

36. Bollenrath, F., *Z. Metallkunde,* **25,** 163 (1933).

37. Borelius, G., *Solid State Physics,* **6,** 15 (1963).

38. Borkovskii, Y. Z., and V. V. Parusov, *Zavod. Laborat.,* **31,** 749 (1965).

39. Bowman, A. L., G. P. Arnold, and N. H. Krikorian, *J. Appl. Phys.,* **41,** 5080 (1970).

40. Bowmann, A. L., N. H. Krikorian, and N. G. Nereson, *Proc. of Thermal Expansion,* 1971, p. 119 [of Ref. (257)].

41. Bowmann, A. L., and N. G. Nereson, *Proc. of Thermal Expansion,* 1973, p. 34 [of Ref. (258)].

42. Boyer, R. F., and R. S. Spencer, *J. Appl. Phys.,* **15,** 398 (1954).

43. Buessem, W. R., in W. W. Kriegel and H. Palmour, Ed., *Mechanical Properties of Engineering Ceramics,* Interscience, New York, 1961, Vol. 3, p. 127.

44. Bunton, G. V., and S. Weintroub, *Cryogenics,* **8,** 354 (1968).

45. Burgers, W. G., and L. J. Groen, *Dis. Farad. Soc.,* **23,** 183 (1957).

46. Caillere, S., and S. Henin, *Mineralogy of Clays,* Masson, Paris, 1963 (in French).

47. Carel, C., *C. R. Hebd. Seances Acad. Sci.,* Sect. C, **262,** 1627 (1966).

48. Carel, C., and P. Vallet, *C. R. Hebd. Seances, Sci.,* Sect. C, **258,** 3281 (1964).

49. Carr, R. H., and C. A. Swenson, *Cryogenics,* **4,** 76 (1964).

50. Carr, R. H., R. D. McCammon, and G. K. White, *Proc. Roy. Soc.,* **A280,** 72 (1964).

51. Case, C. R., K. O. McLean, C. A. Swenson, and G. K. White, *Proc. of Thermal Expansion,* 1971, p. 183 [of Ref. (257)].

52. Chalmin, R., *Genie Civil,* Fr., **130,** 236 (1953).

53. Chaye-D'Albissin, M., and P. Morlier, *Bull. Soc. Fr. Mineral. Cristallogr.* **93,** 488 (1970).

54. Chevenard, P., *Rev. Metallurgie,* **14,** 610 (1917).

55. Chevenard, P., *Rev. Metallurgie,* **22,** 362 (1925).

56. Chevenard, P., *Rev. Metallurgie,* **23,** 92 (1926).

57. Chevenard, P., *Dilatometric Analysis of Materials,* Dunod, Paris, 1929.

58. Chevenard, P., *Rev. Metallurgie,* **47,** 805 (1950).

59. Chmielecki, W., B. Monko, and A. Szymanski, *Thermal Analysis,* Akademiai Kiado, Budapest, 1975, Vol. 3, p. 679.

60. Cizeron, G., Thesis, Paris, 1957.

61. Cizeron, G., "Dilatometrie," in *Techn. Ingr. Mesures et Analyses*, Paris, 1962, p. 870.

62. Cizeron, G., "Dilatometrie," Adamel-Lhomargy Co. Technical Notice.

63. Cizeron, G., *Mem. Sci. Rev. Metallurgie*, **60**, 195 (1963).

64. Cizeron, G., and P. Lacombe, *Rev. Metallurgie*, **62**, 179 (1960).

65. Claudel, C., C. Comel, B. Mentzen, and M. Murat, *C. R. Hebd. Seances Acad. Sci.*, Sect. C, **275**, 215 (1972).

66. Clusener, G. R., *Proc. of Thermal Expansion*, 1971, p. 51 [of Ref. (257)].

67. Cowder, L. R., R. W. Zocher, J. F. Kerrisk, and L. L. Lyon, *J. Appl. Phys.*, **41**, 5118 (1970).

68. Criado, E., S. de Aza, and D. A. Estrada, *Boln. Soc. Esp. Ceram.*, **14**, 271 (1975).

69. Criado, E., D. A. Estrada, and S. de Aza, *Boln. Soc. Esp. Ceram.*, **15**, 319 (1976).

70. Cubaud, J. C., M. Murat, and C. Eyraud, *C. R. Hebd. Seances Acad. Sci.*, Sect. C, **262**, 977 (1966).

71. Cubaud, J. C., M. Murat, and C. Eyraud, *Rev. Mat. Constr.*, **609**, 239 (1966).

72. Dannis, M. L., *J. Appl. Poly. Sci.*, **1**, 121 (1959).

73. Deportes, C., and M. Gauthier, *C. R. Hebd. Seances Acad. Sci.*, Sect. C, **273**, 1605 (1971).

74. Dessieux, R., J. P. Thevenin, and G. Cizeron, *C. R. Hebd. Seances Acad. Sci.*, Sect. C. **275**, 1173 (1972).

75. Drozdz, M., and W. Wolek, Conference on Refractory Concrete, Karlovy Vary (C. S. S. R.), 1974. *Bull. Soc. Fr. Ceram.*, **107**, 39 (1975).

76. Duchene, J., "Dilatometry applied to Catalysts and Catalyst Supports," Lyons, 1965 (in French, unpublished).

77. Duclot, M., and C. Deportes, *J. Thermal Anal.*, **1**, 329 (1969).

78. Engberg C J., and E. H. Zehms, *J. Am. Ceram. Soc.*, **42**, 300 (1959).

79. Evans, D. J., and C. J. Winstanley, *J. Sci. Instrum.*, **43**, 772 (1966).

80. Fahmy, A. A., and A. N. Ragai, *J. Appl. Phys.*, **41**, 5108 (1970).

81. Fahmy, A. A., and A. N. Ragai, *J. Appl. Phys.*, **41**, 5112 (1970).

82. Fahmy, A. A., and A. N. Ragai-Ellozy, *Proc. of Thermal Expansion*, 1973, p. 231 [of Ref. (258)].

83. Faivre, R., and G. Chaudron, *C. R. Hebd. Seances Acad. Sci.*, **219**, 29 (1944).

84. Fenner, C. N., *Am. J. Sci.*, 4th Series, **36**, 331 (1913).

85. Ferry, J. D., Viscoelastic Properties of Polymers, John Wiley, 1961, p. 316.

86. Fitzer, E., and S. Weisenburger, *Proc. of Thermal Expansion*, 1971, p. 25 [of Ref. (257)].

87. Fizeau, M., *Ann. Chim. Phys.*, **2**, (1864).

88. Fizeau, M., *Ann. Chim. Phys.*, **8**, 335 (1866).

89. Fizeau, M., *C. R. Hebd. Seances Acad Sci.*, **62**, 1101 (1866).

90. Fizeau, M., *C. R. Hebd. Seances Acad. Sci.*, **62**, 1133 (1866).

91. Fizeau, M., *C. R. Hebd. Seances Acad. Sci.*, **66**, 1005 (1868).

92. Fizeau, M., *C. R. Hebd. Seances Acad. Sci.*, **66**, 1072 (1868).

93. Forest, J., *Bull. Soc. Fr. Mineral. Crystallogr.*, **94**, 118 (1971).

94. Gaal, P. S., *Symposium on Thermal Expansion of Solids*, 1968.

95. Gaal, P. S., *Proc. of Thermal Expansion*, 1973, p. 102 [of Ref. (258)].

96. Gibbons, D. F., *Phys. Rev.*, **112**, 136 (1958).

97. Green, J. L., and J. A. Leary, *J. Appl. Phys.*, **41**, 5121 (1970).

98. Griffin, R. E., and S. G. Demou, *Proc. of Thermal Expansion*, 1971, p. 302 [of Ref. (257)].

99. Grüneisen, E., *Ann. d. Physik,* **39,** 297 (1912).

100. Grüneisen, E., *Handbuch der Physik,* Springer Verlag, Berlin, 1926, Vol. 10, p. 1.

101. Gschneidner, K. A., Jr., *Solid State Physics,* Academic, 1964, Vol. 16, p. 313.

102. Guillaume, P., J. Senevat, A. Defresne and P. Gilles, *J. Therm. Anal.,* **7,** 317 (1975).

103. Gulati, S. T., and W. A. Plummer, *Proc. of Thermal Expansion,* 1973, p. 196 [of Ref. (258)].

104. Hahn, T. A., *J. Appl. Phys.,* **41,** 5096 (1970).

105. Hahn, T. A., NBS Certificate: SRM 731 Borosilicate Glass—Thermal Expansion, July 31, 1972.

106. Hahn, T. A., and R. K. Rigby, *Proc. of Thermal Expansion,* 1971, p. 13 [of Ref. (257)].

107. Hahn, T. A., and R. K. Kirby, *Proc. of Thermal Expansion,* 1971, p. 87 [of Ref. (257)].

108. Heide, K., R. Haft, J. Paulik and F. Paulik, *J. Therm. Anal.,* **4,** 83 (1972).

109. Henglein, F. A., *Z. Phys. Chem.,* **115,** 91 (1925).

110. Henglein, F. A., *Z. Phys. Chem.,* **117,** 837 (1925).

111. Henry, J. L., and G. G. Thompson, *Am. Ceram. Soc. Bull.,* **55,** 281 (1976).

112. Herpin, A., *Theorie du Magnétisme,* Presses Universitaires de France, Paris, 1968, p. 265.

113. Wiley, Campbell, Ed., High Temperature Technology, 1956.

114. Hocheid, B., A. Tanon and F. Miard, *Mem. Sci. Rev. Metallurgie,* **62,** 683 (1965).

115. Hollenberg, G. W. and R. Ruh, *Proc. of Thermal Expansion,* 1973, p. 241 [of Ref. (258)].

116. Hostache, G., Thesis, Lyons, 1966.

117. Houldsworth, H. S., and J. W. Cobb, *J. Soc. Glass Technol.* (Trans), **5,** 16 (1921).

118. Houska, C. R., *J. Am. Ceram. Soc.,* **47,** 310 (1964).

119. Hummel, F. A., *J. Am. Ceram. Soc.,* **32,** 320 (1949).

120. Hummel, F. A., *J. Am. Ceram. Soc.,* **33,** 102 (1950).

121. Huntz, A. M., G. Cizeron, and P. Lacombe, *Symp. on Powder Technology, Paris,* June 1964, Ed. Metaux, p. 141.

122. Huzan, E., C. P. Abbiss, and G. O. Jones, *Phil. Mag.,* **6,** 277 (1961).

123. Jacobs, S. F., J. W. Berthold III, and J. Osmudsen, *Proc. of Thermal Expansion,* 1971, p. 1 [of Ref. (257)].

124. James, B. W., and B. Yates, *Cryogenics,* **5,** 68 (1965).

125. Jansson, B., and C. A. Sjöblom, *Z. Naturforsch.,* **A25,** 1115 (1970).

126. Jeppesen, M. A., *J. Opt. Soc. Am.,* **48,** 629 (1958).

127. Jones, R. V., *J. Sci. Instr.,* **38,** 37 (1961).

128. Jouenne "Dilatométrie," in Septima, Ed., *Traité de Céramique et Matériaux Minéraux,* Paris, 1975, p. 266–280.

129. Kerner, E. H., *Proc. Phys. Soc.,* **69B,** 808 (1956).

130. Keyser, W. L., *Chim. Anal.,* Fr., **39,** 229 (1957).

131. Kiefer, C., *Proc. of the IIIrd Int. Ceramic Symp.,* Paris, May 1952, p. 55.

132. Kiefer, C., *Bull. Soc. Fr. Ceram.,* **17,** 14 (1952).

133. Kiefer, C., *Bull. Soc. Fr. Ceram.,* **35,** 94 (1957).

134. Kiefer, C., *Bull. Gr. Fr. Argiles,* **18,** 33 (1966).

135. King, B. W., and L. L. Suber, *J. Am. Ceram. Soc.,* **38,** 306 (1955).

136. Kingery, W. D., *J. Am. Ceram. Soc.,* **40,** 351 (1957).

137. Kirby, R. K., and T. A. Hahn, NBS Certificate of Analysis: SRM 739 Fused-Silica Thermal Expansion, May 12, 1971.

138. Kirby, R. K., and T. A. Hahn, NBS Certificate: SRM 736 Copper-Thermal Expansion, August 5, 1975.

139. Kirby, R. K., and T. A. Hahn, NBS Certificate: SRM 737 Tungsten-Thermal Expansion, May 19, 1976.

140. Kirchner, H. P., *Proc. of Thermal Expansion,* 1971, p. 269 [of Ref. (257)].

141. Klemm, W., *Z. f. Elektrochem.,* **34,** 523 (1928).

142. Kollie, T. G., D. L. McElroy, J. T. Hutton, and W. M. Ewing, *Proc. of Thermal Expansion,* 1973, p. 129 [of Ref. (258)].

143. Kovacs, A., *J. Polym. Sci.,* **30,** 131 (1958).

144. Kracek, F. C., *J. Phys. Chem.,* **34,** 225 (1930).

145. Kreider, K. G., and V. M. Patarini, *Met. Trans.,* **1,** 3431 (1970).

146. Krishna Rao, K. V., *Proc. of Thermal Expansion,* 1973, p. 219 [of Ref. (258)].

147. Latapie, J. P., Thesis, Lyons, 1970.

148. Lehr, P., "Les Phénomènes de Dilatation des Matériaux et leurs Conséquences," in *Les Méthodes d'Analyse Thermique et leurs Applications,* C. N. R. S., Paris, 1977, p. D-1 to D-31.

149. Lehr, P., C. E. A. Report (Fr.), 1958, N. 800.

150. Lehr, P., and J. P. Langeron, *Rev. Metallurgie,* **54,** 257 (1957).

151. Lejus, A. M., and R. Collongues, *C. R. Hebd. Seances Acad. Sci.,* Sect. C, **254,** 2005 (1962).

152. Lenel, F. V., *Symp. on Powder Metallurgy,* Paris, June 1964, Ed. Métaux, p. 105.

153. Liebermann, A. and W. B. Crandall, *J. Am. Ceram. Soc.,* **11,** 304 (1952).

154. Lindemann, F. A., *Physikal. Ztschr.,* **11,** 609 (1910).

155. Linseis, M., *Thermal Analysis,* Akademiai Kiado, Budapest, 1975, Vol. 3, p. 913.

156. Majumdar, A. D., and R. Roy, *J. Phys. Chem.,* **63,** 1858 (1959).

157. Manz, W., and J. P. Creedon, *Thermal Analysis,* Birkhäuser Verlag, Basel, 1972, Vol. 3, p. 141.

158. Marchese, B., G. L. Valenti and C. Piccioli, *Il Cemento* (Ital.), **69,** 237 (1972).

159. Marks, S. D., and R. C. Emmanuelson, *Am. Ceram. Soc. Bull.,* **37,** 193 (1958).

160. Mauer, F. A., and C. H. Bolz, WADC Tech. Rep. 55–473, Suppl. 1, June 1957.

161. Mauer, F. A., and T. A. Hahn, *Proc. of Thermal Expansion,* 1971, p. 139 [of Ref. (257)].

162. MacDonald, R. R. and R. P. Pinkley, Report RFP 668, 1966, Dow Chemical Co., Rocky Flats Division, Golden, Colo.

163. Megaw, H. D., *Proc. Roy. Soc.,* **142A,** 198 (1933).

164. Megaw, H. D., *Z. f. Krist.,* **100,** 58 (1939).

165. Meinke, P. P. M., and G. M. Graham, *Can. J. Phys.,* **43,** 1853 (1965).

166. Mentzen, B., and C. Comel, *J. Solid State Chem.,* **9,** 214 (1974).

167. Merard, R., *Bull. Inform. C. E. A.,* **205,** 73 (1975).

168. Mercier, M., *Silic. Industr.* (Belg.), **XL,** 85 (1975).

169. Merz, K. M., W. R. Brown, and H. P. Kirchner, *J. Am. Ceram. Soc.,* **45,** 531 (1962).

170. Meyer, R., P. Potet, M. Czarnul, and J. Leseur, *Symp. on Powder Technology,* Paris, June 1964, Ed. Métaux, p. 27.

171. Miccioli, B. R., and P. T. B. Shaffer, *J. Am. Ceram. Soc.,* **47,** 351 (1964).

172. Miehr, W., J. Kratzert, and H. Immke, *Tonind. Ztg.,* **51,** 417 (1927).

173. Mignot, B., Thesis, Paris, 1972.

174. Miller, R. L., "Crystallinity," in *Encyclopedia of Polymer Science and Technology*, John Wiley, New York, 1966, Vol. 4, p. 449.

175. McKinstry, H. A., *J. Appl. Phys.*, **41**, 5074 (1970).

176. McMillan, P. W., *Glass Ceramics*, Academic, London and New York, 1964.

177. Mondange-Dufy, H., Thesis, Paris, 1958; *Ann. Chimie*, **1**, 107 (1960).

178. Munier, P., *Silic. Industr.* (Belg.), **15**, 67 (1950).

179. Murat, M., *C. R. Hebd. Seances Acad. Sci.*, Sect. D., **270**, 1657 (1970).

180. Murat, M., *C. R. Hebd. Seances Acad. Sci.*, Sect. C, **272**, 1392 (1971).

181. Murat, M., *Thermal Analysis*, Birkhäuser Verlag, Basel, 1971, Vol. 3, p. 467.

182. Murat, M., *Bull. Soc. Fr. Minéral. Crystallogr.*, **95**, 603 (1972).

183. Murat, M., *Rev. Génér. de Thermique*, Fr., **136**, 331 (1973).

184. Murat, M., *Powder Technology*, **10**, 171 (1974).

185. Murat, M., A. Bachiorrini, and A. Negro, *Rev. Phys. Appl.* (Fr.), **12**, 653 (1977).

186. Murat, M., and P. Barriac, *Rev. Mater. Constr.*, **631**, 161 (1968).

187. Murat, M. and P. Barriac, *Thermal Analysis*, Birkhäuser Verlag, Basel, 1971, Vol. 3, p. 483.

188. Murat, M., and M. Charbonnier, *C. R. Hebd. Seances Acad. Sci.*, Sect. C, **274**, 221 (1972).

189. Murat, M., and M. Charbonnier, *J. Therm. Anal.*, **7**, 203 (1975).

190. Murat, M., F. Chatelut, and C. Bardot, *Bull. Soc. Chim. Fr.*, **9**, 3201 (1972).

191. Murat, M., and G. Didier, *Journées d'Analyse Thermique*, Soc. Chim. Fr., Lyons, Oct. 11, 1968.

192. Murat, M., and O. Lahodny, *C. R. Hebd. Seances Acad. Sci.*, Sect. D, **274**, 1601 (1972).

193. Murat, M., and O. Lahodny-Sarc, *Proceed. of the Third Intern. Symp. of I.C.S.O.B.A.*, Nice, France, Sept. 17–21, 1973, Sedal Ed., Paris, 1973, p. 317.

194. Murat, M., and A. Negro, 2°Congr. Nazional sulle Argille, Bari, Italia, Oct. 16, 1976.

195. Nakamura, H. H., and D. C. Larsen, *Proc. of Thermal Expansion*, 1973, p. 117 [of Ref. (258)].

196. Naum, R. G., and C. K. Jung, *J. Appl. Phys.*, **41**, 5092 (1970).

197. Naum, R. G., C. K. Jun, and P. T. B. Shaffer, *Proc. of Thermal Expansion*, 1971, p. 279 [of Ref. (257)].

198. Neel, L., *Ann. Phys.*, **8**, 237 (1937).

199. Negro, A., and A. Bachiorrini, Intern. Symp. on the Valorisation of Slags, Mons, Belg., Oct. 28–29, 1976; *Silic. Ind.* (Belg.), **42**, 121 (1977).

200. Negro, A. and M. Murat, *Thermal Analysis*, Akademiai Kiado, Budapest, 1975, Vol. 3, p. 635.

201. Nicol, A., *Rev. Mat. Constr.*, **34** (Oct. 1950).

202. Nicol, A., and M. Domine-Berges, *C. R. Hebd. Seances Acad. Sci.*, **235**, 1021 (1953).

203. Nielsen, T. H., and M. H. Liepold, *J. Am. Ceram. Soc.*, **46**, 381 (1963).

204. Nielsen, T. H., and M. H. Liepold, *J. Am. Ceram. Soc.*, **48**, 164 (1965).

205. Pampuchova, S., and Z. Szmal, *Bull. Soc. Fr. Ceram.*, **55**, 37 (1962).

206. Pannetier, G., and M. Gaultier, *Bull. Soc. Chim. Fr.*, **3**, 1069 (1966).

207. Paquin, P., Thesis, Paris, 1962. *Suppl. Bull. Soc. Fr. Ceram.*, **58**, 1 (1963).

208. Paulik, F., and J. Paulik, *Conf. Appl. Phys. Chem. Methods in Chem. Anal.*, Budapest, 1966, p. 333.

209. Pearce, J. H., and P. G. Mardon, *J. Sci. Instrum.*, **36**, 457 (1959).

210. Pereira, F. N. D. D., C. H. Barnes, and G. M. Graham, *J. Appl. Phys.*, **41**, 5050 (1970).

211. Pereira, F. N. D. D., and G. M. Graham, *Proc. of Thermal Expansion*, 1971, p. 65 [of Ref. (257)].

212. Pirel, C., Thesis, Lyons, 1976.

213. Plummer, W. A., First Int. Symp. on Thermal Expansion of Solids, Gaithersburg, Maryland, U.S.A., 1968.

214. Plummmer, W. A., *Proc. of Thermal Expansion*, 1971, p. 36 [of Ref. (257)].

215. Plummer, W. A., *Proc. of Thermal Expansion*, 1973, p. 147 [of Ref. (258)].

216. Plummer, W. A., and H. E. Hagy, *Appl. Opt.*, **7**, 825 (1968).

217. Prime, R. B., E. M. Barral, J. A. Logan, and P. J. Duke, *Proc. of Thermal Expansion*, 1973, p. 72 [of Ref. (258)].

218. Pruna, M., R. Faivre, and G. Chaudron, *Bull. Soc. Chim. Fr.*, 1949, D-204.

219. Quenum, B. M., P. Berticat, and G. Vallet, *Polymer J.*, **7**, 300 (1975).

220. Ranby, B. G., K. S. Chan, and H. Brumberger, *J. Polymer Sci.*, **58**, 545 (1962).

221. Regourd, M., and A. Guinier, *Rev. Mat. Constr.*, **695**, 201 (1975).

222. Rencker, E., and P. Dubois, *C. R. Hebd. Seances Acad. Sci.*, **203**, 185 (1936).

223. Revcolevschi, A., J. Hubert, and R. Collongues, *C. R. Hebd. Seances Acad. Sci.*, Sect. C, **269**, 265 (1969).

224. Rigby, G. R., *Trans. Br. Ceram. Soc.*, **50**, 175 (1952).

225. Rigby, G. R., and A. T. Green, *Trans. Br. Ceram. Soc.*, **42**, 95 (1943).

226. Roberts, D E., and L. Mandelkern, *J. Am. Chem. Soc.*, **77**, 781 (1955).

227. Rötger, H., *Silikattechnik, Dtsch.*, **20**, 404 (1969).

228. Rothrock, B. D., and R. K. Kirby, *J. Res. NBS*, **71C**, 85 (1967).

229. Rubens, L. C., and R. E. Skochdopole, "Dilatometry," in *Encyclopedia of Polymer Science and Technology*, John Wiley, New York and London, 1966, Vol. 5, p. 83.

230. Rubin, T., H. W. Altman, and H. L. Johnson, *J. Am. Chem. Soc.*, **76**, 5289 (1954).

231. Ruffino, G., A. Rosso, L. Coslovi, and F. Righini, *Proc. of Thermal Expansion*, 1973, p. 159 [of Ref. (258)].

232. Sarver, J. F., *J. Am. Ceram. Soc.*, **46**, 195 (1963).

233. Saunders, J. B., *J. Res. NBS*, **35**, 157 (1945).

234. Schapery, R. A., *J. Composite Mater.*, **2**, 380 (1968).

235. Schlosser, W. F., E. Latal, P. P. M. Miencke, and G. M. Graham, *Proc. of Thermal Expansion*, 1971, p. 195 [of Ref. (257)].

236. Schoknecht, W. E., and R. O. Simmons, *Proc. of Thermal Expansion*, 1971, p. 169 [of Ref. (257)].

237. Scholze, H., "Le Verre," 2nd Ed., Institut du Verre, Paris, 1974, p. 75.

238. Schossberger, F. V., "High-Temperature Camera Techniques," in E. F. Kaelble, Ed., *Handbook of X-Rays for Diffraction, Emission, Absorption, and Microscopy*, McGraw-Hill, New York, 1967.

239. Schueller, K. H., and H. Groschwitz, "Science of Ceramics," Vol. 8: *Dilatometermessungen zur Kennzeichnung von keramischen Rohstoffen und Massen und zur Texturerfassung.*, British Ceram. Soc. Ed., Stoke-on-Trent, 1976, p. 317.

240. Schulz, J. P., and G. Henning, *Thermal Analysis*, Akademiai Kiado, Budapest, 1975, Vol. 3, p. 1061.

241. Sereda, P. J., and R. F. Feldman, "Mechanical Properties and The Solid-Gas Interface" in E. Alison Flood, Ed., *The Solid Gas Interface*, Marcel Dekker, New York, 1967, Vol. 2, p. 729.

242. Shaffer, T. B., and S. D. Mark, *J. Am. Ceram. Soc.*, **46**, 104 (1963).

243. Shapiro, J. M., D. R. Taylor, and G. M. Graham, *Can. J. Phys.*, **42**, 835 (1964).

244. Shelley, D. L., *Proc. of Thermal Expansion*, 1971, p. 295 [of Ref. (257)].

245. Shrivastava, R. S., and D. S. Joshi, *Proc. of Thermal Expansion*, 1971, p. 44 [of Ref. (257)].

246. Simmons, R. O., and R. W. Ballufi, *Phys. Rev.*, **117**, 52 (1960).

247. Sinha, A. K., R. A. Buckley, and W. Hume-Rothery, *J. Iron Steel Inst.*, **2**, 191 (1967).

248. Sosman, R. B., *Properties of Silica*, Amer. Chem. Soc. Monograph Ser. No. 37, Chem. Catalog Co., Inc., New York, 1927, 856 pages.

249. Southard, J. C., *Bur. Mines Tech. Paper*, **625**, 25 (1941).

250. Sparks, P. W., and C. A. Swenson, *Phys. Rev.*, **163**, 779 (1967).

251. Stamenkovic, I., and F. Sigulinski, *Ceramurgia Int.*, **3**, 25 (1977).

252. Steger, W., *Ber. deut. keram. Ges.*, **23**, 46 (1942).

253. Straumenis, M. E., *J. Appl. Phys.*, **21**, 936 (1950).

254. Terpstra, J., *Appl. Sci. Res.*, Sect. B, **4**, 434.

255. Thermal Expansion 1968. Int. Symp. on Thermal Expansion of Solids, Gaithersburg, Maryland, Sept. 18–20, 1968. (Proceed. not available.)

256. Thermal Expansion 1970. Int. Symp. on Thermal Expansion of Solids. Santa Fe, New-Mexico, U.S.A., June 10–12, 1970. Proceedings in *J. Appl. Phys.*, **41**, 5043 (1970).

257. Thermal Expansion 1971. Int. Symp. on Thermal Expansion of Solids, Corning N.Y., Oct., 1971. Proceedings, *A.I.P. Conference Proceedings* No. 3, M. G. Graham and H. E. Hagy, Ed., American Institute of Physics, New York, 1972, 312 pages.

258. Thermal Expansion 1973. Int. Symp. on Thermal Expansion of Solids, Lake of the Ozarks, Missouri, U.S.A., Nov. 7–9, 1973. Proceedings, *A.I.P. Conference Proceedings* No. 17, R. E. Taylor and G. L. Denman, Ed., American Institute of Physics, New York, 1974, 304 pages.

259. Thermal Expansion 1975. Int. Symp. on Thermal Expansion of Solids, Storrs, Connecticut, U.S.A., 1975. (Proceed. not published.)

260. Thermal Expansion 1978. Int. Symp. on Thermal Expansion of Solids, Winnipeg, Canada, August 29–31, 1977. (Proceed. not yet published.)

261. Thirumalai, K., and S. G. Demou, *J. Appl. Phys.*, **41**, 5147 (1970).

262. Thirumalai, K., and S. G. Demou, *Proc. of Thermal Expansion*, 1973, p. 60 [of Ref. (258)].

263. Thomas, G. H., and I. M. Stephenson, Int. Symp. on Slags and Wastes, Mons, Belg., Sept. 1975.

264. Thomas, J. P., AD 287–826 (General Dynamics, Fort Worth, Tex.).

264b. Thompson, A. M., *I.R.E. Trans. Instr.*, **I-7**, 278 (1958).

265. Thormann, P., and P. Buchmayer, *Tonind. Ztg.*, **91**, 218 (1967).

266. Tranquard A., G. Coffy, and M. J. Boinon, *Bull. Soc. Chim. Fr.*, **8**, 2608 (1969).

267. Travers, A., and D. Goloubinoff, *Rev. Métallurgie*, 27 (1926).

268. Tummala, R. R., and A. L. Friedberg, *J. Appl. Phys.*, **41**, 5104 (1970).

269. Turner, S., *J. Res. NBS*, **37**, 239 (1946).

270. Vergnon, P., F. Juillet, J. Elston, and S. J. Teichner, *Rev. Hautes Temp. et Refract.*, **1**, 27 (1964).

271. Vernay, A. M., Thesis, Lyons, 1972.

272. Wakelin, J. H., and H. J. White, U.S. Dpt. Comm. Office Techn. Serv. PB Rept 147, 170, 47 (1960).

273. Waterhouse, N., and B. Yates, *Cryogenics*, **8**, 267 (1968).

274. Weigle, J., and H. Saini, *Helv. Phys. Acta,* **7,** 257 (1934).

275. Westgren, Λ., *J. Iron Steel Inst.,* **103,** 306 (1921).

276. Weyl, W. A., *Silic. Ind., Belg.,* **34,** 73 and 113 (1969).

277. White, G. K., *Cryogenics,* **1,** 151 (1961).

278. White, G. K., *Proc. of Thermal Expansion,* 1971, p. 59 [of Ref. (257)].

279. White, G. K., *Proc. of Thermal Expansion,* 1973, p. 1 [of Ref. (258)].

280. Wiig, E. O., and A. J. Juhola, *J. Am. Chem. Soc.,* **71,** 561 (1949).

281. Willemsen, H. W., E. Vittoratos, and P. P. M. Meincke, *Proc. of Thermal Expansion,* 1971, p. 72 [of Ref. (257)].

282. Woolfrey, J. L., *J. Austr. Ceram. Soc.,* **12,** 20 (1976).

283. Yates, B., *Thermal Expansion,* Plenum Press, New York and London, 1972, 121 pages.

284. Yates, B., and C. H. Panter, *Proc. Phys. Soc.,* **80,** 373 (1962).

285. Zwetsch, A., *Int. Congress of Ceramics,* Wien, 1956, p. 333.

Chapter 8

ELECTROTHERMAL ANALYSIS

By Donald A. Seanor, *Joseph C. Wilson Center for Technology, Xerox Corporation, Rochester, New York*

Contents

I. INTRODUCTION

The study of the electrical properties of materials has formed an important branch of physics and materials science. Measurement techniques have been developed that are capable of detecting the motion of as few as 10^7 electrons/sec. Techniques of such sensitivity can aid greatly in detecting and understanding phenomena that are not easily probed by other methods. Electrothermal analysis is such a sensitive electrical technique; it probes the types of information that can be gained by studying changes in electrical conductivity of a material as its temperature is continuously changed. Such changes are the related to physical or chemical changes known or thought to be taking place within the solid.

Strictly speaking, the term used should be thermoelectrometry (98). However, in much of the literature the term "electrothermal" (ETA) is used. Other terms are "thermally stimulated current analysis" (TSC), "conductivity glow curve analysis," "thermally stimulated discharge" (TSD), "thermal depolarization" (TD), "thermally stimulated depolarization" (TSD), and "ionic thermoconductivity" (ITC).

The electrical phenomena observed as the temperature of the solid is continuously changed relate to more than the normally observed change in equilibrium electrical conductivity with temperature. Changes in equilibrium electrical conductivity can be well characterized and related to the equilibria of temperature-sensitive processes as, for example, in the case of doped semiconductors (30,89). However, in the sense in which the term electrothermal analysis is used here, the emphasis will be placed on systems in which thermodynamic equilibrium is not attained.

The specific electrical conductivity of a solid, σ, $ohm^{-1} \cdot cm^{-1}$, is the current, i, amps, flowing through a centimeter cube of the material under unit electrical potential, i.e.,

$$\sigma = i \frac{L}{AV} \tag{1}$$

where the sample length is L (cm), its area A (cm^2), and the potential is V (volts).

The specific conductivity, σ, is related to two basic parameters, the charge carrier density, η cm^{-3}, and the charge carrier mobility, μ $cm^2 \cdot volt^{-1} \cdot sec^{-1}$ ($\equiv$ cm sec^{-1}/volt cm^{-1}), i.e.,

$$\sigma = \sum_i q_i \eta_i \mu_i \tag{2}$$

where q_i is the charge on the ith species. Each of the parameters, η_i and μ_i is ambient sensitive; each may be potential sensitive (82,92) (although this aspect will not be discussed).

By ambient it is meant that, in addition to temperature, both the number and the mobility may be sensitive to the precise experimental conditions. This may mean sensitivity to lattice spacing, sample preparation, and ambient atmosphere (such as moisture and the electron-accepting and -donating properties of the surrounding gas). The mobility, μ, is a vector and is therefore direction sensitive. The number of charge carriers is a pure number, i.e., we should write:

$$\begin{aligned} \eta_i &= \eta_i(V, T, A) \\ \mu_i &= \mu_i(V, T, A, z), \end{aligned} \tag{3}$$

where A indicates ambient and z direction. It should be noted that the system is defined ony for good single crystal samples. The influence of grain boundaries, particle size, and interfaces in general, may lead to large deviations from ideality.

Thus, any nonequilibrium phenomena that can be created under a continuously changing temperature which affects parameters such as lattice spacing, adsorption equilibrium, potential distribution, dipole orientation, or molecular species can be studied by the technique of electrothermal analysis. Typically, processes such as first- and second-order phase transitions, solid-state chemical reactions, chemical degradation, dipole alignment and molecular motion, charge-carrier trapping and detrapping, the influence of impurities, and the nature of reactive sites in catalysts have been studied by this method. However, the results of different experimental measurements must, in general, be used in combination with theoretical interpretation to ensure the correct assignments.

It is also necessary to define the parameters that should be controlled in order to specify the system uniquely. This also helps in understanding the behavior under nonequilibrium conditions. In some applications of electrothermal analysis, the phase rule may help define the system and aid in interpretation. The phase rule relates the number of independent variables (the number of degrees of freedom), F, the number of components, C, and the number of phases, P, by the equation:

$$F = C - P + 2 \qquad (4)$$

For example, at a phase transition for a single component material, $P = 2$, $C = 1$, and $F = 1$. Hence, the temperature completely defines the system. On the other hand, for a metal oxide in equilibrium with its vapor, $P = 2, C = 2$, and $F = 2$. Thus, the stoichiometry is defined only upon fixing both temperature and oxygen partial pressure. Such dependencies must be recognized in order to make meaningful experiments.

In general, the number and the mobility of the charge carriers change with temperature. In nonmetals there is an equilibrium between free and localized charge carriers (89) that gives rise to a relationship:

$$n = N_0 \exp - E_n/kT \qquad (5)$$

The relationship of mobility to temperature is more complex (82,89) and depends upon the type of material. The mobility may be controlled by scattering as in metals and wideband seminconductors in which case

$$\mu = \mu_0 T^{-n} \qquad (6)$$

where $1 < n < 2$. Alternatively, the effective mobility may have an activation energy and increase with temperature as

$$\mu = \mu_0 \exp - E\mu/kT \qquad (7)$$

The overall temperature dependence will be to a first approximation:

$$\sigma = \sigma_0 \exp - E_\sigma/kT \qquad (8)$$

where

$$E_\sigma = E_n(+ E_\mu) \qquad (9)$$

i.e., the current measured normally increases exponentially with increasing temperature. Information about the intrinsic conduction processes can be obtained from such measurements [and indeed much of the fundamental understanding of the role of impurities and ions in semiconductors was obtained by studying the change in conductivity with temperature (89,105)]. However, this review is concerned not so much with the isothermal equilib-

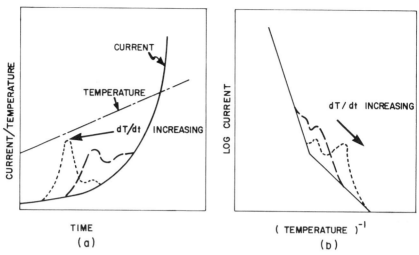

Fig. 8.1. Idealized thermally stimulated current-versus-time curves. (*a*) Current versus time with a linearly increasing temperature; (*b*) log current plotted versus (absolute temperature^{-1}). (————) Equilibrium current; (---------) nonequilibrium current.

rium behavior but with the nonequilibrium nonisothermal behavior. The differences are indicated in Fig. 8.1. Typical isothermal behavior of a semiconductor is shown in which there is a well-defined, characteristic relationship between current and temperature.

In the nonequilibrium measurement, the isothermal (equilibrium) behavior still underlies the transient behavior. The thrust of the discussion is toward the excess current that arises from the nonequilibrium phenomena intrinsic to the material or may be induced by preconditioning. The excess current is normally seen as a peak that shifts to lower temperatures as the heating rate is increased. By suitable analysis, the magnitude of the peak and its rate of appearance can be used to calculate parameters basic to the phenomena occurring.

The information that may be obtained from such curves depends upon factors such as the possible molecular relaxation processes; it may relate to a more precisely defined transition temperature; it may indicate the extent of reaction or be used to estimate the activation energy of reaction. If charge carrier trapping is under study, the number and the activation energy for release from traps can be calculated. Similarly, the activation energy for diopole reorientation can be studied as can the onset of molecular relaxation. In many areas, a number of peaks are observed. Such peaks can be isolated and the parameters corresponding to each peak determined. The technique by which this is accomplished is referred to as peak "cleaning." The sample is preconditioned, cooled under potential, and the temperature is allowed to rise. At some temperature, the sample is cooled without the applied potential and then allowed to warm up again. In this way, broad peaks can be resolved and the individual peaks analyzed. The ability to

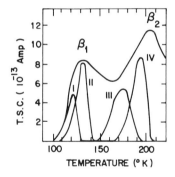

Fig. 8.2. Thermally stimulated discharge current from polycarbonate (4). The broad β_1 and β_2 peaks have been resolved into components by repeated depolarization steps.

"sample" the peak is of particular value since no other method enables the depolarization process to be probed in such detail. Figure 8.2 shows how this technique has been used to study molecular processes in polycarbonate (4). The two broad peaks labeled β_1 and β_2 have been resolved into four peaks with characteristic activation energies of 0.24, 0.27, and 0.36 and 0.46 eV. 1 eV $\equiv$ 23.3 Kcal $\cdot$ mole^{-1} $\equiv$ 97.0 KJ $\cdot$ mole^{-1}. Each peak has been related to local molecular motions involving the methyl groups, carbonate groups restricted by the phenyl groups, a complex relaxation of carbonate/phenyl groups of the main chain, and movement of the phenyl groups attached to carbonate groups, respectively. Figure 8.3 shows the effect of pre-irradiating lead oxide at 90 K with 5500 Å light for different periods of time (78). In this case, four peaks corresponding to trapping levels, I, II, and IV were found corresponding to energies 0.23, 0.33, and 0.38 eV above the valence band. Other experiments under conditions of varying oxygen pressure or illumination identified a further peak, III, 0.26 eV below the conduc-

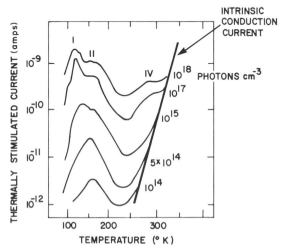

Fig. 8.3. Thermally stimulated discharge current from ultrapure lead oxide as a function of 5500 Å illumination intensity at 90 K (78).

tion band and also established that the trap corresponding to peak II was very sensitive to oxygen pressure.

The beauty of the electrothermal technique lies in its sensitivity to small electrical currents and to changes in small currents. Many current detectors are capable of detecting currents as low as 10^{-13} amps. Since the current flowing in an external circuit reflects the behavior of the charge within the solid, small changes in distribution of charge carriers or internal fields are detected quite readily. For example, polymers have carrier mobilities in the range 10^{-7} cm$^2 \cdot$ volt$^{-1} \cdot$ sec^{-1} (cm $\cdot$ sec^{-1}/volt $\cdot$ cm^{-1}). Thus, a current of 10^{-13} amps corresponds to observing the movement of 10^{13} carriers, or in a reasonably pure semiconductor where μ may be 10^2 cm$^2 \cdot$ volt$^{-1} \cdot$ sec^{-1} the movement of as few as perhaps 10^7 charge carriers can be observed. Thus, problems lie not so much in the detection of the small currents themselves, but in first ensuring that the currents are intrinsic to the solid and not artifacts of the measuring system, noise within the measuring system, or some other extrinsic source. Second, the observed changes should be related to phenomena known or believed to be occurring within the solid. In general, the sensitivity of the electrothermal technique demands cross-correlating experiments. In the course of the review, some specific cross-correlations will be made.

Typically the experimental measurements can be made in three modes:

1. Isothermal studies at a series of temperatures with or without preconditioning.

2. Dynamic measurements under continuously changing (usually increasing) temperature with no specific pretreatment to create nonequilibrium conditions within the sample.

3. Dynamic measurements under continuously increasing temperature after some type of pretreatment to create a nonequilibrium number of charge carriers, field distribution, or dipole orientation. Pretreatment may involve cooling under potential, irradiation at low temperatures, the injection of charge carriers, or changes in ambient gas pressure or species.

Each mode of study yields different but complementary information. For example, studies in mode 1 can show the presence of impurities in semiconductors and can be used to study the kinetics of phase changes or chemical reactions and to study isothermal depolarization and charge carrier trapping or detrapping.

The second mode can be used analytically to distinguish between similar materials, to determine the onset of reaction, as well as to note phase changes.

The third mode can yield information on dipole orientation, charge carrier detrapping, and spatial distribution. Measurements in modes 1 and 3 can be made with or without the imposition of an electric field. It should be again emphasized that mode 1 studies the thermal equilibrium condition or the

TABLE 8.I
Processes Studied by Electrothermal Analysis

Dipole alignment	Phase changes; molecular motions	Cool under potential
Excess space charge	Phase changes, charge carrier trapping, trap states	Cool under high potential. Preirradiation
Excess diffusion drift	Phase changes, charge carrier trapping, trap states	Measure without field; preirradiation under potential
Ohmic conduction	Phase changes, reactions, decomposition	None required
Maxwell Wagner depolarization	Interfaces, composites, phase changes	None required

approach to thermal equilibrium. This is never quite the case in modes 2 and 3. In mode 3, the equilibrium is deliberately distorted, and changes between mode 2 and 3 can be studied. The isothermal approach to equilibrium of a deliberately perturbed system can also be studied. Pulse heating or intermittent studies do not appear to have been made, but, obviously, such a technique could be developed.

Typical phenomena and the process causing them are listed in Table 8.I.

II. EXPERIMENTAL METHODS

Most studies by the electrothermal method have been carried out on insulators rather than conductors. The emphasis has been on dc techniques. Isothermal ac techniques have been extensively described in the literature but little adaption to the dynamic method has been described (45). The experimental methods described for high-resistance insulators using high-impedance electrometers are usually more appropriate than the bridge techniques used with more conductive materials. There are many papers dealing with the electrical properties of such insulators as ceramics and polymers. Detailed references can be found in reviews of these topics to which the reader interested in greater detail is referred (15,66,108,129,150).

Since the electrothermal method involves measuring the current flowing through a sample under continuously changing temperature, the following units are required (Fig. 8.4):

1. The sample holder consisting of sample, electrodes, leads, and means of securely locating the sample.

2. Means of controlling ambient atmosphere.

3. A heater capable of giving a variable but linear rate of temperature increase over a broad range of temperatures.

4. Current detector capable of measuring currents from as small as 10^{-13} amps to as high as, perhaps, 10^{-4} amps. This will usually be a high-impedance electrometer.

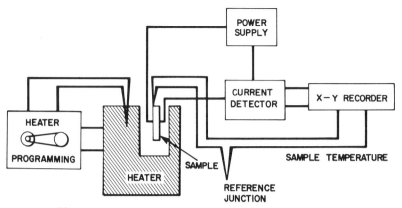

Fig. 8.4. Block diagram of an electrothermal analysis system.

5. Source of controlled low-ripple dc potential.

6. Means of recording current and temperature. This may be an XY recorder or a computer interface.

7. Additional means of preconditioning the sample, such as a source of irradiation, the energy of which depends upon the nature of the experiment.

Different materials may demand different capabilities. For example, the temperature range required for polymers is not likely to exceed 300°C. However, a study of ceramics may require temperatures above 1500°C, thus requiring specialized equipment.

Requirements for isothermal measurements are essentially the same as those for dynamic measurements. Apart from the need to maintain constant temperature over long periods of time, the problems of sample preparation, electrodes, ambient atmosphere control, and current measurement are the same.

A. SAMPLE PREPARATION AND ELECTRODES

Sample preparation and the electrode system can play a critical role in determining the electrothermal behavior. It is necessary to ensure that the observed behavior is characteristic of the material, not of the way in which the sample is made and electroded. Cross-checking experiments using different electrodes and methods of sample preparation are required.

The sample may be in the form of fiber, film, single crystal, block, or compacted disc of powdered material. Each presents its own peculiar problems.

In the case of fibers, electrical contact presents a serious problem, particularly if bundles of yarn are involved as opposed to monofilaments (129).

Powders present a number of problems. Among these can be listed effects of particle size, contacts, interfacial effects, mixing, and compacting. The

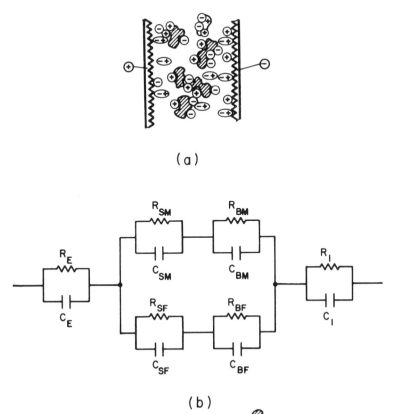

(a)

(b)

Fig. 8.5. Representation of an inhomogeneous sample. ▨ , filler particles; ⊕, ⊖, free charges; + − , dipoles. (a) An array of irregular filler particles (F) in a continuous matrix (M). These may be crystalline regions in a semicrystalline polymer. (b) The equivalent circuit: E represents the electrode and measuring circuit contributions to capacitance and resistance, B, the bulk, and I, the interface contributions.

samples are often prepared using KBr pellet presses or other types of die. The compacting pressure may affect behavior. The range of effective compacting pressure will vary from material to material. Under some conditions of relative humidity, the sample may stick to the die face. The use of mold releases is not recommended as the surface of the sample may be changed. Thin Teflon® sheets cut to size often help to eliminate sticking of the sample to the die face. It is not recommended that the sample be compacted between thin metal foil electrodes as voids may easily be created at the sample foil interface.

Composite materials can show a marked dependence upon processing parameters. For example, Norman (108), in discussing the behavior of conductive rubbers, devotes two chapters to the influence of mixing, type of carbon black, volume loading, and curing on the electrical properties of filled rubbers.

In many respects, thin films, single crystals or slabs cut from larger blocks present the least complicated systems. Even here factors such as residual solvent and morphology may be important. Certainly in case of polymers, the presence of residual solvent, changes in crystalline/amorphous ratio, the sample cooling rate (by its influence on glass transition temperature or crystallite size) and ambient atmosphere all affect electrical behavior.

The electrical system of an inhomogeneous sample can be represented as a resistance/capacitance network to which the factors discussed may each contribute as shown in Fig. 8.5. Depending upon the sample and precautions taken, the equivalent electrical circuit can be simplified and analyzed. For example, some catalyst samples can be simplified to single circuit elements in which the surface component dominates the electrical properties. Unfortunately, simplification is not always possible particularly if interfacial polarization occurs. In addition, care must be taken to minimize and to allow for capacitance effects in the measuring system, leads, and sample holder.

The formation of the electrode is of paramount importance. In laying down the conductive material, the surface should not be chemically changed and there should be no gaps between the electrode and sample. Vacuum deposition of a metal such as silver, aluminum, or gold has commonly been employed. Use of conductive paints is acceptable provided that the solvent does not affect the substrate to be studied. Foils have also been employed, but may create interfacial voids. For experiments involving light, one electrode must be semitransparent. Frequently, Nesa glass has been used for the semitransparent electrode as have been thin layers of gold or grids of metal (evaporated or embedded). Table 8.II lists the more commonly used electrode materials.

TABLE 8.II
Electrode Materials

Hg—Liquid metal, not recommended. Easily oxidized; toxic vapors.

Al—Vacuum deposited. May oxidize.

Ag—Vacuum deposited, silver paint, silver epoxy cement; may react with some substrates. May diffuse into sample.

Au—Vacuum deposited, paint, paste. Inert and not oxidized.

Cu—Vacuum deposited. Reactive in many gases.

Ni—Vacuum deposited, from solution. Reactive at high temperature.

Pd—Vacuum deposited. Inert except in certain atmospheres where chemisorption may occur. Froms Pd/H$_2$ alloys which can be used as proton-injecting electrodes.

Pt—Vacuum deposited with difficulty, paint, paste. May be catalytically active.

Ir—For very high temperatures.

C or Graphite—Applied from colloidal suspension. Requires nonoxidizing atmosphere.

Nesa—A semitransparent conductive tin oxide coating often applied to quartz or glass for contact electrodes. For photoconduction experiments.

Hg/ln/Sn/Gn—Low melting metals used alone or as alloys. Easily oxidized. Vapors may be toxic.

The electrode should be ohmic, that is, it should not perturb the potential distribution across the sample by injecting charge carriers or by creating potential barriers at the surface, It is also advisable to use guard-ring electrodes to minimize surface leakage currents.

B. THE SAMPLE HOLDER

The design of the sample holder can present quite serious problems. Because of the wide range of criteria required for each specific situation, individual workers usually design their own. There are certain common criteria; they all must support and locate the sample and the electrodes, there must be means for easy thermal and ambient control, and sources of stray electrical capacitance must be minimized. The nature of the specific sample frequently dictates the type of sample holder to be used. Typical cells are shown in Figs. 8.6–8.

Figure 8.6 shows a typical glass cell used to study powders (65). The powder is lightly compressed by the spring-loaded, upper platinum electrode in a sintered glass tube with a corresponding electrode at the bottom. Both electrodes are provided with shielded electrical and thermocouple leads. A platinum or gold-plated platinum screen surrounds the sintered glass tube to act as a guard electrode. The holder can be connected to a vacuum system for control of the ambient while the sintered glass reaction tube allows easy access of reactant gas.

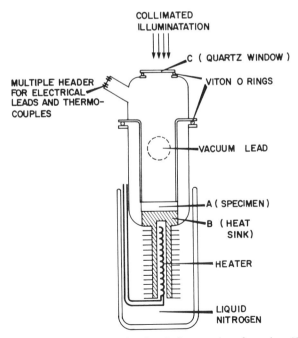

Fig. 8.6. Cell for studying the electrical properties of powders (66).

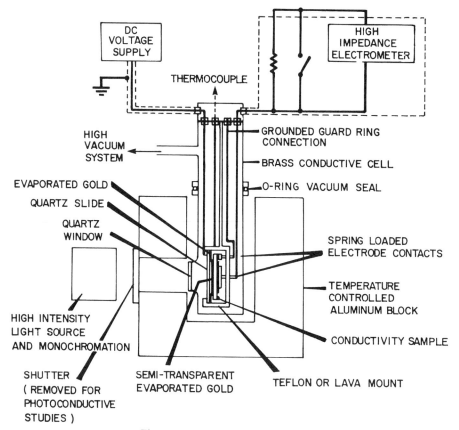

Fig. 8.7. Photoconductivity cell (122).

Figure 8.7 shows a metal photoconductivity cell (122) which can easily be used for electrothermal analysis if preirradiation of the sample is required. The high-intensity monochromater can be readily calibrated and allowance made for window and electrode absorption.

A versatile system which allows a number of other measurements to be made on the sample is shown in Fig. 8.8 (127). Two similar samples are prepared. One sample has electrodes evaporated onto it and is used for electrical and spectroscopic studies; the second is placed on the microbalance and is used for adsorption/desorption studies. By this means, gas uptake or desorption and the attendant changes in electrical or photoconduction can be studied simultaneously as can photoadsorption or desorption. In addition, if the conductivity cell windows are IR-transmitting, attendant changes in IR absorption of the solid can be studied. This system has been used to study effects of gas and water absorption on photoconductors in which surface oxidation can take place and to study the effect of programmed desorption.

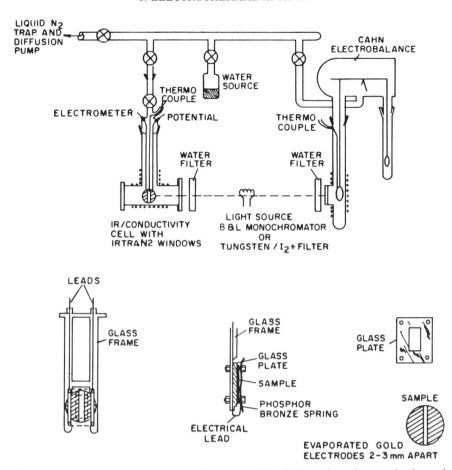

Fig. 8.8. Experimental system for simultaneous studies of adsorption, photoconduction and surface species (127).

Chiu (44) has described a technique for the simultaneous thermogravimetric analysis (TGA), derivative thermogravimetric analysis (DGA), differential thermal analysis (DTA), and electrothermal analysis. The system described was based on the Dupont 900 thermal analyzer in which two specimens are utilized. Both samples are placed in the same thermal environment; one is used for TGA-DTG and the other for DTA-ETA. Such a system can be used to characterize complex systems in which simultaneous or consecutive thermal events occur. Typical of these results is Fig. 8.9 which shows the type of information obtained during the heating of pyromellitic acid dihydride (PMA · 2 H$_2$O). The initial heating cycle shown in Fig. 8.9a contains considerable fine structure. The fine structure is absent in the reheating cycle (Fig. 8.9b). The results are explained in terms of the following transitions:

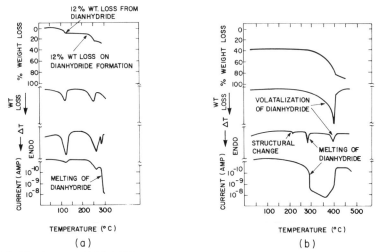

Fig. 8.9. Simultaneous TGA-DTG-DTA and ETA of pyromellitic acid dihydrate. (*a*) Initial heating; (*b*) reheating (44).

$$\text{PMA 2 H}_2\text{O} \xrightarrow[125°\text{C}]{70°-130°\text{C}} \text{PMA} \xrightarrow[250°\text{C}]{180°-270°\text{C}} \text{PMDA} \xrightarrow[289°\text{C}]{}$$
solid

$$\text{PMDA} \xrightarrow[400°\text{C}]{} \text{PMDA}$$
liquid   vapor

The small endotherm at 225°C shown in Fig. 8.9*b* is attributable to a structural rearrangement of the pyromellitic dianhydride (PMDA) created during the initial heating cycle.

In a later paper (45), Chiu described the development of a dynamic elec-thermal analysis technique based on the analysis of dielectric constant and dielectric loss tangent. In this system, the sample was placed in one arm of an automatic capacitance bridge. The capacitance and dissipation (loss) factor was then plotted directly as a function of temperature and frequency. Other methods of making dielectric measurements and their correlation with other thermal properties of polymers have also been described (75,76,101,124,158).

Recently, the use of an electrical system, referred to as the Electric Thermal Analyser manufactured by Toyo Seiki Seisaku-Sho, Ltd., has been described (138). A versatile system with components capable of simultaneously or independently measuring dynamic mechanical, dynamic dielectric, dynamic depolarization and other thermal properties is also commercially available.*

* The Universal Relaxation Spectrometer (Unirelax) marketed by Tetrahedron Associates, San Diego, California.

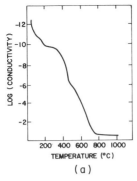 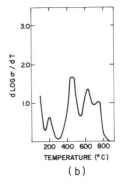

Fig. 8.10. Electrothermal analysis of Markham black shale (116). (*a*) Plot of log σ versus temperature; (*b*) a plot of the derivative *d*(log σ)/*dT* plotted versus temperature.

In passing, it should be noted that it is more usual to record the current as a function of time while simultaneously recording the temperature. In some instances, particularly where comparisons of materials are involved, use of the current/time derivative, *di/dt*, renders small differences more visible. This manipulation can be carried out electronically with no difficulty. The advantage of this technique in defining peaks is clearly shown in Fig. 8.10 which plots the dynamic conductivity of a coal versus temperature (116). The presence of different chemical reactions is clearly indicated in the derivative curve. Subsequent chemical treatment of the base shale allowed direct comparison of coals and, when used in conjunction with other techniques, enabled a good interpretation of the mechanism of carbonation (117) to be made.

III. THEORETICAL BASIS

Electrothermal analysis may be used quantitatively or qualitatively. Qualitative analysis has been used mainly to study phenomena such as phase transitions, curing of polymers, chemical reactions, and to infer the presence of dipoles or impurities. Extensive use of quantitative analysis has been made in studying semiconductor materials, ionic crystals, and, more recently, in studying molecular relaxations, particularly in polymers.

A. ANALYSIS OF THERMALLY STIMULATED CURRENTS

The electrothermal technique in many ways resembles the more familiar thermoluminescence or glow curve analysis techniques that were originally developed for studying detrapping and recombination of charge carriers in semiconductors (121). After cooling and irradiation or after the prolonged application of potential, many semiconductors emit radiation as the temper-

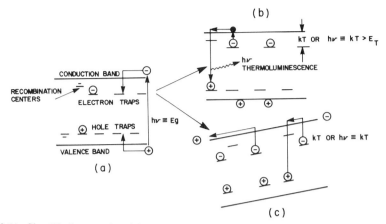

Fig. 8.11. Simplified energy band for an insulator showing: (*a*) The absorption of a photon of energy $h\nu > E_g$ to create an electron and a hole, either of which may be trapped. (*b*) Thermoluminescence as an electron acquires sufficient energy to escape from the trap and recombines with a hole to emit a photon with energy characteristic of the trapping parameter. (*c*) Transient thermally stimulated current as the electron escapes from the trap, moves under the applied electrical potential, and either leaves the sample or recombines.

ature is raised. The emission results from the recombination of a detrapped (thermally stimulated) charge carrier with a charge carrier of opposite sign. The rate of emission depends upon characteristics of the detrapped charge carrier such as the number and depth of the traps; the energy of the photon emission is characteristic of the initial and final states of the recombination event (as shown in Fig. 8.11). Experimentally, there are close correspondences between thermoluminescence and electrothermal analysis. Frequently, however, the thermoluminescence peaks occur at lower temperatures than do the electrothermal peaks as shown in Fig. 8.12 (16).

It is assumed that, in preconditioning the sample, optically created or injected charge carriers are trapped within the solid. As the solid is warmed up, a fraction of the charge carriers acquire sufficient thermal energy to escape from the trap and contribute to the current measured in the external circuit. Note that it is not possible to trap more than one CV of charge (C = sample capacitance; V = applied voltage) and that the current in the external circuit reflects the distribution and magnitude of the moving charge within the sample. It is not required that the detrapped charge exit the sample; only that it move within the sample. Similarly, changes in potential distribution, caused by dipole orientation, can be determined.

The fate of the charge carrier will depend upon the number and the state of traps remaining in the solid. It may be swept out of the solid, be retrapped, or recombine with a charge carrier of opposite sign. The information which can be gained from the thermally stimulated current relates to:

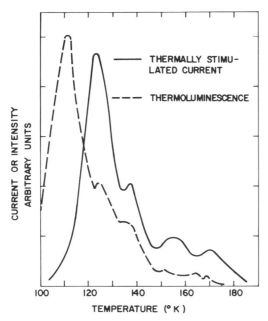

Fig. 8.12. Thermally stimulated properties of irradiated $Ca(NbO_3)_2$ (16). (———), Thermo-luminescence after cooling to 90 K under potential; (---------) thermally stimulated current after X-ray irradiation at 90 K.

1. The trap depth that can be obtained from either the temperature at which the peak in thermally stimulated current is observed or from the initial rate of current increases with temperature.

2. The number of trapped charges from the area under the current-time curve.

3. The capture cross-section of the trap which is obtained from the de-tailed thermally stimulated current versus temperature curve.

However, the precise interpretation of thermally stimulated current curves is seldom unambiguous. Various models have been developed such as that in which no recombination occurs (121), a model in which there is equal probability for recombination and retrapping (60), and a fast retrapping model (73). While most models correctly predict the general shape of the "glow" curve, unambiguous analyses based on thermally stimulated cur-rents have seldom, if ever, been made (23).

Analysis for the thermally stimulated current follows that developed to described thermoluminescence. The escape probability, p, at a temperature, $T°K$, for an electron from a trap depth, E_T, below the conduction level is given by

$$p = v \exp - E_T/kT \qquad (10)$$

where v is the attempt frequency. If there are N trapped electrons, then the instantaneous rate of release is, in the absence of retrapping (121),

$$\frac{dN}{dt} = -Np$$

$$= -Nv \exp(-E_T/kT)$$

(11)

Rewriting dN/dt as $(dN/dT) \cdot (dT/dt)$ and assuming a rate of temperature increase, $dT/dt = \beta$

$$\frac{dN}{N} = -\frac{v \exp - E_T/kT}{\beta} dt$$

(12)

Therefore,

$$\frac{N}{N_0} = \exp\left[-\frac{v E_T}{\beta k} \exp - E_T/kT\right]$$

(13)

The electrical conductivity, σ, is given by

$$\sigma = qN_C\mu$$

(14)

and

$$N_C = -\frac{dN}{dt} \tau$$

(15)

where N_C is the density of free charge carriers which have an average lifetime, τ. Therefore,

$$\sigma = -q\mu\tau\beta \frac{dN}{dT}$$

$$= q\mu\tau v N_0 \exp\left(-\frac{E_T}{kT}\right) \exp\left[-\int_{T_0}^{T} \frac{v}{\beta} \exp\left(-\frac{E_T}{kT} dT\right)\right]$$

(16)

For small values of T, the second exponential term approaches 1, therefore

$$\sigma = q\mu\tau v N_0 \exp(-E_T/kT)$$

in the absence of retrapping.

Garlick and Gibson (60) have treated the case where retrapping and recombination occur rapidly, Assuming there are N traps in the solid of which n are filled at any time, there are $N-n$ empty traps and n recombination centers.

The current at any time is determined by the rate of escape minus the fraction which is retrapped or which recombines, i.e.,

$$i = -\frac{dn}{dt}$$

$$= v \frac{n^2}{N} \exp(-E_T/kT)$$

(17)

Again, putting

$$\frac{dn}{dt} = \frac{dn}{dT} \frac{dT}{dt} \qquad (18)$$

$$= \beta\, dn/dT$$

$$i = n_0^2\, v \exp\left(-E_T/NkT\right)\left[1 + \frac{n_0}{N}\int \frac{v}{\beta} \exp\left(-E/RT\right)dT\right]^2 \qquad (19)$$

where n_0 is the initial trap filling.

The characteristic parameters of the thermally stimulated current curves analyzed by this method show:

1. The maximum current, for fixed β, N_0, and v, occurs at a temperature proportional to the trap depth, E_T.

2. The maximum current for fixed E_T and N_0 occurs at a temperature proportional to β^{-1} and v, i.e., the greater the heating rate the lower the peak.

3. The area under the stimulated current versus temperature curve is proportional to N_0, assuming a single trapping level only.

4. The initial current increase is given by $i = $ constant $\cdot \exp\left(-E_T/kT\right)$, from which the E_T can be readily obtained.

It is to be emphasized that many models show the same general behavior. The above analyses were developed for thermoluminescence: there is close analogy with thermally stimulated currents. For example, peaks in thermoluminescence and thermally stimulated current should coincide. Simultaneous measurements such as those shown in Fig. 8.12 (16) show this not to be the case; thermoluminescence peaks generally occur at lower temperatures. The theoretical reasons for this are discussed by Chen (43).

There have been many electrothermal analyses made upon semiconductors. The usual analysis involves a discussion of trap depths and the number of traps; there has been little discussion of the chemical nature of the traps or the processes which lead to their creation. Some specific examples will be discussed in Section V.

B. THERMALLY STIMULATED DEPOLARIZATION CURRENTS

If an electrical potential is applied to a material in which the dipoles are free to rotate, the dipoles become oriented. As the material is cooled, still under potential, the oriented dipoles become "frozen in" as the dipolar groups lose thermal energy and are no longer free to rotate. Conversely, on reheating those polar groups that can rotate most easily are the first to become nonoriented in the absence of an external field. The change in the number of oriented dipoles results in a change in the internal field distribution. Such changes in the internal field distribution are reflected by current flow in the external circuit. If there are a number of polar groups, the dipoles become free to rotate at rates and temperatures characteristic of the particular molecular relaxation process (43,41,49,109). In this respect, the technique is a low-frequency analog of a dielectric measurement.

The technique is thus ideally suited for the study of dielectric materials and molecular relaxations since it is characterized by a very low (10^{-2}–10^{-4} Hz) equivalent frequency compared to the normal dielectric ac method (1–10^6 Hz). The sensitivity and the ability to "clean" the current–temperature curves has enabled many broad relaxation peaks to be resolved into more than one peak with consequent clarification of the particular relaxation modes (4,42,93,134,148,150). In particular, the method is receiving great attention with respect to the relaxation modes of polymers (101).

The technique was originally applied by Bucci, Fieschi, and Guidi to alkali halides in which the dipoles were formed from polar impurities (34). It has been variously referred to as ionic thermoconductivity (ITC), depolarization thermocurrent (DTC), thermostimulated current (TSC), and thermostimulated depolarization current (TDC).

The temperature-dependent relaxation time, τ, of a dipole within a dielectric is given by

$$\tau(T) = \tau_0(n) \, T^{-n} \exp(\Delta U/kT) \tag{20}$$

where T is the temperature and ΔU is an activation energy. $\tau_0(n)$ is constant and $n = 0, \frac{1}{2}, 1$ according to the particular relaxation theory used. In the phenomenological (Arrhenius) theory, $n = 0$ and $\tau_0(0)$ is not related to the microscopic parameters of the system. In the Eyring relaxation theory (54), $n = 1$ and

$$\tau_0(1) = \frac{h}{k} \exp(-\Delta S/k) \tag{21}$$

where h is Planck's constant and ΔS is the change in entropy between the normal and activated state. $n = \frac{1}{2}$ corresponds to the Bauer theory in which $\tau_0(\frac{1}{2})$ is approximately $(2\pi I/k)^{\frac{1}{2}}$ where I is the moment of inertia of the relaxing unit (8).

In exactly the same way that the thermally stimulated current was derived for detrapping, the depolarization current, $J(T)$, may be calculated (33,110)

$$J = \frac{P_0}{\tau(T)} \exp\left[-\int \frac{dt}{\tau(T)}\right] \tag{22}$$

where P_0 is the initial polarization. Let $dT/dt = \beta$, then

$$J(T) = \frac{P_0 T^n}{\tau_0(n)} \exp\left(-\frac{\Delta U}{kT}\right) \exp \frac{1}{\beta \tau_0(n)} \int_{T_0}^{T} T^n \left(\exp - \frac{\Delta U}{RT}\right) dt \tag{23}$$

The temperature of maximum current, T_m, given by

$$\frac{dJ\,(T)}{dT} = 0$$

is expressed by

$$\tau_0(n) = \frac{T_m^{2n} \exp\left(-\dfrac{\Delta U}{RT}\right)}{\beta\, n\!\left(T_m^{n-1} + T_m^{n-2}\,\dfrac{\Delta U}{R}\right)} \tag{24}$$

The initial rise in current is

$$J(T)T^{-n} = \exp\left\{\text{constant} - \frac{\Delta U}{kT}\right\} \tag{25}$$

In other words, the phenomenon can be described in exactly the same way as the detrapping and the parameters obtained by the same analysis.

The preceeding analysis applied to systems in which there is one characteristic relaxation time. In practice, many depolarization peaks may be composites of a number of "pure" processes. "Pure" is used in the sense that the relaxation process has a unique activation energy, hence a unique relaxation time. The value of the cleaning technique can now be clearly understood—a broad peak may be resolvable into a sequence of "pure" relaxations (Fig. 8.2). On the other hand, no pure relaxation may be involved; the process may be characterized by a continuous distribution of activation energies and relaxation times. Depending upon the precise conditions of polarization, a range of activation energies would be obtained on "cleaning" the peaks by partial depolarization (148).

IV. ELECTROTHERMAL ANALYSIS OF POLYMERS

The use of electrothermal thermal analysis in the study of polymers is of increasing current interest. The focus is twofold; the technique has been used extensively in the study of electrets (131,150), piezo and pyroelectricity in polymers (24,25), and as a sensitive means of investigating molecular motions and relaxations in bulk polymers (4,42,93,148).

Much of the literature on polymeric electrets is contained in two Electrochemical Society publications (9,111), which are collections of papers given in Symposia in 1967 and 1972. Van Turnhout (150) has published a book *Thermally Stimulated Discharge of Polymer Electrets* which extensively reviews the literature to 1972. Seanor (129) has some applications of electrothermal analysis to polymers.

A. POLYMERIC ELECTRETS

Polymeric electrets are materials that exhibit long-lived permanent electrical polarization, i.e., the relaxation time is very long and may be as long as

10^{13} sec (at 25°C). Polarization arises from a number of sources which include interfacial polarization, dipole orientation, ion displacement, and charge injection. Polarization may arise spontaneously, but it is more usual to precondition the sample by a thermal treatment in which the polymer is subjected to "poling" or polarization at a temperature higher than its glass transition temperature and then cooled. This type of polarization is referred to as "heterocharge" since the surface of the dielectric is polarized oppositely to that of the poling electrode. However, if a beam of ionized particles such as an electron beam or corona is used to charge the dielectric, then the injected charge is of the same sign as the polarizing electrode and is referred to as "homocharge." The homocharge is trapped within the solid where it may interact with, or even neutralize, the heterocharge.

Typical of polymeric electret materials is polyvinylidene fluoride (pVF$_2$), which has the composition–CH$_2$–CF$_2$–. This material is tough, easily fabricated into thin films and can be made to exhibit unusual, nonlinear electrical and optical effects (13,24,57,87,102,106,112,114). In the solid state, pVF$_2$ can exist in three different forms referred to as α, β, and γ phases (113,119) in which the relative amounts of each phase determine the unusual electrical properties. Prest (119) has reviewed the relationship of the electret-forming ability with pretreatment. The thermally stimulated current response can be directly related to the γ phase content as revealed by the infrared spectra of differently treated samples (113) as is shown in Fig. 8.13. It was concluded that the γ form (all *trans* configuration) and its orientation were the important factors in controlling the charge storage. On the other hand, the pyro- and piezoelectric behavior seem to depend on both field-induced dipole alignment of the β phase (57) and space-charge injection (112). Detailed studies of the morphology and charging (39,88,97,135) suggest there are several aspects associated with the poling process. Initially, there is a rapid activation of the film as a result of charge injection and/or the inversion of the β-phase dipoles (but not rotation of the entire crystal). At significantly longer times, some of the α phase is destroyed (135). Subsequently, some of the resulting free chains recrystalize into the β form.

Similarly, polyacrylonitrile has also been studied as an electret material (133,138,139,140). In much the same way that the electret activity of polyvinylidene fluoride can be altered by stretching (133) so can that of polyacrylonitrile (140). The processes, however, appear to be complicated by the presence of ionic impurities. Figure 8.14 shows the thermally stimulated current from an unpoled film of polyacrylonitrile (138). Three current peaks noted at 90°C, 140–160°C, and 180–205°C are referred to respectively as γ, β, and α peaks. The γ peak is opposite in sign to that of the α and β peaks and is seen only in unpoled films. It is attributed to the redistribution of ionic impurities creating an internal electric field which causes the polar nitrile groups to adopt a preferred orientation in response to the ioninduced electrical asymetry (138). The β peak is associated with chain segment mobility as

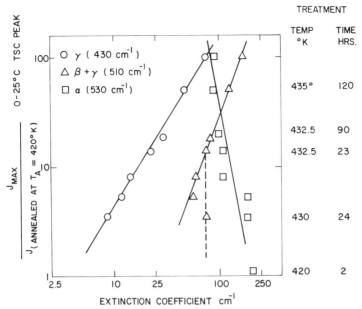

Fig. 8.13. Correlation between thermally stimulated current and crystallinity in polyvinylidene fluoride revealed by IR spectroscopy (119). (○) γ content (430 cm⁻¹), (△) β + γ content (510 cm⁻¹), (□) α content (530 cm⁻¹).

compared with rotation of the nitrile groups which is hypothesized for the γ peak. The α peak has been attributed to the earliest stages of thermal degradation (139). Spectroscopic studies (139) suggest that the mechanism of degradation involves the creation of unsaturation (–C=C–), conjugation, the formation of imine (–C=N–) groups and possibly the creation of CN⁻ ions. Earlier studies (64) had suggested a mechanism

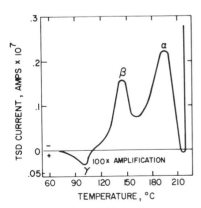

Fig. 8.14. Thermally stimulated current of an unpolarized polyacrylonitrile film cast from dimethylformamide (139).

On the other hand, there is also experimental evidence for dehydrogeneration of the polymer backbone as the initial step in the thermooxidative degradation polyacrylonitrile (47). On the basis of the thermally stimulated current-analysis combined with the spectroscopic analysis, it is suggested that unsaturated groups ($-C=C-$ and $-C=N-$) can act as trapping sites that would have different affinity for the different charge carriers. This is in agreement with earlier investigations which suggested that unsaturated groups in substituted polyolefins (50) and polyethylene (51) could behave as traps for either positive or negative charge carriers.

B. POLYMER CHARACTERIZATION STUDIES

In addition to studies of polymer electrets, the electrothermal technique has been used to characterize polymer structure and molecular motions. In these studies, use is made of the different molecular environments in which side chains, small chain segments, and large segments are found after different pretreatments (which may involve polarization at high temperatures). A limited number of specific examples will be used to illustrate the main ideas involved.

In the simplest case, the sample is simply polarized at a suitably high temperature, under the applied potential and the discharge measured. However, in the case of polymers interpretation is not easy. For example, polyethylene, ideally the simplest organic macromolecule, exhibits one of the most subtle and complex behaviors found among high polymers (86,101). The complexity arises from several sources:

1. Molecular structure (36): the method of synthesis can yield different material. The low-pressure (Phillips) synthesis yields a high density, essentially linear polymer. Ziegler polyethylene contains ~5 ethyl branches per 1000 carbon atoms, and free radical polymer is highly branched having as many as 50 short branches per 1000 carbon atoms.

2. Degree of crystallinity, which depends upon the synthetic process as well as the preparation technique. Free-radical polymer has reduced crystallinity and low density whereas linear polyethylene can be highly crystalline

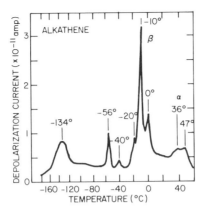

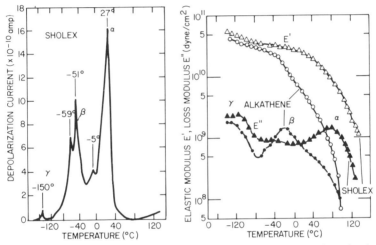

Fig. 8.15. Electrothermal analysis of polyethylenes (142). (*a*) Alkathene (low density); (*b*) Sholex (high density); (*c*) dynamic modulus (ϵ') and dynamic loss modulus (ϵ'') at 30 Hz for Alkathene and Sholex.

and has a high density. Either polymer can be quenched from the melt to yield totally amorphous solid.

3. The character of the noncrystalline fraction, which may be influenced by crystal defects, chain folding, chains ends, and branch points.

4. In addition, peaks may overlap and be composed of more than one peak.

Thus, the complexity of the thermally stimulated current spectra shown in Fig. 8.15 (142) for two polyethylenes is not entirely unexpected. Comparing the mechanical properties of low density polyethylene (Alkathene) and high density polyethylene (Sholex) in Fig. 8.15*c*, the peaks are labeled α, β, and γ

in order of decreasing temperature. These peaks can be compared with the electrothermal peaks in the temperature regions 20–40°C, 0–20°C, and < – 120°C (allowing for the lower equivalent frequency of the electrothermal technique).

Detailed studies of the dielectric (99) and mechanical properties (86,101) of polyethylene and its polar copolymers have led to the following assignments for the α, β, and γ peaks. The α peak is ascribed to relaxation of the noncrystalline material between crystal lamellae as they slip past each other and to crystal defects within the crystal lamellae. The β relaxations are associated with both the amorphous regions of the polymer and the number of branch points. The temperature appears to represent the T_g of the branched polymer. The γ peak is a composite relaxation related to both the crystalline and amorphous phases. The crystalline component is thought to be related to the motion of chain ends or defects within a chain folded crystal (80); the amorphous contribution is thought to arise from a crankshaft motion of the main chain (126). Thus, the amorphous contribution represents the T_g of linear polyethylene.

From the foregoing analysis, the electrothermal peaks can be ascribed with the exception of the peaks in the – 50°C to – 60°C range which are observed in both low-density and high-density polymer. These peaks do not appear to have an analog in either the mechanical or dielectric spectra. However, the possibility that they are related to injected charge still remains.

For the purposes of characterizing molecular relaxations in polymers, a single polarization step is inadequate to give detailed understanding of the relaxation process. However, repeated polarization/depolarization (sampling) experiments at different temperatures yield more useful information. This point was illustrated earlier in Fig. 8.2, which showed how two β relaxation peaks in polycarbonate could be resolved (4). In this respect, the technique is particularly valuable for there is no analog in dielectric or mechanical measurements that allows the polarization or relaxation to be sampled.

Typical of the manner in which research has progressed are the curves in Fig. 8.16 for polymethyl methacrylate which progress from a single polarization through multistage thermally stimulated depolarization (109) to fractional polarization during cooling (148). The multistage measurements (109) were made cooling under continuous potential to the temperatures (T_{ms}) noted, held at T_{ms} for 1 hr and then heating. The activation energy measured increased steadily from 4.2 Kcal · mol^{-1} for T_{ms} = – 140°C to 160 Kcal · mol^{-1} when T_{ms} was equivalent to the temperature at which $J(T)$ was maximum for the saturated discharge. This measurement indicates that there is a range of activation energies involved in the β relaxation mode. However, the maximum value of the activation energy (11.0 Kcal · mol^{-1}) does not agree with that obtained from dielectric measurements (18.0 Kcal · mol^{-1}). The reason for this was suspected to be a consequence of the range vibrational

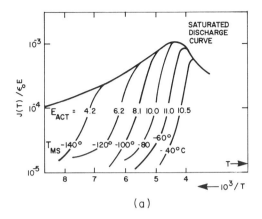

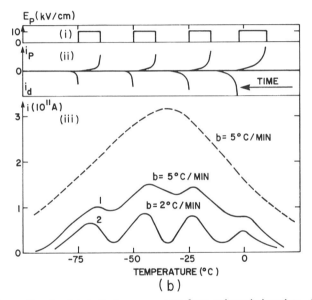

Fig. 8.16. Thermally stimulated discharge currents from polymethylacrylate. (*a*) Multistage thermally stimulated discharge as a function of the preheat temperature at which the electret was maintained for 1 hr before heating. [$T_p = 20°C$, $E = 8.0 \, kV \cdot cm^{-1}$ for 1 hr, heating rate = $3°C^{-1} \cdot mm^{-1}$] (109). (*b*) Fractional polarization: (i) polarizing steps during linear cooling, (ii) polarization and depolarization currents observed during cooling, (iii) partial TSC curves obtained during linear heating at rates of 5°C min^{-1} (1) and 2°C mm^{-1} (2). Dashed curve corresponds to the saturated TSC curve. [$T_p = 40°C$, $T_o = -196°C$, $E_p = 10 \, K\dot{V}cm^{-1}$, $t_p = 30$ min] (148).

frequencies associated with the dipole group involved in the relaxation. This conclusion was confirmed by fractional polarization studies (148). In these experiments, the potential was applied intermittently during cooling with the sample being short-circuited between potential applications. Thus, the molecular motions were "sampled" and selectively frozen in during cooling. Each peak corresponding to partial polarization was analyzed by both the Gibson-Garlick initial rise method (60) and the Bucci-Freschi-Guidi method of graphical integration (34). In all cases, which covered the range of polymers indicated in Fig. 8.17, there was continuous increase in activation energy with the difference between the temperature of maximum current (T_m) and temperature at which the sample was short circuited (T_c). Thus, it was concluded that the β relaxation peaks of many polymers were characterized by a continuous distribution in activation energy. This conclusion agrees with the hypothesis that the β relaxation involves the motion of relatively small polar groups enjoying a local interaction with their immediate environment.

In the work previously referred to on polycarbonate (Fig. 8.2) (4), two broad TSC peaks referred to as $β_1$ and $β_2$ peaks were each resolved into two peaks. The $β_1$ peak, consisted of two peaks with activation energy 5.6 Kcal · mol^{-1} and 6.3 Kcal · mol^{-1} which occurred at 121°C and 131°C, respectively, and were relatively unchanged on annealing. On the other hand, the two peaks resolved from the $β_2$ peak at 175°C and 193°C changed in activation energy from 8.2 Kcal · mol^{-1} and 11.0 Kcal · mol^{-1} to 10.25 Kcal · mol^{-1} and 14.0 Kcal · mol^{-1}, respectively; however, temperature changes were small. Annealing the polycarbonate caused an increase in the magnitude of the $β_1$ peak and reduced that of the $β_2$ peak. On this basis, it was proposed that

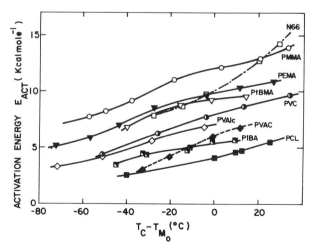

Fig. 8.17. Variation of apparent activation energy of the β peaks of various polymers as a function of $[T_c - T_{m_0}]$ (148).

motion of the carbonate groups restricted by the phenyl groups, thought to be responsible for the β_1 peak (3,96) was made easier by annealing. Motion of the phenyl groups with cooperative motion of the carbonate groups (the β_2 relaxation) was made more difficult. Behavior of the resolved peaks was rationalized by ascribing peaks I, II, and IV to motions involving the methyl group, the motion of carbonate groups attached to a phenyl group of the main chain, and the movement of phenyl groups attached to moving carbonate groups. Peak III was attributed to a complex relaxation of carbonate groups restricted by phenyl groups on the polymer backbone. Calculation of the activation parameters also suggsted that chains involving movements of the phenyl groups were more ordered after annealing below T_g. Annealing appeared to affect mainly the relaxation processes relating to phenyl group motion. Therefore, the decreased impact strength and improved strength after annealing below T_g were attributed to the reduced mobility of the phenyl groups and ordering of the polymer chains (63). These assignments are also in agreement with the assignments made on the basis of dynamic mechanical measurements (83,95).

Polyamides also show multiple peaks in their mechanical and dielectric spectra. These peaks have been referred to as α, β, γ, and δ peaks in order of descending temperature (22). The thermally stimulated current curve for poly (hexamethylene adipamide) (Nylon 66) between 90 K and its melting point at ~258°C is shown in Fig. 8.18a (41). The 66°C peak has been analyzed (41) and assigned to the α' process (120). This peak was resolved into four distinct peaks each having a constant activation energy and corresponding to a unique relaxation mode. The peaks were: α_1' ($E_{act} = 12.8$ Kcal $\cdot$ mol^{-1}) tentatively assigned to end-group relaxation; α_2' ($E_{act} = 19.80$); and α_3' ($E_{act} = 30.3$ Kcal $\cdot$ mol^{-1}) attributed to self-dissociation of two amide groups (53):

$$2 \quad \begin{array}{c} \diagdown \\ \diagup \end{array}\!\!\!C\!=\!O \quad \begin{array}{c} \diagdown \\ \end{array}\!\!\!C\!=\!O \\ \quad \overset{+}{\underset{\diagdown}{N}}\!-\!H\!\rightleftharpoons\!H\!-\!\underset{|}{N}\!-\!H \;+\; \begin{array}{c} \diagdown \\ \diagup\diagup \end{array}\!\!\!C\!-\!O^{\ominus} \\ \qquad\qquad\qquad\qquad \underset{\diagdown}{N}$$

The value of 30.3 Kcal $\cdot$ mol^{-1} is the activation energy of proton conduction in the amorphous regions of the polymer (128,160). The α_4' ($E_{act} = 48.7$ Kcal $\cdot$ mol^{-1}) is consistent with the activation energy of electronic charge transfer in the amorphous polymer which is related to rotation of ~10 amide groups (6,128). Thus, all the contributions to the α' process may be considered as relating to transitions of the hydrogen-bonded amide protons.

The peak at -15°C was very moisture sensitive, and, therefore, was attributed to the relaxation of nonhydrogen-bonded amide groups (42). The high temperature peak at 207°C was associated with protonic space charge

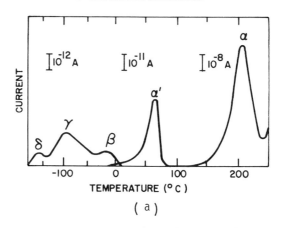

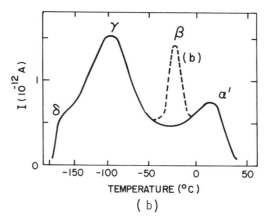

Fig. 8.18. Thermally stimulated currents in polyamides. (a) Polyhexamethylene adipamide (Nylon 66) (41) [T_p = 220°C (α peak), 70°C (α' peak) and 20°C (β, γ, δ peaks), T_o = 90 K, E_p = 5 × kV·cm^{-1}, t_p = 2 min]. (b) Polyethylene isophthalimide (93) [T_p = 20°C, T_o = 90 K, E_p = 7.5 kV·cm^{-1}, T_p = 2 min.] Dashed curve corresponds to a damp sample.

between the crystalline and amorphous regions of the polymer and termed the α peak. The γ and δ peaks were unassigned.

In later studies of polyethylene isophthalamide (93), the thermally stimulated discharge current curves shown in Fig. 8.18b were obtained. Similarities to the curve of Nylon 66 can be readily noted—a moisture-sensitive β peak, a large γ peak, and small δ peaks. Using the stepwise polarization and depolarization technique, the γ peak was resolved into a series of peaks, each characterized by a unique relaxation time and having a unique activation energy. Thus, the whole γ peak could be described in terms of a discrete but continuous range of relaxation processes. Cross-correlation with dielectric measurements was excellent.

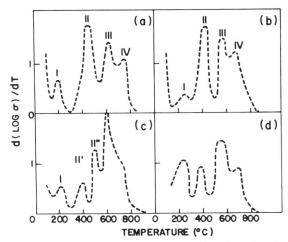

Fig. 8.19. Rate of change of conductivity (in arbitrary units) plotted against temperature of carbonization for Markham black shale which was: (*a*) untreated; (*b*) methylated; (*c*) dehydrogenerated; and (*d*) brominated (116).

The electrothermal technique can also be used to characterize materials and differentiate small differences between materials. The earlier discussion of the differences in the thermally stimulated current of low- and high-density polyethylenes (Fig. 8.15) illustrates one use in an analytical mode.

The technique was used to differentiate between coals treated in different ways (116,117) and to study the carbonization process. The curves shown in Fig. 8.19 are the derivative, $d \log \sigma/dT$. Similar to those shown in Fig. 8.10, they show considerable fine structure but are in themselves insufficient to give a clear picture of the processes occurring in the thermal degradation of coal. However, with the aid of other techniques, a reasonable interpretation of the carbonization process was possible (117).

The initial leveling off of conductivity (Fig. 8.10*a*) at ~150°C was related to loss of moisture. Between 150°C and 350°C, little structural change occurred, but the grain structure was disrupted leading to peak *I*. Primary carbonization began between 400° and 500°C and was accompanied by a marked increase in conductivity (peak II). In this regime, free radical concentration reaches a maximum, the extent of conjugation increases, each process contributing to increased conductivity. Secondary carbonation involves loss of peripheral material from aromatic clusters and growth of the carbon in the form of small sheets similar to, but smaller than, graphite. The accompanying increase in conduction leads to peak III and IV.

The effect of chemical treatment of the coals led to the curves illustrated in Figs. 8.19*b*, *c*, and *d*. The changes in intensity and temperature at the peaks was discussed in similar terms and were reinforced by other experiments. For example, bromination (Fig. 8.19*d*) led to a considerable enhancement of the 200°C derivative peak. TGA showed the evolution of hydrogen

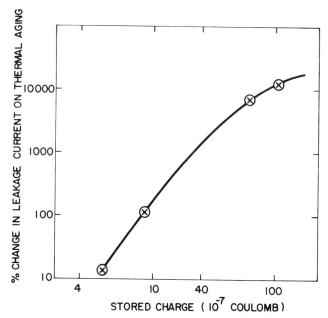

Fig. 8.20. Relationship between changes in leakage current and area under the thermally stimulated current discharge curve of epoxy encapsulant resins (157).

bromide at this temperature which resulted in a residue containing a considerable–C=C–character.

The technique was used by Warfield (151,152,153) to study glass transitions and curing of various resins under static conditions. As is usually observed, the curves showed a distinct break at the glass transition temperature with the activation energy being lower above T_g. Such an observation has frequently been used to suggest that conduction was ionic. However, such a conclusion should only be made after more direct evidence has been obtained (128) since charge transport involving localized sites and electrons (or positive holes) would also be easier above T_g, as many of the experiments discussed earlier indicate.

Thermally stimulated discharge currents have been used to predict the life of epoxy resin-encapsulated, semiconductor devices (156). As shown in Fig. 8.20, the change in leakage current on heat aging is strongly related to the area under the TSD curve. Further work (15) related the area under the TSD curve to such factors as resin/hardener ratio, postcure times, and catalyst concentration. In general, it was noted the least charge storage occurred with 1:1 stoichiometry and complete curing. The addition of catalyst, in addition to hastening the cure, reduced the area of the TSD curve. It was suggested that while surface species associated with unreacted fragments of the base resin were involved in loss of device performance, moisture also was shown to have a detrimentional effect on life and to introduce new peaks into the TSD curve. The new peaks were associated with side chain rotation (100).

The thermal depolarization of composites of epoxy resin and mica flakes has also been studied (143) (Fig. 8.21). Mica alone exhibited a single broad peak at 148°C, the resin showed peaks of 132°C (α_R) and 106°C (β_R), whereas the composite had peaks at 84°C (γ_c), 106°C (β_c), and 152°C (α_c). The α_R peak was considered due to electrode polarization of injected electrons. The β_R and β_c peaks were considered to arise from dipole orientation. The γ_c peak, which in common with the α_c peak was dependent upon mica concentration, was assigned to interfacial dipole orientation. Associated with the α_c peak was a large dielectric dispersion. Accordingly, the α_c peak was attributed to charge carriers trapped in the boundary regions between the resin and mica flakes. In view of the proposed ionic conductivity of these resins (104,153), interfacial polarization (Maxwell-Wagner polarization) is considered to be related to the presence of ions trapped in the interfacial regions of the composite.

Attempts have been made to determine the relationship between thermally stimulated currents and the structure of poly(2,6 dimethyl-1,4-phenylene oxide) (PPO) blends with polystyrene-related polymers (1). While the individual polymers each showed a peak characteristic of the glass transition, another peak associated with space charge drift was observed with polystyrene (PS), poly(p-chlorostyrene), and a random copolymer of styrene and p-chlorostyrene. This latter peak was attributed to ionic impurities.

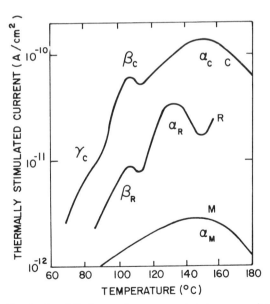

Fig. 8.21. Thermally stimulated discharge of an epoxy resin/mica composite. (143) R, resin, M, mica, and C, composite [$E_p = 18$ kV · cm^{-1}, $T_p = 183$°C, $t_p = 20$ min.]

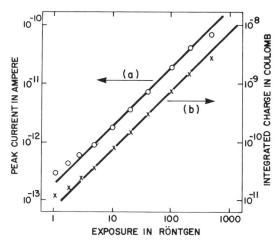

Fig. 8.22. The relationship between thermally stimulated current response of a PTFE foil and radiation exposure (69). (*a*) Peak current; (*b*) charge under the TSC peak.

The TSD of PPO and its blends with the above polymers showed large and sometimes variable peaks. In the case of polystyrene-rich blends, the peaks corresponded to the T_g' values of the individual polymers. However, in the case of 50/50 PS/PPO and poly(*p*-chlorostyrene) blends, neither peak corresponded clearly to a calorimetrically determined T_g.

Radiation dosimetry using thermally stimulated discharge currents from polymers has been an ongoing topic of research (19–21,56,58,70,74,132). For a material to be useful for dosimetry, there are a number of requirements to be made of the material. It must be an electrical insulator in which any conduction electrons are due entirely to absorbed radiation. Such charge carriers must be promptly trapped and stored for an appreciable time. For practical purposes, this means that release (thermal stimulation) should occur above 100°C. The material should not be light sensitive. To avoid corrections, the dosemeter material should have the same absorption characteristics as the medium in which it is required to know the dose.

Of the materials studied on a systematic basis, a synthetic silica (Spectrosil) and polytetrafluoroethylene showed significant promise (20). In particular, Fluon G163 showed a marked thermally stimulated current response which varied linearly with dose over a 2–500-Roentgen range (69) as shown in Fig. 8.22. Studies of electron and γ-ray-irradiated Teflon foils (132) showed the thermally stimulated current to decay with a time constant of many years. The consequence of charge decay from the shallowest traps was to move the TSC peaks to higher temperatures. Later studies (70) using penetrating electrons and γ rays showed the presence of two peaks in Teflon at ~70°C and 200°C when the thermally stimulated current was measured under an applied bias. This observation is compared to the single TSC peak observed when measured under short circuit. Measurement using the applied bias increased the current sensitivity of both peaks. However, the 70°C peak tended to saturate with increasing bias and was sensitive to storage conditions. The 200°C peak was smaller and did not saturate with bias potential during measurement but did saturate with dose. While suggesting that the higher temperature peak was related to a smaller number of deep traps, there was no indication of the nature of the chemical nature of the deeper shallow traps.

V. ELECTROTHERMAL ANALYSIS AS APPLIED TO SEMICONDUCTORS

Many studies have been carried out on semiconductor materials with the purpose of identifying trapping levels. In most cases, there has been no attempt to identify specific chemical species with the trapping level. Therefore, only a few cases are discussed, to illustrate particular uses of the technique.

Cadmium sulfide is a photoconducting material in which copper ions play an important part in improving its photosensitivity (30,31). Deliberate incorporation of the impurities (doping) into pure cadmium sulfide is accomplished by the controlled addition of copper and chlorine in the form of cuprous or cupric chloride. By this means, an insulating but more photosensitive material is created. The stability of the doped material is of obvious importance. The stability of the Cu^+ and CU^{2+}-doped material has been investigated by thermally stimulated current analysis with the results shown in Fig. 8.23 (125). The Cu^+-doped material originally contained shallow traps in the range 0.15–0.25 eV, which emptied over the 100–180 K range, and deep traps in the 0.6–0.8 eV range, which emptied in the 300–400 K range. After illumination for 6 hr, a threefold increase in the shallow-trap density and a corresponding decrease in the density of deep traps was observed. By contrast, the Cu^{2+}-doped material was far less stable: the high initial concentration of shallow traps was reduced by a factor of three and the deep traps showed an increase in density. While these experiments show the superior stability of the cuprous ion-doped material, they shed little light on the chemical processes occurring under intense illumination.

Selenium is also an important photoconductor which can exist in the amorphous or crystalline state. Amorphous selenium polarized at 25°C shows two peaks at 173 K and 273 K and a small homocharge at 213–233 K (71). Partial polarization allowed the resolution of each peak. The upper temperature peak had characteristics of a single relaxation process and an activation energy of 0.79 eV similar to that of the ac conductivity in amorphous selenium (94). It was ascribed to the field-independent mobility of the charge carriers (positive holes) measured in this temperature regime. The low temperature peak had a characteristic activation energy of 0.54 eV. However, this peak had characteristics of a continuous distribution of acti-

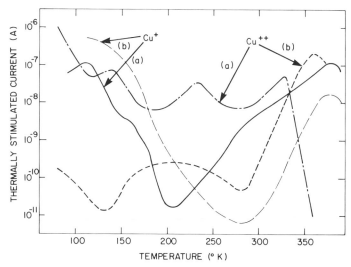

Fig. 8.23. Thermally stimulated discharge of Cu^+ and Cu^{2+} doped CdS and the effect of prolonged irradiation (125). (*a*) Initial curves; (*b*) after 6 hr irradiation. (————) Cu^+; (—————) Cu^{2+}.

vation energies similar to the distribution of activated mobilities observed (141). This conclusion was reinforced by the potential dependence of both the peak temperature and mobility.

There have been several studies on crystalline selenium grown by different techniques (5,7,52,79,90,123,137). At least four peaks have been reported by different authors occurring over a temperature range from 115 K to 273 K. The nature of the species or sites responsible for the peaks has not been discussed in great detail (Table 8.III).

The same situation holds for other similar materials. Despite a number of studies on chalcogenide glasses, many of which are photoconductors, little information has been generated concerning the specific mechanisms or

TABLE 8.III
Thermally Stimulated Current Peaks in Selenium

Sample	Peak temperature (K) and [activation energy (eV)]
Monocrystalline selenium [bulk (5), vapor grown (7)]	115 (0.1), 165 (0.14), 180 (0.17)
Polycrystalline hexagonal selenium (90)	(0.12), (0.16), (0.23)
Single crystal (high pressure and preirradiation) (79)	(0.45), (0.54),
Single crystal (123)	130 (0.065), 162, 200, 250 (0.43)
Strained vapor grown (137)	170 (0.065–0.15)
Epitaxially grown (52)	145 (0.06–0.16)
Amorphous selenium (71)	173 (0.54), 273 (0.79)

TABLE 8.IV
Thermally Stimulated Current Peaks in ZnO (68)

Temperature (K)	E_{act}(eV)	Species
88–100	0.12–0.15	Zn/Zn$^+$
114–130	0.4 –0.5	Cu impurity
	0.6	O_2 deficiency
150–180	0.8	O^-
190–20	1.0	
275	1.1 –1.2	Ionic conduction

chemical species responsible for the thermally stimulated current (2,14,18, 55,91,103,130,136,144,145). The attribution usually made is to trapping centers or localized sites in the bandgap when referring to amorphous materials although an ionic contribution has also been suggested (144,145). The formation of clusters of arsenic or arsenic oxide has been invoked for As_2Se_3 (136) which is known to decompose under the influence of light (14). The lack of precise understanding is related to the complex nature of the conduction process in amorphous materials which is the subject of intense on-going discussion.

Zinc oxide is a semiconducting oxide that has been used in electrophotography and catalysis (77). There have been a number of electrothermal analysis studies on this material (67,68,115). Gray (68) made a very detailed study of the effects of gases, pressure, and light on the thermally stimulated currents from zinc oxide and titanium dioxide with the results shown in Tables 8.IV and 8.V. Correlation of the energy level with pretreatment in vacuum, oxygen, and light enabled an association to be made between the low energy

TABLE 8.V
Thermally Stimulated Current Peaks in TiO$_2$ (anastase) (67)

Pure TiO$_2$			1% Nb modified TiO$_2$	
Evacuation at 300–350 K and oxygen adsorbed at 300°K (eV)	Reoxidation at 775°K with PO$_2$ 100 mm (eV)	Photoelectret measurements after evacuation at 700 K (eV)	Reoxidized sample (775 K; PO$_2$ = 100 mm) (eV)	Photoelectret measurements after evacuation at 700 K (eV)
—	.07	—	.01	—
.04	—	.04	.04	.04
.12	.12	.12	—	.12
—	.29–.32	.29–.30	.27	.24–.27
.35–.38	.35–.38	—	.34–.36	.31–.32
—	.40–.42	—	.42	—
.50	.47	—	.46	—
.53–.54	—	—	.50	—

peak and donor levels, Zn or Zn^+. The 0.6-eV level was associated with an oxygen deficiency and the 0.8-eV level with O_2^- ion adsorption. The high energy peak was associated with water absorption and may be related to ionic conduction. The 0.4-eV level was associated specifically with the presence of residual copper impurities. Table 8.V shows that as many as seven trapping levels could be isolated in niobium-doped anastase (TiO_2), although no assignments were made.

VI. ALKALI HALIDES

Whenever a divalent impurity is introduced into an alkali halide crystal it enters substitutionally into the lattice. The charge neutrality of the crystal is maintained by the simultaneous creation of vacancies equal in number to the number of impurity ions. Such vacancies have a charge equal to but of opposite sign to that of the impurity ion. Because of the coulombic attraction, the impurity and vacancy occupy adjacent or next nearest neighbor lattice sites, and the ion/vacancy pair possess a dipole moment. The properties of the impurity/vacancy dipole have been extensively studied by the electrothermal technique. The term "ionic thermocurrent" (33) has often been used to describe the experimental technique. In addition to studying the dipolar properties of the ion/vacancy pair, it has also been possible to study their agglomeration and precipitation (34).

Potassium chloride doped with strontium chloride was one of the first systems reported by Bucchi and Fieschi in their first publication on ionic thermoconductivity (33). Typical results for the $KCl/SrCl_2$ system are shown in Fig. 8.24. Associated with the Sr^{2+}/vacancy dipole is a strong peak at 235

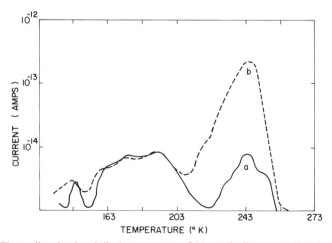

Fig. 8.24. Thermally stimulated discharge curves of (a) nominally pure KCl and (b) KCl containing 0.06% $SrCl_2$ (33) [T_p = 237 K, E_p = 305 V for 3 min, heating rate = 0.1 K sec^{-1}].

TABLE 8.VI

Thermally Stimulated Current Peaks in KCl Doped with Divalent Metals (27)

Ion	Atomic mass	Radius	T_{max} (K)	E (eV)	τ_0^{-1} (sec^{-1})	XCl$_2$ concentration in the melt (mol%)	Reference
K$^+$	39	1.33	—	—	—	—	
Ba^{2+}	137	1.34	233	0.71 ± 0.01	3.1 × 10^{13}	0.03	29
Pb^{2+}	207	1.20	233.5	0.65 ± 0.01	4.8 × 10^{12}	0.1	29
Sr^{2+}	87	1.12	222.5	0.63 ± 0.01	2.7 × 10^{12}	0.01	29,33,34
			235	0.66	1 × 10^{13}		
Sm^{2+}	150	1.11	227	0.7 ± 0.03	3.7 × 10^{13}	0.1	147
Eu^{2+}	152	1.09	224	0.68 ± 0.01	2.7 × 10^{13}	0.1	147
Ca^{2+}	40	0.99	210.5	0.61 ± 0.01	3.3 × 10^{12}	0.05	29
Cd^{2+}	112	0.97	—	0.64 ± 0.01	2.06 × 10^{12}	—	40
Yb^{2+}	173	0.93	215	0.67 ± 0.01	9.1 × 10^{13}	0.1	147
Mn^{2+}	55	0.8	192	0.49 ± 0.01	1.6 × 10^{11}	0.3	28
Co^{2+}	59	0.72	—	0.83 ± 0.06	1.0 × 10^{12}	—	84
			—	0.85 ± 0.06	1.3 × 10^{13}		
Ni^{2+}	58	0.69	186	0.46 ± 0.01	8.6 × 10^{10}	0.3	27
Mg^{2+}	24	0.66	189	0.49 ± 0.03	2.6 × 10^{11}	0.3	27
Be^{2+}	9	0.35	161	0.45	2.0 × 10^{12}	0.1	32
			133	0.24	1.0 × 10^8		

K which has an activation energy of 0.79 eV and a single relaxation time. It is significant that four other peaks, undetectable by any other method, are observed at 178 K (0.60 eV), 168 K (0.56 eV), 153 K (0.51 eV), and 135 K (0.46 eV). These bands were always present in nominally pure crystals and were probably caused by small amounts of different divalent impurities at levels in the 10^{16}/cm^3 range (34). After polarization at higher temperatures, broad peaks attributed to space charge relaxation were seen (35). In a later paper (34), the activation energy of the dipole relaxation was given as 0.66 eV. More recently (81), two peaks have been observed after different thermal treatments. The one peak with an activation energy to 0.74 eV is seen only in an annealed and quenched solid. If the annealling was extended, two relaxation processes with activation energies of 0.68 eV and 0.74 eV were seen. The 0.68-eV process was related to an M^{2+} cation vacancy complex and the 0.74-eV peak to dimers of the M^{2+} cation vacancy complex.

Impurities other than strontium chloride have been studied in potassium chloride (27,28,29,32,40,84,147). Table 8.VI lists the impurities, the peak temperature, the activation energy, and preexponential terms. The relationship between activation energy and ion radius is shown in Fig. 8.25. Analysis (27) leads to the conclusion that the most favored vacancy site is the next nearest neighbor site and that this site becomes increasingly stable as the ionic radius of the impurity ion increases. However, in the case of beryllium, the smallest ion, the nearest-neighbor site predominates.

Not only have divalent cation impurities been studied in alkali halides but also divalent anions. For the case of potassium iodide doped with S^{2-} (118),

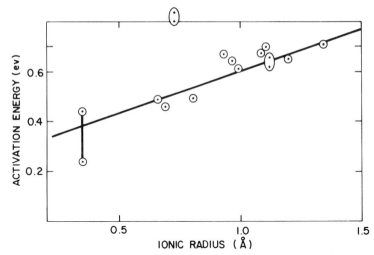

Fig. 8.25. The relationship between ionic radius and activation energy of reorientation of divalent impurities in KCl (27).

a single peak at 204 K was observed with an activation energy of 0.61 eV. On irradiation at $-30°C$, a photochemical reaction, in which the S^{2-} impurity decomposes into S^- and an F center (an electron trapped in an anion vacancy), results in a decreased peak current. On the other hand, irradiation under potential at temperatures where the dipole is already frozen in results in a higher peak current attribute to realignment of the dipoles.

VII. PHASE CHANGES AND CHEMICAL REACTIONS

The study of electrical resistivity or dielectric constant during a phase transformation has been an on-going topic of research, particularly when combined with other techniques such as differential thermal analysis. It should be noted that this type of measurement does not involve any prepolarization. The current passing through the sample is measured as the temperature is continuously changed. At the phase change or as reaction occurs, both the number and mobility of the charge carrier(s) may change. This is reflected in the measured current.

For example Fig. 8.26 shows the well-defined peaks observed at phase transitions in crystalline barium titanate and niobate (61). The peak may be either an increase in resistivity as with $BaTiO_3$ or an increase in conductivity as in $KNbO_3$. The peak is not observed in mixed $KNbO_3/KTaO_3$, but changes in the resistivity/temperature profile (Fig. 8.26c) indicate the onset of a phase transition. Such measurements have been used to construct phase

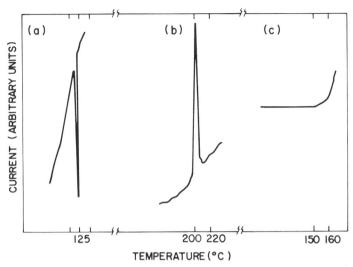

Fig. 8.26. Current changes during phase transformations (61) (a) $BaTiO_3$; (b) $KNbO_3$ single crystal; (c) $KNb_{0.9}Ta_{0.1}O_3$.

diagrams (62) of systems such as the $KNbO_3/KTaO_3$ system where, for example, differential thermal analysis is insufficiently sensitive to give clearly defined transitions.

Dielectric measurements using ac bridge techniques have also been used to study phase transitions particularly with high dielectric constant materials. However, these measurements fall beyond the scope of this article.

Studies have also been made of phase transitions such as amorphous-crystalline transformations in semiconductors and semiconductor–metal transitions. An example of each is given.

The crystallization of amorphous semiconductors and metals is of interest in switching and other electronic devices. The process can be initiated by various energy sources such as radiation, electric field, heat transfer, and electron beam. However, the mechanism of crystallization and its influence on electronic behavior is still incompletely explained. For example, the crystallization of amorphous Cd_3As_2 films was studied by a variety of techniques including electrical resistivity (107). Fig. 8.27 shows the changes in electrical resistivity as a function of temperature (Fig. 8.27a) and time at temperature (Fig. 8.27b). From these and other measurements, it was concluded that a series of transformations occurs; namely, amorphous phase, a′ → intermediate crystalline phase, c′ → a second metastable amorphous film, a″ → stable crystalline film c.

Studies of the microstructure and kinetics allowed the authors to conclude that the initial amorphous phase consisted of clusters of atoms in which the transition to a crystalline phase was initiated. The resistivity

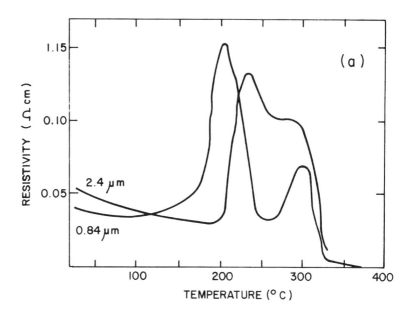

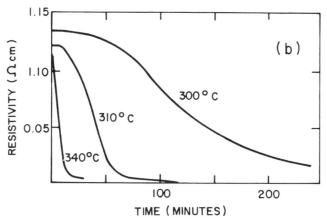

Fig. 8.27. Resistivity of amorphous Cd_3As_2 films (102). (a) As a function of continuously changing temperature; (b) isothermal measurements at a series of temperatures.

changes were then attributed to a sequence of reactions in which the formation of the crystalline clusters from amorphous clusters led to the first minimum in the resistivity-temperature curve. The second maximum which was observed in both amorphous and crystalline films was attributed to the removal of crystalline defects with heat treatment.

The sintering of metal powders has also been studied by a variety of thermal methods (72).

Typical of the use of electrical resistivity and cross-checking experiments is the investigation of the semiconductor-metal transition in doped vanadium oxide (26). Plots of resistivity, thermoelectric power and magnetic susceptibility are shown in Fig. 8.28. In addition, X-ray investigations showed that phases MI (monoclinic), T (Triclinic), M2 (Monoclinic), and (R) metallic Rutile (tetragonal) exist. The transitions T-M2 and M2-R are first-order, can be determined readily electronically (Fig. 8.28a), and are also seen in the dependence of thermoelectric power (Fig. 8-28b) and magnetic susceptibility (Fig. 8.28c) on temperature (72).

In Fig. 8.29 is shown the simultaneous electrothermal current curve and the water adsorption curve for a synthetic zeolite (Union Carbide 13) (85). The increase in weight and the increase in current around 5°C was associated with adsorption and desorption on the wall of the pyrex apparatus. Since adsorbed water freezes at a much lower temperature than pure water and its vapor pressure is higher than that of ice, it was suggested that, on warming, water desorbed from the walls on the sample. The large increase in current above 200°C was caused by Na^+ ions migrating within the zeolite.

Another application of electrical analysis has been to the detection of quadruple points in metal salt-hydrate systems (154). The dehydration of hydrated copper sulfate shows three endothermic peaks in the DTA curve (17). The electrothermal curve shows only one peak which corresponds to the first DTA peak (Fig. 8.30). It was considered that the electrothermal peak was caused by the liberation of liquid water which was followed by rapid evaporation. No electrical peaks were noted for the other DTA curves. Thus, it was concluded that the follow reactions occurred.

$$CuSO_4 \cdot 5\ H_2O(s) \rightarrow CuSO_4 \cdot 3\ H_2O + H_2O(l)\text{–Peak I}$$
$$H_2O(l) \rightarrow H_2O(g) \qquad\qquad\qquad\text{–Peak II}$$
$$CuSO_4 \cdot 3\ H_2O \rightarrow CuSO_4 \cdot H_2O + H_2O(g) \quad \text{–Peak III}$$

In addition, since the first two reactions are in equilibrium, a liquid and gaseous phase coexist in the presence of two metal salt hydrates. This is a quadruple point, i.e., a temperature at which four phases exist. Quadruple points were also postulated in the $BaCl_2 \cdot 2\ H_2O$ and $BaBr_2 \cdot H_2O$ systems.

There have been several applications of the electrothermal technique to chemical reactions (as opposed to phase changes) in recent years. Within this category one might also include the catalyst–gas studies and dehydration experiments previously discussed as well the polymer decompositions.

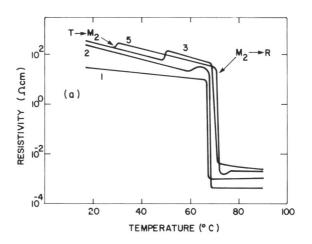

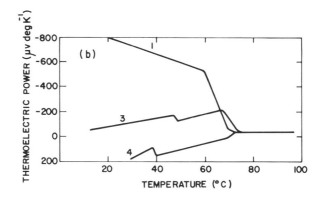

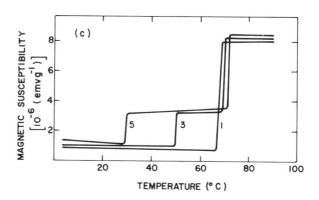

Fig. 8.28. Temperature behavior of gallium-doped vanadium dioxide, $V_{1-x}Ga_xO_2$(26). (a) Electrical resistivity, (b) thermoelectric power, (c) magnetic susceptibility. (1) $x = 0$, (2) $x = 0.0019$, (3) $x = 0.0039$, (4) $x = 0.0078$, (5) $x = 0.0099$.

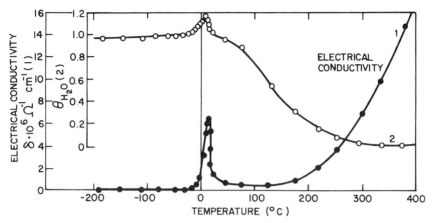

Fig. 8.29. Thermogravimetric analysis and electrical conductivity versus temperature for system water-zeolite 13X. Curve 1, electrical conductivity; Curve 2, TG curve (85).

In addition to these, we should list several papers by Burmistrova and co-workers (10,11,37,38) which deal with simultaneous DTA and ETA (10) of salt mixtures (37,38) and reactions in mixtures of salts (11). Simultaneous DTA, ETA, and evolved gas detection were also used to study the thermal behavior of various metals sulphides (12). Good correlations between ETA and DTA were obtained for the decomposition of ammonium metavanadate,

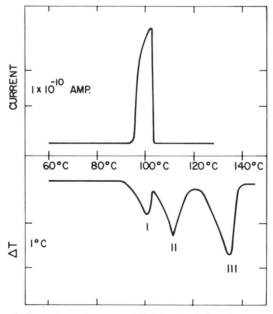

Fig. 8.30. DTA and electrothermal curves of $CuSO_4$, $5H_2O$ [heating rate 2–5° mm^{-1}] (154).

including evidence for the formation of anhydrous ammonium divanadate (146).

VIII. SUMMARY

The electrothermal analytical technique is capable, in many cases, of detecting directly or more sensitively small changes that can only be detected with difficulty using other more conventional techniques. These changes are related to factors that change the mobility and number of the charge carriers or are related to the internal polarization of the sample. It is of particular value in studying complex molecular relaxations, since no other technique allows the relaxation process to be probed in such detail. The technique has broad applicability, but, in general, other cross-correlating experiments are required before the causes can be unambiguously assigned.

REFERENCES

1. Alexandrovich, P., F. E. Karasz, and W. J. MacKnight, *J. Applied Phys., 47,* 4251 (1976).
2. Andriesh, A. M., and B. T. Kolomiets, *Bull. Akad. Sci. USSR. Phys Ser., 28,* 1193 (1964).
3. Aoki, Y., and J. O. Brittain, *J. Polymer Sci. (Physics), 14,* 1297 (1976).
4. Aoki, Y., and J. O. Brittain, *J. Polymer Sci. (Physics), 15,* 199 (1977).
5. Arkedeva, E. N., and S. M. Rivkin, *Soviet Physics (Solid State), 1,* 1264 (1959).
6. Baird, M. E., G. T. Goldsworthy, and C. J. Creasey, *J. Polymer Sci., B6,* 737 (1968).
7. Bakirov, M. Ya., and N. Z. Dzalilov, *Soviet Physics (Solid State), 9,* 968 (1967).
8. Bauer, E., *Cah. Phys., 20,* 1, (1944).
9. Baxt, L. M., and M. M. Perlman, Eds., *Electrets and Related Electrostatic Charge Storage Phenomena,* The Electrochemical Society Inc., 1968.
10. Berg, L. G., and N. P. Burmistrova, *Russ. J. Inorganic Chem., 5,* 326 (1960).
11. Berg, L. G., N. P. Burmistrova, and N. I. Lisov, *J. Thermal Anal., 7,* 111 (1975).
12. Berg, L. G., and N. I. Shlyapkina, *J. Thermal Anal., 8,* 417 (1975).
13. Bergman, J. G., Jr., J. H. McFee, and G. R. Crane, *App. Phys. Lett., 18,* 203 (1971).
14. Berkes, J. S., S. W. Ing, and W. J. Hillegas, *J. App. Phys., 42,* 4908 (1971).
15. Blumenthal, R. N., and M. A. Seitz, in N. M. Tllen, Ed., *Electrical Conductivity in Ceramics and Glass,* Marcel Dekker, New York, 1974, p. 35.
16. Böhm, M., and A. Scharmann, *Phys. Stat. Sol.,* (a) *22,* 143 (1974).
17. Borchart, H. J., and F. Daniels, *J. Phys. Chem., 61,* 917 (1974).
18. Botila, T., and H. K. Henisch, *Phys. Stat. Sol.,* (a) *38,* 331 (1976).
19. Bowlt, C., *J. Phys. (D), 6,* 616 (1973).
20. Bowlt, C., and D. J. Waggett, *Phys. Med. Biol., 19,* 534 (1974).
21. Bowlt, E., *Contemp. Phys., 17,* 461 (1976).
22. Boyd, R. H., *J. Chem. Phys., 30,* 1276 (1959).
23. Braunlich, P., and A. Scharmann, *Phys. Stat. Sol.,* (a) *18,* 307 (1967).

24. Broadhurst, M. G., *Proceedings of the Piezoelectric and Pyroelectric Symposium Workshop*, U.S. National Bureau of Standards, Publication NBSIR 75-760, 1975.

25. Broadhurst, M. G., and G. T. Davies, *Piezoelectric and Pyroelectric Properties of Electrets*, U.S. National Bureau of Standards, Publication NBSIR 75-787, 1975.

26. Brückner, W., U. Gerlach, W. Moldenhauer, H. P. Brückner, N. Mattern, H. Oppermann, and E. Wolf, *Phys. Stat. Sol.*, (a) **38**, 93 (1977).

27. Brun, A., and P. Dansas, *J. Phys.* (*c*), **7**, 2593 (1974).

28. Brun, A., P. Dansas, and F. Beniere, *J. Phys. Chem. Sol.*, **35**, 249 (1974).

29. Brun, A., P. Dansas, and P. Sixov, *Sol. State Comm.*, **8**, 613 (1970).

30. Bube, R. H. *Photoconductivity in Solids*, Wiley, New York, 1960.

31. Bube, R. H., and S. M. Thomsen, *J. Chem. Phys.*, 23, 15 (1955).

32. Bucci, C., *Phys. Rev.*, **164**, 1200 (1967).

33. Bucci, C., and R. Fieschi, *Phys. Rev. Lett.*, **12**, 16 (1964).

34. Bucci, C., R. Fieschi, and R. Guidi, *Phys. Rev.*, **148**, 876 (1966).

35. Bucci, C., and S. Riva, *J. Phys. Chem. Sol.*, **26**, 363 (1965).

36. Bunn, C. W., in A. Renfrew and P. W. Morgan, Eds., *Polythene*, Iliff Press, London, 1960, p. 87.

37. Burmistrova, N. P., and R. G. Fitzeva, *J. Thermal Anal.*, **4**, 161 (1972).

38. Burmistrova, N. P., E. G. Volozhanina, and N. V. Dubkova, *J. Thermal Anal.*, **4**, 323 (1972).

39. Cessac, G. L., and J. G. Curro, *J. Polymer Sci.* (*Physics*), **12**, 695 (1974).

40. Chaney, R. E., and W. J. Fredericks, *J. Sol. State Chem.*, **6**, 240 (1973).

41. Chatain, D., P. Gautier, and C. Lacabanne, *J. Polymer Sci.* (*Physics*), **11**, 1631 (1973).

42. Chatain, D., C. Lacabanne, and M. Maitrot, *Phys. Stat. Sol.*, (a) **13**, 303 (1972).

43. Chen, R., *J. Appl. Phys.*, **42**, 5899 (1971).

44. Chiu, J., *Anal. Chem.*, **39**, 861 (1967).

45. Chiu, J., *Thermochimica Acta*, **8**, 15 (1974).

46. Chudleigh, P. W., R. E. Collins, and G. D. Hancock, *Appl. Phys. Lett.*, **23**, 211 (1973).

47. Conley, R. T., and J. F. Bieron, *J. Appl. Polymer Sci.*, **7**, 1757 (1963).

48. Cortilli, G., and G. Zerbi, *Spectrochim. Acta*, **23A**, 285 (1967).

49. Cresswell, R. A., and M. M. Perlman, *J. Appl. Phys.*, **41**, 2365 (1970).

50. Cresswell, R. A., M. M. Perlman, and M. Katayama, in F. Karasz, Ed., *Dielectric Properties of Polymers* Plenum, New York, 1975, p. 215.

51. Davies, D. K., and P. J. Locke, in M. M. Perlman, Ed., *Electrets, Charge Transfer and Storage in Dielectrics*, Electrochemical Soc., Princeton, N.J., 1973.

52. El-Azab, M. I., and C. H. Champness, *Appl. Phys. Lett.*, **31**, 295 (1977).

53. Eley, D. D., and D. I. Spivey, *Trans. Farad. Soc.*, **57**, 2280 (1961).

54. Eyring, H., *J. Chem. Phys.*, **4**, 283 (1936).

55. Fagan, E. H., and H. Fritzsche, *J. Non-Cryst. Sol.*, **4**, 480 (1970).

56. Fowler, J. F., *Proc. Royal Soc.*, **A236**, 464 (1956).

57. Fukada, E., and S. Takashita, *Jpn. J. Appl Phys.*, **8**, 960 (1969).

58. Fullerton, G. D., and P. R. Morgan, *Med. Phys.*, **1**, 161 (1974).

59. Garlick, G. F., *Luminescent Materials*. Oxford University Press, 1960.

60. Garlick, G. F., and A. F. Gibson, *Proc. Phys. Soc.*, **60**, 574 (1948).

61. Garn, P. D., and S. S. Flaschen, *Anal. Chem.*, **29**, 268 (1957).

62. Garn, P. D., and S. S. Flaschen, *Anal. Chem.*, **29**, 271 (1957).

63. Golden, J. H., B. L. Hamment, and E. A. Hazell, *J. Appl. Polymer Sci.*, **11**, 1571 (1967).

64. Grassie, N., and J. N. Hay, *J. Polymer Sci.*, **56**, 189 (1962).

65. Gray, T. J., *Actes 2nd Congress Int. Catalyse, Paris,* Editions Technip, Paris, 1961, p. 1561.

66. Gray, T. J., in R. Anderson, Ed., *Experimental Methods in Catalysis,* Vol. 2, 1969.

67. Gray, T. J., *J. Can Ceram. Soc.*, **38**, 103 (1969).

68. Gray, T. J., and P. Amigues, *Surface Sci.*, **13**, 103 (1969).

69. Grinter, M., and C. Bowlt, *J. Phys. (D)*, **8**, L159 (1975).

70. Gross, B., G. M. Sessler, and J. E. West, *J. Appl. Phys.*, **47**, 968 (1976).

71. Guillaud, A., J. Fornazero, M. Maitrot, D. Chatain, and C. Lacabanne, *J. Appl. Phys.*, **48**, 3428 (1977).

72. Guillaume, P., J. Senevat, A. DeFresne, and P. Gilles, *J. Thermal Anal.*, **7**, 317 (1975).

73. Haering, R. R., and E. N. Adams, *Phys. Rev.*, **117**, 451 (1960).

74. Harper, M. W., and B. Thomas, *Phys. Med. Biol.*, **18**, 409 (1973).

75. Hedvig, P., and Czuikouski, *Angew. Makromol. Chem.*, **27**, 79 (1972).

76. Hedvig, P., and M. Kisbenyi, *Angew. Makromol. Chem.*, **7**, 198 (1969).

77. Heiland, G., E. Molvo, and F. Stockman, *Electronic Processes in Zinc Oxide, Solid State Physics,* Vol 8, Academic, New York, 1959.

78. Heijne, L., *Phillips Research Reports,* Suppl. 4 (1956).

79. Henisch, H. K., and M. H. Engineer, *Phys. Lett.*, **A26**, 188 (1968).

80. Hoffman, J. D., G. Williams, and E. Pasaglia, *J. Polymer Sci.*, **C14**, 173, 1966.

81. Hor, A. M., P. W. M. Jacobs, and K. S. Moodie, *Phys. Stat. Sol.*, (a) **38**, 293, 1976.

82. Hughes, R. C., in A. V. Patsis and D. A. Seanor, Eds., *Photoconductivity in Polymers— an Interdisciplinary Approach,* Technomic Press, 1976, Chapter 6.

83. Illers, K. H., and H. Breur, *Kolloid Zeit.*, **176**, 110 (1961).

84. Jain, S. C., K. Lal, and U. Mitra, *J. Physics (c)*, **3**, 2420 (1970).

85. Juranic, N., D. Karaulic, and D. Vucelic, *J. Thermal Anal.*, **7**, 119 (1975).

86. Kambour, R. P., and R. E. Robertson, in A. D. Jenkins, Ed., *Polymer Science,* North Holland Publishing Co., Amsterdam, 1972, p. 720.

87. Kawai, H., *Jpn. J. Appl. Phys.*, **8**, 975 (1969).

88. Kepler, R. C. *U.S. National Bureau of Standards,* Publication NBSIR 75-760, 46, 1975.

89. Kittel, C., *Introduction to Solid State Physics,* John Wiley, Inc., New York, 1967, Chapters 7–10.

90. Kolomiets, B. T., and P. K. Khodosevich, *Soviet Phys. (Solid State)*, **6**, 2556 (1965).

91. Kolomiets, B. T., and T. F. Mazets, *J. Non-Cryst. Sol.*, **3**, 46 (1970).

92. Kryszewski, M., and A. Szysmanski, *Macromol. Reviews*, **4**, 245 (1970).

93. Lacabanne, C., D. Chatain, J. Guillet, G. Seytre, and J. F. May, *J. Polymer Sci. (Physics)*, **13**, 445 (1975).

94. Lakatos, A. I., and M. Abkowitz, *Phys. Rev.*, **B3**, 1791 (1971).

95. LeGrand, D. G., and P. F. Ehrhardt, *J. Appl. Polymer Sci.*, **13**, 1707 (1969).

96. Locati, G., and A. B. Tobolski, *Adv. Mol. Relax. Proc.*, **1**, (1970).

97. Luongo, J. P., *J. Polymer Sci. (A2)*, **10**, 1119 (1972).

98. MacKenzie, R. C., in P. J. Elving, I. M. Koltoff, and C. Murphy, Eds. *Treatise on Analytical Chemistry,* Wiley-Interscience, New York, Vol. III, p.

99. MacKnight, W. J., in K. Frisch and A. V. Patsis, Eds., *Electrical Properties of Polymers,* Technomic Press, Westport, 1972, p. 85.

100. May, C. A., and Y. Tanaka, *Epoxy Resins*, Marcel Dekker, New York, 1973, p. 339, 688.

101. McCrum, N. G., B. E. Read, and A. Williams, *Anelastic and Dielectric Effects in Polymeric Solids*, John Wiley, London, 1967.

102. McFee, J. H., J. Bergman, and A. R. Crane, *Ferroelectrics*, **3**, 305 (1972).

103. Minami, T., A. Yoshida, and M. Tanuka, *J. Non-Cryst. Solids*, **7**, 328 (1972).

104. Miyamoto, T., and T. Sugano, *Polymer J.* **6**, 451 (1974).

105. Mott, N. F., and R. W. Gurney, *Electronic Processes in Ionic Crystals*, Oxford University Press, 1940.

106. Nakamura, K., and Y. Wada, *J. Polymer Sci.*, **A29**, 161 (1971).

107. Niedzwiedz, M., and L. Zdanowicz, *J. Non-Cryst. Solids*, **23**, 167 (1977).

108. Norman, R. H., *Conductive Rubbers and Plastics*, Elsevier Publishing Company, 1970.

109. Ong, P. H., and J. van Turnhout, in M. M. Perlman, Ed., *Electrets, Charge Storage and Transport in Dielectrics*, The Electrochemical Soc., Princeton, N.J., 1973, p. 214.

110. Perlman, M. M., *J. Appl. Phys.*, **42**, 2685 (1971).

111. Perlman, M. M., Ed. *Electrets, Charge Storage and Transport in Dielectrics*. The Electrochemical Soc., Princeton, N.J., 1973.

112. Pfister, G., M. Abkowitz, and R. G. Crystal, *J. Appl. Phys.*, **44**, 2064 (1973).

113. Pfister, G., W. M. Prest, D. J. Luca, and M. Abkowitz, *Appl. Phys. Lett.*, **27**, 486 (1975).

114. Phelan, R. J., Jr., R. L. Peterson, C. A. Hamilton, G. W. Day, and L. O. Mullen, *Ferroelectrics*, **7**, 375 (1974).

115. Pillai, P. K. C., R. Nath, and P. K. Nair, *Phys. Lett.*, **58A**, 474 (1976).

116. Pope, M. I., *Polymer*, **8**, 49 (1967).

117. Pope, M. I., *Proc 2nd Conf. on Carbon and Graphite*, Soc. Chem. Ind., London, 1965, p. 474.

118. Prakush, J., and F. Fischer, *Phys. Stat. Sol.*, (a) **39**, 499 (1977).

119. Prest, W. J., and D. J. Luca, *Soc. Plast. Engineers, 35th Annual Tech. Conf.*, 376, 1977.

120. Prevorsek, D. C., R. H. Butler, and H. K. Reimschussel, *J. Polymer Sci. (A2)*, **9**, 867 (1971).

121. Randall, J. T., and M. H. F. Wilkins, *Proc. Royal Soc.*, **A181**, 366 (1945).

122. Reucroft, P. J., H. Scott, and F. L. Serafin, *J. Polymer Sci.*, **C30**, 261 (1970).

123. Roberts, A. A., *J. Phys. (c)*, **4**, 1348 (1971).

124. Sacher, E., *J. Polymer Sci. (A2)*, **10**, 1179 (1972).

125. Saleh Menshadi, M. A., and J. Woods, *Phys. Stat. Sol.*, (a) **40**, K43 (1977).

126. Schatzki, T. F., *Polymer Preprints*, American Chemical Society, Polymer Division, 6, 646, 1965.

127. Seanor, D. A., Unpublished.

128. Seanor, D. A., *J. Polymer Sci.*, **C17**, 195 (1967).

129. Seanor, D. A., in P. H. Slade, Jr., and L. Jenkins, Eds. *Thermal Methods of Polymer Analysis*, Vol. 2, 293, 1969.

130. Servini, A., and A. K. Jonscher, *Thin Solid Films*, **3**, 341 (1969).

131. Sessler, G. M., and J. E. West. *J. Acoust. Soc. Am.*, **63**, 1589 (1973).

132. Sessler, G. M., and J. E. West, *Phys. Rev.*, **B10**, 4488 (1974).

133. Shuford, R. J., A. F. Wilde, J. J. Ricca, and G. R. Thomas, *Polymer Engineering and Science*, **16**, (1976).

134. Solunov, Ch. A., and Ch. S. Ponevsky, *J. Polymer Sci. (Physics)*, **15**, 969 (1977).

135. Southgate, P. D., *Appl. Phys. Lett.*, **28**, 250 (1976).

136. Street, R. A., and A. D. Yoffe, *Thin Solid Films*, **11**, 161 (1972).

137. Stuke, J., *Phys. Stat. Sol.*, (a) **6**, 441 (1964).

138. Stupp, S. I., and S. H. Carr, *J. Appl. Phys.*, **46**, 4120 (1975).

139. Stupp, S. I., and S. H. Carr, *J. Polymer Sci.* (*Physics*), **15**, 485 (1977).

140. Stupp, S. I., R. J. Comstock, and S. H. Carr, *J. Macromol. Sci.* (*Physics*), **B13**, 101 (1977).

141. Tabak, M., *Phys. Rev.*, **B2**, 2104 (1970).

142. Takamatsu, T., and E. Fukada, *Jpn. Polymer J.* **1**, 101 (1970).

143. Tanaka, T., S. Hayashi, and K. Shibayama, *J. Appl. Phys.*, **48**, 3478 (1977).

144. Thurso, I., D. Barancok, E. Mariani, and J. Janci, *J. Non-Cryst. Sol.*, **18**, 129 (1975).

145. Thurso, I., and J. Doupovec, *J. Non-Cryst. Sol.*, **22**, 205 (1976).

146. Trau, J., *J. Thermal Anal.*, **6**, 355 (1974).

147. Unger, S., and M. M. Perlman, *Phys. Rev.*, **B6**, 3973 (1972).

148. Vanderschoeren, J., *J. Polymer Sci.* (*Physics*), **15**, 873 (1977).

149. Van Turnhout, J., *Polymer J.*, **2**, 173 (1971).

150. Van Turnhout, J., *Thermally Stimulated Discharge of Polymer Electrets*, Elsevier, Amsterdam, p. 475.

151. Warfield, R. W., *Soc. Plast. Engineers J.*, **14**, 39 (1958).

152. Warfield, R. W., *Soc. Plast. Engineers J.*, **17**, 364 (1961).

153. Warfield, R. W., and M. C. Petrie, *Macromol. Chem.*, **58**, 139 (1962).

154. Wendtland, W. W., *Thermochimica Acta*, **1**, 11 (1970).

155. Wiseal, B., and J. E. Willard, *J. Chem. Phys.*, **45**, 4387 (1967).

156. Woodard, J. B., *I.E.E.E.—14th Annual Proceeding—Reliability Physics*, 234, 1976.

157. Woodard, J. B., *J. Electron. Mat.*, **6**, 145 (1977).

158. Yalof, S., and W. Wrasidlo, *J. Appl. Polymer Sci.*, **16**, 2159 (1972).

159. Yamafugi, K., and Y. Ishida, *Koll. Zeit.*, **221**, 63 (1967).

160. Yu, L. T., *J. Phys.*, **24**, 677 (1963).

Chapter 9

THERMOACOUSTIMETRY

By Pronoy K. Chatterjee, *Personal Products Co.*,
A Johnson & Johnson Co., *Milltown, New Jersey*

Contents

I. INTRODUCTION

The characterization of materials by acoustical techniques is an old art. Centuries ago people learned to differentiate dissimilar metals by striking them with another object and listening to the characteristic frequency and amplitude of the resulting sound waves. The technique, however, emerged as a scientific tool when the subjective listening was changed into the objective measurements by modern instrumentations. In an acoustical technique for characterization of materials, the sound waves of known frequencies and amplitudes are transmitted to the test material at different environmental conditions, and the characteristic changes of these waves are observed.

Thermoacoustimetry, a relatively new technique, is defined as follows: "A technique in which characteristics of imposed acoustic waves are measured as a function of temperature after passing through a substance whilst the substance is subjected to a controlled temperature programme." There is another thermal acoustic technique which is known as thermosonimetry. Thermoacoustimetry and thermosonimetry are two different techniques involving different principles and theories. In the former case, the material characterization is done by analyzing the imposed sound waves whereas in the latter case the characterization is done by analyzing the spontaneously emitted sound from the material.

The present chapter describes the thermoacoustimetry principle for the determination of chemical relaxations and transitions in solids, primarily in polymers. A comprehensive description on chemical relaxation phenomena and transitions in solids and their influence on transmitted acoustical waves is beyond the scope of this chapter. Therefore, the theoretical description is limited to a few fundamental aspects only to introduce the reader to the basic principles of thermoacoustimetry in different analytical areas.

Since this chapter is dealing with sound waves, a few notes on the fundamentals of sound waves are pertinent. Sound waves are generally audible to human ear if the vibration frequency is in a range defined as 20–20,000 Hz (cycles/sec). The frequency range below 20 Hz is the infrasonic and that above 20,000 is the ultrasonic. In thermoacoustimetry, the selection of frequency which ranges from a few hundred hertz (Hz) to several thousand megahertz is dependent on the particular phenomena to be investigated. For the investigation of chemical relaxations, a high frequency, i.e., ultrasonic frequency, is necessary whereas in the case of the polymer characterization a high frequency ultrasonic wave is not desirable because it may initiate the chain scission of the polymer molecules.

Whatever might be the frequency, a sound wave is a periodic disturbance in a material in which the molecules in certain regions are momentarily displaced from their equilibrium position and experience a restoring force due to the elasticity of the medium. This restoring force is responsible for the propagation of the disturbance wave in the form of an oscillation of the molecules about their mean positions and its magnitude influences the velocity with which the wave is propagated. Sound waves in fluids (gases and liquids) are only compressional or longitudinal, and the molecule displacements and restoring forces are along the direction of the propagation of the wave. In solids, other types of sound waves are possible, such as shear waves or torsional waves. The propagation of the wave involves a continual interchange between the translational kinetic energy and the potential energy of the molecules, and also between the translational energy and intramolecular energy such as vibrational and rotational, if the time period of temperature cycling is long enough to permit equilibrium. As in any wave, there is a simple relationship, $f\lambda \propto c$ between the frequency f, the wavelength λ, and the velocity c of the sound wave.

II. THEORY AND PRINCIPLE

Thermoacoustimetry deals with the determination of the physicochemical properties of materials subjected to high frequency (sonic or ultrasonic) stresses at varying temperatures. Through the determination of those properties, one can study the chemical relaxations, physical transitions, or chemical transformations of materials.

A. CHEMICAL RELAXATION (1–4)

The application of ultrasonic waves of different frequencies (20 kHz–1 GHz) to the investigation of molecules and their interactions is known as ultrasonic spectrometry. This technique has been employed to study fast chemical interactions having relaxation times of the order of 10^{-10} sec. A compound may undergo several physical or chemical reactions with different time constants that will influence the absorption of the sonic energy. The equilibrium of one type of reaction may be perturbed by a certain frequency range whereas the equilibrium of another type may be perturbed by another frequency range. As a result of this perturbation of the equilibrium condition, the velocity and the energy of the sound waves will be affected. A study of sonic properties as a function of frequency determines the time constants for reactions. If this study includes the variation of environmental temperatures in a controlled manner, the technique is called thermoacoustimetry.

To further illustrate this principle, when a longitudinal high-frequency sound wave passes through a liquid, it creates local pressure and temperature variations in the liquid. These variations shift at a rate that is proportional to the sound frequency. The sound frequencies are extremely high compared with the heat flow; therefore, there will be no heat flow from the compressed part of the liquid to its surroundings. Consequently, the compression and decompression take place adiabatically. The local temperature and pressure variations affect a given chemical equilibrium in the system. The pressure, P, and the temperature, T, dependence of the equilibrium constant, K, is given by the following equation:

$$dK = K\{\Delta H^\circ / RT^2)_P\ dT - (\Delta V^\circ / RT)_T\}dP \tag{1}$$

where ΔH° and ΔV° are the reaction enthalpy and reaction volume respectively.

A necessary condition for the equilibrium constant to change with pressure and temperature is that the coefficient K on the right-hand side of equation 1 not be zero. However, this condition alone is not sufficient to affect the absorption of the propagating sound wave. The actual magnitude of the equilibrium constant is also important for the detection of the change of the sound amplitude.

Let us assume an equilibrium system between two different conformational states of a molecule in a volume element of a liquid through which the sound wave propagates. When the sound wave propagates through the system, the volume element oscillates between the compressed and expanded situation resulting in a shifting of the chemical equilibrium in one direction in the compressed stage and in the other direction in the expanded stage. At sufficiently low oscillation frequencies, the energy transferred to the chemical equilibrium is returned to the sound wave and consequently the sound wave does not register the energy shift as taking place. By increasing the oscillating frequency, a frequency region is reached in which the chemical equilibrium is not able to adjust itself fast enough to return the transferred energy, thus the sound wave registers as a loss of energy. The frequency at which this happens obviously is related to the rates involved in the chemical reaction. Thus, at a given chemical equilibrium, the sound absorption coefficient increases in a limited frequency range only and this frequency range contains information about the rates of the chemical reaction.

The phenomenon described here is called a chemical relaxation. The frequency at which the ratio of the absorption coefficient to the sound frequency attains a maximum value is called the relaxation frequency.

If this acoustical technique is carried out for a reaction system which consists of a single elementary reaction step of the type

$$a_1A_1 + a_2A_2 + \ldots a_nA_n \underset{k_{-1}}{\overset{k_1}{\rightleftharpoons}} b_1B_1 + b_2B_2 + \ldots b_nB_n \qquad (2)$$

the following relaxation equation is obtained

$$\frac{\alpha}{f^2} = \frac{A}{1 + (\omega T)^2} + B \qquad (3)$$

where A is the sound amplitude, and α is the absorption coefficient of the sound amplitude, f is the frequency (ω is $2\pi f$), B is a constant, defined as α/f^2 for the solvent and T is the relaxation time which can be expressed as follows:

$$\frac{1}{T} = k_1 \chi \sum_{i=1}^{i=n} [A_i]^\circ a_i \qquad (4)$$

where

$$\chi = \sum_{i=1}^{i=n} (a_i/[A_i]^\circ + b_i^2/[B_i]^\circ) \qquad (5)$$

If n different elementary reaction steps contribute to the relaxation, the relaxation equation can be written as

$$\frac{\alpha}{f^2} = \sum_{i=i}^{i=n} \frac{A_i}{1 + (\omega T_i)^2} + B_i \tag{6}$$

This experiment results in values of α/f^2 at certain frequencies for different solute concentrations and temperatures. The analysis of the data involves finding the minimum value of i required for equation 6 to describe the experimental data within the experimental error. The analysis thus results in values of i, T_i, A_i, and B_i. To make such an analysis, it is necessary to define a criterion that decides whether or not a certain value of i makes equation 6 describe the experimental data. This criterion must depend on the experimental error in α/f^2, the actual number of measured points on the relaxation spectrum, the frequency range covered, the location of the relaxation frequency in this range, and the size of the amplitudes involved.

The acoustical measurements described above do not give information about the type of phenomena that actually cause the observed relaxation. The first step in such an interpretation is to formulate all possible phenomena that may give rise to the observed relaxation. Theoretical consideration and results obtained by means of this and other experimental techniques very often give rise to several plausible mechanisms.

B. VISCOELASTIC PROPERTIES AND FINE STRUCTURES

The predominant characteristic of polymer molecules is the thread-like character (45) caused by the joining of a large number of repeating elements by carbon bonds. The properties change materially with the degree of cross-linking between the chains, and on the formation of closely packed segments of the neighboring chains to form crystals. The thread-like character allows the molecule to take up a large number of forms which continually change due to thermal agitation. The mode or the degree of change of the molecular forms of the polymer is also influenced by an externally applied stress or heat.

There is a wide range of frequencies over which the properties of polymers are similar to a rubber where only segments of chains can undergo motion. In the case of a very low frequency stress, i.e., a long time application of the stress, the whole polymer chain can eventually move. But, in the case of very high-frequency stress, i.e., a very short time application of a force, even the shortest complete segments of the polymer can no longer respond in the time of a cycle.

Therefore, the motion of selected segments of polymer chains will depend upon the application of specific stress frequencies. The stress, with fixed or variable frequencies, can be applied by transmitting plane acoustic waves through a polymetric material. The quantities measured are the attenuation and velocity of wave propagation or resonance frequency. The desired quantities for a theoretical interpretation, are the real and imaginary parts of the stiffness (93). In the case of the thermal acoustical analysis those parameters are determined as a function of temperature.

1. Modulus of Elasticity and Internal Viscosity of Polymers

As suggested by Ballou and Smith (11), the dynamic modulus of elasticity (E) and the coefficient of internal viscosity (γ) can be obtained according to the following equation derived by Nolle (103)

$$E = \rho C^2 \omega^2 \, (\omega^2 - \alpha^2 C^2)/(\omega^2 + \alpha^2 C^2)^2 \qquad (7)$$

$$\gamma = 2\rho\alpha C \, (\omega C)^2/(\omega^2 + \alpha^2 C^2)^2 \qquad (8)$$

where ρ = density in g/cc, C = velocity of sonic pulse in cm/sec, α = attenuation in nepers per cm, and ω = angular frequency in rad/sec. For the case of small damping (11) these expressions reduce to

$$E = \rho C^2 \qquad (9)$$

$$\gamma = 2\rho\alpha C^3/\omega^2 \qquad (10)$$

In addition, the relaxation time is given by

$$T = \gamma/E \qquad (11)$$

The above equations have been widely used to characterize different polymers. Also, equation 9 had been the basis of the development of a number of fiber characterization techniques (33,23,24,28).

2. Molecular Orientation

If a sound pulse is sent across an array of parallel molecules, sonic energy is presumed to be transmitted from one molecule to another by the stretching of intermolecular bonds (98). If sound is sent along the length of a bundle of parallel polymer molecules, the sonic energy is presumed to be transmitted principally by stretching of the chemical bonds in the backbone of the polymer chain. In the case of partially oriented molecules, the molecular motion due to sound transmission is presumed to have perpendicular components along and across the direction of the molecular axis. The magnitude of either of these two components is taken to be a function of the angle between the molecular axis and the direction of sound propagation, θ. Mosley and coworkers (98,21) derived the following relationship between the angle of molecular orientation and sound velocity:

$$\cos^2\theta = 1 - \frac{2C_u^2}{3C^2} \qquad (12)$$

or

$$C^2 = \tfrac{2}{3}\frac{C_u^2}{\sin^2\theta} \qquad (13)$$

where C is the sonic velocity along a fiber axis and C_u is the sonic velocity of the sample compound but unoriented molecular structure.

The Mosley equation has been extensively used to determine the orientation of many crystalline and noncrystalline polymers at different temperatures. However, Grechiskin et al. (56) concluded that the Mosley equation is valid only for amorphous polymers and therefore there was always a disparity between the values obtained from the Mosley equation and that obtained from X-ray measurements (52). It was shown that the velocity of sound measured in polyethylene terephthalate below the glass transition temperature (T_g) was higher in amorphous than in crystallized samples.

3. A General Relationship between Pulse Propagation and Viscoelasticity of Polymers

By combining Nolle's (103) equation for pulse propagation and Tobolosky and Eyring's viscoelastic equation (130) based on Maxwell-Wiechert model of a polymer (141), one can obtain the following equation in the case of a small damping (26):

$$\rho C^2 = \sum_{i=1}^{m} \frac{E_i \omega^2 T_i^2}{1 + \omega^2 T_i^2} \tag{14}$$

where, E_i and T_i are the modulus and relaxation time, respectively, of the i-th element of the model where i ranges from 1 to m.

Since a polymer block contains elements of different T values ranging from very low to very high values, it was assumed (26) that a fraction of elements had T values so large that their relaxation was not responsive to sonic waves of frequency ω, and the rest of the elements had relatively smaller T values and the relaxation of those elements was responsive to sonic waves. By assigning the latter fraction as β and by applying Urick's approximation theory (136) and Moseley's relationship (98) between velocity of sound and molecular orientation, Chatterjee (26,27,25) derived the following equation:

$$\mu = 0.82 \chi \sin \theta \left[\frac{\rho \beta}{\sum_{i=1}^{m\beta} E_i \sin^2 \delta_i} + \frac{\rho (1 - \beta)}{\sum_{j=1}^{m(1-\beta)} E_j} \right]^{1/2} \tag{15}$$

where μ is the pulse propagation time for a fixed length χ of a sample, δ_1 is the phase angle between the stress and the strain and i and j represent the elements of low and high relaxation times, respectively.

For simplicity and practical considerations, the above equation can be written as

$$\mu = 0.82 \chi (\sin\theta) \epsilon_s \tag{16}$$

where ϵ_s is defined as a sonic viscoelastic function of a polymer at a constant frequency ω.

Equation 16 has been used in studying the viscoelastic nature of many synthetic fibers under programmed temperature changes. This technique was termed as dynamic thermoacoustical technique.

4. Influence of Crystallinity on Velocity of Sound

A correlation between the velocity of sound and the degree of crystallinity in a polymer has been proposed by Perepechko (108). The degree of crystallinity is defined as $x = V_1/V$ where V_1 is the volume of the crystalline part of the polymer and V is the total volume of the polymer. The degree of crystallinity is related to the velocity of sound as described below:

$$\frac{1}{\rho C^2} = \frac{x}{E_1} + \frac{(1 - x)}{E_2} \tag{17}$$

or

$$\frac{A}{\rho C^2} = \frac{xA_1}{\rho_1 C_1^2} + \frac{(1 - x)A_2}{\rho_2 C_2^2} \tag{18}$$

where

$$\frac{1}{\rho_i C_i^2} = \frac{1}{E_{iq}} + \int_0^\infty \frac{\phi_i(T)dT}{i + \omega^2 T^2} \tag{19}$$

and

$$A_i = \tfrac{1}{2}\rho_i C_i^2 \int_0^\infty \frac{\phi_{i(T)}\omega T dT}{1 + \omega^2 T^2} \tag{20}$$

The subscript i is either 1 or 2 where 1 and 2 indicate the crystalline and amorphous regions respectively. E_{iq} represents modulus of the spring elements of the generalized Kelvin-Voigt model and A_1 and A_2 represent the amorphous and crystalline volume fractions respectively. It follows from equations 17 and 18 as $x \to 0$, $C \to C_2$ (decreasing velocity of sound), $A \to A_2$; as $x \to 1$ (increasing crystallinity), $C \to C_1$ and $A \to A_1$.

Further, at sufficiently low temperature, below Tg, $T_i \to \infty$ and equation 19 becomes:

$$\frac{1}{\rho_i C_i^2} = \frac{1}{E_{iq}} \tag{21}$$

Therefore, according to Perepechko, below T_g the velocity of sound becomes independent of the frequency and crystallinity. The crystallinity influences the velocity of sound only at temperatures in excess of the glass transition temperature.

Moseley (98) and Ward's (139) theory on orientation and velocity of sound in a single-phase system has been further extended by Samuels (119) to a two-phase system consisting of crystalline and noncrystalline regions. Assuming that certain polymers, such as isotatic polypropylene, consist of an ideal mixture of amorphous and crystalline phases, Samuels (119) derived the following equation:

$$\frac{1}{\rho C_{or}^2} = \frac{1}{E_{or}} = (x/E_{t,1}^0)(1 - \cos^2\theta) + [(1 - x)/E_{t,2}^0](1 - \cos^2\theta_2) \quad (22)$$

where E_t^0 is the intrinsic lateral modulus, θ is the angle between molecular axis and the direction of sound propagation, the subscripts 1 and 2 stand for the crystalline and amorphous regions, respectively, C_{or} and E_{or} represent the velocity and sonic modulus, respectively, of the oriented sample, ρ represents the density of the polymer, x is the fraction of crystalline material and $(1 - x)$ is the fraction of amorphous material.

The above equation is further reduced to the following form:

$$\frac{3}{2(\Delta E^{-1})} = (xf_1/E_{t,1}^0) + [(1 - x)f_2/E_{t,2}^0] \text{ and}$$

$$(\Delta E^{-1}) = (E_u^{-1} - E_{or}^{-1}) \quad (23)$$

where, E_u is the sonic modulus of an unoriented system, and f_1 and f_2 are defined as orientation fractions for the crystalline and amorphous phases, respectively.

III. TECHNIQUE AND INSTRUMENTATION

The basic types (18) of sonic (or ultrasonic, i.e., frequency over 20 KHz) systems used in an acoustical measurement are based on pulse propagation or sonic resonance techniques. The pulse propagation technique can be subdivided into two categories: pulse-through-transmission and pulse-echo system. The pulse-through-transmission system places continuously pulsed or modulated waves through a transducer coupled to one side of the material with a pick-up on the other side. The pulse-echo system uses a radio frequency pulse ranging in length (time) from a fraction of a microsecond to several microseconds and an amplitude from 50–250 V across the transducer. The pulse travels in the material and is reflected by an interface; the time of travel is measured. The resonance system uses a single acoustical transducer and varies the frequency applied to it. The applied frequency at which the sample attains the resonance frequency is measured for various calculations.

In this chapter, the techniques have been broadly classified into two categories: (1) pulse propagation technique and (2) resonance frequency technique.

A. PULSE PROPAGATION TECHNIQUES

1. Liquid Samples

Measurements of the attenuation and velocity of sound are made by converting an electrical signal to a sonic signal of the same frequency and measuring the decrease in amplitude and time of flight of this signal as it passes through the specimen of interest.

To illustrate the principle of a through-transmission system, a continuous sinusoidal electrical signal (approximately 50-V peak) is generated by an oscillator and is passed through the pulse modulator whereby the signal is converted to a pulsed sinusoidal signal of the desired frequency and length, which is then amplified by a power amplifier and applied to a transmitting transducer usually made of quartz or other piezoelectric crystals. The transducer converts the electrical signal to a sound signal of similar wave shape. After passing the transmitted pulsed sonic waves through the test specimen at a selected temperature, the sonic wave is converted to an electrical signal by a receiving transducer similar to the transmitting transducer. This electrical signal is then amplified and displayed in an oscilloscope or recorded by an appropriate device.

For chemical relaxation spectrometry, the pulse-echo technique has been widely used for measuring the sound absorption coefficient in the frequency range 10–800 MHz. The technique was originally developed by Pellam and Galt (106), Teetar (127) and Pinkerton (109). Briefly, the procedure is as follows (106): A quartz crystal transducer radiates a short train of waves (periodically repeated) into the liquid and later serves as a receiver for the waves returned from a plane parallel reflector, which can be moved with respect to transducer, as shown in Fig. 9.1. Loss is determined by the decrease of the received pulse amplitude as the distance traveled by the waves is increased. Velocity of propagation is determined by a direct measurement of the transit time for a given pulse. The temperature equilibrium is maintained by balancing the effects of dry ice in the bath against the warming action of a thermostatically controlled electric heater. A water pumping system improves the temperature uniformity outside the measuring tank and a motor-driven stirrer maintains constant temperature within the liquid sample itself.

The above method can yield accuracies to within a few percent for attenuation and few parts in 10^4 for velocities. However, there are other arrangements that have been used. For example, Rapuano (115,116) used a short rod of fused silica to delay one of the received waves with respect to the signal applied to the quartz transducer as shown in Fig. 9.2b. This modification made it possible to make measurements with small liquid path lengths, since there was ample time for the receiving amplifier to recover from the high overload voltage of the applied pulse. Attenuation measured over a frequency range extending to about 300 MHz were reported for a number of liquids and to 520 MHz for water. Heasell and Lamb (66) used two fused silica rods in a double ended arrangement also shown in Fig. 9.2c. This method is also described in detail by Andreae et al. (55). An excellent discussion of the problem of measuring attenuation by the pulse technique is given in their paper. Also, experimental arrangements and apparatus are described. The two-transducer method (Fig. 9.2d) was used by Brooks (15).

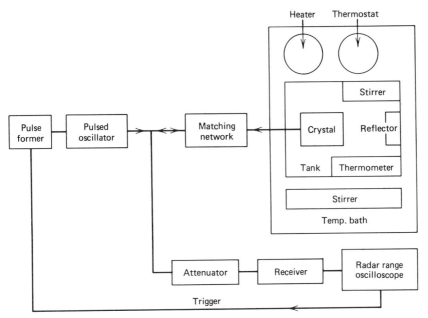

Fig. 9.1. The pulse-echo technique for chemical relaxation spectrometry (106).

He used this form with a variable and fairly long (40–80 in.) path length, for accurate measurements on water.

In a method termed "sing around," separate crystals are used for transmitting and receiving signals, as shown in Fig. 9.3. The short wave train propagated through the specimen is detected and used to trigger another pulse. The repetition rate of the series of pulses thus obtained can be accu-

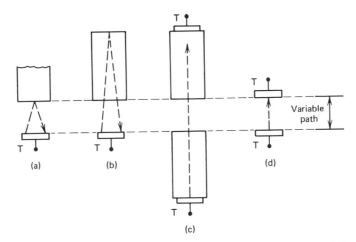

Fig. 9.2. Variable path arrangements for high-frequency pulse technique (Ref. 93, p. 289).

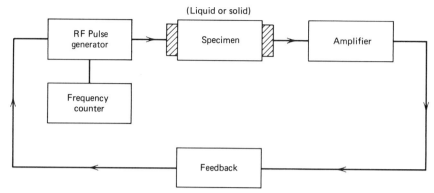

Fig. 9.3. "Sing-around" acoustical method with separate crystals for transmitting and receiving signals (Ref. 93, p. 290).

rately measured and used to determine the delay time and velocity. Careful considerations must be given to the time delay associated with electrical circuits for accurate measurement of the absolute velocity. Nevertheless, this method has good potential for measuring very small changes in velocity. A detailed description of this method is given by Holbrook (69), Cedrone and Curran (20), and Ficken and Hiedemann (47). For measuring attenuation, a method has been devised which depends on the temperature rise at a given point in the liquid as caused by the passage of a train of waves (48). A very small thermocouple is used for this purpose.

A number of other variations of the above methods have also been used. A comprehensive review of the subject is written by McSkimin (94). Many of these methods can be used for making measurements over a wide range of pressure and temperature. For the latter, the pulse-through-transmission or acoustic interferometer and the pulse-echo method have been used extensively. A detailed description of the various adaptations for moderate temperature ranges is beyond the scope of this chapter. However, it should be noted that measurements have been made on such materials as liquid helium (107) and other liquified gases, liquid metals (82), molten sulfur (112), and molten salts (68).

The determination of time lag in ionic reactions from sound absorption results requires measurements over a wide frequency range because of the relatively large half width (41). Such measurements have been carried out by Kurtze and Tamm (87) and were discussed (43) with respect to physicochemical problems involved. As the sound absorption increases with the second power of the frequency, any single device can only be applied to a limited range of frequencies. Kurtze and Tamm used five different apparatus including resonance (126), reverberation, optical (87), and impulse methods by which they measured the sound absorption of a great number of aqueous electrolytic solutions as a function of concentration, temperature, pH, etc.

The pulse technique in general has the big advantage of being a technique that gives absolute values of the sound absorption coefficients. The basic requirements of the technique are, that (a) the sound wave is planar, (b) the loss of energy due to scattering from the transducer and the reflector is negligible, and (c) the distance between transducer and reflector can be changed without affecting the leveling of the transducer and reflector surfaces. However, the pulse technique is sometimes applied in situations where the above conditions are not fulfilled. The method then requires a calibration procedure. In the case of a relaxation spectrometry, at frequencies lower than 5 MHz the pulse technique definitely requires a precise calibration procedure and in this situation a different technique is preferred, for instance, the resonance method.

2. Solid Samples

One of the major experimental problems in sonic measurements on solids is to find a way to introduce the sound wave into the specimen and detect the sound wave leaving the specimen in a reproducible manner. With either the transducers directly touching the specimen or delay rods between the specimen and the transducers, a more reproducible result is expected if a bonding material is used (63). This bonding material could be a grease or, if possible, a more permanent bond, like epoxy. The bond must provide a good and reproducible transmission of the sound wave while maintaining its characteristics over the temperature range of the measurements. There are other variables that would also influence the absorption measurements, such as bond thickness, specimen parallelism, etc. In an attempt to overcome these problems, Hartman and Jarzynski (63,64) proposed an immersion apparatus in which, according to those authors, both longitudinal and shear measurements can be made and a good, reproducible coupling between specimen and transducers is obtained due to the intimate contact of the liquid in which the specimen and transducers are immersed. However, there are some limitations to this approach too. The liquid must be very carefully chosen so that the liquid is completely inert to the specimen within the temperature range of interest.

The magnitude of the problem of transferring the sound waves from transducer to the specimen or vice versa is very much dependent on the physical characteristics of the sample. There has been no standard procedure available to date. The problem has been tackled in different manners by different scientists.

The following are some of the measurement techniques used in different laboratories.

a. Sonic Interferometer with Variable Sonic Frequency

Ballou and Smith (11) described an experimental procedure for measuring sonic velocity and attenuation with a sonic interferometer. A signal genera-

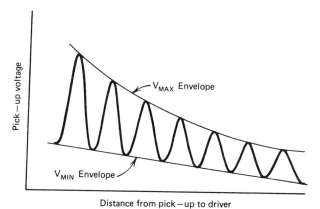

Fig. 9.4. General type of curve obtained with the sonic interferometer used by Ballou and Smith (11).

tor provides a voltage to a driver consisting of a phonograph cutter head. Longitudinal waves are excited in a continuous filament yarn or monofil samples which are attached to the cutter head needle. The waves are received by a Rochelle salt crystal in contact with the sample. The in-phase positions of the received wave with respect to the input wave provide a measure of the wavelength and therefore the velocity.

Figure 9.4 shows a typical curve obtained from the interferometer. Ballou and Smith (11) mathematically derived that the attenuation factor is the slope of the plot of the arc hyperbolic tangent of the ratio of points on the upper envelope versus the distance.

To determine the effect of temperature on the sonic properties, the sample is enclosed in a controlled temperature chamber and the temperature is measured with a thermocouple. In many cases (63), the entire acoustic system is placed inside a resistance heating oven. Heat was transferred to the acoustic system by radiation and natural convection. Specimen temperature is controlled by a rheostat that changed the current in the heating element. Specimen temperature is measured with a chromel-Alumel thermocouple connected to a precision potentiometer.

A general set-up of an interferometer with sample immersion technique (113,70,74,144,92,121) is shown in Fig. 9.5. A piezoelectric crystal is mounted at the base of a vertical steel chamber containing the liquid. At the top of the chamber is a reflecting steel plate which is moved parallel to the direction of propagation of the sound beam by a micrometer screw. The crystal is energized by a constant frequency source, which is coupled to it. Loose coupling is provided to eliminate the effects of the variable load on the frequency of the source. A frequency meter is used to measure the frequency of the exciting emf. These ultrasonic waves are produced in the liquid column by the crystal and the presence of the reflector plate produces a standing-wave system. As the micrometer screw moves the reflector plate,

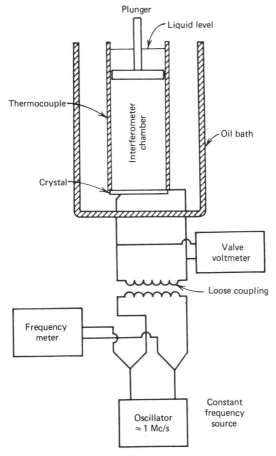

Fig. 9.5. An interferometer with sample immersion technique (113).

a voltmeter indicates voltage maxima at regular half-wave intervals, due to the reduction of the standing-wave system on the crystal. The movement of the micrometer screw is calibrated and gives the half wavelength directly. Hence, on measuring the frequency with a frequency meter, the velocity of sound may be calculated. In actual practice, the velocity in the liquid is measured over the temperature required and then the velocity measurement is repeated after immersing the sample (solid) into the liquid. At constant temperature and frequency, if a plane parallel slab of the solid sample is immersed in the liquid with its major axis parallel to the plane wavefront, the antinodal planes are displaced and the displacement ΔS can be measured (113). The velocity of sound in the solid is calculated using the relationship:

$$\Delta S = d(\mu - 1)/\mu$$

Where d is the thickness of the solid and μ is given by

$$\mu = \frac{V_s}{V_1} = \frac{\text{velocity of sound in solid}}{\text{velocity of sound in liquid}} \qquad (24)$$

The temperature control is provided by a thermostatically controlled heated oil bath.

Hartman and Jarzynski (64) described a special type of apparatus capable of measuring both longitudinal and shear sound speeds and absorptions in polymers. When the specimen is held in an immersion chamber perpendicular to the path of the sound beam, longitudinal waves are developed in the specimen. A variation of this method is to hold the specimen at an angle to the sound beam. In this way, both longitudinal and shear waves are generated in the specimen. If the angle at which the specimen is held is greater than the critical angle, the longitudinal wave is internally reflected and only the shear wave is propagated. In a typical immersion apparatus, when making shear measurements, the specimen is held vertically and is rotated with respect to the transducers to obtain shear waves. This arrangement is not suitable, however, for making measurements through the melting point because it is difficult to mount and rotate the molten specimen. In a special immersion apparatus, the specimen is held horizontally and the transducers are rotated to obtain shear waves. This arrangement allows the specimen to be mounted in such a manner as to minimize the deformation of the specimen in the melting region.

The technique of determining velocity by measuring the time required for pulses to traverse samples of different thicknesses was preferred by Arnold and Guenther (6) to eliminate complicating considerations such as transducer and circuitry delay times.

Most of the velocity determinations are made through three samples of different thicknesses, x_1, x_2, and x_3. The corresponding transit times t_1, t_2, and t_3 are determined and then they are correlated. Velocity calculations are made according to the following formulas:

$$V_{12} = (x_1 - x_2)/(t_1 - t_2); \; V_{13} = (x_1 - x_3)/(t_1 - t_3); \qquad (25)$$

$$V_{23} = (x_2 - x_3)/(t_2 - t_3) \qquad (26)$$

The arithmetic mean of three values is computed and the resultant value is plotted as a point on the velocity–temperature curve. One can also compute the velocity by taking the slope of an x versus t plot.

For determining sound velocities as a function of pressure and temperature, Asav et al. (7) used an experimental technique originally developed by Williams and Lamb (142). As shown in Fig. 9.6, a continuous wave oscillator drives a pulsed amplifier. The pulse amplifier incorporates a gating circuit which allows the amplification of the input CW signal into two phase coherent rf pulses for every repetition of the amplifier. The rf pulse is applied through a 93 Ω coaxial cable to one of two transducers closely matched in resonant frequency and located on flat and parallel surfaces of the specimen. The transmitting transducer excites acoustic vibrations which are received

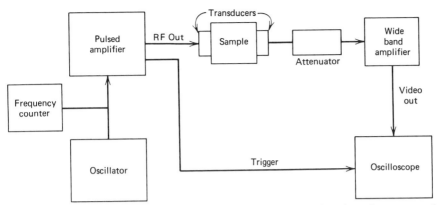

Fig. 9.6. An experimental set up to measure sound velocity as a function of pressure and temperature (7).

by the second crystal, amplified, and displayed on the oscilloscope. An attenuator is used to compare the amplitude of the acoustic signal which has traversed a single length of the specimen with one or more of the echos resulting from that signal.

The first acoustic pulse generates a series of signals on the second trans-ducer, which corresponds to the first transmitted signal and the corre-sponding echos which result from reverberation of this pulse within the specimen. The second applied signal generates a similar wave train on the receiving transducer. The amplitude of the second applied pulse is adjusted so that the amplitude of the straight transmitted signal is equal to that of the first echo from the first pulse. By adjusting the delay of the second pulse, these two signals can be made to overlap and a null condition is obtained. The transit time t is given by the relation

$$\omega_n t - \varphi_R = (2n + 1)\,\pi/2 \tag{27}$$

where ω_n is the carrier frequency corresponding to the null, n is an integer closely approximating the number of acoustic wavelengths in the specimen, and φ_R is the phase shift resulting from acoustic reflection at the transducer-bond-sample boundary. Experimentally n is determined by measuring the null frequencies and φ_R is approximated as π. The transit time t is then calculated from equation 27.

b. Pulse Propagation Meter with Constant Sonic Frequency

A simple instrument which measures only pulse propagation time through a sample within a specified distance has been widely used in characterizing fibers and sheets (33,23,24,28,9). Back et al. (9) used this instrument for measuring thermal softening of paper products and the influence of thermal auto-cross-linking reactions. Briefly, the instrument consists of a ceramic piezoelectric transducer which converts the electrical pulse and propagates a longitudinal sonic wave in the sample at a frequency sufficiently low to ensure complete damping before the succeeding wave is generated. A receiving transducer, at a measured distance from the transmitter, reconverts the sonic pulse into an electronic signal. From the time elapsed between transmission and reception of the signal, the sonic pulse velocity through the sample and thus Young's modulus of elasticity is obtained. The samples examined by Back (9) were thermally treated in a range between $-80°C$ and $400°C$ prior to the measurement of the sound velocity.

c. THERMOACOUSTIMETRIC SCANNING MEASUREMENT TECHNIQUE

Chatterjee (25,26,27,28) developed a procedure whereby a differential thermal analysis and a pulse propagation technique were combined to obtain a programmed temperature thermoacoustical analysis of textile fibers.

In principle, the method consists of a continuous measurement of the propagation time of sonic pulses of constant frequency (7 kHz) through a sample which is being held under light tension while being heated at a programmed rate. During the investigation the samples were heated from ambient temperatures to the respective melting points or decomposition.

The experimental set-up is shown in Fig. 9.7. In the figure, A represents the heating block of a Dupont 900 DTA where H represents the heating element and T_2 and T_3 represent the reference and sample wells, respectively. Two additional holes (T_1 and T_4) were drilled all the way from one end to the other through the heating block. The heating block was thoroughly insulated by putting asbestos caps on both ends and covering the rest with asbestos tape. Melting point tubes, open on both ends, were inserted into the holes T_1 and T_4. The heating block was mounted horizontally and thermocouples were inserted in T_1, T_2, and T_3. These thermocouples were connected to different terminals of the Dupont DTA cell as shown in Fig. 9.7. The fiber sample was tied at one end to a clamp S, passed under a pulley P_1 and on a notched ceramic piezoelectric crystal transducer Z_1, through the hole T_4, and supported by another identical piezoelectric crystal Z_2 and a set of pulleys P_2 and finally terminated at a suspended weight of 5 grams. The piezoelectric crystals were connected to an electrical pulse generating and recording device, R. The distance between Z_1 and Z_2 was 3.6 cm, which was kept constant throughout the experiment. The piezoelectric transducers and the pulse generating and recording devices were the parts of an instrument known as Pulse Propagation Meter, Model PPM-5R, manufactured by H. M. Morgan and Co., Cambridge, Massachusetts. The maximum range switch of the instrument was modified to obtain a time response as high as 2000 μsec.

As soon as the pulse propagation meter is activated, the electrical pulses of 7 kHz frequency are transmitted to the piezoelectrical crystal Z_1. The crystal lattice of Z_1 then begins to vibrate at the same frequency as the electrical pulses, and thus the electrical pulses are converted to sound pulses. Piezoelectricity is a means of converting electrical energy into

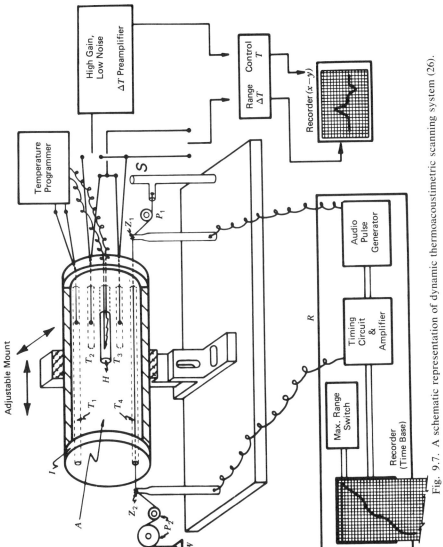

Fig. 9.7. A schematic representation of dynamic thermoacoustimetric scanning system (26).

mechanical energy and vice versa. The sound pulses are transmitted through the fiber sample to the crystal Z_2, which reconverts the sound pulses back to electrical pulses. As soon as this reconversion takes place, an internal timing circuit in the instrument is closed and the recorder instantaneously records the time required for pulses to propagate from Z_1 and Z_2, one can calculate the velocity of sound through the sample. However, in the present technique it is necessary to record the pulse propagation time only; the velocity conversion is not required.

For thermoacoustimetric scanning the sample was heated in air atmosphere at a programmed rate of 20°C/min. The system temperature was continuously recorded on a DTA chart whereas the sonic response was simultaneously recorded on a time-base recorder provided with the pulse propagation meter. The abscissa of the original sonic chart was later converted to a temperature scale.

In the case of simultaneous differential thermal analysis and thermoacoustimetric scanning analysis, the fibers were cut into small pieces by using a Wiley mill with 60-mesh screen, and 5 mg of this sample was poured into a melting point tube. The tube was then placed into the sample cavity T_2; and a thermocouple, as shown in Fig. 9.7, was inserted into the sample. Similarly, in the reference cavity T_3, a melting point tube containing reference glass beads was inserted. The reference thermocouple was then embedded into the glass beads. For thermoacoustimetric scanning, the set-up was the same as that described in the preceding paragraphs.

On heating the metal block A, the DTA curve of the sample was obtained on the X-Y recorder of the DTA instrument and a thermoacoustical curve of the sample was obtained on the time base recorder of the PPM-5R.

A sketch of a hypothetical thermoacoustic scanning curve is shown in Fig. 9.8. The pulse propagation time for a distance of x cm of the sample at room temperature is represented by the horizontal portion of the curve AB. As long as the distance x is kept constant, AB remains parallel to the abscissa. The velocity of sound through the material at 25°C is equal to $(x/120) \times 10^6$ km/sec. The temperature programming of the DTA apparatus is initiated at B. As long as the sample remains physically and chemically unchanged, the curve continues to indicate a horizontal line. At C the sample begins to transform to a different phase and the curve deviates from the base line. A change toward the upward direction indicates the lowering of the sound velocity. It is known that the velocity of sound is highest in solid, lowest in gas, and intermediate in liquid. Therefore, one may assume that the upward trend of the curve would indicate the change of polymer from compact form to relatively fluid form or, in other words, an increase of molecular motion of polymers. An opposite phenomenon is indicated by the downward trend of the curve FG. The sample at G is certainly in a less fluid state than that at F. Again G to H shows no physical or chemical change in the sample. At H the sample reveals the premelting behavior. As the melting starts there is a sharp upward trend of the curve until the sample breaks at I

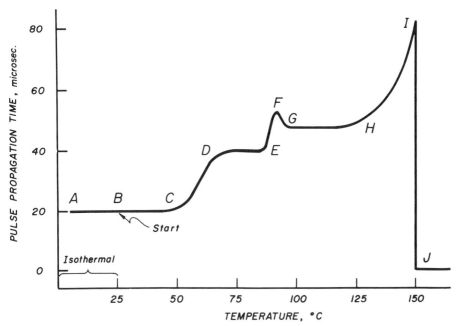

Fig. 9.8. A hypothetical dynamic thermoacoustic scanning curve (26).

due to the actual melting. The recorder pen drops immediately to zero, indicating thereby a discontinuity of the pulse propagation path.

In the case of a metal wire, the interpretation of the thermoacoustical curve is much simpler. For a metal, the sonic velocity is related to Young's modulus according to

$$C = (Y/D)^{1/2} \qquad (28)$$

where C is the sonic velocity, D is the density of the metal, and Y is the Young's modulus. Therefore, the entire curve can be interpreted as the change of Young's modulus or the density. In the case of polymers, the curve should be interpreted with respect to equation 16, i.e., in terms of the change of the viscoelastic function.

B. RESONANCE FREQUENCY TECHNIQUES

1. Liquid Samples (117)

The resonance technique makes use of resonator constructed from two x-cut crystals with liquid sample introduced between them. The technique measures the quality factor, Q, of the liquid at different resonance frequencies. The quality factor can be calculated from the resonance curve shown in Fig. 9.9 and is given by the frequency, f_0 at which the resonance curvature

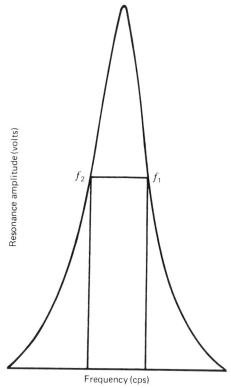

Fig. 9.9. Resonance amplitude in volts versus frequency plot for a specimen.

has a maximum divided by the width of the curvature at the 3 db point, $f_2 - f_1$.

$$Q = \frac{f_0}{f_2 - f_1} \tag{29}$$

The resonance curve is obtained experimentally by exciting the transmitting crystal with a frequency signal over a suitable band width. The resulting sound waves produced in the system are detected by the second crystal and recorded as a voltage against frequency on a display oscillograph.

The quality factor is related to the sound absorption, α, according to the following equation

$$\frac{1}{Q} = \alpha u / f_0 \tag{30}$$

where u is the sound velocity in quartz and Q is the quality factor of the liquid column only. Experimentally, however, the measured value of α is always larger than expected due to the fact that the resonator cell does not have absolutely rigid walls. Therefore calibration of the system is essential.

2. Solid Samples

The quality factor of a solid sample, Q, determined from the resonance curve as explained above, is related to the flexural resonance and energy loss (tanδ) of the material as follows (40,132):

$$\frac{1}{Q} = \tan\delta = \frac{(f_2 - f_1)}{\sqrt{3}f} \tag{31}$$

where f is the flexural resonance frequency; $(f_2 - f_1)$ is the half width, and δ is the angle between the stress and response vectors and tanδ is the ratio of the loss modulus to the storage modulus. The storage modulus, E', of a specimen can be calculated as follows (123):

$$E' = R_1 f^2 \rho \tag{32}$$

where ρ is the density of the specimen (grams/cm^3), f is the resonance frequency (Hz), and R_1 is the shape factor. For a rectangular base specimen

$$R_1 = \frac{0.94642L^4}{t^2} \tag{33}$$

where L is the length of the specimen and t is its thickness.

The loss modulus, E'', and dynamic viscosity η' can be calculated as follows:

$$E'' = \tan\delta\, E' = \frac{1}{Q} E' \tag{34}$$

and

$$\eta' = \frac{E''}{\omega} = \frac{E''}{2\pi f} \tag{35}$$

where ω is angular the frequency (rads/sec). The shear storage modulus (G') is calculated by

$$G' = \rho(2Lf)^2 R_0$$

where R_0 is the specimen shape factor. For a rectangular specimen

$$R_0 = \frac{1 + (a/t)^2}{4 - 2.521\frac{(t)}{(a)}} \tag{36}$$

where a is the width of the specimen. The bulk modulus (K') is calculated from

$$K' = \frac{E'}{3(1 - 2\mu)} \tag{37}$$

where the Poission's ratio is

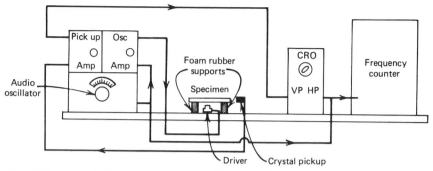

Fig. 9.10. A sketch diagram of a set up used for dynamic resonance measurements (123).

$$\mu = \frac{E'}{2G'} - 1 \tag{38}$$

The principle of the equipment (123) is to excite and detect resonance in the specimen at a controlled temperature and measure the frequency of the resonance. Fig. 9.10 represents a schematic of the instrumentation.

The output of the audio oscillator is amplified and fed to the driver, whose mechanical energy in turn is transmitted to the specimen. While the driver is being energized, the oscillator output also goes to the horizontal plate of the scope. As the oscillator frequency is scanned, it eventually reaches one of the mechanical resonance frequencies of the specimen. The predominant characteristic of a system in resonance is the large increase in the amplitude of its vibrations. The resonance frequency, f, and quality factor Q are then determined from the resonance–frequency curve represented in Fig. 9.9. For thermal acoustical measurements, those parameters were determined as a function of temperature.

By the resonance frequency technique, change in modulus, with temperature may be followed (11) by noting how the fundamental resonance frequency changes with temperature and computing the modulus at any temperature T from

$$E_T = E_0 \, (f_T/f_0)^2 \tag{39}$$

where E_T = modulus at temperature T, E_0 = modulus at room temperature, and f_T = resonance frequency at temperature T and f_0 = resonance frequency at room temperature.

IV. APPLICATIONS

A. LIQUIDS

The sonic velocity and attenuation at different temperatures were determined for a number of liquids including various alkyl halides, alkyl acetates, alcohols, alkanes, benzene, carbon tetrachloride, etc. by Pellam and Galt

(106) for the calculation of chemical relaxation parameters. The measurement of sound velocity and absorption was made at 15 MHz. The technique used could measure the velocity and attenuation with accuracies of 0.05% and 5%, respectively. Aqueous solutions of alcohol and various organic liquids were also studied by thermal acoustical methods for determining the activation energy of viscous flow, the relaxation for the organic liquid–water mixtures (100,125), the structure of water around nonpolar groups (16,101, 102,111) and the hydrogen-bonding characteristics (91). A correlation between ultrasonic propagation velocity and spin–lattice relaxation time in toluene and α-picoline was reported by Linde et al. (91,143).

Ultrasonic absorption studies of lithium nitrate solutions in tetrahydrofuran at 25°C were carried out (138) in the frequency range 10–150 MHz. The observed relaxation was attributed to a process, possibly a dissolution, characteristic of the nitrate ion within an ion pair. The ultrasonic absorption of $LiCLO_4$ in tetrahydrofuran in the frequency range 5–330 MHz at 25°, −15°, and −30°C was also measured (76). The ultrasonic relaxation data were interpreted as being due to the process of formation of triple ions, the barrier of activation energy for the process being comparable to the one for viscous flow.

B. AQUEOUS SOLUTIONS

In general ionic reactions in aqueous solutions of the type $A^+ + B^- = AB$ proceed so rapidly that they cannot be investigated by usual experimental methods. The half-times of these reactions are often as short as 10^{-10} sec. However, a sound absorption method can be used to study such a reaction system. The method is based on measurements of the chemical relaxation of an electrolytic dissociation equilibrium effected by rapid variation of pressure, electric field density, and temperature. The results indicated that a biomolecular reaction in which protons and hydroxide ions take part has rate constants of the order of 10^{10}–10^{11} per mol sec (41).

Kinetics of the nickel carboxylate complex formation in aqueous solutions were studied using acoustical techniques by Harada et al. (59,60). Rate constants and activation parameters suggest that the rate determining step of the complex formation is the dehydration process of the nickel ion. The MHz frequency range relaxation phenomenon was studied by sound absorption as a function of temperature and initial concentration of nickel ion. An interferometric technique was used to study the relaxation in aqueous solution of urea and thiourea in the temperature range 60–80°C (114). The result was discussed in terms of the structure breaking property of urea and thiourea in water. The temperature dependence of ultrasound absorption in alkali metal chlorides and alkali metal bromides was studied (14) to calculate the relaxation times for the transitions between the different structural forms.

Ultrasonic attenuation measurement of poly (L-glumatic acid) in 0.2 M NaCl H_2O-dioxide (2:1) solution were carried out over a frequency range of 6–175 MHz using a pulse-echo method (17). According to the reported results, the estimated average relaxation time (T) characteristic of the helix-coil transition is $5 \times 10^{-8} < T < 10^{-5}$ sec.

C. LIQUID CRYSTALS

Thermal acoustical studies on various liquid crystals are reported (77,19, 10,95). The ultrasonic velocity and attenuation were studied on all well-known liquid crystal symmetries (nematic, cholesteric, and smectic A, B, and C) of N-p-cyanobenzylidene-p-octyloxyaniline, ethyl-p-[(p-methoxy-benzylidene) amino] cinnamate, p-methoxybenzylidene-p-(n-butylaniline), p-azoxyanisole-azoxybenzene, etc., and the results were interpreted in terms of the various equilibrium changes near certain phase transitions. A temperature-dependent change in the nonlinear coupling coefficient in the nematic phase on approaching the nematic-isotropic transition is reported (19).

D. GLASS AND CERAMICS

Ultrasonic attenuation coefficients were measured in tellurite glass by using 10–200 MHz ultrasonic waves in the range -200–300°C and the results were interpreted in terms of phonon–phonon interactions (75). For both longitudinal and transverse polarization, the variation of the velocity of sound was measured in borosilicate glass at 0.28–4.2 K to calculate the thermal condition at low temperatures (72). The temperature dependence of the shear and longitudinal sound velocities were measured for Mn aluminosilicate and glass at 2–100 K. At about 10 K, small dips appeared in the velocities which correlated with peaks in the magnetic susceptibility (65). Shear moduli for Si_3N_4 and SiC ceramics and LiAl silicate glass–ceramic at 20–1000°C at 10 MHz are reported (44). To explain the various anomalous properties of glasses observed at low temperature, including unusual temperature dependence near 1 K, ultrasonic measurements were carried out on glasses and different crystals (71). The effects of γ-ray induced defects and their subsequent annealing on ultrasonic attenuation in quartz and tourmaline were studied in the temperature range 4.2–38 K and at about 150 K (128). It was confirmed that irradiation introduces scattering centers and lowers the thermal phonon relaxation time. The longitudinal and transverse sound velocities and hardness of a number of metallic glass consisting of palladium, nickel, iron, platinum, and phosphorous were measured and the results were interpreted to explain the transitions and the ductile behavior of the glass (35). Sound-velocity anomalies in cobalt and manganese and aluminosilicate glasses were used to study the domain structure and the magnetic susceptibility of the glass (96).

E. METALS AND OTHER ELEMENTS

The temperature dependence of ultrasound velocity and its characteristic changes near paramagnetic-ferromagnetic transitions have been widely studied. Ultrasonic attenuation and phase velocity measurements in the frequency range 10–500 MHz were performed in the vicinity of the 290 K paramagnetic-ferromagnetic transition of MnP (46,53). A study on the ultrasonic (390 MHz) attenuation for a linear antiferromagnetic C_5NiCl_3 in the temperature range 2–180 K was performed to explain the magnetic susceptibility of the compound at various transition temperatures (4,3). The temperature dependence of elastic stiffness constants and ultrasound velocity of triglycine selenate single crystals were measured near the ferroelectric phase transition to study the anomalous part of the elastic stiffness constants (131). The attenuation was studied below and above the Curie point. The paramagnetic-ferromagnetic transition of nickel-chromium-iron alloys was studied ultrasonically at 4–300 K (140). The critical behavior of sound propagation above the Curie point in ferromagnetic metals was explained by an electron model with the generalized random phase approximation by Kim (80).

According to a recent publication of Kostial (83), in sample of amorphous Se containing 0.5–10% Ge inflection points in the temperature dependences of the electrical conductivity and ultrasonic attenuation can be correlated. It was shown that the points correspond to the beginning of the viscous flow of the materials. Anomalies in the transverse electromagnetic response in niobium were studied by measuring the attenuation of transverse acoustic waves as a function of temperature (90). To elucidate the mechanism of attenuation of sound oscillations in films at the superconductive transition Postnikov et al. (110) measured the internal friction of polycrystalline V and Ta superconductors by ultrasonic measurements. Chi and Sladek (36) reported that ultrasonic attenuation in single crystals of Ti_2O_3 at various frequencies and at temperatures of 298–525 K indicated that phonon viscosity losses were mainly responsible for the intrinsic attenuation. Ultrasonic attenuation in superconductors has been reviewed by Tittman (129). The review included theories and derivations to explain the amplitude dependence associated with the motion of dislocation and the effect of vortex lines on the magnetic field dependence.

There have been various studies reported in the literature (49,135,120,81, 89,99,13) concerning the structure and transitions in metal alloys using thermal acoustical measurements. Those studies include the investigation of martensitic transition in gold-cadmium alloy (49), the structure relaxation of thallium-tellurium alloys (135), cubic to rhombohedral transition temperature in termanium telluride-tin telluride alloy single crystals (99), and elastic properties of wrought and annealed aluminum alloys.

The temperature dependence of the electronic mean free path in molybdenum was measured by using ultrasonic attenuation methods (2). The temperature dependences of the acoustic absorption in silicone were also measured at less than 30 K and 0.5–2 GHz (73). Ultrasonic measurements were made at 17 MHz while quasi-static stress cycles were applied to aluminum plastically deformed (1%) at 90 K (137). Changes in attenuation versus stress measured at low temperature after annealing and during a linear temperature increase were found to be strongly dependent on the previous treatments. The calcite 1-2 transition in single and polycrystals of calcite was studied at high and low temperatures (137). The elastic constants of the double hexagon close-packed allotrope of neodymium were measured at 4.2 to 300 K by ultrasonic measurement techniques (58).

F. POLYMERS AT CONTROLLED TEMPERATURES

The resonance frequency measurement technique was applied to observe the modulus changes with temperature of polyethylene terephthalate (PET) in various states (11). As shown in Fig. 9.11, the orientation had a marked influence on the modulus. The first break in the curve (concave downward) occurred in the region of 70°C for the unoriented state, whereas it was near 90°C for the oriented. The second or concave upward break occurred at approximately 130°C and 150°C depending on orientation again. The latter type of break had been referred to as a "cold point" (97). Besides a small increase in modulus which was attributed to increased orientation, the crystalline-oriented sample had about the same characteristics as the amorphous-oriented states. The crystallinity had a very small effect on the dynamic behavior of the particular polymer.

A study of polyethylene terephthalate by pulse propagation technique was published by Charch and Mosley (21). The temperature-sonic velocity curves of three forms of PET are shown in Fig. 9.12. The curves for the crystalline samples were notably similar in shape, but their differences in position were attributed to the supramolecular structure of the polymer. The

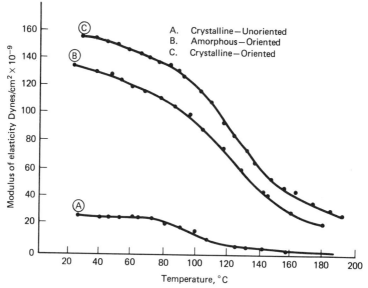

Fig. 9.11. High-frequency (sonic) modulus versus temperature for polyethylene terephthalate in various physical tests (11).

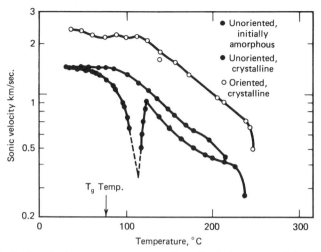

Fig. 9.12. Sonic velocity versus temperature for polyethylene terephthalate fibers (21).

bottom curve represented an initially unoriented amorphous fiber. It started at room temperature along with the unoriented crystalline sample but as the glass-transition temperature of about 80°C was approached, relaxation peaks in the amorphous sample shifted and the sonic velocity dropped sharply as more sound energy was absorbed or converted into heat. As the environmental temperature crossed the transition temperature, crystallization was complete. The curve then showed essentially the same trend with temperature as though one had started initially with unoriented crystallized sample. Perepechko (108) showed theoretically that the sonic velocity is independent of the frequency and crystallinity below the glass transition temperature. However, Charch and Mosley's (21) experimental data demonstrates that sonic velocity or sonic modulus is independent of crystallinity only appreciably below the glass transition temperature. In the region of transition temperature, sonic data can be used to follow both rate and the extent of crystallinity.

The low temperature (4.2–240 K) thermal acoustical studies were conducted on a series of Nylon-6 fibers, Nylon-7, Nylon 6, 10, Nylon 11, and Nylon 12, by Golub and Perepechko (54). Near 4.2 K, the sound velocity was found to be independent of both temperature and frequency. An interesting fact was noted in the case of Nylon-6 and Nylon 6, 10, viz. an anomalous dependence of the velocity of ultrasound on the degree of crystallinity in the very low temperature interval, where the sound velocity decreased as the degree of crystallinity increased, which was contrary to the phenomena at higher temperature.

Polymers and other dielectric materials are frequently used for many purposes in the construction of cryogenic apparatus. The determination of thermal properties of those materials below 4 K is extremely important.

Cieloszyk et al. (32) studied a polycarbonate sample, an easily measured transparent polymer useful in the manufacturing of cryogenic apparatus. Measurements of the velocity of propagation of polarized 10 MHz to 4 KHz sound waves were used to determine the Debye contribution to the heat capacity and to calculate the elastic moduli of the material at temperatures approximately 0 K.

The dependence of the velocity and absorption of sound in polyformaldehyde fibers on their structures was investigated with thermal acoustical measurements by Genina et al. (50). The correspondence between the optical and acoustical measurements of the molecular orientation of fibers showed deviations when a change of orientation was accompanied by a structural rearrangement at the supramolecular level in this polymer. During the study of molecular orientation of a series of fibers at different temperatures, Genina et al. observed an abrupt change of the temperature coefficient of the sound velocity which was not due to the change in molecular orientation but due to the inception of relaxation processes or phase transitions.

The sonic modulus and internal friction were also observed for the melting, supercooling, and recrystallization of polyethylene by measuring sonic pulse propagation between $-80°C$ and $260°C$ (39). The internal friction was found to have maxima at $-52°C$, $60°C$, and $100–130°C$. The maximum at $100–130°C$ was found to be associated with the presence of crystals. At about $130°C$, the sonic modulus increased linearly with the logarithm of recrystallization time and the internal friction decreased. The effect of melting and pressure on the α-relaxation process in polyethylene was studied using sonic technique by Boyd and Biliyar (12) and Kijima et al. (79). The sonic velocity versus temperature curves at 1, 1000, and 3100 atm pressure, indicated dispersions in the sound velocity at $96°C$, $120°C$, and $149°C$, respectively. These dispersions correspond to the α_2-relaxation which has been identified as being associated with the crystalline phase of polyethylene. The pressure coefficient of the relaxation temperature at 1 atm was semiempirically evaluated using the thermodynamical properties of polyethylene crystals and was found to be in reasonable agreement with the experimental value.

Polytetrafluoroethylene has two phase transitions: at $20°C$ and $30°C$. Kravtsov (85) studied temperature dependence of the velocity of sound near those transitions. An ultrathermostat which had a thermal regulation precision of $\pm 0.02°C$ was used to control the temperature. A small velocity minimum was observed in the vicinity of the phase transition at $20°C$, and the phase transition at $30°C$ was marked by a change in the temperature coefficient of the velocity.

The cooperative and group motions in poly[3,3-*bis*(chloromethyl)-oxetane] between 100 K and 450 K were studied by sound velocity and resonance frequency methods (8). The experimental data are plotted in Fig. 9.13. At the glass transition temperature, at about 270 K, the velocity and damping factor both undergo a significant change of slope. The damping factor versus temperature curve showed two maxima, at 230 K and at 330 K.

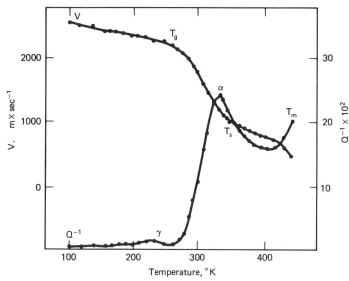

Fig. 9.13. Sound velocity and damping factor in poly[3,3-bis(chloromethyl)oxetane] (8).

The latter was attributed to the relaxation process accompanying the excitation of the cooperative motions of chain segments in the amorphous region (T_g) and the low-temperature damping maxima was assigned to a particular motion of the $-CH_2O-$ group. Sonic dynamic mechanical properties were also determined for poly(2,6-dimethyl-1,4-diphenylene oxide) over a temperature range of 80–500 K (37). The data indicated two mechanical relaxation effects, one, α, at a temperature above 480 K and another β, below the glass transition point, T_g, between 290 K and 370 K. The α relaxation effect was attributed to the thermal excitation of cooperative motions in the chain while the secondary β relaxation was interpreted as due to oscillation of aromatic rings around the C–O–C bond.

The thermal acoustical properties of polymers containing methyl methacrylate including the effect of tacticity on segmental motions of the polymer were widely studied (118,105,55,62,51,38,124,134). The glass temperature of a series of copolymers of methyl methacrylate and methacrylic acid were determined by the measurement of sound velocity. It was shown by Pavlinov et al. (105) that the increase in the proportion of methacrylic acid in the copolymer and also in the polymerization pressure increased the velocity of sound. This increase in velocity was attributed to the increased number of interchain hydrogen bonds which increase the rigidity of the chain backbone. The velocity of longitudinal and transverse ultrasonic waves in polymethyl methacrylate (PMMA) in the temperature interval from 2.1 K to 240 K and at frequencies of 1 and 5 MHz was studied by Golub and Perepechko (55). Those authors concluded that when ultrasound propagates in the polymer at low temperature, only the methyl groups whose tunneling frequencies are close to the experimental driving frequencies would interact with the ultrasonic wave. With the change of ultrasonic frequency, other methyl groups having different tunneling frequency would begin to react with the sound wave. The entropy of the interior and of the surface layer of a PMMA sample was determined by means of the transition function of ultrasonic vibrations by Tsarev and Lipatov (134). It was found that the entropy of the surface layer was less than that of the interior because of surface tension effects. The dependence of the entropy of the surface layer of PMMA on frequency and temperature was found. Measurements are reported on the acoustic attenuation and velocity of dry and wet samples of PMMA over a temperature range of 5°C and 70°C and over a frequency range of 5–35 MHz. The acoustic attenuation and infrared spectra of wet and dry samples of PMMA at different temperatures suggested that water could act as plasticizer for the cooperative motion of the pclymer backbone.

Thermal acoustical measurements of wood pulp sheet was reported by Back et al. (9) where the sheets were preheated and the sonic modulus was measured to investigate softening and autocrosslinking of cellulose. However, the data obtained by Back et al. do not unequivocally prove that the changes in sonic response was due to the change in state of the individual fibers in the sheet rather than the changes in the characteristic fiber to fiber bonding (31).

In a rheological study of aqueous solutions of polyethylene glycol at ultrasonic frequencies within a temperature range of 10–70°C, Arakawa and Takenaka (1) determined the apparent activation energy of the polymer relaxation process as $1 \ K \cdot cal/mol$. A study of sound velocity as a function of temperature with poly(oxyethylene-6-laurylether) in aqueous solution revealed that the sonic technique could be used to determine the cloud point (104).

While studying the velocity of sound of polyvinyl chloride plasticized with dialkylphthalates, Karyakin (78) determined relationships among the density, velocity of sound, adiabatic compressibility, and glass transition temperature, Tg, of the polymer. The glass transition temperature for polystyrene was determined to be 90°C from the slope change of the sound speed versus temperature plot (122).

Hartman (61) followed the curing of phenolic and polyphenylquinoxaline polymers by thermal acoustical measurements. Their findings indicated that the sound speed was not dependent on the degree of cross-linking but the sound absorption was inversely related to the degree of cross-linking.

Thermal acoustic studies of various fluoropolymers are available in the literature (88,84). Kracheva and Perepechko (84) published the acoustical studies of polytetrafluorethylene in the temperature range -150–200°C. Six transition temperatures were observed. Those transitions were at $-108°C$, $-83°C$, 19°C, 30°C, 48°C, and 118°C. The two lowest transition temperatures, $-108°C$ and $-83°C$, were attributed to the glass transition temperature, Tg, of the amorphous domains of the polymer and the mobility of four or more CF_2 groups (γ relaxation), respectively.

The acoustical techniques under controlled temperatures were also applied to study acoustical decoupling properties of Corprene DC-100 (a polychloroprene-cork composite) (67), rheological properties of dental restorative materials (133) and to calculate Debye temperature of rotational and nonrotational forms of solid *dl*-camphene, cyclohexane, and cyclohexanol (57).

G. POLYMERS AT PROGRAMMED HEATING

According to equation 16 derived by Chatterjee (26) the pulse propagation time measured at constant frequency with fixed propagation path length (fixed distance between transmitting and receiving transducers) is directly proportional to the product of an orientation factor ($\sin\theta$) and the viscoelastic function of the polymer sample (ϵ_s). In a series of experiments, it was also shown that in the case of fiber samples, under the experimental conditions described, the heating of fibers did not change the orientation of the polymer molecule. Therefore, any change of pulse propagation time would indicate simply the change in viscoelastic functions of the polymer. Chatterjee applied this principle to monitoring the changes in viscoelastic properties of polymers (fibers) at a programmed temperature ranging from room temperature to about 500°C.

The thermoacoustical curve of nylon 610 filament drawn to a ratio of 3.5:1 was studied. The curve showed a distinct upward deviation from the baseline at 45°C. It leveled off again at 55°C. This deviation was attributed to the glass transition temperature, *Tg*, of Nylon 610. Premelting behavior of the polymer was revealed by the upward trend of the curve and then an instantaneous drop to the zero line. It is important to note that above the glass transition temperature the pulse propagation time increased continuously with the rise of temperature. Prior to melting, however, the propagation time increased at an accelerating rate. The behavior of undrawn nylon 610 and drawn Nylon 610 at different draw ratios were also compared. The increase of pulse propagation time near *Tg* was more pronounced in the case of the undrawn fiber. The curves shifted downward as the draw ratio was increased. This shift toward the lower pulse propagation time or, in other words, toward the higher sonic velocity was attributed to the increase of the degree of orientation of the polymer molecules with the increased draw ratio.

Thermoacoustical curves of a variety of synthetic fibers are also reported (26). Those curves were all obtained under the same condition and at the same scale sensitivity. The curves of Nylon 4, Nylon 66, and Nylon 610 all showed different characteristic natures. Dacron and Fortrel polyester fibers behaved differently at temperatures below 100°C, but above 100°C both of them behaved identically. Teflon behaved similar to Nylon 4 between 100 and 175°C.

Cellulose fibers, such as cotton and rayon, do not melt but decompose at

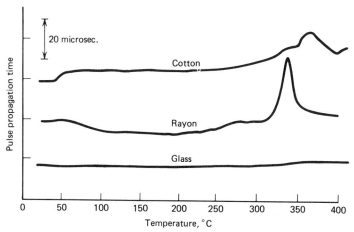

Fig. 9.14. Thermoacoustic scanning curves of cellulose in air (26).

temperatures above 300°C. Cotton and rayon are chemically the same (polymer made up of cellobiose units) but differ in their morphological structure, degree of crystallinity, and degree of polymerization. They are hard to differentiate by differential thermal analysis or by thermogravimetric analysis. The thermoacoustic scanning curves of cotton and rayon are shown in Fig. 9.14. They were distinctly different, particularly above 250°C. Rayon showed a distinct peak at about 340°C, whereas cotton showed a series of overlapping peaks at higher temperatures. Those peaks could be attributed to the decomposition of cellulose fibers. Because of a variety of chemical changes at the decomposition temperature, such as polymer scission and end-group unzipping (31,29,30,22), the velocity of pulses was slowed down, resulting in a peak. The right-hand side of the peak indicated the resumption of the original speed as the chemical changes were over and the cellulose molecule was converted to a stable carbonized form. The curve of a glass fiber has been included as a reference.

A simultaneous thermoacoustical analysis and differential thermal analysis was also done with Nylon 610 by the technique described in the experimental section. For the thermoacoustical analysis the sample was mounted as shown in Fig. 9.7, and for DTA the sample was cut into small pieces. Analytical plots were obtained simultaneously on two separate chart recorders.

REFERENCES

1. Arakawa, K., and Takenaka, N., *Bull. Chem. Soc. Jpn.,* **40,** 2063 (1967).

2. Almond, D. P., D. A. Dewiler, and J. A. Rayne, *Phys. Lett. A,* **54A,** 229 (1975).

3. Almond, D. P., and J. A. Rayne, *Proc. Int. Conf. Low Temp. Phys.,* 14th, **5,** 433 (1975).

4. Almond, D. P., and J. A. Rayne, *Phys. Lett. A.*, **54A**, 295 (1975).

5. Andreae, J. H., R. Bass, E. L. Heasell, and J. Lamb, *Acustica*, **8**, 131 (1956).

6. Arnold, N. D., and A. H. Guenther, *J. Appl. Polymer Sci.*, **10**, 731 (1966).

7. Asav, J. R., D. L. Lamberson, and A. H. Guenther, *J. Appl. Phys.*, **40**, 1768 (1969).

8. Baccaredda, M., E. Butta, and V. Frosini. *Eur. Polymer J.*, **2**, 423 (1966).

9. Back, E. L., M. T. Htung, M. Jackson, and F. Johanson, *TAPPI*, **50**, 542 (1967).

10. Bahadur, B., *Acustica*, **33**, 277 (1975).

11. Ballou, J. W., and J. C. Smith, *J. Appl. Phys.*, **20**, 493 (1949).

12. Boyd, R. H., and K. Biliyar, *Polymer Preprints* **14**, 329 (1973).

13. Boyle, W. F., and R. J. Sladek, in D. Lenz and K. Luecke, Eds., *Intern. Frict. Ultrason. Attenuation Crystal Solids Proc. Int. Conf.*, 5th, Springer, Berlin, Germany, 1975, Vol. 1, p. 98.

14. Breitschwerdt, K. G., in W. P. Luck, Ed., *Struct. Water Aqueous Solutions, Proc. Int. Symp.*, 1973, p. 473.

15. Brooks, R., *J. Acoust. Soc. Am.*, **20**, 590 (1960).

16. Bruan, S. G., G. P. Soerensen, and A. Hvidt, *Acta Chem. Scand.*, Ser. A, **A28**, 1047 (1974).

17. Burke, J. J., G. G. Hammes, and T. B. Lewis, *J. Chem. Phys.*, **42**, 3520 (1965).

18. Carlin, B., in *Mark's Standard Handbook of Mechanical Engineering*, 7th Edition, McGraw-Hill, New York, 1967, Sect. 12, P. 177.

19. Castro, C. A., A. Hikata, and C. Elbaum, *Mol. Cryst. Liq. Cryst.*, **25**, 167 (1974).

20. Cedrone, N. P. and D. R. Curran, *J. Acoust. Soc. Am.*, **26**, 963 (1954).

21. Charch, W. H., and W. W. Moseley, Jr., *Text. Res. J.*, **29**, 525 (1959).

22. Chatterjee, P. K., *J. Appl. Polymer Sci.*, **12**, 1859 (1968).

23. Chatterjee, P. K., *TAPPI*, **52**, 699 (1969).

24. Chatterjee, P. K., *Svensk Pepperstid.*, **74**, 503 (1971).

25. Chatterjee, P. K., *Nuova Chimica*, **49**, 91, (1973).

26. Chatterjee, P. K., *J. Macromol. Sci.-Chem.*, **A8**, 191 (1974).

27. Chatterjee, P. K., *Proc. Int. Conf. Therm. Anal., 4th, 1974,* Budapest, **3**, 835 (1975).

28. Chatterjee, P. K., ACS Symposium Series, No. 48, *Cellulose Chemistry & Technology*, J. C. Arthur, Ed., 1977, p. 173.

29. Chatterjee, P. K., and C. M. Conrad, *Text. Res. J.*, **36**, 487 (1966).

30. Chatterjee, P. K., and C. M. Conrad, *J. Appl. Polymer Sci.*, **6**, 3217 (1968).

31. Chatterjee, P. K., and R. F. Schwenker, Chapter 5 in R. T. O'Connor, Ed., *Instrumental Methods in the Study of Oxidation, Degradation and Pyrolysis of Cellulose*, Marcel Dekker, New York, 1972, p. 273.

32. Cieloszyk, G. S., M. T. Cruz, and G. L. Salinger, *Cryogenics*, **13**, 718 (1973).

33. Craver, V. K., and D. L. Taylor, *TAPPI* **48**, 142 (1965).

34. Czerlininski, G. H., *Chemical Relaxation*, Marcel Dekker, New York, 1966, p. 4.

35. Chen, H. S., J. T. Krause, and E. Coleman, *J. Non-Cryst. Solids*, **18**, 157 (1975).

36. Chi, T. C., R. J. Sladek, in D. Lenz and K. Luecke, *Int. Frict. Attenuation Crystal Solids Proc. Int. Conf.*, 5th Ed., 1975, p. 127.

37. DePetris, S., V. Frosini, E. Butta, and M. Baccaredda, *Makromol. Chem.*, **109**, 54 (1967).

38. Dunbar, J. H., A. M. North, R. A. Pethrick, and D. B. Steinhauer, *J. Chem. Soc., Faraday Trans.*, **2:71**, 1478 (1975).

39. Eby, R. K., *J. Acoust. Soc. Am.*, **36**, 1485 (1964).

40. Eggers, F., *Acustica,* **19,** 323 (1968).

41. Eigen, M., *Disc. Faraday Soc.,* **17,** 194 (1954).

42. Eigen, M., in A. Weissberger, Ed., *Techniques of Organic Chemistry,* Vol. VIII, pt. 2, Wiley (Interscience), New York 1963, p. 793.

43. Eigen, M., G. Kurtze, and K. Tamm, *z. Electrochem.,* **57,** 103 (1953).

44. Fate, W. A., in J. J. Burke et al., Eds., *Ceram. High-Perform. Appl. Proc. Army, Mater. Technol. Conf.,* 2nd, Brook Hill Publ. Co., Chestnut Hill, Mass., 1974, p. 687.

45. Ferry, J. D., *Viscoelastic Properties of Polymers,* John Wiley, New York, 1970, p. 3.

46. Ferry, B., and B. Golding, *AIP Conf. Proceeding* **24,** 290 (1975).

47. Ficken, G. W., Jr., and E. A. Hiedemann, *J. Acoust. Soc. Am.,* **28,** 921 (1956).

48. Fry, W. J., and R. B. Fry, *J. Acoust. Soc. Am.,* **26,** 311 (1954).

49. Gefen, Y. and M. Rosen, *J. Phys. Chem. Solids,* **37,** 669 (1976).

50. Genina, M. A., Z. A. Golik, Yu F. Zabastha, K. A. Zubovich, and M. Yu. Kuchinka, *Akust. Zh.,* **21,** 33 (1975); Eng. Trans. in *Sov. Phys. Acoust.,* **21,** 19 (1975).

51. Gilbert, A. S., R. A. Pethrick, and D. W. Phillips, *J. Appl. Polym. Sci.,* **21,** 319 (1977).

52. Goikhman, A. S., L. A. Osipina, S. G. Osinin, and M. P. Nosov, *Vysokomolek. Soed.* **8,** 94 (1966).

53. Golding, B., *Phys. Rev. Lett.,* **34,** 1102, (1975).

54. Golub, P. D. and I. I. Perepechko, *Acoust. Zh.* **20,** 38 (1974), Eng. Trans. in *Soviet Phys. Acoust.* **20,** 22 (1974).

55. Golub, P. D. and I. I. Perepechko, *Acoust. Zh.,* **19,** 619 (1973), Eng. Trans. *Sov. Phys. Acoust.* **19,** 391 (1974).

56. Grechishkin, V. A., L. G. Kazaryan, and I. I. Perepechko, *Soviet Physics-Acoustics,* **16,** 187 (1970); Translated from *Akusticheskii Zhurnal,* **16,** 223 (1970).

57. Green, J. R. and C. E. Scheie, *J. Phys. Chem. Solids,* **28,** 383 (1967).

58. Greiner, J. D., D. M. Schlader, O. D. McMasters, K. A. Gschneidner, and J. F. Smith, *J. Appl. Phys.,* **47,** 3427 (1976).

59. Harada, S., *J. Sci. Hiroshima Univ., Ser. A, Phys. Chem.* **39,** 183 (1975).

60. Harada, S., T. Yasunaga, K. Tamura, and N. Tatsumoto, *J. Phys. Chem.,* **80,** 313 (1976).

61. Hartman, B., *J. Appl. Polym. Sci.,* **19,** 3241 (1975).

62. Hartman, B. and J. Jarzynski, *J. Appl. Phys.,* **43,** 4304 (1972).

63. Hartman, B. and J. Jarzynski, Report of Naval Ordnance Laboratory, White Oak, Silver Spring, Maryland, Report numbers NOLTR70-248 (1970) and NOLTR72-73 (1972).

64. Hartman, B. and J. Jarzynski, *J. Acoust. Soc. Am.* **56,** 1469 (1974).

65. Hayes, D. J., M. D. Rechtin, and A. R. Hilton, in J. Deklerk, *Ultrason. Symp. Proc.,* IEEE, New York, 1974, p. 502.

66. Heasell, E. L. and J. Lamb, *Proc. Phys. Soc.* (London), **B69,** 861 (1956).

67. Higgs, R. W. and L. J. Eriksson, *J. Acoust. Soc. Am.,* **46,** 1254 (1969).

68. Higgs, R. W. and T. A. Litovitz, *J. Acoust. Soc. Am.,* **32,** 1108 (1960).

69. Holbrook, R. D., *J. Acoust. Soc. Am.,* **20,** 590 (1948).

70. Hubbard, J. C. and A. L. Loomis, *Phil. Mag.,* **5,** 1177 (1928).

71. Huklinger, S., in J. Klerk, Ed., *Ultrason. Symp. Proceeding,* IEEE, New York, 1974, p. 493.

72. Huklinger, S. and L. Piche, *Solid State Commun.,* **17,** 1189 (1975).

73. Ishiguro, T. and H. Tokumoto, *J. Phys. Soc. Jpn,* **37,** 1716 (1974).

74. Ivey, D. G., B. A. Mrowca, and E. Guth, *J. Appl. Phys.,* **20,** 486 (1949).

75. Izumitani, T. and I. Masuda, *Dig. Tech. Pap. Int. Qua. Electron Conf. 8th,* IEEE, New York, 1974, p. 11.

76. Jagodzinski, P. and S. Petrucci, *J. Phys. Chem.,* **78,** 917 (1974).

77. Karatha, C. G. and A. R. K. L. Padmini, *Mol. Cryst. Liq. Cryst.,* **29,** 243 (1975).

78. Karyakin, N. V., I. B. Rabinovich, and V. A. Ulynov, *Vysokomol. Soyed, A11,* No. 12, 2779 (1969).

79. Kijima, T., K. Koga, and J. Takayanagi, *Macromol. Sci-Phys.,* **B10,** 709 (1974).

80. Kim, D. J., *J. Phys. Soc. Jpn.,* **40,** 1250 (1976).

81. King, P. J. and S. G. Oates, *J. Phys. C,* **9,** 389 (1976).

82. Kleppa, O. J., *J. Chem. Phys.,* **18,** 1331 (1950).

83. Kostial, P., *Czech. J. Phys.,* **B26,** 835 (1976).

84. Kracheva, L. A. and I. I. Perepechko, *Akusticheskii Zhurnal,* **18,** 409 (1972), Eng. Trans. in *Sov. Phys.-Acoust.,* **18,** 343 (1973).

85. Kravtsov, V. M., *Akust. Zh.,* **11,** 400 (1965), Eng. Trans. in *Sov. Phys. Acoust.,* **11,** 335 (1966).

86. Kurtze, G., *Nachr. Ges. Wiss. Gittingen,* 57 (1952).

87. Kurtze, G., and K. Tamm, *Acustica,* **3,** 33 (1953).

88. Kwan, S. F., F. C. Chen, and C. L. Choy, *Polymer,* **16,** 481 (1975).

89. Lassmann, K., *Phys. Lett. A,* **56A,** 409 (1976).

90. Leibowitz, J. R., T. L. Francavilla, and E. M. Alexander, *Phys. Rev. B,* **11,** 3362 (1975).

91. Linde, B., H. Szmacinski, and A. Sliwinski, *Acta Phys. Pol. A,* **A46,** 635 (1974).

92. Maeda, Y., *J. Polymer Sci.,* **18,** 87 (1955).

93. Mason, W. P., *Proc. Int. School Phys.* ''*Enrico Fermi,*'' **27,** 223 (1963).

94. McSkimin, H. J., in W. P. Mason, Ed., *Physical Acoustics,* Vol. 1, Part A, Chapter 4, Academic, New York, 1964.

95. Miyano, K. and J. B. Ketterson, *Phys. Rev. A,* **12,** 615 (1975).

96. Moran, T. J., N. K. Batra, R. A. Verhelst, and A. M. De Graaf, *Phys. Rev. B.,* **11,** 4436 (1975).

97. Moll, H. W. and W. J. LeFevre, *Ind. Eng. Chem.,* **40,** 2172 (1948).

98. Moseley, W. W., Jr., *J. Appl. Polymer Sci.,* **III,** 266 (1960).

99. Naimon, E. R., H. M. Ledbetter, and W. F. Weston, *J. Mater. Sci.,* **10,** 1309 (1975).

100. Narasimham, A. V., and B. Manikiam, *Ind. J. Phys.,* **48,** 1068 (1975).

101. Nishikawa, S., M. Mashima, and T. Yasunaga, *Bull. Chem. Soc. Jpn.,* **48,** 661 (1975).

102. Nishikawa, S., M. Mashima, M. Maekawa, and T. Yasunaga, *Bull. Chem. Soc. Jpn.,* **48,** 2353 (1975).

103. Nolle, A. W., *J. Acous. Soc. Am.,* **19,** 194 (1947).

104. Nomoto, O. and H. Endo, *Bull. Chem. Soc. Jpn.,* **43,** 3722 (1970).

105. Pavlinov, I. I., I. B. Rabinovich, V. Z. Pogorelko, and A. V. Ryabov, *Vysokomol. Soyed.,* **A(10),** 1270 (1968).

106. Pellam, J. R. and J. K. Galt, *J. Chem. Phys.* **14,** 608 (1946).

107. Pellam, J. R. and C. F. Squire, *Phys. Rev.,* **72,** 1245 (1947).

108. Perepechko, I. I., *Sov. Phys. Acoust.* **13,** 118 (1967); Trans. from *Akusticheskii Zhuranl,* **13,** pp. 143 (1967).

109. Pinkerton, J. M. M., *Nature,* **160,** 128 (1947).

110. Postnikov, V. S., I. V. Zolotukhin, V. E. Miloshenko, and G. E. Shumin, in D. Lenz and K. Luecke, Eds., *Intern. Frict. Ultrason. Attenuation Crystal Solids Proc. Int. Conf., 5th,* Springer, Berlin, Germany, 1975, Vol. 1, p. 137.

111. Prakash, S., S. B. Srivastava, and O. Prakesh, *Ind. J. Pure Appl. Phys.*, **13**, 191 (1975).

112. Pryor, A. W. and E. G. Richardson, *J. Phys. Chem.* **59**, 14 (1955).

113. Pullen, W. J., J. Roberts, and T. E. Whall, *Polymer*, **5**, 471 (1964).

114. Rao, N. P. and K. C. Reddy, *Z. Phys. Chem.*, **100**, 133 (1976).

115. Rapuano, R. A., *Phys. Rev.*, **72**, 78 (1947).

116. Rapuano, R. A., *MIT Res. Lab. Electronics Tech. Rept.* No. 151 (1950).

117. Rassing, J., *NATO Adv. Study Inst. Serv. Ser. C.*, **18**, 1 (1975).

118. Reese, W., *J. Appl. Phys.*, **37**, 3959 (1966).

119. Samuels, R. J., *J. Polym. Sci.*, Part A, **3**, 1741 (1965).

120. Seddon, T., J. M. Farley, and G. A. Saunders, *Solid State Commun.*, **17**, 55 (1975).

121. Smith, R. E., *J. Appl. Phys.* **43**, 2555 (1972).

122. Smith, D. M. and T. A. Wiggins, *Applied Optics,* **11**, 2680 (1972).

123. Spinner, S. and W. E. Tefft, *Proc. ASTM,* **61**, 1221 (1961).

124. Sutherland, H. J. and R. Lingle, *J. Appl. Phys.*, **43**, 4022 (1972).

125. Takagi, K. and K. Negishi, *Jpn. J. Appl. Phys.* **14**, 953 (1975).

126. Tamm, K., *Nachr. Ges. Wiss. Gottingen,* 81 (1952).

127. Teeter, C. E., Jr., *J. Acoust. Soc. Am.,* **25**, 1124 (1953).

128. Thuraisingham, M. S., and R. W. B. Stephens, in D. Lenz and L. Kurt, Eds., *Intern. Frict. Ultrason. Attenuation Cryst. Solids,* Proceeding Int. Conf. 5th, Springer, Berlin, Germany, 1975, p. 308.

129. Tittmann, B. R., D. Lenz and K. Luecke, Eds., *Intern. Frict. Ultrason. Attenuation Crystal Solids Proc. Int. Conf.,* 5th, Springer, Berlin, Germany, 1975, Vol. 1, p. 52.

130. Tobolsky, A. V., and H. Eyring, *J. Chem. Phys.*, **11**, 125 (1943).

131. Todo, I., *J. Phys. Soc.* (Japan), **39**, 1538 (1975).

132. Torgalkar, A. M., *J. Dent. Res.*, **52**, 476 (1973).

133. Torgalkar, A. M., *J. Dental Res.*, **52**, 1216 (1973).

134. Tsarev, P. K. and Yu, S. Lipatov, *Vysokomol. Soyed* **A17**, 717 (1975).

135. Turner, R., *J. Phys. C.,* **7**, 3686 (1974).

136. Urick, R. J., *J. Appl. Phys.*, **18**, 983 (1947).

137. Vincent, A., and J. Perez, *Nuovo Cimento So. Ital. Fis. B,* **33**, 147 (1976).

138. Wang, H. and P. Hammes, *J. Am. Chem. Soc.*, **95**, 5115 (1973).

139. Ward, I. M., *Text. Res. J.,* **34**, 806 (1964).

140. Weston, W. F., H. M. Ledbetter, and E. R. Naimon, *Mater. Sci. Eng.*, **20**, 185 (1975).

141. Wiechert, E., *Wied. Ann. Phys.*, **50**, 335, 546 (1893).

142. Williams, J. and J. Lamb, *J. Acoust. Soc. Am.*, **30**, 308 (1958).

143. Zana, R. and J. Lang, *Adv. Mol. Relaxation Process,* **7**, 21 (1975).

144. Zosel, A., *Kolloid,* **213**, 121 (1966).

SUBJECT INDEX